Second California Climate Scenarios Assessment

Springer Atmospheric Sciences

For further volumes:
http://www.springer.com/series/10176

Daniel R. Cayan • Susanne Moser
Guido Franco • Michael Hanemann
Myoung-Ae Jones
Editors

Second California Climate Scenarios Assessment

Previously published in *Climatic Change*
Volume 109, Supplement 1, 2011

Springer

Editors

Daniel R. Cayan
Scripps Institution of Oceanography
University of California
San Diego, La Jolla, CA, USA

Guido Franco
Public Interest Energy Research
California Energy Commission
Sacramento, CA, USA

Myoung-Ae Jones
Scripps Institution of Oceanography
University of California
Sacramento, CA, USA

Susanne Moser
Stanford University and Susanne
Moser Research and Consulting
Santa Cruz, CA, USA

Michael Hanemann
University of California
Berkeley, CA, USA

ISBN 978-94-007-4013-6   ISBN 978-94-007-4014-3 (eBook)
DOI 10.1007/978-94-007-4014-3
Springer Dordrecht Heidelberg New York London

Library of Congress Control Number: 2012934280

© Springer Science+Business Media Dordrecht 2013
This work is subject to copyright. All rights are reserved by the Publisher, whether the whole or part of the material is concerned, specifically the rights of translation, reprinting, reuse of illustrations, recitation, broadcasting, reproduction on microfilms or in any other physical way, and transmission or information storage and retrieval, electronic adaptation, computer software, or by similar or dissimilar methodology now known or hereafter developed. Exempted from this legal reservation are brief excerpts in connection with reviews or scholarly analysis or material supplied specifically for the purpose of being entered and executed on a computer system, for exclusive use by the purchaser of the work. Duplication of this publication or parts thereof is permitted only under the provisions of the Copyright Law of the Publisher's location, in its current version, and permission for use must always be obtained from Springer. Permissions for use may be obtained through RightsLink at the Copyright Clearance Center. Violations are liable to prosecution under the respective Copyright Law.
The use of general descriptive names, registered names, trademarks, service marks, etc. in this publication does not imply, even in the absence of a specific statement, that such names are exempt from the relevant protective laws and regulations and therefore free for general use.
While the advice and information in this book are believed to be true and accurate at the date of publication, neither the authors nor the editors nor the publisher can accept any legal responsibility for any errors or omissions that may be made. The publisher makes no warranty, express or implied, with respect to the material contained herein.

Printed on acid-free paper

Springer is part of Springer Science+Business Media (www.springer.com)

Table of Contents

Second California Assessment: integrated climate change impacts
assessment of natural and managed systems. Guest editorial 1
G. Franco, D.R. Cayan, S. Moser, M. Hanemann and M. Jones

Projecting long-run socioeconomic and demographic trends in California
under the SRES A2 and B1 scenarios 21
A.H. Sanstad, H. Johnson, N. Goldstein and G. Franco

Current and future impacts of extreme events in California 43
M.D. Mastrandrea, C. Tebaldi, C.W. Snyder and S.H. Schneider

Potential increase in floods in California's Sierra Nevada under future
climate projections 71
T. Das, M.D. Dettinger, D.R. Cayan and H.G. Hidalgo

Simulating cold season snowpack: Impacts of snow albedo and multi-layer
snow physics 95
D. Waliser, J. Kim, Y. Xue, Y. Chao, A. Eldering, R. Fovell, A. Hall, Q. Li,
K.N. Liou, J. McWilliams, S. Kapnick, R. Vasic, F. De Sale and Y. Yu

Human-induced changes in wind, temperature and relative humidity during
Santa Ana events 119
M. Hughes, A. Hall and J. Kim

Adapting California's water system to warm vs. dry climates 133
C.R. Connell-Buck, J. Medellín-Azuara, J.R. Lund and K. Madani

Climate change impacts on two high-elevation hydropower systems in
California 151
S. Vicuña, J.A. Dracup and L. Dale

The impact of price on residential demand for electricity and natural gas 171
F.V. Lavín, L. Dale, M. Hanemann and M. Moezzi

Simulating the impacts of climate change, prices and population on
California's residential electricity consumption 191
M. Auffhammer and A. Aroonruengsawat

Effects of climate change and wave direction on longshore sediment transport patterns in Southern California.. 211
P.N. Adams, D.L. Inman and J.L. Lovering

Potential impacts of increased coastal flooding in California due to sea-level rise 229
M. Heberger, H. Cooley, P. Herrera, P.H. Gleick and E. Moore

A methodology for predicting future coastal hazards due to sea-level rise on the California Coast 251
D.L. Revell, R. Battalio, B. Spear, P. Ruggiero and J. Vandever

Estimating the potential economic impacts of climate change on Southern California beaches 277
L. Pendleton, P. King, C. Mohn, D.G. Webster, R. Vaughn and P.N. Adams

Modifying agricultural water management to adapt to climate change in California's Central Valley 299
B.A. Joyce, V.K. Mehta, D.R. Purkey, L.L. Dale and M. Hanemann

California perennial crops in a changing climate 317
D.B. Lobell and C.B. Field

Effect of climate change on field crop production in California's Central Valley.. 335
J. Lee, S. De Gryze and J. Six

Climate extremes in California agriculture 355
D.B. Lobell, A. Torney and C.B. Field

Economic impacts of climate change on California agriculture 365
O. Deschenes, and C. Kolstad

Economic impacts of climate-related changes to California agriculture 387
J. Medellín-Azuara, R.E. Howitt, D.J. MacEwan and J.R. Lund

Case study on potential agricultural responses to climate change in a California landscape 407
L.E. Jackson, S.M. Wheeler, A.D. Hollander, A.T. O'Geen, B.S. Orlove, J. Six, D.A. Sumner, Santos-Martin, J.B. Kramer, W.R. Horwath, R.E. Howitt and T.P. Tomich

The impact of climate change on California timberlands 429
L. Hannah, C. Costello, C. Guo, L. Ries, C. Kolstad, D. Panitz and N. Snider

Climate change and growth scenarios for California wildfire.................... 445
A.L. Westerling, B.P. Bryant, H.K. Preisler, T.P. Holmes, H.G. Hidalgo, T. Das and S.R. Shrestha

The impact of climate change on California's ecosystem services 465
M.R. Shaw, L. Pendleton, D.R. Cameron, B. Morris, D. Bachelet,
K. Klausmeyer, J. MacKenzie, D.R. Conklin, G.N. Bratman,
J. Lenihan, E. Haunreiter, C. Daly and P.R. Roehrdanz

The climate gap: environmental health and equity implications of
climate change and mitigation policies in California—a review of
the literature 485
S.B. Shonkoff, R. Morello-Frosch, M. Pastor and J. Sadd

Climate change-related impacts in the San Diego region by 2050 505
S. Messner, S.C. Miranda, E. Young and N. Hedge

Index 533

Springer

Addendum to "Simulating cold season snowpack: Impacts of snow albedo and multi-layer snow physics": Waliser, D., J. Kim, Y. Xue, Y. Chao, A. Eldering, R. Fovell, A. Hall, Q. Li, K. N. Liou, J. McWilliams, S. Kapnick, R. Vasic, F. De Sale, and Y. Yu (2011), *Climatic Change*, 109 (Suppl 1):S95–S117, DOI 10.1007/s10584-011-0312-5

Duane E. Waliser · Bin Guan · Jui-Lin F. Li · Jinwon Kim

Received: 20 June 2012 / Accepted: 21 June 2012
© U.S. Government 2012

Using simulations and projections by 16 global climate models (GCMs) contributing to the Coupled Model Intercomparison Projection (CMIP) 3, and thus in turn the Intergovernmental Panel on Climate Change's (IPCC's) 4th Assessment Report (AR4), Waliser et al. (2011) showed that 80 % of the mass of the Sierra Nevada snowpack will be lost by year 2050 under the middle-of-the-road greenhouse gas concentration scenario, based on the multi-model ensemble mean (Fig. 1b in Waliser et al. 2011). The uncertainty of the projection is notably large, with individual models projecting 20–95 % loss of the snowpack.

Here, the above results are updated with CMIP5 data, which are slated to form the model and projection basis for IPCC AR5. Model output from 14 GCMs that have snowpack data available as of this writing are utilized. In this case, the multi-model ensemble mean indicates 70 % loss of the Sierra Nevada snowpack by year 2050, which is very similar to the previous result based on CMIP3 (i.e., 80 %). The degree of uncertainty is also similar to the previous result, as indicated by the relatively large model spread. It is concluded that ensemble projections of the Sierra Nevada snowpack for the 21st century are largely the same from CMIP3 to CMIP5. The results again call for adaptation considerations for the western US, in this case California in particular, that rely so heavily on water reserves that derive from the annually recurring snowpack. Improvements in GCMs, particularly in terms of snow pack physics and higher resolution to resolve topography are needed to reduce the inter-model uncertainty in the projected snowpack properties and regional climate.

D. E. Waliser · B. Guan · J.-L. F. Li
Jet Propulsion Laboratory, California Institute of Technology, Pasadena, CA, USA

J. Kim (✉)
Joint Institute for Regional Earth System Science and Engineering, University of California, Los Angeles, CA, USA
e-mail: jkim@atmos.ucla.edu

Published online: 18 July 2012

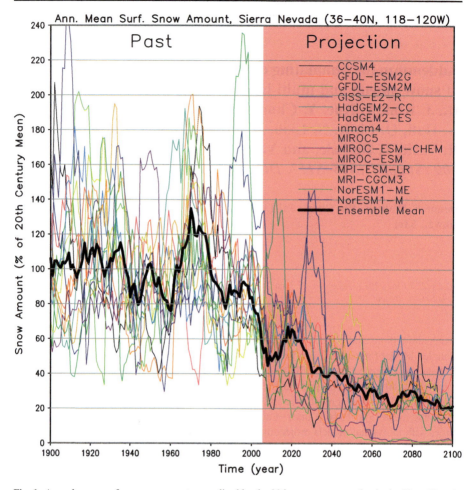

Fig. 1 Annual mean surface snow amount normalized by the 20th century mean value in the Sierra Nevada area (36–40 N, 118–120 W) during the 20th and 21st centuries simulated by 14 GCMs (model names indicated in the legend) contributing to CMIP5. The *thick black line* represents the average over the 14 GCMs

Acknowledgements This study has been supported in part by NASA NCA (ID 11-NCA11-0028) and performed on behalf of the Jet Propulsion Laboratory, California Institute of Technology, under a contract with the National Aeronautics and Space Administration.

Reference

Waliser DE, Kim J, Xue Y, Chao Y, Eldering A, Fovell R, Hall A, Li Q, Liou K-N, McWilliams J, Kapnick S, Vasic R, De Sale F, Yu Y (2011) Simulating cold season snowpack: impacts of snow albedo and multi-layer snow physics. Clim Chang 109:S95–S117. doi:10.1007/s10584-011-0312-5

Second California Assessment: integrated climate change impacts assessment of natural and managed systems. Guest editorial

Guido Franco · Daniel R. Cayan · Susanne Moser · Michael Hanemann · Myoung-Ae Jones

Received: 6 September 2011 / Accepted: 21 September 2011 / Published online: 1 December 2011
© Springer Science+Business Media B.V. 2011

Abstract Since 2006 the scientific community in California, in cooperation with resource managers, has been conducting periodic statewide studies about the potential impacts of climate change on natural and managed systems. This Special Issue is a compilation of revised papers that originate from the most recent assessment that concluded in 2009. As with the 2006 studies that influenced the passage of California's landmark Global Warming Solutions Act (AB32), these papers have informed policy formulation at the state level, helping bring climate adaptation as a complementary measure to mitigation. We provide here a brief introduction to the papers included in this Special Issue focusing on how they are coordinated and support each other. We describe the common set of downscaled climate and sea-level rise scenarios used in this assessment that came from six different global climate models (GCMs) run under two greenhouse gas emissions scenarios: B1 (low emissions) and A2 (a medium-high emissions). Recommendations for future state assessments, some of which are being implemented in an on-going new assessment that will be completed in 2012, are offered.

G. Franco (✉)
Public Interest Energy Research, California Energy Commission, Sacramento, CA, USA
e-mail: Gfranco@Energy.state.ca.us

D. R. Cayan
Scripps Institution of Oceanography, University of California, San Diego, La Jolla, CA, USA

S. Moser
Stanford University, Santa Cruz, CA, USA

M. Hanemann
Economics Department, Arizona State University, Tempe, AZ, USA

M. Jones
Scripps Institution of Oceanography, University of California, San Diego, Sacramento, CA, USA

D. R. Cayan
U.S. Geological Survey, La Jolla, CA, USA

S. Moser
Susanne Moser Research & Consulting, Santa Cruz, CA, USA

1 Introduction

The Second California Scenarios Assessment originates from Governor Arnold Schwarzenegger's Executive Order S-3-05, which charges the Secretary of the California Environmental Protection Agency to report to the Governor and the State Legislature by January 2006 and periodically thereafter on the impacts of global warming to California. The 2009 assessment builds upon previous climate model-based studies of possible climate change impacts on various sectors in the California region, including a broad assessment of possible ecological impacts by Field et al. (1999); an assessment of a range of potential climate changes on ecosystems, health, and economy in California described by Wilson et al. (2003); a study of how a "business-as-usual emissions scenario simulated by a low sensitivity climate model would affect water resources in the western United States" by Barnett et al. (2004); a multisectoral assessment of the difference in impacts arising from high versus low greenhouse gas (GHG) emissions in Hayhoe et al. (2004); and the First 2006 California Assessment (e.g., Franco et al. 2008; Cayan et al. 2008a).

Key elements of the 2009 Assessment presented in this Special Issue are the use of a common set of climate and sea-level rise scenarios that were provided to sectoral researchers in conjunction with demographic and urban projections for California counties for the rest of this century (Sanstad et al. 2011). The sectors investigated include: 1) water supply; 2) agriculture; 3) coastal resources; 4) ecosystem services; 5) forestry; 6) public health; and 7) energy demand and hydropower generation. Additional studies report on extreme events and potential environmental justice issues associated with climate change policies and impacts. Finally, a regional study for San Diego County demonstrated how state-wide studies could inform and coordinate with local or regional assessment efforts.

The Second Assessment included 39 individual studies, of which 25 are presented in this Special Issue. In this overview we present integrated summary findings from the combined effort and refer to some of the studies not included in this issue. Our emphasis will be on findings that either confirm previous conclusions, thus strengthening scientific confidence, or on those that deviate from previous studies, thus raising interesting questions for further study.

2 Climate scenarios

This section presents an overview of the climate and sea-level-rise scenarios used for the Second Assessment. As these are not included in this Special Issue, the reader should consult Cayan et al. (2009) for more information. In view of the uncertainty of the climate responses to greenhouse gases and other forcings and the variability amongst models in representing and calculating key Earth system processes, it is important to consider results from several climate models rather than rely on just a few. For this Assessment, the set of global climate models (GCMs) has been expanded compared to previous assessments for the region to include more GCMs that contributed to the Intergovernmental Panel on Climate Change (IPCC) Fourth Assessment (IPCC 2007) using *Special Report on Emissions Scenarios* (SRES) A2 and B1 emission scenarios (Cayan et al. 2009).[1]

---

[1] The GCMs selected for the assessment include: the National Center for Atmospheric Research (NCAR) Parallel Climate Model (PCM); the National Oceanic and Atmospheric Administration (NOAA) Geophysical Fluids Dynamics Laboratory (GFDL) model version 2.1; the NCAR Community Climate System Model (CCSM); the Max Plank Institute ECHAM5/MPI-OM; the MIROC 3.2 medium-resolution model from the Center for Climate System Research of the University of Tokyo and collaborators; and the French Centre National de Recherches Météorologiques (CNRM) models.

The two emissions scenarios considered are the same ones that were used for the 2006 California Climate Change Assessment (Cayan et al. 2008b). The A2 emissions scenario represents a differentiated world in which economic growth is uneven and the income gap remains large between the now-industrialized and developing parts of the world, and people, ideas, and capital are less mobile so that technology diffuses more slowly. The B1 emissions scenario presents a future with a high level of environmental and social consciousness, combined with a globally coherent approach to a more sustainable development. Each GCM differs, to some extent, in its representation of various physical processes from other GCMs, and so the different models contain different levels of warming, different patterns and changes of precipitation, and so on. The result is a set of model simulations having different climate characteristics, even when the models are driven by the same GHG emissions scenario. Consequently, climate projections from these simulations should be viewed as possible outcomes, each having uncertainties that stem from imperfect model representations that differ between climate models, uncertain future emissions, and unpredictable internal climate variability (Hawkins and Sutton 2009). In short, these modeling results provide a set of scenarios of plausible futures, but they are not detailed probabilistic predictions.

The six GCMs employed were selected on the basis of their ability to provide a set of relevant monthly, and in some cases daily, data. Another rationale was that the models provided a reasonable representation, from their historical simulation, of the following elements: seasonal precipitation and temperature, the variability of annual precipitation, and El Niño/Southern Oscillation (ENSO). It should be noted, however, that the historical skill criteria is probably not very well founded, since it has been shown that model historical skill is not well related to model climate change performance (Coquard et al. 2004; Brekke et al. 2008; Pierce et al. 2009).

Two downscaling methods were employed in the Second Assessment. These are (1) constructed analogues (CA), and (2) bias corrected spatial downscaling (BCSD). Maurer and Hidalgo (2008) and Maurer et al. (2010) compare the two methods and find that they both perform reasonably well, but they contain some noteworthy differences. Both methods have been shown to be skillful in different settings, and BCSD (Wood et al. 2004) has been used extensively in hydrologic impact analysis. The BCSD and CA methods both use coarse scale precipitation and temperature from reanalysis as predictors of the desired fine scale fields. The CA (Hidalgo et al. 2008) method provides downscaled daily large-scale data directly, and BCSD downscaled monthly data, with a random resampling technique based on historical patterns to generate daily values. Both methods yield reasonable and similar levels of skill in their resultant downscaled precipitation and temperatures for monthly and seasonal aggregated time scales. Daily precipitation is more problematic, wherein both methods produce about the same level of limited skill in simulating observed wet and dry extremes(Maurer and Hidalgo 2008). In the selected examples shown here, the results were obtained either by the BCSD or the CA method or both.

2.1 Warming

Overall, the six models' warming projections in mid-century range from about 1°C to 3°C (1.8°F to 5.4°F), rising by end-of-twenty-first century, from about 2°C to 5°C (3.6°F to 9°F). The upper part of this range is considerably greater than the historical rates estimated from observed temperature records in California (Bonfils et al. 2008).

There is considerable variability between the six GCMs, but the lower sensitivity model (PCM) contains the lowest temperature rise in both cool and warm seasons. The models do contain decade-to-decade variability, but this decadal component is not too large, and overall there is a steady, rather linear increase over the 2000–2100 period. All of the model

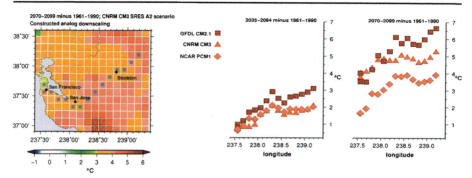

Fig. 1 Amount of warming in July, (2035–2064 minus 1961–1990) and (2070–2099 minus 1961–1990), along a coast-to-interior transect for three GCMs under A2 simulation downscaled via Constructed Analogues (Maurer and Hidalgo 2008) to the region from San Francisco through the interior region of Central California. The transect is shown in the map on the left, which illustrates the amount of warming for July for the CNRM CM3 A2 simulation. Source: Cayan et al.(2009)

runs result in a loss of spring snowpack in California, confirming previous findings (e.g., Hayhoe et al. 2004; Cayan et al. 2008b), and the models produce substantial warming during the hydrologically sensitive spring period.

There is considerable asymmetry, both seasonally and spatially, in the amount of warming. Winter (January–March) temperature changes range from 1°C to 4°C (1.8°F–7.2°F) in the six GCMs, under A2 and B1 GHG emissions scenarios, averaged over 30 years at the end of the twenty-first century relative to the 1961–1990 climatology. The noteworthy trend here is that there is greater warming in summer than in winter. Summer (July–September) temperature changes range from 1.5°C to 6°C (2.7°F–10.8°F) over the six GCMs, under both emissions scenarios. During summer, the models suggest that climate warming of land surface temperatures is amplified in the interior of the California, resulting in a gradient of temperature changes along a coast-interior transect through the San Francisco Bay region. A distinct Pacific Ocean influence occurs, wherein warming is more moderate in the zone of about 50 kilometer (km) from the coast, but rises considerably—as much as 4°C (7.2°F) higher—in the interior landward areas as compared to the warming right along the coast, as shown in Fig. 1.

2.2 Precipitation changes

Precipitation in most of California is characterized by a strong Mediterranean pattern wherein most of the annual precipitation falls in the cooler part of the year between November and March. The climate change simulations from these GCMs indicate that California will retain its Mediterranean climate with relatively cool and wet winters and hot dry summers. Another important aspect of the precipitation climatology is the large amount of variability, not only from month to month but from year to year and decade to decade. This variability stands out when mapped across the North Pacific and western North America complex, and it is quite well represented by models in comparison to the observed level of variability from global atmospheric data, via the NOAA National Centers for Environmental Prediction (NCEP) Reanalysis. The climate model-projected simulations indicate that the high degree of variability of annual precipitation will also prevail during the next century, which would suggest that the region will remain vulnerable to drought. The example presented here (Fig. 2), oriented on Sacramento, do not capture the magnitude

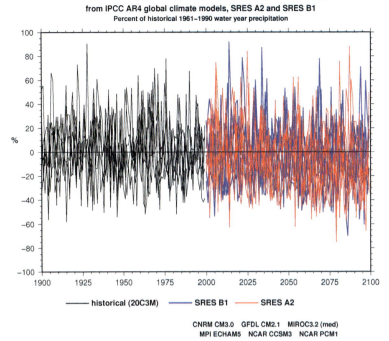

Fig. 2 Precipitation, by water year, 1901–1999 historical period (*black*) and 2000–2100 climate change period for SRES B1 (*blue*) and SRES A2 (*red*) GHG emission scenarios from six GCMs. The values plotted are taken directly from the GCMs from the grid point nearest to Sacramento. Source: Cayan et al. (2009)

of precipitation in the heaviest key watersheds in California. However, because winter precipitation in Sacramento is well correlated to that in the Sierra Nevada, these measures are representative of precipitation variability in the watersheds of the central Sierra Nevada and coast regions.

In addition to strong interannual-decadal variability contained within the climate simulations, there is a decided drying tendency. By mid- and late-twenty-first century, all but one of the simulations has declined relative to its historical (1961–1990) average. For the B1 simulation in mid-twenty-first century, two of the six simulations have a 30-year mean precipitation in Sacramento that is more than 5% drier than its historical average, and by late twenty-first century, three of the six have 30-year averages that decline to more than 10% below their historical average. By the late twenty-first century, the differences of 30-year mean precipitation from its historical average in three of the B1 simulations and four of the A2 simulations reaches a magnitude exceeding the 95% confidence level, as gauged from a Monte Carlo exercise that establishes the distribution of a historical samples. By the mid- and late-twenty-first century, only one of the simulations has 30-year mean precipitation that is wetter (slightly) than the historical annual average. Changes are stronger and more consistent in the southern part of the state than in the northern part of the state.

Cayan et al. (2009) used the Variable Infiltration Capacity (VIC) model driven from the outputs from the BCSD and CA downscaling techniques to estimate changes in river flows at representative stream gauges in California. As before, they report an accelerated early melting of snow and a shift in the hydrograph towards more flows in the winter and less in the spring and summer months (see also Vicuna et al. 2011; Das et al. 2011).

Mastrandrea et al. (2011) examine how an interwoven set of extreme meteorological and hydrological events would change in each county in California using the climate projections described above. In general they find consistent increases in extreme heat such as, for example, the July 2006 heat wave would become an annual event by the of this century under the high emissions scenario.

2.3 Sea-level rise

Over the past several decades, sea-level measured at tide gages along the California coast has risen at a rate of about 17–20 centimeters (cm) per century, a rate that is nearly the same as that from global sea-level rise estimates (Church and White 2006). In 2007, Rahmstorf demonstrated with his semi-empirical method that over the last century observed global sea-level rise can be linked to global mean surface air temperature. This provides a methodology to estimate global sea-level using the surface air temperature projected by the global climate model simulations, and also leads to larger rates of sea-level rise than those produced by other recent estimates (Cayan et al. 2008c). The estimates presented in the Second California Assessment include those using Rahmstorf's method, assuming that sea-level rise along the California coast will be the same as the global estimates. Also, the projections here include a second set of estimates that are a modification of Rahmstorf's method that attempts to account for the global growth of dams and reservoirs, which have artificially changed surface runoff into the oceans (Chao et al. 2008), in addition to the effects of climate change. In the simulations here, the sea-level estimates were adjusted so that for year 2000 their value was set to zero—this allows for comparison across the simulations of the amount of projected sea-level rise over the twenty-first century. By 2050, sea-level rise, relative to the 2000 level, ranges from 30 cm to 45 cm. By 2100, sea-level rise ranges from 0.5 to 1.4 m. As sea-level rises, there will be an increased rate of extreme high sea-level events, which occur during high tides, often accompanied by winter storms and periodically exacerbated by El Niño occurrences (Cayan et al. 2008c). It is important to note that, as decades proceed, these simulations also contain an increasing tendency for heightened sea-level events to persist for longer hours, which would imply a greater threat of coastal erosion and other damages. Hourly sea-levels simulations using the method of Cayan et al. (2008c) were updated with the new secular global sea-level projections described above (Fig. 3).

3 Impacts

As explained in section 2, the impacts researchers had at their disposal a relatively large set of climate and sea-level rise scenarios with at least daily temporal resolution and geographical resolution of about 12 km. However, only a handful of the researchers were able to make use of all of the scenarios, given the resource demands for their own impacts models. For this reason, as needed, the discussions in this section identify the specific scenario(s) used for the impacts being described.

3.1 Water supply

The Second California Assessment used two approaches to estimate potential impacts of climate change to the supply of water to different sectors of the economy. A group of

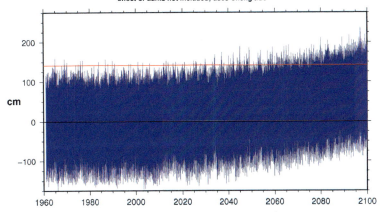

Fig. 3 Hourly sea level simulated for San Francisco (Fort Point) location, using secular change estimated using the Rahmstorf (2007) method. Hourly sea-level model from Cayan et al. (2008c) includes this secular rise and superimposes predicted astronomical tides, barometric pressures winds, and ENSO from GFDL A2 simulation. Sea-level values are referenced to the long-term mean historical average. Source: Cayan et al. (2009)

researchers associated with the California Department of Water Resources utilized a simulation water supply model known as CalSim II (Chung et al. 2009) driven by 12 climate BCSD scenarios. The authors implicitly assumed that current water rights, regulations, laws, and management practices would not change for the rest of this century. CALSIM simulates the two major water supply infrastructures in California designed to transfer water from Northern California to Central and Southern California via the Sacramento-San Joaquin Delta (Chung et al. 2009). For the second approach, Connell-Buck et al. (2011) investigated potential impacts of changes in streamflows using the CALVIN model, which is an economic-engineering optimization model of the vast California's intertied water supply system. The authors simulated water demand conditions in 2050 under a warm-dry scenario (GFDL CM 2.1 for the A2 global emission scenario) to explore the system response in California's Central Valley to severe drought in the midst of a warmer climate. In contrast to the projected warm dry conditions, a second run considered historical hydrological conditions, and a third run examined only warming without changes in total annual amount of streamflows from historical values but with a shift of the peak streamflows to earlier parts of the year to consider increased rain vs. snow and hastened snowmelt.

The CalSim II simulation model found substantial reductions in annual exports of water through the Sacramento-San Joaquin Delta. These reductions amount to 7–10% by mid–century and 21–25% at the end of the century resulting in increased annual Sacramento Valley groundwater pumping to supplement surface water supplies by 5–9% by mid–century and by 13–17% at the end of the century. Water shortage worse than the 1977 drought could occur in one out of every 6–8 years by mid–century and one out of every 3–4 years at the end of the century.

Connell-Buck et al. (2011) using CALVIN estimated difference of scarcity costs (cost associated with unmet target demand for water) by 2050. For the warming-only scenario (no change in total annual streamflows), the difference of scarcity costs by 2050 is almost indistinguishable from those incurred with no climate change. However, the run which included declines in precipitation from the warm-dry GFDL simulation produced increased scarcity costs of about $1.3 billion per year by 2050. More realistic economic losses could be much higher because the CALVIN model assumes perfect foresight, perfect water markets without the limitations of existing water rights, and a perfect operation of reservoirs and well-coordinated management of surface and groundwater resources. The modeling results suggest an important adaptation under dry conditions, wherein the drawdown and refill for reservoirs should advance by about 1 month earlier than in historical practice. Additionally, CALVIN demonstrated that statewide economic losses can be reduced substantially by transferring water from agricultural uses to consumptive uses in urban areas.

Together, these two studies suggest that without changes in California's present system of fresh water deliveries, serious water shortages would take place, but that technical solutions are theoretically possible. Moving to the idealized system represented by CALVIN, however, is challenged by serious structural, institutional, and political hurdles (Hanak et al. 2011).

3.2 Agriculture

Water supply and agriculture are very closely connected in California, given the copious amount of irrigation in the state. The agricultural sector consumes about 80% of the water withdrawals (DWR 2005). Prior studies have examined the connection between water supply and agriculture production (Wilson et al. 2003; Schlenker et al. 2007) but potential changes in crop yields have been estimated based on econometric relationships using average monthly temperature data during the growing season (Adams et al. 2003). This may be problematic because crop quality and yields, especially for perennial crops, also depend on weather conditions outside the growing season such a minimum number of hours below a threshold temperature required for dormancy for certain nuts and fruits in the cold winter season (Baldocchi and Wong 2008). Lobell and Field (2011) and Lee et al. (2011) attempted to address these limitations using, among other parameters, maximum and minimum temperatures through the year. They also increased the number of crops analyzed from prior studies.

Lobell and Field (2011) used county records for perennial crop harvests and weather from 1980 to 2005 which was complemented by Lee et al. (2011) using a well calibrated process-based crop model with daily time steps known as DAYCENT (Del Grosso et al. 2005) to simulate annual crops. Medellin et al. (2011) used all these results together with findings of prior studies on weather and yields for California as an input in their Statewide Agricultural Production Model (SWAP) to estimate how the agricultural sector would respond to both the changes in the availability of water estimated using the CALVIN model and a general warming that would be experienced by 2050. The authors performed an extensive sensitivity analysis of several assumptions in the SWAP model, concluding that their results seem to be robust. Medellin et al. assumed that California will maintain its role as a major provider of certain agricultural products in the United States, such as tree nuts, some fruits, and vegetables. For these crops, SWAP internally estimated changes in prices but for global commodities such as rice, grain, and corn, prices are provided exogenously to SWAP (California is a price taker). SWAP also considered the amount of land that would no longer be available for agriculture production due to its conversion to urban dwellings given the urban projections reported by Sanstad et al. (2011).

Results from Medellin et al. suggested a general shift to higher-value, less water intensive agricultural crops which would reduce the overall economic damages to the agricultural sector. Nevertheless, revenues would fall by about $3 billion a year or about 10% from what would be expected in 2050 without climate change. These results are in general agreement with those from a version of the SWAP model produced in 2003 (Howitt et al. 2003). The economic losses from the 2009 Assessment seem larger than Howitt et al. had produced, but the comparison is uneven given the differences in climate scenarios and model assumptions.

As reported before, using other climate scenarios with less drastic reductions in water supply would have significantly reduced economic impacts (Wilson et al. 2003) while relaxing other assumptions in CALVIN-SWAP, such as perfect adaptation in the water and agricultural sectors, should increase costs considerably. For example, these models used monthly time steps which do not allow the consideration of potential economic losses due to inland flooding which Lobell et al. (2011) reported as historically one of the main drivers for costly climatic extremes in the agricultural sector in California. In addition, Das et al. (2011) used the Variable Infiltration Capacity (VIC) macroscale hydrological model forced by downscaled GCM output to estimate how daily flood flows on the western slopes of the Sierra Nevada might change over the remainder of this century. Three-day averaged streamflows have been shown to be very well correlated with flooding events in California (Roos 1998; Florsheim and Dettinger 2007) and Das et al. (2011) used this metric to estimate the probability of floods in the rest of this century. By the second half of this century, all three models (NRM CM3, NCAR PCM, and GFDL CM 2.1) selected by the research team produced increases in the magnitude of floods for all the scenarios while the frequency of floods increased in the CM3 and PCM models but not in GFDL CM2.1. These increases are caused by multiple factors, including an increase in storm intensity and frequency and climate warming-related shifts in precipitation toward more rain rather than snow. This suggests that not considering impacts of flooding to the agricultural sector would underestimate economic losses. Nevertheless, the CALVIN-SWAP modeling studies conducted here provide useful insights about potential adaptation options. More detailed feasibility studies would need to investigate these options, including an analysis of how the regulatory and legal structure governing the agricultural and water sectors would have to change to allow for the implementation of technically promising adaptation options.

3.3 Coastal resources

The 2006 California Climate Change Assessment did not consider coastal impacts and only a few previous studies (e.g., Gleick and Maurer 1990; Newmann et al. 2003) had considered potential impacts of climate change on coastal resources in California. The Second California Assessment improved prior studies in various ways, such as going beyond simple static inundation estimates, using modern GIS tools, and considering coastal erosion of cliffs and other similar topographic features well above sea levels that in the past were assumed not to be affected by sea-level rise.

In the 2009 Assessment four studies investigated the potential impacts of sea-level rise on coastal resources. Adams et al. (2011) investigated how longshore sediment transport on the beaches in Southern California would change with changes in deep water wave direction in the Pacific Ocean. They reported that specific impacts on individual beaches depend on the direction of the waves originating in the open ocean far from California. This is important because other studies have reported a northward shift in cyclonic activities in the Pacific Ocean (Salathé 2006; Bender et al. 2011) which some GCMs suggest being a climate change signal (Yin 2005).

Revell et al. (2011) applied new methodologies using statewide data sets to evaluate potential erosion hazards on open ocean coastlines of California for a 1.4 m sea-level rise scenario. They also estimated future 100-year coastal flood elevations along open ocean and bay/estuarine shorelines extrapolating from current FEMA coastal flood maps. Although this is an exploratory analysis, this type of work is essential in providing coastal managers with information potentially useful in developing adaptation strategies. Knowles (2010) tackled the issue of potential inundation in the interior part of San Francisco Estuary and Bay using the highest resolution topographic data available in 2008 with the hourly sea-level rise projections of Cayan et al. (2009) driving a hydrodynamic model. In mapping potential inundation, Knowles (2010) found that the current 100-year peak flood events would become a yearly occurrence by the middle of this century. Heberger et al. (2011) used the geographical information system (GIS) provided by Revell et al. (2011) and Knowles (2010) to determine the resources and assets (using data cataloging infrastructure currently in place) along the open coast and estuarine shoreline of San Francisco Bay that would be affected. They reported that by the end of this century coastal flooding could threaten areas that currently are home to approximately half a million people and $100 billion in property and assets. Assuming no additional protective measures, their maps show critical infrastructure, including roads, hospitals, schools, emergency facilities, wastewater treatment plants, airports, and power plants, currently at risk from flooding exacerbated by a 1.4 m sea-level rise. Widely publicized in the local media, their results engendered widespread recognition of the need for adaptation and initiated various planning activities in the San Francisco Bay region.

Finally, Pendleton et al. (2011) examined economic impact on beach recreation in Southern California associated with permanent beach loss caused by inundation due to sea-level rise of 1 m, and an extreme storm event such as occurred during the strong El Niño year of 1982/1983. Pendleton et al. used a model of beach visitation in Southern California to estimate how beach attendance would be impacted by the consequent beach closures in each case and to calculate the resulting economic cost in terms of lost revenue and lost consumer's surplus. They compared these costs with the costs of beach nourishment as an adaptation measure. In the case of sea-level rise of 1 m, the changes in attendance at the 51 beaches in Los Angeles and Orange Counties generated a loss of consumer's surplus amounting to about $63 million per year. The costs are considerably higher than estimated costs of beach nourishment of about $4 million per year suggesting that, in general, beach nourishment would be a cost effective adaptation option.

3.4 Ecosystem services

Even under current conditions it is extremely difficult and controversial to value, in economic terms, ecosystem services (Serafy 1998; McCauley 2006). Shaw et al. (2011) examined how climate change would affect two ecosystem services: 1) carbon sequestration in natural terrestrial ecosystems, and 2) non-irrigated forage production for livestock. Shaw et al. used the MC1 dynamic vegetation model together with the urban expansion reported in Sanstad et al. (2011) to estimate changes in carbon stocks in natural ecosystems. Using social cost of carbon figures reported in the literature by Tol (2007), Watkiss and Downing (2008), and Nordhaus (2008), they concluded that if a relatively mild form of climate change (climate scenarios from the PCM global change model) becomes a reality, carbon sequestration would result in a net benefit between $38 million annually in the period from 2005 to 2034 and about $22 billion annually by 2070. On the other hand, if a hotter and drier climate change scenario materializes, the social costs would range from approximately $600 million to $5.2 billion annually for the period 2005–2034 and $62 billion annually by 2070–2099. They estimated natural forage production to decline dramatically by the end of this century in all future climate projections, with potentially significant impacts on ranching agriculture and costs for adaptive measures.

3.5 Timber industry and wildfires

Westerling et al. (2011) applied an enhanced statistical model, originally developed for seasonal forecasts of fire risks (Westerling et al. 2003), to estimate how climate change would affect wildfires in California. The novel approach in their 2009 Assessment study was the use of projections of human settlement in the wildland-urban interface provided by the U.S. Environmental Protection Agency (Bierwagen et al. 2010). This is important because human population expanding into wildland areas is one of the explanatory variables traditionally used to estimate fire risks. The urban projections by Sanstad et al. (2011) were of little use for the study by Westerling et al., however, because their projections are only for core urban centers. Westerling et al. found increases in burned area that go up with time (i.e., increases as the magnitude of climate change increases), with estimates in burned area by the end of the 21st Century exceeding 100% of the historical area burned in much of the forested areas of Northern California in all A2 runs. The resulting risk in this new study is greater than previously reported (Westerling and Bryant 2008), most likely because the new study models a broader range of climate-vegetation-fire relationships, uses additional climate scenarios, employs multiple thresholds for defining the wildland-urban interface, and explores a range of population and development scenarios rather than assuming that human settlements in the future would be unchanged from current conditions. A companion study by Bryant and Westerling (2009) estimated costs from losses of property due to both increased wildfire risk and human encroachment in forested areas, which implied that alternative growth patterns would reduce the probability of property damage.

Another study by Hannah et al. (2011) modeled productivity of different forest trees coupled with economic models of landowner adaptation. Hannah et al. took into account that timber is a commodity with a global market and made use of global timber prices projected for this century as estimated by Sohngen et al. (2001). Hannah et al. reported, in general, increased timber production with climate change and a decrease in timber values, in line with Sohngen et al.'s findings that climate change increases global timber production, resulting in lower timber prices. In addition, they found that losses in California are not geographically homogeneous and driven mainly by assumptions about the price of timber in global markets. The findings of Hannah et al. agree with another study in the 2009 Assessment that used a new empirical model of timber production and found increased timber yields in California with climate change (Battles et al. 2009).

Hughes et al. (2011) downscaled the outputs from the NCAR CCSM3 global climate model using the Weather Research and Forecast (WRF) model to estimate how climate change would change dry, windy fall and winter Santa Ana events that have historically proceeded some of Southern California's largest wildfires which in turn produced the most substantial economic and property losses. The authors reported approximately a 20% decrease of Santa Ana events in the mid-21st century from historical conditions with an accompanying decrease in relative humidity, but with a simultaneous increase in temperature. The latter would normally favor the occurrence of more wildfires. Future work will need to include a fire behavior model to account for the effect of changes in wind regimes, relative humidity, and temperatures to reach more robust conclusions about the potential effects of climate change on fire risks in Southern California due to Santa Ana events.

3.6 Public health

Studies on public health that were part of the Second Assessment have been published elsewhere. In general they showed that the association between elevated temperatures and human mortality is independent of air pollution (Basu et al. 2008); that mortality effects are differentiated by age and ethnic group but not by gender or educational level (Basu and Ostro 2008); and that high temperatures have important morbidity effects measured by hospital admission data (Green et al. 2010). Furthermore, they showed that more days with conditions conducive to high tropospheric ozone levels will increase with climate change (Mahmud et al. 2008). This will result in an "air quality penalty" in the sense that more than the anticipated reduction of emissions of ozone precursors will have to be realized to be able to continue to improve air quality in California and eventually comply and maintain compliance with state and federal air quality standards. Assuming current control costs of nitrogen oxide and volatile organic compound, this air quality penalty will result in $8 billion per year of additional expenditures by the middle of this century (Motallebi, personal communication).

Finally, Cayan et al. (2009) indicated that hot daytime and nighttime temperatures (heat waves) are increasing in frequency, magnitude, and duration from the historical period. Within a given heat wave, there is an increasing tendency for multiple hot days in succession, and the spatial footprint of heat waves is more and more likely to encompass multiple population centers in California. These findings heighten public health concerns in the coming decades.

3.7 Energy demand and hydropower generation

Auffhammer and Aroonruengsawat (2011) expanded upon previous studies (Miller et al. 2008; Franco and Sanstad 2008) in the 2009 Assessment by evaluating climate impacts on electricity demand. They used a unique data set consisting of household level residential electricity consumption to estimate potential changes in electricity demand in the residential sector at the U.S. mail zip code level. Their estimated impacts are much higher than what has been reported in the past (Miller et al. 2008). This may be due to the fact that the residential sector is more responsive to temperature than other sectors (e.g., industrial sector). Auffhammer and Aroonruengsawat (2011) reported increases in electricity demand in the residential sector of up to 55% by the end of this century if future climatic conditions are superimposed on the current stock of homes and their electricity consuming devices. The actual impacts could be much lower due to the implementation of new aggressive energy efficiency programs and by consumer responses to potential increases in electricity rates. At the same time, more urban development is projected to take place in the California Central Valley and other inland areas that are already experiencing higher summer temperatures and are expected to warm at a faster rate than coastal areas. Increases in temperatures will also favor an increased penetration of air conditioning units (Sailor and Pavlova 2003) or better building practices (insulation). Net electricity demand, however, will not necessarily grow in strict proportion to the increased penetration of air conditioning because new air conditioning units are likely to be more efficient and better suited to California's dry summer conditions (Buntine et al. 2008). Electricity expenditures in the residential sector were about $13 billion in 2009 (EIA State Energy Data System 2011) so that even fractional increases in demand will represent non-trivial economic losses.

The future amount of hydropower generation depends heavily on how climate change affects overall precipitation amounts. If a drier climate becomes reality, hydropower generation would decline, but go up if precipitation increases. Connell-Buck et al. (2011)

reported results from the CALVIN model suggesting a 4.5% decrease ($20 million a year) in hydropower benefits in their warm-dry scenario from the low-elevation hydropower units associated with relatively large reservoirs. However, historically about 74% of the hydropower generated in California comes from high-elevation hydropower units that use snow as their main water reservoir (Aspen Environmental Group 2005). Madani and Lund (2009) reported reductions on the order of 14% for these units due to changes in runoff patterns, lower snowpack volumes, and limited storage capacity, under the warm-dry-scenario. Regardless of how climate change will affect precipitation amounts, a reduction of electricity generation during the hot summer months, a period traditionally relied on for hydroelectricity to satisfy peak cooling demand, is projected for all climate scenarios.

3.8 Differential vulnerability to climate change and policy

Shonkoff et al. (2011) reviewed literature on climate change and environmental justice issues to draw some preliminary conclusions about potential disproportional impacts to low income and minority groups with relatively limited resources available to adapt or to relocate if needed. They also discussed how efforts to reduce greenhouse gas emissions might preferentially benefit certain segments of California's population, but generally not low income and minority groups. For example, they argued, cap-and-trade program will not necessarily result in reduction of co-pollutants such as volatile organic compounds and hazardous air pollutants in areas that are currently considered environmental justice hotspots in California, i.e., areas where low-income and ethnic minority populations experience undue burdens from environmental pollutants and nuisances. An example of this outcome is the expected lack of change in emissions near oil refineries in Southern California because they were able to purchase emissions offsets generated from the destruction of old, polluting vehicles in lieu of installing equipment reducing emissions at these refineries. This type of argument was persuasive to the state judicial system in halting, at least temporarily, implementation of the cap-and-trade program designed by the California Air Resources Board to comply with the state's Global Warming Solutions Act of 2006.

3.9 Regional impacts foci

Finally, the 2009 Assessment included a regional climate impacts study conducted for San Diego County by local scientists. Messner et al. (2011) estimated climate change impacts by 2050 for that county covering the sectors included in the statewide Assessment. They paid special attention to water supply issues for the San Diego region, given the fact that San Diego County relies heavily on imported water from Northern California and from the Colorado River. Due to climate change and increased drought tendency, the region's reliance on imported water will significantly increase in the coming decades. As discussed extensively in scientific journals (e.g., MacDonald 2010), both sources of imported water are threatened by climate change.

4 Statewide assessment and its linkage to adaptation policies in California

An important goal of the recurrent production of assessments is to inform policy decisions with the best available science. As such, while it was an independent scientific endeavor, the Second Assessment has had a strong connection to state climate policy evolution in California. A steering committee formed by senior technical managers in different state agencies was involved from the start to help shape the overall design of the study. They also participated in meetings with the researchers organized to discuss preliminary results. In several cases the members of the steering committee made substantial contributions by identifying important government data sets, providing insightful comments from their in-depth knowledge of California issues, and helping to produce research products that are pertinent to and useable in—as much as possible—actual resource management decisions. These interactions were very fruitful and, in some cases, have influenced long-term planning activities in California. For example, the California Department of Forestry and Fire Protection used the ecological model enhanced and used by Hannah et al. (2011) for their 2010 Forest and Range Assessment (CalFire 2010), which this agency is required to prepare and submit to the Governor and the Legislature every 5 years. For the first time, this agency was able to quantitatively consider climate change in their long-term management plan of resources under its jurisdiction.

Another important policy development for which the Second Assessment was instrumental was the preparation of the first statewide climate adaptation strategy for California. Mandated by Executive Order S-13-08, which was issued by Governor Schwarzenegger in November 2008, state agencies in charge of the management of natural resources, infrastructure and public health were directed to identify adaptation measures for those assets and populations likely to be affected by climate change. One of the motivating factors for the Executive Order was the release of a short synthesis of scientific findings between the First and Second California Assessment, which made the case that even if strong mitigation measures were implemented at a global scale, California would see substantial changes in its climate and impacted physical, natural, and social systems (Moser et al. 2008; Moser et al. 2009). The Governor released the strategy in December 2009 (California Natural Resources Agency 2009), which is seen by state authorities as a starting point in a long journey of continuously updating and refining the measures California must take to adapt to a changing climate. The close interaction between the scientists involved in this assessment and senior technical managers in state government enabled the consideration of the research findings presented in this Special Issue in the preparation of the 2009 California Climate Adaptation Strategy (CAS). One of the clear recognitions articulated in the CAS was that climate-scenario driven (top-down) impacts assessments alone are insufficient to fully understand the challenges to be expected from climate change. Thus, as one of its overarching recommendations, the CAS proposed an integrated, top-down and bottom-up vulnerability and adaptation assessment to inform ongoing adaptation policy developments. That set of studies is currently (2011) underway—building on, yet also significantly expanding on, past investigations—with results expected in the first quarter of 2012.

More recently, the California Ocean Protection Council issued interim guidelines on the assumptions that governmental agencies should use with regard to sea-level rise when issuing permits or for long-term planning work (OPC 2010). Results from the Second Assessment contributed to the preparation of these guidelines and some of the scientists involved in the second Assessment provided expert advice.

5 Discussion

Several lessons can be drawn from the Second California Assessment but we focus here on lessons related to research management and policy related issues. Papers included in this issue can be referred to for future scientific directions.

Some key aspects of regional climate change are still quite uncertain, as evidenced by the range that is contained across the scenarios that have been included. In particular, these include the direction and magnitude of precipitation changes, the rate of sea-level rise and the intensity and frequency of future storminess. The impacts of these regional climate changes will cascade through a network of sectors and ecosystems and thus will require continued scrutiny from ongoing observational and model assessments.

In general, the studies indicate that climate change impacts in California will be distributed unevenly across social groups, industries and regions. While some may benefit from climate change in a relative sense, others will have to cope with predominantly negative impacts. For example, Pendleton et al. (2011) estimated increases in beach attendance and economic benefits for some beaches and economic losses at other beaches, as well as an overall negative economic impact for Southern California. Heterogeneous impacts imply the need for site-specific local studies informed by regional and/or statewide studies to be most useful to local and state-level decision makers. This point takes on a special urgency in light of the unequal distribution of socioeconomic impacts reported by Shonkoff et al. (2011). Given the relative paucity of literature on this topic and the important policy implications of this type of work, the study on equity implications of climate change and climate mitigation policies is an important research area in need of further development.

The San Diego study (Messner et al. 2011) and other similar regional/local efforts demonstrate that regional/local entities are willing and in some cases eager to engage with the research community. For example, this regionally focused study has been influential in motivating local adaptation planning efforts, and in shaping follow-up assessment work in other regions of California (e.g., the focus on San Francisco Bay in the 2011 assessment currently underway). How to collaborate with them without overburdening the scientists involved in the California Assessments and how to increase mutual understanding for the needs and limitations of both sides are issues that need to be continuously addressed in future endeavors.

Some of the statewide studies already contain information relevant for the local level. For example, the climate projections (Cayan et al. 2009) have a geographical resolution of about 12 km and the electricity demand estimates are resolved at the U.S. mail zip code level. Scenarios and findings of the impacts studies have begun to be made available at a website known as Cal-Adapt (http://cal-adapt.org/) through visually approachable, interactive tools that allow users to inspect research results at smaller scales and access output data for further use by local/regional decision makers. Ongoing monitoring and critical evaluation of the use of this website, its tools, and the available data is needed to assess their usefulness and to ensure appropriate use.

In some cases, economic impacts in California will strongly depend on forces from national and international markets. For example, the California timber industry, even if yields increase, may be negatively affected if world prices for timber decline as suggested by Sohngen et al. (2001). For this reason, it would be desirable to coordinate international and national studies in such a way that they inform each other. Moreover, more research into such "teleconnections" (Adger et al. 2009) is required to more realistically assess impacts from and societal responses to climate change. Effects on California will also derive from climate change effects in other regions and the responses in these other regions to them. An IPCC-level of effort that is purposefully designed to both review the state of the science and the creation of new knowledge through the coordination of regional and national studies is highly desirable. The recently started new U.S. National Climate Assessment is a good start in this direction (NCA 2011).

Many policymakers and agency personnel in California have recognized the value of concerted research efforts, regular updates on the state of climate change science, and the growing attention to the salience and relevance of research findings to decision making. This is presenting new opportunities for novel research and also for ongoing learning about mutual needs and capabilities between researchers and practitioners. Tight state budgets and some political challenges to climate change impacts research threaten to narrow the breadth of ongoing assessments, but the demand for location-specific, decision-relevant climate and adaptation research is only growing.

For future research to continue to be policy-relevant, the identification of technically sound adaptation strategies is only one part of a long process that must come to grips with regulatory, legal, institutional and other non-technical barriers (Ekstrom et al. 2011). These technical and societal components of the adaptation strategy must be explored and addressed in an ongoing interaction with decision makers. Studies with models such as CALVIN and SWAP, while important, leave a number of unanswered questions on how to get closer to their idealized adaptation scenarios. Alternatively, to be more decision-relevant, studies must reflect the more realistic circumstances that decision making face, and help them develop practically feasible adaptation options.

In closing, we and others have found that an ongoing, periodic assessment involving technical staff from state agencies is highly beneficial for both the scientists and for state agency decision makers. Despite the effort at the state and federal level, it is clear from the rapidly growing interest in adaptation planning, and the significant lag in adaptation science (NRC 2010), that a substantial amount of adaptation will take place at the local level without adequate scientific information to inform it. Continued scientific effort, sustained research investment, strategic science policy and research priority-setting, rapid building of the necessary science-practice bridging capacity, and proactive advice-seeking by decision makers, are essential to inform adaptation planning and implementation with the current scientific understanding of our rapidly changing climate, environment, and society.

Disclaimer This paper reflects the views of the authors and does not necessarily reflect the views of the California Energy Commission or the state of California.

References

Adams R, Wu J, Houston L (2003) The effects of climate change on yields and water use of major California crops. In: Wilson, T, Williams L, Smith J, Mendelsohn R (eds) Global climate change and California: potential implications for ecosystems, health, and the economy. Appendix IX. California Energy Commission, Sacramento. CEC 500-03-058

Adams P, Inman D, Lovering J (2011) Effects of climate change and wave direction on longshore sediment transport patterns in Southern California. Clim Change (Suppl 1), doi:10.1007/s10584-011-0317-0

Adger W, Eakin H, Winkels A (2009) Nested and teleconnected vulnerabilities to environmental change. Front Ecol Environ 7(3):150–157

Aspen Environmental Group (2005) Potential changes in hydropower production from global climate change in California and the Western United States. California Energy Commission, Sacramento. CEC-700-2005-010

Auffhammer M, Aroonruegsawat A (2011) Simulating California's future residential electricity demand under different scenarios of climate change, electricity prices and population electricity demand. Clim Change 109 (Suppl 1), doi:10.1007/s10584-011-0299-y

Baldocchi D, Wong S (2008) Accumulated winter chill is decreasing in the fruit growing regions of California. Clim Chang 87(Suppl1):S153–S166

Barnett T, Malone R, Pennell W, Stammer D, Semtner A, Washington W (2004) The effects of climate change on water resources in the west: introduction and overview. Clim Chang 62:1–11

Basu R, Ostro B (2008) A multicounty analysis identifying the populations vulnerable to mortality associated with high ambient temperature in California. Am J Epidemiol. doi:10.1093/aje/kwn170

Basu R, Wen-Ying F, Ostro B (2008) Characterizing temperature and mortality in nine California counties. Epidemiology. doi:10.1097/EDE.0b013e31815c1da7

Battles J, Robards T, Das A, Stewart W (2009) Projecting climate change impacts on forest growth and yield for California's Sierran mixed conifer forests. California Energy Commission, Sacramento. CEC-500-2009-047-F

Bender F, Ramanathan V, Tselioudis G (2011) Changes in extratropical storm track cloudiness 1983–2009: observational support for a poleward shift. Clim Dyn. doi:10.1007/s00382-011-1065-6

Bierwagen B, Theobald D, Pyke C, Choate A, Groth P, Thomas JV, Morefield P (2010) National housing and impervious surface scenarios for integrated climate impact assessments. PNAS. doi:10.1073/pnas.1002096107

Bonfils C, Duffy P, Santer B, Wigley T, Lobell DB, Phillips TJ, Doutriaux C (2008) Identification of external influences on temperatures in California. Clim Chang 87:43–55

Brekke LD, Dettinger MD, Maurer EP, Anderson M (2008) Significance of model credibility in projection distributions for regional hydroclimatological impacts of climate change. Clim Chang. doi:10.1007/s10584-007-9388-3

Bryant B, Westerling A (2009) Potential effects of climate change on residential wildfire risk in California. California Energy Commission, Sacramento. CEC-500-2009-048-F

Buntine C, Proctor J, Knight B (2008) Energy performance of hot, dry optimized air-conditioning systems. California Energy Commission, Sacramento. CEC-500-2008-056

CalFire (2010) California's forests and rangelands: 2010 Assessment. California Department of Forestry and Fire Protection. http://frap.cdf.ca.gov/assessment2010/pdfs/california_forest_assessment_nov22.pdf. Accessed 25 July 2011

California Natural Resources Agency (2009) 2009 California climate adaptation strategy. http://www.energy.ca.gov/2009publications/CNRA-1000-2009-027/CNRA-1000-2009-027-F.PDF. Accessed 20 July 2011

Cayan D, Bromirski P, Hayhoe K, Tyree M, Dettinger M, Flick R (2008a) Climate change projections of sea level extremes along the California coast. Clim Chang 87(Suppl 1):S57–S73

Cayan D, Luers A, Franco G, Hanemann M, Croes B, Vine E (2008b) Overview of the California climate change scenarios project. Clim Chang 87(Suppl 1):S1–S6

Cayan D, Maurer E, Dettinger M, Tyree M, Hayhoe K (2008c) Climate change scenarios for the California region. Clim Chang. doi:10.1007/s10584-007-9377-6

Cayan D, Tyree M, Dettinger M, Hidalgo M, Das T, Maurer E, Bromirski P, Graham N, Flick R (2009) Climate change scenarios and sea level rise estimates for the California 2009 climate change scenarios Assessment. California Energy Commission, Sacramento, CEC-500-2009-014-F

Chao B, Wu Y, Li S (2008) Impact of artificial reservoir water impoundment on global sea level. Science 320 (5873):212–214

Chung F, Anderson J, Arora S et al (2009) Using future climate projections to support water resources decision making in California. California Energy Commission, Sacramento. CEC-500-2009-052-F

Church J, White N (2006) A 20th century acceleration in global sea-level rise. Geophys Res Lett. doi:10.1029/2005GL024826

Connell-Buck c, Medellín-Azuara J, Lund J, Madani K (2011) Adapting California's water system to warm vs. dry climates. Clim Change 109 (Suppl 1), doi:10.1007/s10584-011-0302-7

Coquard J, Duffy P, Taylor K, Iorio J (2004) Present and future surface climate in the western USA as simulated by 15 global climate models. Clim Dyn 23(5):455

Das T, Dettinger M, Cayan D, Hidalgo H (2011) Potential increase in floods in California's Sierra Nevada under future climate projections. Clim Change 109 (Suppl 1), doi:10.1007/s10584-011-0298-z

Delgrosso S, Mosier A, Parton W, Ojima D (2005) DAYCENT model analysis of past and contemporary soil N2O and net greenhouse gas flux for major crops in the USA. Soil Tillage Res 83:9–24

DWR (2005) The California water plan update. Bulletin 160–05. California Department of Water Resources. http://www.waterplan.water.ca.gov/docs/public_comments/prdcomments/(07-22-05)_-_IID_-_Anisa_Divine_2_attach1.pdf

EIA. State energy data system. http://205.254.135.24/state/seds/. Accessed 31 July 2011

Ekstrom J, Moser S, Torn M (2011) Barriers to Climate Change Adaptation: A Diagnostic Framework. California Energy Commission, Sacramento. CEC-500-2011-004 http://www.energy.ca.gov/2011publications/CEC-500-2011-004/CEC-500-2011-004.pdf

Field C, Daily B, Davis F, Gaines S, Matson P, Melack J, Miller N (1999) Confronting climate change in California: ecological impacts on the golden state. Union of Concerned Scientists and Ecological Society of America. http://www.ucsusa.org/assets/documents/global_warming/calclimate.pdf. Accessed 20 July 2011

Florsheim J, Dettinger M (2007) Climate and floods still govern California levee breaks. Geophys Res Lett. doi:10.1029/2007GL031702

Franco G, Sanstad A (2008) Climate change and electricity demand in California. Clim Chang 87(Suppl 1): S139–S151

Franco G, Cayan D, Luers A, Hanemann M, Croes B (2008) Linking climate change science with policy in California. Clim Chang 87(Suppl 1):S7–S20

Gleick P, Maurer E (1990) Assessing the costs of adapting to sea-level rise: a case study of San Francisco Bay. Pacific Institute, Oakland

Green R, Basu R, Malig B, Broadwin R, Kim J, Ostro B (2010) The effect of temperature on hospital admissions in nine California counties. Int J Public Health. doi:10.1007/s00038-009-0076-0

Hanak E, Lund J, Dinar A, Gray B, Howitt R, Mount J, Moyle P, Thompson B (2011) Managing California's water: from conflict to reconciliation. Public Policy Institute of California

Hannah Lee, Costello C, Guo C, Ries L, Kolstad C, Panitz D, Snider N (2011) The impact of climate change on California timberlands. Clim Change 109 (Suppl 1), doi:10.1007/s10584-011-0307-2

Hawkins E, Sutton R (2009) The potential to narrow uncertainty in regional climate predictions. Bull Am Meteor Soc 90:1095–1107

Hayhoe K, Cayan D, Field C et al (2004) Emissions pathways, climate change, and impacts on California. PNAS 101(34):12422–12427

Heberger M, Cooley H, Herrera P, Gleick P, Moore E (2011) Potential Impacts of Increased Coastal Flooding in California Due to Sea-Level Rise. Clim Change 109 (Suppl 1), doi:10.1007/s10584-011-0308-1

Hidalgo H, Dettinger M, Cayan D (2008) Downscaling with constructed analogues: daily precipitation and temperature fields over the United States. California Energy Commission, Sacramento. CEC-500-2007-123

Howitt R, Tauber M, Pienaar E (2003) Impacts of global climate change on California's agricultural water demand. In: Wilson, T, Williams L, Smith J, Mendelsohn R (eds) Global climate change and California: potential implications for ecosystems, health, and the economy. Appendix X. California Energy Commission, Sacramento. CEC 500-03-058

Hughes M, Hall A, Kim J (2011) Human-induced changes in Wind, Temperature and Relative Humidity during Santa Ana events. Clim Change 109 (Suppl 1), doi:10.1007/s10584-011-0300-9

IPCC (2007) Climate change 2007: the physical science basis. contribution to the fourth assessment report by working group I. Cambridge University Press, Cambridge, UK, New York, NY. www.ipcc.ch/ipccreports/ar4-wg1.htm

Knowles N (2010) Potential inundation due to rising sea levels in the San Francisco Bay Region. San Franc Estuary Watershed Sci 8(1) http://escholarship.org/uc/jmie_sfews?volume=8;issue=1

Lee J, De Gryze S, Six J (2011) Effect of climate change on field crop production in California's Central Valley. Clim Change 109 (Suppl 1), doi:10.1007/s10584-011-0305-4

Lobell D, Field C (2011) California Perennial Crops in a Changing Climate. Clim Change 109 (Suppl 1), doi:10.1007/s10584-011-0303-6

Lobell D, Torney A, Field C (2011) Climate Extremes in California Agriculture. Clim Change 109 (Suppl 1), doi:10.1007/s10584-011-0304-5

MacDonald GM (2010) Water, climate change, and sustainability in the southwest. PNAS 107(50):21256–21262

Madani K, Lund J (2009) Modeling California's high-elevation hydropower systems in energy units. Water Resour Res. doi:10.1029/2008WR007206

Mahmud A, Tyree M, Cayan D, Motallebi N, Kleeman M (2008) Statistical downscaling of climate change impacts on ozone concentrations in California. J Geophys Res. doi:10.1029/2007JD009534

Mastrandrea M, Tebaldi C, Snyder C, Schneider S (2011) Current and future impacts of extreme events in California. Clim Change (Suppl 1), doi:10.1007/s10584-011-0311-6

Maurer E, Hidalgo H (2008) Utility of daily vs. monthly large-scale climate data: an intercomparison of two statistical downscaling methods. Hydrol Earth Syst Sci 12:551–563

Maurer E, Hidalgo H, Das T, Dettinger M, Cayan D (2010) The utility of daily large-scale climate data in the assessment of climate change impacts on daily streamflow in California. Hydrol Earth Syst Sci 14 (6):1125–1138

McCauley D (2006) Selling out on nature. Nature 443:26–27

Medellin-Azuara J, Howitt R, MacEwan D, Lund J (2011) Economic Impacts of Climate-Related Agricultural Yield Changes in California. Clim Change 109 (Suppl 1), doi:10.1007/s10584-011-0314-3

Messner S, Miranda S, Young E, Hedge N (2011) Climate change-related Impacts in the San Diego region by 2050. Clim Change 109 (Suppl 1), doi:10.1007/s10584-011-0316-1

Miller N, Hayhoe K, Jin J, Auffhammer M (2008) Climate, extreme heat, and electricity demand in California. J Appl Meteorol Climatol 47:1834–1844

Moser S, Franco B, Cayan D (2008) The future is now: an update on climate change science impacts and response options for California. California Climate Change Center. CEC-500-2008-077

Moser S, Franco G, Pittiglio S, Chou W, Cayan D (2009) The future is now: an update on climate change science impacts and response options for california. California Climate Change Center. CEC-500-2008-071 http://www.energy.ca.gov/2008publications/CEC-500-2008-071/CEC-500-2008-071.PDF

NCA (2011) National climate assessment. Strategy—summary. http://www.globalchange.gov/images/NCA/nca-summary-strategy_5-20-11.pdf. Accessed on 8 July 2011

Newmann J, Hudgens D, Herr J, Kassakian J (2003) Market impacts of sea level rise on California coasts. In: Wilson T, Williams L, Smith J, Mendelsohn R (eds) Global climate change and California: potential implications for ecosystems, health, and the economy. Appendix VIII. California Energy Commission, Sacramento. CEC 500-03-058

Nordhaus W (2008) A question of balance: weighing the options on global warming policies. Yale University Press, New Haven

NRC (2010) America's climate choices: adapting to the impacts of climate change. National Academies Press, Washington, DC

OPC (2010) State of California sea-level rise interim guidance document. California Ocean Protection Council

Pendleton L, King P, Mohn C, Webster D, Vaughn R, Adams P (2011) Estimating the potential economic impacts of climate change on Southern California beaches. Clim Change 109 (Suppl 1), doi:10.1007/s10584-011-0309-0

Pierce D, Barnett T, Santer B, Gleckler P (2009) Selecting global climate models for regional climate change studies. Proc Natl Acad Sci. doi:10.1073/pnas.0900094106

Rahmstorf S (2007) A semi-empirical approach to projecting future sea-level rise. Science 315(5810):368–370

Revell D, Battalio R, Spear B, Ruggiero P, Vandever J (2011) A Methodology for Predicting Future Coastal Hazards due to Sea-level Rise on the California Coast. Clim Change 109 (Suppl 1), doi:10.1007/s10584-011-0315-2

Roos M (1998) The great new year's flood of 1997 in Northern California. Proc. Annual PACLIM Workshop. Interagency Ecology Program Technical Report 57:107–116. http://cepsym.info/Sympro1997/roos.pdf

Sailor D, Pavlova A (2003) Air conditioning market saturation and long-term response of residential cooling energy demand to climate change. Energy 28(9)

Salathé E (2006) Influences of a shift in North Pacific storm tracks on western North American precipitation under global warming. Geophys Res Lett. doi:10.1029/2006GL026882

Sanstad A, Johnson H, Goldstein N, Franco G (2011) Projecting long-run socioeconomic and demographic trends in California under the SRES A2 and B1 scenarios. Clim Change 109 (Suppl 1), doi:10.1007/s10584-011-0296-1

Schlenker W, Hanemann M, Fisher A (2007) Water availability, degree days, and the potential impact of climate change on irrigated agriculture in California. Clim Chang 81(1):19–38

Serafy S (1998) Pricing the invaluable: the value of the world's ecosystem services and natural capital. Ecol Econ 25:25–27

Shaw R, Pendleton L, Cameron D, Morris B, Bachelet D, Klausmeyer K, MacKenzie J, Conklin D, Bratman G, Lenihan J, Haunreiter E, Daly C, Roehrdancz P (2011) The impact of climate change on California's ecosystem services. Clim Change 109 (Suppl 1), doi:10.1007/s10584-011-0313-4

Shonkoff S, Morello-Frosch R, Pastor M, Sadd J (2011) The Climate Gap: Environmental Health and Equity Implications of Climate Change and Mitigation Policies in California - A Review of the Literature. Clim Change 109 (Suppl 1), doi:10.1007/s10584-011-0310-7

Sohngen B, Mendelsohn R, Sedjo R (2001) A global model of climate change impacts on timber markets. Am J Agric Econ 26(2):326–343

Tol R (2007) The social cost of carbon: trends, outliers and catastrophes. Economics Discussion Papers. Discussion Paper 2007–44

Vicuna S, Dracup J, Dale L (2011) Climate change impacts on two high-elevation hydropower systems in California. Clim Change 109 (Suppl 1), doi:10.1007/s10584-011-0301-8

Watkiss P, Downing T (2008) The social cost of carbon: valuation estimates and their use in UK policy. Integr Assess J 8:85–105

Westerling A, Bryant B (2008) Climate change and wildfire in California. Clim Chang 87(1):231–249

Westerling A, Gershunov A, Cayan D (2003) Statistical forecasts of the 2003 Western wildfire season using canonical correlation analysis. Exp Long-Lead Forecast Bull 12(1,2):49–53

Westerling A, Bryant B, Preisler H, Holmes T, Hidalgo H, Das T, Shrestha S (2011) Climate Change and Growth Scenarios for California Wildfire. Clim Change 109 (Suppl 1), doi:10.1007/s10584-011-0329-9

Wilson T, Williams L, Smith J, Mendelsohn R (2003) Global climate change and California: potential implications for ecosystems, health, and the economy. California Energy Commission, Sacramento. CEC 500-03-058. pp 1–138

Wood A, Leung LR, Sridhar V, Lettenmaier D (2004) Hydrologic implications of dynamical and statistical approaches to downscaling climate model outputs. Clim Chang 62:189–216

Yin J (2005) A consistent poleward shift of the storm tracks in simulations of the 21st century climate. Geophys Res Lett 32:L18701

ERRATUM

Erratum to: Second California Assessment: integrated climate change impacts assessment of natural and managed systems. Guest editorial

Guido Franco · Daniel R. Cayan · Susanne Moser · Michael Hanemann · Myoung-Ae Jones

Published online: 22 June 2012
© Springer Science+Business Media B.V. 2012

Erratum to: Climatic Change (2011) 109 (Suppl 1):S1–S19
DOI 10.1007/s10584-011-0318-z

In the original study included in this special issue, Auffhammer and Aroonruengsawat (2011) reported a 55 % increase in household electricity consumption by 2100. The authors later discovered that the large increase was due to a coding error in the simulation and ran the models again with corrected codes. The corrected results now show that the household electricity consumption rate by 2100 one order of magnitude smaller than the original estimation. Our discussion in section 3.7 on the energy demand and hydropower generation covering their study should now read as below:

The online version of the original article can be found at http://dx.doi.org/10.1007/s10584-011-0318-z.

G. Franco (✉)
Public Interest Energy Research, California Energy Commission, Sacramento, CA, USA
e-mail: Gfranco@Energy.state.ca.us

D. R. Cayan
Scripps Institution of Oceanography, University of California, San Diego, La Jolla, CA, USA

S. Moser
Stanford University, Santa Cruz, CA, USA

M. Hanemann
Economics Department, Arizona State University, Tempe, AZ, USA

M. Jones
Scripps Institution of Oceanography, University of California, San Diego, Sacramento, CA, USA

D. R. Cayan
U.S. Geological Survey, La Jolla, CA, USA

S. Moser
Susanne Moser Research & Consulting, Santa Cruz, CA, USA

Auffhammer and Aroonruengsawat (2011) expanded upon previous studies (Miller et al. 2008; Franco and Sanstad 2008) in the 2009 Assessment by evaluating climate impacts on electricity demand. They used a unique data set consisting of household level residential electricity consumption to estimate potential changes in electricity demand in the residential sector at the U.S. mail zip code level. Their estimated impacts are consistent with what has been reported in the past (Miller et al. 2008). What the new study adds is high spatial resolution showing that annual increases in electricity demand would grow at a faster rate in the Central Valley and the southeastern parts of the state.

Projecting long-run socioeconomic and demographic trends in California under the SRES A2 and B1 scenarios

Alan H. Sanstad · Hans Johnson · Noah Goldstein · Guido Franco

Received: 9 March 2010 / Accepted: 21 September 2011 / Published online: 9 December 2011
© Springer Science+Business Media B.V. 2011

Abstract The State of California is developing and implementing a new generation of environmental policies to transition to a low-carbon economy and energy system in order to reduce the risks of future damages from global climate change. At the same time, it is increasingly clear that climate change impacts are already occurring and that further effects cannot be completely avoided. Thus, anticipating and planning for emerging and potential future climate change impacts in California must complement the state's greenhouse gas mitigation efforts. These impacts will depend substantially on the future evolution of the state's social structure and economy. To support impact studies, this report describes socioeconomic storylines and key scenario elements for California that are broadly consistent with the global "A2" and "B1" storylines in the 2000 *Special Report on Emissions Scenarios* of the Intergovernmental Panel on Climate Change, including qualitative socioeconomic context as well as quantitative projections of key variables such as population, urbanization patterns, economic growth, and electricity prices.

1 Introduction

Over the past decade, the State of California has developed and begun implementing a portfolio of energy and environmental policies aimed at effecting a shift to a low-carbon

---

This paper is adapted from a longer report, Sanstad et al. (2009), which was funded by the California Energy Commission and the California Environmental Protection Agency. Mr. Sanstad's work was supported through the U.S. Department of Energy under Contract No. DE-AC02-05CH11231.

A. H. Sanstad (✉)
Lawrence Berkeley National Laboratory, #1 Cyclotron Rd., Berkeley, CA 94720, USA
e-mail: ahsanstad@lbl.gov

H. Johnson
Public Policy Institute of California, Sacramento, CA, USA

N. Goldstein
Lawrence Livermore National Laboratory, Livermore, CA, USA

G. Franco
California Energy Commission, Sacramento, CA, USA

state economy and energy system. Prominent among these is AB 32, the California Global Warming Solutions Act of 2006, which established the goal of returning statewide greenhouse gas (GHG) emissions to 1990 levels by 2020 and created a regulatory system to achieve this goal.[1] In addition, other U. S. states as well as the federal government have in the past several years accelerated their efforts to achieve significant, large-scale reductions of GHG emissions. These efforts and concurrent initiatives around the world are aimed at the United Nations Framework Convention goal of preventing "dangerous anthropogenic interference" with the global climate system.

There is increasing evidence, however, that climate change impacts are already occurring and that further effects in the near term cannot be avoided. Moreover, our increasing but still-limited understanding of the immensely complex climate system implies that even the establishment of an internationally coordinated, worldwide GHG abatement regime can be expected to reduce but not eliminate the risks of future impacts of climate change on the global environment and human society. Accordingly, anticipating and planning for emerging and potential future climate change impacts in California must complement the state's GHG mitigation efforts.

The impacts of climate change on California's natural and human environment will depend substantially on the future evolution of the state's social structure and economy. Climate change, particularly as manifest in localized changes in weather and precipitation, will have complex, differentiated effects across sectors of the economy, income groups, and geographical regions. For example, climate change-driven increases in summer temperatures may weigh most heavily on low-income households in the inland region, while precipitation changes will affect agriculture far more than service industries. At the same time, expectations of California's future socioeconomic configuration will influence the manner in which adaptation to climate change is conceived and carried out. This socioeconomically mediated nature of potential climate change impacts, and the interactions among vulnerabilities and response strategies, are the motivations for this paper.

As noted by Meehl et al. (2007), variations in assumptions regarding future trends in social and economic variables, and the consequent range of potential paths of anthropogenic GHG emissions, are by far the biggest source of uncertainty in numerical projections of future climate change. As described by Franco et al. (2011), the quantitative projections of global climate change conducted under the auspices of the Intergovernmental Panel on Climate Change (IPCC), and applied in the Second California assessment of which this paper is one element, are driven by modeled simulations of two sets of projections of twenty-first century social and economic development around the world, the so-called "A2" and "B1" storylines in the 2000 *Special Report on Emissions Scenarios* (SRES) (IPCC 2000). The SRES study was conducted as part of the IPCC's Third Assessment Round, released in 2001. The A2 and B1 storylines and their quantitative representations represent two quite different possible trajectories for the world economy, society, and energy system, and imply divergent future anthropogenic emissions, with projected emissions in the A2 being substantially higher.[2]

The rationale for using the SRES A2 and B1 scenarios in the Second California assessment is given by Franco et al. (*op. cit.*). This paper describes socioeconomic storylines and key scenario elements for California that are broadly consistent with these global scenarios, and that were developed to support several of the assessment's impact analyses. In contrast to the A2 and B1-driven regional climate projections, we do not

---

[1] Assembly Bill 32 (Nuñez), Chapter 488, Statutes of 2006.

[2] In SRES terminology, "storyline" refers to a qualitative description of a pattern of global and regional socioeconomic development, including the characteristics of energy systems and implications for greenhouse gas emissions. A "scenario" is a simulation of a storyline by a specific numerical model.

formally downscale the scenarios or conduct economic simulation modeling of twenty-first century California. Instead, we provide a general, qualitative socioeconomic context as well as quantitative projections of key variables, including population, urbanization patterns, and economic growth, that reflect the main elements of the global scenarios.

In the past several years, increasing attention has been devoted to developing and applying new, general methods for creating scenarios of this type, in part to meet the growing demand for climate change impacts and adaptation analysis. This topic is the subject of, for example, a recent report of the U. S. National Research Council (NRC 2010). By contrast, this paper can be seen as a case study in the pragmatics of generating relatively localized policy-relevant scenario information, drawing upon available techniques and data resources.

The paper is organized as follows. Section 2 briefly reviews the SRES framework and the A2 and B1 storylines and quantitative scenarios. Scenarios of economic growth, population, urbanization and other key drivers are presented in Section 3. Concluding remarks are given in Section 4.

2 SRES background and selected details

Understanding the factors that determine anthropogenic GHG emissions is both critical for climate science and policy making, and extremely challenging. Both the importance and the difficulty increase when attempting to project possible paths of these emissions a century into the future. It is widely understood that emissions on this time scale cannot be predicted with any plausibility or reliability. Basic data requirements for applying conventional forecasting methods can probably not be met in the case of global GHG emissions; if they could, the statistical errors associated with such methods would substantially dominate their predictions. The need for some means of quantitatively and defensibly projecting long-run GHG emissions in the face of such hurdles has led over the last several decades to the use of various "scenario" techniques for this purpose.

There is no single or standard definition of "scenario" in energy and environmental analysis and modeling, but the term is generally used to convey the idea of a "plausible projection," whether qualitative or quantitative, of future events, variables, or system behaviors. As summarized in the SRES,

> Scenarios are images of the future, or alternative futures. They are neither predictions nor forecasts. Rather, each scenario is one alternative image of how the future might unfold. A set of scenarios assists in the understanding of possible future developments of complex systems...many physical and social systems are poorly understood, and information on the relevant variables is so incomplete that they can be appreciated only through intuition and are best communicated by images and stories...Scenarios can be viewed as a linking tool that integrates qualitative narratives or stories about the future and quantitative formulations based on formal modeling.[3]

Of the four SRES scenario families, the A2 and B1 families constitute a "high" and "low" set of CO2 emissions projections, respectively. (By contrast with the so-called "FI" or "Fossil Intensive" subset of the A1 family, however, the A2 does not reflect a continued or increased reliance on fossil fuels across regions.) This motivated their selection for the current study.

The foundation of both the qualitative storylines and the quantitative scenarios in SRES is the recognition of economic development, demographics, and technological change as

---

[3] IPCC 2000, Chapter 1, p. 62.

primary drivers of GHG emissions. The storylines summarize the SRES view of how these drivers might evolve. The A2 and B1 are summarized as follows.

A2: Slower economic growth and technological change, higher population growth, and less socioeconomic convergence (relative to the A1 scenario).

As described in the SRES, A2 represents a "differentiated world," with respect to demographics, economic growth, resource use and energy systems, and cultural factors (although the latter are not represented, per se, in the quantitative scenarios). There is what in current terminology could be called a de-emphasis on globalization, reflected in heterogeneity of economic growth rates and rates and directions of technological change. Globally, slow decline of fertility rates results in very high population, 15 billion by 2100. Lack of convergence of economic growth rates, among other factors, results in substantial differences in per capita income across regions. As noted above these assumptions, when quantified, result in continued growth in global CO_2 emissions, which reach nearly 30 gigatons of carbon (GtC) annually in the marker scenario by 2100.

B1: Intermediate economic growth and low population growth, global convergence emphasizing environmental priorities and sustainability.

The B1 storyline is in essence the reflection of the A2 across the key dimensions. It can be characterized as a "global sustainability" scenario. Worldwide, environmental protection and quality and human development emerge as key priorities, and there is an increase in international cooperation to address them as well as convergence in other dimensions. A demographic transition results in global population peaking around mid-century and declining thereafter, reaching roughly 7 billion by 2100. Economic growth rates are higher than in A2, so that global economic output in 2100 is approximately one-third greater. The combination of these economic and demographic trends results in a much richer world, in terms of per capita income, in B1 than A2.

No numerical probabilities or likelihoods were assigned to any of the storylines or their quantified associated scenarios. Each is regarded as a plausible future that might unfold under certain conditions; none is a "base case" or "business as usual" case. Equally important, the storylines and scenarios are all "no GHG policy" cases, meaning specifically that it is assumed that no coordinated global emissions mitigation such as the Kyoto Protocol is enacted. As noted in the report, however, this assumption does not preclude in principle either other environmental policy actions or underlying economic or technological developments, that would result in lower GHG emissions than might otherwise occur.

The quantification of these storylines was conducted using six different numerical models, representing a range of approaches to energy and emissions modeling.[4] This multi-model approach was used among other reasons to capture the uncertainties associated with the representation of economic dynamics, technological change, demographic effects, and other key factors. Multiple models were used to simulate each storyline, with the aforementioned uncertainties resulting in varying emissions projections even for a given storyline. However, for each of A1, A2, B1, and B2 storylines, a single model was selected to provide the "marker" scenario for the given storyline; these markers were selected by the SRES team to illustrate the complete set of storylines in a more manageable form (than the forty scenarios comprising the complete set). The markers are not in any sense "most likely," but these scenarios did receive a higher degree of technical review and scrutiny, in accord with their intended purpose.

---

[4] The quantification of the storylines is described in detail in SRES (IPCC 2000) Chapter 4.

In the B1 marker scenario, a worldwide transition to efficient and renewable energy technologies—although not driven by global GHG emissions policies—results in a peaking and subsequent decline in global emissions comparable to what might be achieved with such policies: Annual emissions reach about 12 GtC in 2040 and decline to about 4 GtC in 2100.

2.1 United States projections in the A2 and B1 marker scenarios

The SRES reported numerical scenario results on a global basis as well as for four large regions.[5] In most cases, however, a finer degree of geographical or geopolitical disaggregation was incorporated in individual models and scenario results, including the models that generated the A2 and B1 markers: The Atmospheric Stabilization Framework (ASF) for the A2, and the Integrated Model to Assess the Greenhouse Effect (IMAGE) for the B1. For this study, we obtained the main U.S.-specific results from these two models/scenarios.

Table 1 contains key results on U. S. economic and population growth from A2–ASF and B1–IMAGE, as well as U. S. historical figures for comparison. The latter are for the second half of the twentieth century (1950–2000 and 1975–2000), as well as the decade ending in 2005. For the SRES scenarios, results are presented for 2000–2020, 2000–2050, and 2050–2100. As the numbers indicate, both scenarios anticipate declines in growth rates from historic levels in all three of the drivers: Real Gross Domestic Product (GDP), national population, and GDP per capita. These can be interpreted as continuations of trends already observed at the time (1990s) that the SRES was written—that is, long-run slowing of both economic and population growth. It is important to emphasize, however, that even these lower-than-previous rates of economic growth imply that the United States economy will continue to expand and that in dollar terms the United States will be substantially wealthier in the future than it is today, in both absolute and per capita terms

3 Projections for California

3.1 Methodological remarks

As we noted in the Introduction, this study took a pragmatic approach, generating simple and plausible projections of a small set of relevant quantities, drawing primarily upon already available information, rather than, e.g., developing new scenario methods or apply more complex modeling tools. This strategy in part reflects the practical constraints under which studies of this type are typically conducted, notwithstanding in this case the relatively high level of usable climate-change-and-impacts information available in, and about, California. The study was part of an effort to provide California policy-makers and other stakeholders with illustrative order-of-magnitude estimates of potential climate change impacts, conditional on a small set of plausible scenario assumptions, that combined existing information from disparate intra-state sources, and could be assembled on a relatively short schedule. These purposes implied among other things that validating the close alignment of the specific numerical projections with the details of the global SRES A2 and B1 scenarios was not a priority.

---

[5] The four regions are: (1) "OECD90," all member countries of the Organization for Economic Cooperation and Development as of 1990; (2) "REF," reforming or in-transition economies, including Eastern European nations and former states of the Soviet Union; (3) "ASIA," all developing economies in Asia, and (4) "ALM," other developing countries including those in Africa, Latin America, and the Middle East.

Table 1 United States economic and population growth: Historical and SRES projections

		Period	Average annual growth rates in percent		
			Real GDP	Population	Real GDP per capita
Historical		1950–2000	3.48	1.25	2.20
		1975–2000	3.35	1.07	2.25
		1985–2005	3.03	1.11	1.90
Projected	A2–ASF	2000–2020	1.91	0.90	1.0
		2000–2050	1.80	0.79	1.0
		2050–2100	1.83	0.82	1.0
	B1–IMAGE	2000–2020	2.63	0.78	1.84
		2000–2050	1.96	0.64	1.31
		2050–2100	1.23	0.36	0.86

Real GDP in chained 2000 dollars. Sources for historical data: U. S. Bureau of Economic Analysis 2008 (GDP); U. S. Energy Information Administration (2007) (population)

It is worth pointing out that in addition to practicality this approach also stems from an implicit weighing of the costs and benefits that would be associated with a more elaborate analysis given the pervasive and irreducible uncertainties intrinsic to long-run projections of this type in conjunction with their essential purpose in the present context. These uncertainties attend both to the particulars of California and to the conformance between the California projections and the original SRES scenarios, and imply that the use of more elaborate quantification tools per se—such as numerical downscaling or integrated modeling, or both—would not have clearly yielded additional value. When the time and expense of developing and/or implementing such tools is taken into account, it is even more questionable whether the benefits of doing so would have exceeded the costs.

Nevertheless, there are at least two potential liabilities of the approach taken here. One is the reliance on expert judgment to a greater extent than might apparently be the case with, for example, formal modeling. It is also true, however, that long-range quantitative scenario modeling also relies on expert judgment to a considerable degree, in a manner that is not always transparent to non-expert users of model output. The other is the possibility of hidden inconsistencies among different estimates. Ensuring internal consistency of joint assumptions and calculations is commonly posited as a key reason to undertake formal, quantitative modeling for scenario analysis, and that capability is foregone in our informal approach.

3.2 Interpreting the A2 and B1 scenarios for California

Quantitative scenario analysis in energy and climate policy applications has generally been organized around "baseline" or "business as usual" cases with "policy" cases represented as fundamentally incremental departures there from. This reflects, among other influences, the static cost-benefit origins of the economic models that are typically the basis of such analysis. The SRES departure from this convention reflects in part the essential impossibility of defending any particular projection of the society and economy as a base or business as usual case a century into the future.[6]

---

[6] As noted in the SRES report, the choice of an even number of broad storylines was deliberate, to prevent selecting a "middle" case as an implicit baseline.

This logic applies *a fortiori* to climate change-related scenarios during the present era of rapid developments in state, national, and international efforts to significantly reduce GHG emissions. While we regard as increasingly likely the implementation of a coordinated global emissions reduction policy in the coming decades, we cannot foretell the timing, magnitude, costs, or other aspects of the ultimate portfolio of abatement strategies that national governments will implement individually and collectively. In California, while AB 32 implementation is underway, the strategy that the state will follow after 2020 to reach longer-term GHG reduction goals is in the earliest stages of determination.

Addressing these issues in California scenarios of key socioeconomic and demographic trends that are broadly consistent with the global SRES A2 and B1 presents a particular challenge in the A2 case. The SRES emphasizes that its blanket assumption of no global GHG abatement regime does not exclude mitigation actions at sub-global levels that may be consistent with individual storylines and scenarios. In the A2 case, however, global, regional, and national emissions continue to rise. Thus, a narrow interpretation requires a scenario in which California achieves a transition to a low-carbon economy and energy system by mid-century while the United States as a whole, as well as the rest of the world, continues to increase GHG output. We regard this outcome as implausible at best. On the one hand, as noted at the outset, efforts to design and implement policies to achieve significant GHG reductions are well underway around the world; successful implementation and execution of AB 32 by California will stimulate these efforts. In the longer term, a demonstration by California of progress toward the mid-century emissions reduction called for in Executive Order S-3-05 (EO S-3-05) would similarly provide support for comparable progress elsewhere.[7] On the other hand, by the same token, any problems California might encounter reaching either its near-term or long-term GHG targets could reasonably be expected to impede the abatement efforts of other governments. We would here draw an analogy with electricity restructuring, a case in which the California experience affected other states' regulatory strategies and decisions.[8]

Thus, with the present state-of-play of GHG reduction policy around the world, the A2 must be interpreted somewhat differently than was appropriate at the time of the SRES report. It now can be viewed as describing a future in which current efforts to reduce GHG emissions within OECD countries as well as sub-national entities (notably U. S. states and Canadian provinces) fail to reach fruition, and the developing world does not itself undertake or achieve significant reductions.

By contrast, despite the assumption of "no global policy," the B1 case is consistent both in general and in key specifics with a storyline in which aggressive worldwide emissions mitigation policy is undertaken and results in a peaking and then decline in emissions globally and regionally in this century. B1 is not a "stabilization" scenario of the type that is increasingly the focus of policy, for example, in the mid-century goal established in EO S-3-05 and contemplated in proposed U. S. Congressional legislation. In the IMAGE B1 marker scenario, OECD and U. S. emissions peak and begin to decline by mid-century, but are not on the order of eighty percent below 1990 or 2000 levels by 2050. Nevertheless, with reasonable flexibility of interpretation regarding exact levels and timing, B1 can be seen as describing a future in which

---

[7] EO S-3-05, issued by the Governor in June 2005, officially recognized the risks of climate change and the necessity for GHG emissions abatement policies, and among other provisions established the goal of reducing California statewide GHG emissions to 80% below 1990 levels by 2050.

[8] It is certainly possible to imagine a scenario in which problems with policy implementation in California yield lessons that other governments apply to *improve* their own processes. But in this case, one would in turn expect a "feedback" to California, so that the overall process becomes in a sense self-correcting. The point is that, in this case also, California's and the rest of the world's GHG emissions trajectories would in the long term be convergent.

Table 2 Comparison of historical economic growth rates, U. S. and California

Period	Units	Average annual growth rates in percent	
		U. S. Gross National Product	California Gross State Product
1963–1997	SIC-based, current dollars	7.94	8.37
1997–2007	NAICS-based, current dollars	5.34	6.07
1990–1997	SIC-based, chained 2000 dollars	2.98	1.6
1997–2007	NAICS-based, chained 2000 dollars	2.89	4.06

U. S. Bureau of Economic Analysis (2008). Series are separated at 1997 due to inconsistencies in state-level data between the SIC (Standard Industrial Classification) and NAICS (North American Industrial Classification System)

California's and the United States' long-term efforts to radically reduce emissions succeed, including the maintenance of robust economic growth and improving living standards while transitioning to a low-carbon energy system. As in the SRES interpretation of B1, a variety of social and market drivers would contribute to this future, but in contrast to that interpretation, in this alternative view of B1 policy plays a fundamental role.

3.3 Historical U. S. and California economic growth

Since the early 1960s, growth in California's gross state product has on average exceeded that of the U. S. GDP, although not in every year or succession of years. This is summarized in Table 2, which presents various measures of state and national economic growth in this period.

3.4 California economic growth projections

The A2 and B1 storylines and U. S. economic growth projections shown in Table 1 suggest corresponding "lower" and "higher" growth scenarios for California, as well as their respective relationships to the U. S. scenarios. In its most recent forecast, the California Energy Commission projected annual state economic growth to average 3.1% from 2006 to 2011, declining to 2.5% from 2011–2018. We use these figures for transition between recent historical and long-run future trends; both lower and higher growth scenarios match these growth rates to 2020. In the lower growth scenario, this rate then declines to 2% until 2050, and then to 1.8% to 2100. In the higher growth scenario, economic growth remains at 2.5% annually to 2050, and is then 2% until 2100. These scenarios and their implications for the size of California's economy over the coming century are summarized in Table 3; note that the California Gross State Product in 2007 was an estimated $1.55 Trillion chained 2000 dollars (U. S. Bureau of Economic Analysis 2008).[9]

These two scenarios are broadly consistent with the SRES A2 and B1 storylines, respectively, in envisioning slower-than-historical rates of economic growth. They also assume a continuation of the observed historical pattern of California's growth being somewhat higher than the national rate, with respect to the SRES United States projections.

---
[9] These scenarios were developed prior to the disruptions in the second half of 2008 in the financial system and the macroeconomy. As a consequence, our near-term growth estimates may be unrealistically high unless there is a fairly rapid—i.e., within several years—global economic recovery. Assuming that such a recovery occurs over the coming decade, however, we do not view current events as materially affecting the plausibility of the longer-term estimates.

Climatic Change (2011) 109 (Suppl 1)

Table 3 Two scenarios of California economic growth

Period	Average annual growth rates in Gross State Product, in percent		Gross State Product at end of period, trillions of chained 2000 dollars	
	Lower growth	Higher growth	Lower growth	Higher growth
2008–2020	2.5	2.5	$2.14T	$2.14T
2020–2050	2.0	2.5	$3.87T	$4.48T
2050–2100	1.8	2.0	$9.44T	$12.01T

While other specific values could be chosen, it is the case that in keeping with the remarks in Section 3.1, above, these qualitative criteria reflect the study's philosophy and purpose, and would not provide a firm basis for ranking among different such choices. (The same statement applies to the numbers reported in subsequent sub-sections.) The end consequence of growth differential in the two scenarios is an approximately 30% difference in the size of the state economy in 2100. Note, however, that in both cases this size is immensely increased from its present-day magnitude. To understand the implications of these two scenarios for per capita state income, we first discuss scenarios of future state population growth.

3.5 California population projections

The manifold uncertainties associated with long-run social and economic scenarios are exemplified by population forecasting. The underlying processes governing population growth are difficult to forecast and include economic, political, environmental, technological, social, and behavioral elements. Social scientists do not fully understand how these elements have shaped fertility, mortality, and migration in the past, let alone how each of these forces might change in the future. Moreover, long-term projections and projections for small areas, including states and particularly counties, are even less certain than projections for large areas. Climate change could affect demographic forces directly, for example by increased mortality associated with high temperatures (or decreased mortality associated with a reduction in cold weather), or indirectly, for example through climate-induced changes in local economies.

For this study, three sets of population projections for California and its counties to 2100 were developed: A low series, a middle series, and a high series. The projections include breakdowns by age, gender, ethnicity, and nativity (U.S. born and foreign born). A cohort component model was used in which the population was aged over time by applying mortality and migration rates. New cohorts were created by applying fertility rates to women of childbearing ages. Projections were developed for every 5 years from 2005 to 2100. Existing national population projections were used to estimate the size of populations providing the sources of migrants to California. Past trends in migration, fertility, and mortality rates in California were used to develop future rates, which are shown in Table 4.

The three sets of projections represent three very different demographic futures. In the low series, population growth slows as birth rates decline to levels similar to those experienced in California during the nadir of the baby bust, migration out of the state accelerates, international migration slows quickly and considerably, and mortality rates show little improvement. In the high series, population growth accelerates as birth rates increase, migration increases, and mortality declines. In this series, total fertility rates increase to 2.6 children per woman, well below the baby boom peak of 3.6 reached in 1961 in California, but much higher than the replacement level of 2.1. This scenario is consistent with strong economic growth in California (recall that the baby boom occurred during a period of rapid economic

Table 4 Components of change assumptions for statewide population projections

		Net international migration (thousands per year)	Net interstate migration (thousands per year)	Total fertility rate[a]	Mortality rate[b]
Low series	2005–2010	161	−113	2.15	1.00
	2020–2025	26	−63	2.06	0.98
	2045–2050	1	−71	1.93	0.96
	2095–2100	0	−1	1.64	0.94
Middle series	2005–2010	190	−90	2.15	0.98
	2020–2025	225	−30	2.09	0.95
	2045–2050	225	−30	2.09	0.90
	2095–2100	50	−25	2.09	0.85
High series	2005–2010	220	7	2.23	0.98
	2020–2025	240	45	2.30	0.92
	2045–2050	250	50	2.46	0.80
	2095–2100	360	100	2.64	0.67

[a] The total fertility rate is the average number of children a woman will have over her reproductive years

[b] Age-specific mortality rates relative to 2005

growth) and continued global disparities in income, as observed in the A2 storyline, that lead to large and sustained flows of international migrants to the state.

The middle series assumes that future growth in California will be similar to patterns observed over the state's recent history, including a moderation of previous growth rates but still large absolute changes in the state's population. In the middle series, international migration flows to California remain strong to mid-century and then subside, net domestic migration remains negative but of small magnitude, fertility levels (as measured by total fertility rates) decline slightly, and age-specific mortality rates continue to improve. The middle series projections could be thought of as consistent with the B1 storyline. Growth rates in California slow, but absolute increases remain large as the state's economy continues to attract international migrants from developing countries that continue to experience strong population growth.

As indicated in Table 4, the key drivers of uncertainty are first migration, and second fertility. For many decades, changes in age-specific mortality rates been relatively stable, and the projections assume that no long-lasting catastrophic events will occur. Migration, in contrast, has been quite volatile, with the state experiencing both large net flows into and out of the state at different times within the past 25 years. Fertility changes have also been notable, with total fertility rates in California falling from 3.6 children per woman during the height of the baby boom to 1.7 children during the nadir of the baby bust.

Table 5 displays aggregate results from the California projections created for this project along with national projections developed by the Census Bureau as well as a state projection developed by the California Department of Finance. The level of uncertainty reflected in these projections is striking. The national projections reflect even greater relative uncertainty for the nation, with the high series four times as great as the low series. To the extent that the U.S. population determines one of the most important pools for potential migrants to California, the widely divergent national projections are necessarily incorporated into the California projections.

Divergence among our California projections is not large in 2020, but increases thereafter. This reflects the compounding effects of the demographic components that drive population change. For perspective, in 1900 California's population was 1.49 million and by 2000 the state's population had grown 23 times larger to 34 million. For one comparison, we

Table 5 Population projections for California and the United States

	U.S. Projections[a]			California Projections			CA-DOF[b]
	Low	Middle	High	Low	Middle	High	
2010	291,413	308,936	310,910	38,862	39,170	39,896	39,136
2020	303,664	335,805	354,642	41,978	44,184	46,863	44,136
2050	313,546	419,854	552,757	44,204	59,283	69,376	59,508
2100	282,706	593,820	1,182,390	43,835	85,264	147,698	

[a] The U. S. "Low" and "High" projections are the Census Bureau's "Lowest" and "Highest" series, respectively, in its 2000 projections (U. S. Census Bureau 2000). The U. S. "Middle" projection to 2050 is based on the Census Bureau's 2004 interim projection series; from 2050 to 2100, the Census Bureau's 2000 series was used but modified in light of the 2004 series

[b] State of California, Employment Development Department (2007)

might turn to Japan, a country of similar geographic size to California. In 1900, Japan was home to 43 million people, somewhat more than California's current population of about 38 million. By 2000, Japan's population had reached 127 million (similar to the high series projection for California for 2100). In contrast to the explosive growth of the twentieth century, projections for Japan to 2050 suggest the country will lose about 20% of its residents, declining to a population of 103 million by 2050.

Global population projections also exhibit considerable uncertainty. Even without the volatile and most difficult to project component of migration, United Nations projections to 2050 place the world's population at 7.8 billion in the low variant to 10.8 billion in the high variant, an increase of 20% over the 2005 global population in the low variant and 65% in the high variant (United Nations 2006, 2007).

Finally, it should be noted that there is uncertainty about the state's current (c. 2008) population, with the California Department of Finance estimating that the state's population is about one million higher than Census Bureau estimates for the state. It is not surprising, then, that population projections to 2100 exhibit such great uncertainty.

3.6 California economic growth in per capita terms

The economic wealth of a state or nation is a function not just of absolute output but also of per capita output. A given sized economy (in dollar or other currency terms) can reflect vastly different standards of living and social conditions depending on the population size as well as the distribution of income. Thus, in the long run economic and population growth are closely related in determining the wealth of a society. Recall (cf. Table 2) that California's growth rate of overall economic output had generally exceeded that of the U. S. in recent decades. Table 6

Table 6 Growth of real personal income, U. S. and California, 1959–2006

	Average annual growth rates in percent	
	U. S.	California
Personal income	3.05	3.9
Per capita personal income	2.40	2.1

California Dept. of Finance

Table 7 Per capita income in California under lower and higher economic growth scenarios, middle population projection

Decade	Average annual growth rate in percent		End-of-decade per capita personal income in thousands of chained 2000 dollars	
	Lower	Higher	Lower	Higher
2010–2020	1.10	1.10	40.3	40.3
2020–2050	1.01	1.5	54.5	63.1
2050–2100	1.1	1.26	92.4	118.0

shows that, in per capita terms, population trends resulted in California's per capita income growth being somewhat lower than the nation as a whole from 1959 through 2006, despite a higher absolute growth rate.

Table 7 presents two scenarios of long-run per capita income growth in California implied by our lower and higher economic growth scenarios jointly with our middle population scenario.

Note that the growth rate of per capita income in the lower growth case approximately matches that of the SRES A2 U. S. projection, while the higher growth case is substantially higher starting in 2020, reaching the recent historic trend in mid-century. As in the case of absolute economic growth, the lower-than-historical future rates still imply a substantial increase in the income of individual Californians over the coming century.

3.7 Labor productivity implications

The population and economic growth projections for this report were developed separately. By contrast, a procedure that is applied in "integrated assessment" models of the type used in the SRES is to project economic growth rates as a function of population growth, labor force participation, and labor productivity (Smith 2008). We here review the long-run California labor productivity rates that are jointly implied by our economic and population growth projections.[10]

Because we have not projected future California labor force growth per se, we use population growth as a proxy. To compare with recent historical trends: According to data of the California Department of Finance and Employment Development Department, from 1981 to 2008 the annual average state population growth rate exceeded the labor force growth rate by 0.06%.[11] Thus, while labor force projections are beyond the scope of this study, assuming that the labor force grows at the rate of population increase in the long term may be considered a reasonable approximation for illustrative purposes.

Table 8 shows the labor productivity rates implied by the population and GSP growth rates for the 2020–2050 and 2050–2100 periods, by the low, middle, and high growth projections, under the assumption that the labor force growth rate equals that of population.

Again by way of historical comparison, Bauer and Lee (2006) estimated California's labor productivity growth rate to have been approximately 1.7 annually from 1977 to 2004. Wilson (2002) noted that California's productivity exceeded that of the rest of the U. S. from 1986 through 2000. Given the derivation of our economic growth scenarios from the SRES, it is also worth comparing the estimates in the table with those reported there. SRES noted the historical U. S. productivity growth rates (in annual average increase in GDP per

---

[10] We are indebted to Steven Smith for highlighting the importance of this issue.
[11] Authors' calculations using State of California, Employment Development Department (2005, 2009) and State of California, Department of Finance (Undated, 2007, 2008).

Climatic Change (2011) 109 (Suppl 1)

Table 8 Labor productivity growth rates implied by GSP and population scenarios, 2020–2050 and 2050–2100

GSP scenario	Average annual labor productivity growth rates in percent by population scenario		
	Low series	Middle series	High series
2020–2050 Low	1.8	1.0	0.7
2020–2050 High	2.5	1.5	1.2
2050–2100 Low	1.6	1.1	0.3
2050–2100 High	2.0	1.3	0.5

hour worked) of 2.3% from 1870 to 1973, and 1.1% from 1973 to 1992.[12] According to the SRES, productivity assumptions were not directly comparable across the numerical models used to project the scenarios, but labor productivity projections ranged from 0.79% to 5.85% across the scenario families and world regions.[13] While no further detail is given, we assume that U. S. rates in the scenarios were on the lower end of that range.

More recent U. S. national data show a return to near the long-term historical level.[14] In addition, the U. S. Energy Information Administration recently projected national labor productivity improvement of 1.9% on average to the year 2030 (U. S. Energy Information Administration 2008b).

We conclude that, while the implicit labor productivity projections shown in the table are generally plausible, those below 1% implied by the High population growth scenario represent a very pessimistic view of future California productivity. By contrast, those implied by the Middle series and High series are consistent with the views that future productivity growth will tend toward the low end or the high end, respectively, of historical experience.

These results highlight the value of incorporating explicit assumptions regarding future productivity trends in any subsequent California climate change scenario studies of this type.

3.8 California urbanization projections

Urbanization projections were developed for this project to provide urban extents and spatially distributed population output for use in sector-specific impact studies. These outputs are twofold for each of the three population scenarios described in Section 3.3: Projections of the spatial urban footprint in California over the coming century, and scenario-derived population projections allocated to the urban footprint. In this section we describe the procedures used for these projections, and highlight the results for several California cities.

3.8.1 Creating urban footprint layers

We built on the methods and results of Landis (2001) and Landis and Reilly (2003), who used a spatial-statistical model to project California urban growth and development—the urban "footprint"—to 2100. Their projections were computed for the years 2000, 2020, 2050, and 2100 at a one-hectare spatial resolution for 38 counties in California, excluding

---

[12] Table 3–3, p. 126, IPCC 2000.
[13] Box 4–5, p. 198, IPCC 2000.
[14] Authors' calculations from non-farm business sector data in Table B-49, "Productivity and related data, business and nonfarm business sectors, 1959–2007," in United States President (2008a, b).

the northernmost and least populated counties for (Landis and Reilly 2003). These results were the starting point for creating urban "seeds"—complete, statewide estimates of urban extent—for 2000, 2020, 2050, and 2100. To extend them to all state counties—particularly those in sparsely populated, rural regions, we drew upon geospatial data from the Nighttime Residential Population dataset produced at Los Alamos National Laboratory (LANL) (McPherson and Brown 2003). This dataset modifies the vector-based U.S. Census dataset, which is itself a nighttime estimate of population, and attributes population using data products from GeoData Technologies, Inc. The LANL dataset attributes grid cells of 250 m with the population from the corresponding Census blockgroup.

To complete the seeds for the projection years, we first added urban extent from the LANL data. Under the assumption that expanding population will locate either in or nearby current urban extent, albeit at different densities than today, we defined the following thresholds for the additions of urban extent from the LANL data set:

For year 2000 "seed" layer, densities >40 people/km^2 are "urban"
For year 2020 "seed" layer, densities >20 people/km^2 are "urban"
For year 2050 "seed" layer, densities >8 people/km^2 are "urban"
For year 2100 "seed" layer, any gridcell with any nighttime population is "urban"

While these criteria define a very low population density, they were chosen to account for the highly dispersed nature of California population because, in rural regions, this low density has an "urban influence" on the natural environment. This threshold approach will, nevertheless, result in under-estimation of the very sparsely populated regions not modeled by Landis and Reilly. We therefore further augmented the Landis and Reilly data by incorporating the spatial centroid of every U.S. Census as an urban cell to account for rural regions with very sparse population.

With this completion of the urban seeds for 2000, 2020, 2050, and 2100, the intervening quinquennial years were created by simply "growing" the seed layers. Under the assumptions that all urban extents grow at constant rates, and that urban regions grow from their edges outward (and not from "spotting," as modeled in Clarke and Gaydos 1998), seeds for the intervening years were created by adding urban pixels to a seed layer in five year increments. This was performed iteratively, until reaching the next seed layer.

3.8.2 Future population allocation

Once the quinquennial future urban footprint layers were created, future population was allocated to create a spatially distributed population map in two steps; this allocation was calculated for each of the three projections described in the previous section. First, population was allocated for each time step by county, which was the scale of the demographic forecasts. We assumed isotropy within counties, i.e., within the urban footprint or urban extent, all population is distributed evenly, at a uniform density, and divided population evenly by the number of urban pixels for each county per time step. The pixel-based population value was then attributed to each pixel so that the sum of all pixels in each county provides the total county population for that time step.

Second, this county-level allocation was refined to provide additional detail at the Census Tract level. Using data from the 2000 U.S. Census, the relative population share of every tract in each county was calculated. Because forecasting the change of these shares over the coming century was beyond the scope of this project, we pragmatically assumed that these shares would be fixed for all future years. Next, the county population for each time step was divided by the geographical size of each tract, according to their relative share. Each pixel was then attributed with the appropriate number of people in each tract, reflecting the county-level population. Finally, the numbers of households were calculated, using the proportion of number of households in each region, based on California Energy Commission earlier work (Abrishami et al. 2005). Household proportions were held constant for the study period; future investigations can incorporate explorations of dynamic household proportions in each region.[15]

3.8.3 Results for selected metropolitan areas

As would be expected, the statewide results reflect significant differences in the overall "footprint" projections. In the low population-based projection, large regions of sparse population in California persist over the coming century, and extant urban regions will shrink in population. In the high population results, existing cities become denser and populations increase in rural regions. The middle population projection, while more "moderate" than the high, nevertheless envisions dramatic changes compared to today's urban landscape.

To better understand these trends in changing urban density, we examined four regions in greater detail. These metropolitan regions are formally described by the U.S. Census Bureau as "Urbanized Areas" that "meet minimum population density requirements, along with adjacent densely settled census blocks that together encompass a population of at least 50,000 people." (ESRI 2007, U.S. Census Urbanized Areas Metadata) They are:

1. City and County of San Francisco
2. City of Fresno
3. Greater Los Angeles region (Los Angeles—Long Beach—Santa Ana)
4. City of San Diego.

The boundaries of these regions, as recognized by the Census Bureau, were used as "bounding boxes" for population tallies. While this assumption does not account for inevitable future changes in city boundaries, it does allow us to compare and validate the relative and absolute densities of the cities as the model progresses through time.

Population density projections for these metropolitan regions are summarized in Table 9. In the low series, San Diego's density remains nearly constant due to a relatively level population growth and a small expansion of urban extent. San Francisco and Los Angeles densities peak mid-century and then decline; indeed, the de-population of San Francisco is rather dramatic in this projection. Fresno's doubling of population is matched with its urban expansion at a fairly even rate.

In the middle series, San Francisco density is almost unchanged in 2100, while San Diego and Los Angeles show comparable increases, of seventy-four and sixty-three percent, respectively. The most dramatic change is in Fresno, where density increases almost four-fold by 2100.

By contrast, San Francisco density nearly doubles in the high series, and both San Diego and Los Angeles densities increase significantly relative to the middle series. The greatest projected percentage change is again in Fresno. It is worth noting that while some of the forecasted urban densities in the high series (San Francisco in

---

[15] This allocation methodology was slightly augmented for the Low and High population scenarios to account for prospective shifts in the density of urban core regions. In essence, the Low series population forecast emphasizes a shift of population to the urban areas, while the high series emphasizes a shift away from the urban core. In addition, to avoid artificially skewing relative rankings in projected population growth, an additional adjustment was made for San Quentin State Prison in Marin County, which appears in U. S. Census Bureau estimates as the densest location in California.

Table 9 Population densities (persons/sq km) for selected cities over time, by population scenario

Year	Scenario	City			
		San Francisco	San Diego	Fresno	Los Angeles
2000		6966	1843	1725	2913
2020	Low	7062	2039	2179	3339
	Middle	7327	2126	2296	3431
	High	8294	2254	2330	3688
2050	Low	4824	1975	2604	3025
	Middle	7285	2568	3494	3941
	High	9636	3069	3677	4804
2100	Low	2790	1986	3229	2466
	Middle	6780	3212	6661	4735
	High	12895	6243	8593	7829

particular) appear extreme, they are comparable to current urban densities elsewhere in the world (Cox 2008).

3.9 Future costs of driving in Southern California

In this section we apply our scenario results to project key inputs—household income and costs-of-driving—for use in the companion study on valuing economic impacts of climate change on Southern California beaches.

3.9.1 Household income

Section 3.4 presented two statewide per capita income growth scenarios based on the middle series population projection presented in Section 3.3. Passing from statewide per capita to regional household income in principle requires additional information and assumptions regarding trends in California household composition that in turn are related to the evolution of other demographic factors. Full accounting for these details is beyond the scope of this study, and we apply the statewide aggregates directly with the following simplifying assumption. According to the California Energy Commission (Energy Commission), California's mean household size has increased in recent decades, and the Energy Commission's most recent energy demand forecast projects that it will reach 3 persons per household in the next decade (Marshall et al. 2007). Yi et al. (2006), in their "medium" projection, posit that average household size nationally will decline from 2.6 in 2000 to 2.4 in 2020 and remain stable thereafter. We adopted the assumption that average household size will remain constant after 2020, so that household income will grow at the same rates as per capita income shown in Table 7.

3.9.2 Cost of driving

The future cost of driving in Los Angeles and Orange counties will of course be determined by a range of local, regional, national, and international factors, both policy and market driven. These will include evolving cultural priorities and consumer preferences, fundamental shifts or stability in the structure of the global energy system, technological advances in vehicles, and local land-use planning and patterns. A comprehensive scenario analysis of these influences and their implications for travel costs is again beyond the scope of this study; we consider two plausible scenarios that relate future possibilities to historical and current trends.

The rapid and dramatic increase and subsequent fall in the world price of petroleum and U. S. gasoline prices in 2008 serve as a reminder of the unpredictability of fundamental determinants of transportation costs and highlight the importance of historical perspective. Despite the oil shocks of the 1970s and early 1980s, transportation expenditures as a percentage of personal consumption expenditures among U. S. consumers were on average flat in real terms from 1970 through 2005, and they never exceeded 14% during this period (compared to 12.6% in 1970 and 12% in 2005).[16] The real cost of driving (nationwide) grew at an average of 0.7% annually from 1985 through 2005, but the share of gas and oil costs fell, and these latter costs on average were also essentially flat.[17]

Oil prices follow quite different trajectories under the SRES A2 and B1 scenarios, as shown in Table 10.

In interpreting the implications of these projections for California scenarios, however, it is important to bear in mind that the relationship between oil prices and driving costs is likely to become more complicated. Comparable future driving costs could be achieved under quite different socioeconomic scenarios and oil price levels. For example, a continued dependence on petroleum and the internal combustion engine could lead to development of new, non-traditional fuel supplies and therefore gasoline prices that are relatively moderate in the long-run, although higher than historical levels.[18] This indeed might be one interpretation of the A2 projection. However, a similar outcome could occur under a radical shift in the transportation system toward alternative technologies and fuels, as might be envisioned under the B1 scenario. The reason is that such a shift could result in lower vehicle life-cycle operating costs for consumers, as has been projected in California (Arthur D. Little, Inc. 2002). Moreover, it is easy to imagine a scenario that began along one path but engendered a shift to another: Just as the increase in driving costs in the United States during the 1970s stimulated the introduction and penetration of fuel efficient vehicles, so might the persistence of current, extremely high gasoline prices lead to a similar shift in the vehicle fleet.

These complications notwithstanding, two cost scenarios are presented in Table 11. The first is a slight permanent increase in the growth in the cost of driving from recent trends; the second, a permanent approximate doubling of this growth. We assume, naturally, that the evolution of the regional transportation infrastructure is such that, while alternative modes may become more available, personal transport in essentially its current form remains readily accessible—that is, people driving vehicles on freeways and surface roads. These projections are based on an estimate of a $0.145 per mile cost in the year 2000 by Pendleton et al. (2008).

3.10 Electricity prices

This section develops several scenarios of future California residential statewide electricity prices for use in the companion study of potential climate change impacts on state residential electricity demand.

---

[16] See Table 10.13 in Davis and Diegel (2007).
[17] Table 10.11, Davis and Diegel, *op cit.*
[18] A supply curve for global petroleum resources including unconventional sources is presented in Farrell and Brandt (2006).

Table 10 Oil price increases in SRES A2 and B1 scenarios

Year	Percentage increases (in real dollars) over year 2000 price	
	A2 – ASF model	B1 – IMAGE model
2020	7%	32%
2050	41%	92%
2100	70%	170%

The SRES provides only limited information on possible future electricity prices, and none for the A2 or B1 scenarios. This reflects in part both methodological differences among the models used in the study and limitations of specific models. For this reason, we draw upon alternative sources.

Attempting to project California electricity prices, even in a scenario context, raises a host of issues that are more complex than those involved in long-term aggregates such as population and economic growth. Projecting these prices amounts to positing, at a minimum, how the state, Western United States, and national electricity systems will evolve under GHG regulation, a situation that remains profoundly uncertain. This is illustrated in a recent assessment by the Congressional Research Service of model-based projected economic impacts, including price impacts, of Senate Bill S. 2191, the Lieberman-Warner Climate Security Act of 2007 (Parker and Yacobucci 2008). Among six models, estimated increases in national average electricity prices under S. 2191 range from under 15% to nearly 130% by 2030. An even wider range of uncertainty was revealed in a recent model analysis of global atmospheric GHG concentration stabilization scenarios conducted by the U. S. Department of Energy. Among three economic models, carbon prices needed to achieve several stabilization targets varied by an order of magnitude (Clarke et al. 2007). It is important to emphasize that such uncertainty in outputs persists after several decades of model development and analysis, and shows no signs of narrowing.[19]

Figure 1 depicts real statewide average residential electricity prices in California from 1970 to 2005.[20] Following an increase over the decade of the 1970s, these prices have been on average flat for most of the past thirty years, regular fluctuations notwithstanding; 1980 and 2005 prices were approximately equal.

A recent analysis of AB 32 compliance within the electric power sector conducted for the California Public Utilities Commission projected an average statewide rate increase of 30% over 2008 levels in 2020 in a scenario of accelerated energy efficiency and renewable energy deployment (Price 2008). This projected increase is comparable to that in the 1970s shown in Fig. 1. It is also interesting to compare this finding with one of the above-mentioned recent national studies of S. 2191, the Lieberman-Warner Climate Security Act of 2007, by the U. S. Environmental Protection Agency (2008). S. 2191 aimed to achieve national GHG annual emissions reductions of approximately 11% below 1990 levels by 2030. A computable general equilibrium model simulation estimated that this would entail national average electricity price increases of 44% over 2010 levels by 2030.[21]

---

[19] The significance and potential sources of the large uncertainty in model-based estimates of the costs to the United States of large-scale GHG abatement policies are discussed by Fischer and Morgenstern (2006).

[20] Nominal prices were obtained from the U. S. Energy Information Administration (2008a). Real prices were calculated using the implicit price deflator for personal consumption expenditures (U. S. Bureau of Economic Analysis 2008).

[21] Results from the Applied Dynamic Analysis of the Global Economy (ADAGE) model (Ross 2008).

Table 11 Projections of the cost of driving, 2020–2100

Year	Cost per mile in 2000 dollars	
	Lower growth ($)	Higher growth ($)
2020	0.18	0.22
2030	0.20	0.26
2050	0.24	0.39
2100	0.39	1.05

Work to-date on AB 32 and other California GHG policies has not encompassed potential price trajectories in the long term, and it can be expected that as this kind of analysis emerges the uncertainties will be comparable to those from national-level studies. To construct plausible scenarios, it is first useful to recall the magnitude of California's 2050 target. The California Air Resources Board has estimated California's 1990 GHG emissions as 427 million metric tonnes of carbon dioxide equivalent (MMTCO2e), and projected baseline 2020 emissions as 596 MMTCO2e (California Air Resources Board 2007, 2008). Thus, reductions of 169 MMTCO2e from baseline levels are required by AB 32. The 2050 goal of 80% below 1990 requires an additional 342 MMTCO2e of reduction, that is, twice again as much. It is not currently known how this target will be met, whether in terms of the mix of sectoral reductions, the specific technologies, or the portfolio of policies and measures.

Additional model results on S. 2191 from the U. S. Environmental Protection Agency are relevant. Beyond the previously noted 2030 target of 11% below 1990 levels, this legislation aimed for national GHG reductions to approximately 25% below 1990 levels by 2050. For this emissions trajectory, electricity prices in the Western U.S. region were projected to increase by 46% over 2010 levels by 2030 but recede to 34% over 2010 levels by 2050 (Ross 2008).

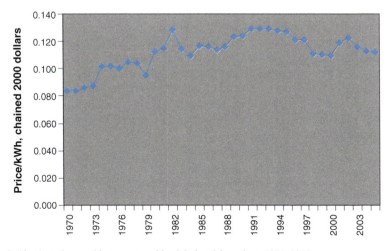

Fig. 1 California real statewide average residential electricity prices, 1970–2005

Against this background, we posit the following two price scenarios: (A) A 30% increase as of the 2020–2040 period, followed by flat prices for the remainder of the century, and (B) The same 30% increase followed by another 60% increase by mid-century and flat prices thereafter. For the coming several decades, the first scenario is consistent with the recent California Public Utility Commission results. A flat trajectory thereafter can be interpreted in several ways. An "optimistic" interpretation is that it reflects technological breakthroughs that allow the widespread deployment of very-low carbon electric power generation technology, as well as demand-side energy efficiency, without further upward price pressures. This might be considered as naturally fitting with the SRES B1 storyline. A "pessimistic" interpretation is that, following initial GHG policy successes, further progress toward the more aggressive mid-century target is stalled, and the status quo as of the 2030 time frame prevails. This might be considered as consistent with the A2 storyline.

The second scenario is again consistent with the existing near-term estimate, and also admits several interpretations for the "out-decades." The assumption of a price increase that is linearly scaled with the percentage emissions reduction can be seen as either optimistic or pessimistic. A pessimistic view might be that technological progress is insufficient to forestall significant price increases, while the optimistic counterpart would be that the unprecedented GHG reduction is achievable at no more than a doubling of prices. Assuming no further change for the remainder of the century reflects both the lack of reasonable information upon which to base any alternative assumption, and our view that in any case only rough orders of magnitude are meaningful in this context.

4 Conclusion

Steadily improving scientific understanding of the evolution of the global climate system has sharpened our knowledge of the potential risks from climate change, both those that can be reduced by GHG mitigation and those that appear to be unavoidable. Consequently, projecting and planning for climate change impacts has emerged in recent years as a necessary complement to developing and implementing mitigation policies. Adaptation to climate change may be the most challenging environmental problem human society has yet faced, for it requires dealing successfully with unprecedented scientific complexity and fundamental uncertainty about the interactions of natural and social systems far into the future.

Scenario studies are a basic tool for approaching these issues. This report has suggested a context for California climate change impact assessment consistent with the SRES scenarios, and provided projections of specific key variables to support specific impact studies. Our hope is that this work and the companion studies in this overall second assessment will contribute to the knowledge base for California's successful response to the climate challenge.

Acknowledgements We are indebted to Alexei Sankovski and William Pepper of ICF International, and Detlef Van Vuuren of the National Institute for Public Health and Environmental Hygiene (RIVM), The Netherlands, for making available detailed results from the SRES A2/ASF and B1/IMAGE scenarios, respectively; Hugh Pitcher of the Joint Global Change Research Institute (JGCRI) for his advice on the methodology and results of the SRES; and Martin Ross of the Research Triangle Institute for providing results from the ADAGE model analysis of S. 2191. We would like to thank Steven Smith of the JGCRI, Gary Yohe, and an anonymous referee for their invaluable comments and suggestions. Finally, we acknowledge and thank Mark Wilson for his outstanding editorial work.

References

Abrishami M, Bender S, Lewis KC, Movassagh N, Puglia P, Sharp G, Sullivan K, Tian M, Valencia B, Vidivar D (2005) Energy Demand Forecast Methods Report: Companion Report to the California Energy Demand 2006–2016 Staff Energy Demand Forecast Report. Staff Report CEC-400-2005-036

Arthur D Little, Inc (2002) Benefits of reducing demand for gasoline and diesel—vol. 3, Task 1 Report. Joint Report to the California Air Resources Board and the California Energy Commission, March 28

Bauer PW, Lee Y (2006) Estimating GSP and labor productivity by state. Federal Reserve Bank of Cleveland Policy Discussion Paper Number 16, March

California Air Resources Board (2007) California 1990 Greenhouse Gas Emissions Level and 2020 Emissions Limit. Staff Report, Jamesine Rogers, Primary Author, November, 16

California Air Resources Board (2008) Climate change draft scoping plan: a framework for change. June 2008 Discussion Draft

Clarke KC, Gaydos LJ (1998) Loose-coupling a cellular automaton model and GIS—long-term urban growth prediction for San Francisco and Washington/Baltimore. Int J Geogr Inf Sci 12(7):699–714

Clarke L, Edmonds J, Jacoby H, Pitcher H, Reilly J, Richels R (2007) Scenarios of greenhouse gas emissions and atmospheric concentrations. Sub-report 2.1A of Synthesis and Assessment Product 2.1 by the U. S. Climate Change Science Program and the Subcommittee on Global Change Research. Department of Energy, Office of Biological & Environmental Research, Washington, D.C. 154 pp

Cox W (2008) Demographia World Urban Areas (World Agglomerations). Wendel Cox Consultancy, Illinois, 91 pp

Davis SC, Diegel SW (2007) Transportation Energy Data Book—Edition 26. ORNL—6978 (Edition 26 of ORNL-5198); report prepared for the Office of Planning, Budget Formulation, and Analysis, Energy Efficiency and Renewable Energy, U. S. Department of Energy by the Oak Ridge National Laboratory

ESRI (2007) U.S. Census Urbanized Areas Metadata. www.esri.com/data/community_data/census/overview.html

Farrell AE, Brandt AR (2006) Risks of the oil transition. Environ Res Lett 1:014004

Franco G, Cayan D, Moser S, Hanemann M, Jones M (2011) Second California Assessment: Integrated Climate Change Impacts Assessment of Natural and Managed Systems. Climatic Change

Fischer C, Morgenstern RD (2006) Carbon abatement costs: why the wide range of estimates? Energy J 27 (2):73–86

IPCC (2000) Special Report on Emissions Scenarios. Working Group III of the Intergovernmental Panel on Climate Change. N. Nakicenovic, Lead Author; N. Nakicenovic and R. Swart, Report Editors. Cambridge University Press

Landis J (2001) CUF, CUF II, and CURBA: a family of spatially explicit urban growth and land-use policy simulation models. In: Brail RK, Klosterman RE (eds) Planning support systems: integrating geographic information systems, models and visualization tools. ESRI Press, Redlands, California, pp 157–200

Landis J, Reilly M (2003) How we will grow: baseline projections of California's urban footprint through 2100. Department of City and Regional Planning Institute of Urban and Regional Development, University of California at Berkeley, October, www.energy.ca.gov/reports/2003-10-31_500-03-058CF_A03.PDF

McPherson T, Brown M (2003) U.S. Day and Night Population Database (Revision 2.0)—Description of Methodology. LA-UR-03-8389. Los Alamos National Laboratory, 30 pp

Marshall L, Gorin T, principal authors (2007) California Energy Demand 2008–2018. Staff Revised Forecast. California Energy Commission Staff Final Report CEC-200-2007-015-SF2, November

Meehl GA, Stocker TF, Collins WD, Friedlingstein P, Gaye AT, Gregory JM, Kitoh A, Knutti R, Murphy JM, Noda A, Raper SCB, Watterson IG, Weaver AJ, Zhao Z-C (2007) Global Climate Projections. Chapter 10 in Climate Change 2007: The Physical Science Basis. In: Solomon S, Qin D, Manning M, Chen Z, Marquis M, Averyt KB, Tignor M, Miller HL (eds) Contribution of Working Group I to the Fourth Assessment Report of the Intergovernmental Panel on Climate Change. Cambridge University Press, Cambridge and New York

National Research Council (NRC) 2010 Describing Socioeconomic Futures for Climate Change Research and Assessment: Report of a Workshop. Panel on Socioeconomic Scenarios for Climate Change Research and Assessment. Committee on the Human Dimensions of Global Change, Division of Behavioral and Social Sciences and Education. Washington, DC: The National Academies Press

Parker L, Yacobucci B (2008) Climate change: costs and benefits of S. 2191. Congressional Research Service Report for Congress. Order Code RL34489, May 15

Pendleton L, King P, Mohn C, Webster DG, Vaughn RK (2008) Estimating the potential economic impacts of climate change on Southern California Beaches. California Energy Commission draft paper

Price S (2008) Electricity & natural gas GHG modeling: revised results and sensitivities. Energy and Environmental Economics, Inc., May, 13

Ross M (2008) Personal communication, August

Sanstad AH, Johnson H, Goldstein N, Franco G (2009) Long-run socioeconomic and demographic scenarios for California. California Energy Commission Final Paper CEC-500-2009-013F, August

Smith SJ (2008) Personal communication, December

State of California, Department of Finance (Undated) E-4 Population Estimates for California Cities and Counties, January 1, 1981 to January 1, 1990. Sacramento, California

State of California, Department of Finance (2007) E-4 Historical Population Estimates for City, County, and the State, 1991–2000, with 1990 and 2000 Census Counts. Sacramento, California, August

State of California, Department of Finance (2008) E-4 Population Estimates for Cities, Counties, and the State, 2001–2008, with 2000 Benchmark. Sacramento, California, May

State of California, Employment Development Department (2005) Historical Civilian Labor Force [Not Seasonally-Adjusted] 1976 to 1989; March 2004 Benchmark. Sacramento, California, April

State of California, Employment Development Department (2007) Population projections for California and its counties 2000–2050, by age gender and race/ethnicity, Sacramento, California, July

State of California, Employment Development Department (2009) Historical Civilian Labor Force [Not Seasonally-Adjusted]; March 2007 Benchmark. Sacramento, California, January

United Nations (2006) World Urbanization Prospects: The 2005 Revision. Population Division of the Department of Economic and Social Affairs of the United Nations Secretariat

United Nations (2007) World Population Prospects: The 2006 Revision. Population Division of the Department of Economic and Social Affairs of the United Nations Secretariat

United States President (2008) Economic Report of the President—Transmitted to the Congress February 2008, together with the Annual Report of the Council of Economic Advisers. Washington, D.C.: U. S. Government Printing Office, February

U. S. Bureau of Economic Analysis (2008) National Income and Product Accounts

U. S. Census Bureau (2000) Table NP-T1 in Annual Projections of the Total Resident Population as of July 1: Middle, Lowest, Highest, and Zero International Migration Series, 1999–2100

U. S. Energy Information Administration (2007) Annual Energy Review. Report No. DOE/EIA-0384(2007)

U. S. Energy Information (2008a) State Energy Data System. February release

U. S. Energy Information Administration (2008b) Annual Energy Outlook 2008. Report #DOE/EIA-0383 (2008), June

U.S. Environmental Protection Agency (2008) EPA Analysis of the Lieberman-Warner Climate Security Act of 2008—S. 2191 in 100th Congress. U. S. EPA Office of Atmospheric Programs, March 14

United States President (2008b) Economic Report of the President Transmitted to the Congress together with The Annual Report of the Council of Economic Advisers. United States Government Printing Office, Washington, DC

Wilson D (2002) Productivity in the Twelfth District. FRBSF [Federal Reserve Bank of San Francisco] Economic Letter Number 2002–33, November 8

Yi Z, Land KC, Wang Z, Gu D (2006) U. s. family household momentum and dynamics: An extension and application of the ProFamy method. Popul Res Policy Rev 25:1–41

Current and future impacts of extreme events in California

Michael D. Mastrandrea · Claudia Tebaldi ·
Carolyn W. Snyder · Stephen H. Schneider

Received: 12 February 2010 / Accepted: 26 September 2011 / Published online: 24 November 2011
© Springer Science+Business Media B.V. 2011

Abstract In the next few decades, it is likely that California must face the challenge of coping with increased impacts from extreme events such as heat waves, wildfires, droughts, and floods. This study presents new projections of changes in the frequency and intensity of extreme events in the future across climate models, emissions scenarios, and downscaling methods, and for each California county. Consistent with other projections, this study finds significant increases in the frequency and magnitude of both high maximum and high minimum temperature extremes in many areas. For example, the frequency of extreme temperatures currently estimated to occur once every 100 years is projected to increase by at least ten-fold in many regions of California, even under a moderate emissions scenario. Under a higher emissions scenario, these temperatures are projected to occur close to annually in most regions. Also, consistent with other projections, analyses of precipitation extremes fail to detect a significant signal of change, with inconsistent behavior when comparing simulations across different GCMs and different downscaling methods.

1 Introduction

Extreme events, such as heat waves, wildfires, droughts, and floods can cause significant damages every year and are responsible for a large fraction of climate-related impacts (Kunkel

---

This paper is adapted from Mastrandrea et al. (2009), PIER Research Report CEC-500-2009-026-D.

M. D. Mastrandrea (✉) · S. H. Schneider
Woods Institute for the Environment, Stanford University, Y2E2, Mail Code 4205,
473 Via Ortega, Stanford, CA 94305, USA
e-mail: mikemas@stanford.edu

C. Tebaldi
Climate Central, 1 Palmer Square, Suite 330, Princeton, NJ 08542, USA

C. W. Snyder
Emmett Interdisciplinary Program on Environment and Resources, Stanford University, Y2E2 Suite 226,
Stanford, CA 94305, USA

S. H. Schneider
Department of Biology, Stanford University, Stanford, USA

et al. 1999; Easterling et al. 2000; Meehl et al. 2000; Tebaldi et al. 2006). Observed changes in extremes and projected impacts from future changes in extremes are summarized in Trenberth et al. (2007) and Parry et al. (2007). Trenberth et al. (2007) find that some extreme events are likely (at 66% probability) to have already changed in frequency and intensity over the past few decades. A human contribution to these changes is at least "more likely than not" (Trenberth et al. 2007).

Changes in extreme events are very likely to be some of the earliest impacts experienced from continuing anthropogenic climate change. Meehl et al. (2007) contend that "the type, frequency and intensity of extreme events are expected to change as Earth's climate changes, and these changes could occur with relatively small mean climate changes" (p. 783). Changes in climate extremes can result from changes in mean climate state, changes in distribution of climate states, or a combination of both. Given the potential for increased variability in the future, the sensitivity of extremes to changes in mean climate may be greater than one would assume from a changed mean alone (Mearns et al. 1984; Katz and Brown 1994; Tebaldi et al. 2006). Observations confirm that changes in extremes are not always proportional to changes in the mean, such as greater changes in minimum than maximum temperatures (Trenberth et al. 2007).

Adaptation to extreme events can be more challenging than adaptation to gradual changes in mean climate states. Extreme events can disproportionately affect vulnerable populations that experience higher exposure (e.g., extreme heat and low-income populations without access to air conditioning or individuals living in flood-prone areas) or higher susceptibility (e.g., extreme heat and elderly individuals) to such events. Changes in climate may not only change the frequency and magnitude of individual extreme events, but may also change the likelihood of extreme climate events occurring concurrently. Vulnerability to and impacts of repeated and coincident extreme events are generally expected to be higher than similar events occurring individually (e.g., Hallegatte et al. 2007).

For these reasons, investigations of extreme events uniquely contribute to impact assessments and adaptation planning. Extreme events are also particularly of interest because they provide near-term, concrete experiences of potential future conditions and impacts related to climate change.

1.1 Defining extreme events

The exact definition of an *extreme event* varies widely in the literature. The Intergovernmental Panel on Climate Change (IPCC) Fourth Assessment Report (AR4) focused on six types of "extreme weather events" in discussions of observed changes in extreme events and projections of future changes: (1) daily maximum and minimum temperatures (coldest and hottest 10% each year); (2) heat waves; (3) heavy precipitation events; (4) droughts; (5) intense tropical cyclone activity; and (6) incidences of extreme high sea levels (Solomon et al. 2007; Parry et al. 2007). The term also is used often to refer to floods and wildfires. The IPCC states, "Extremes refer to rare events based on a statistical model of particular weather elements, and changes in extremes may relate to changes in the mean and variance in complicated ways. Changes in extremes are assessed at a range of temporal and spatial scales" (Trenberth et al. 2007, 299). *Extreme* is generally defined as events occurring between 1% and 10% of the time at a particular location in a particular reference period (Trenberth et al. 2007).

Core to the definition of extreme events is the question: extreme with respect to what?

First, extreme events must be defined temporally. Extreme events encompass variations in physical, weather-related parameters such as temperature or precipitation over days (e.g., heat waves, heavy precipitation) or months to years (e.g., drought).

Second, extremes events must also be defined spatially. Heat waves, for example, can vary widely in their spatial extent, affecting the magnitude and nature of impacts. Different levels of spatial resolution may be examined depending on the desired output and impacts of interest. However, the resolution of the relevant observations or projections provides a fundamental constraint.

Third, the thresholds used to define what is extreme must be defined. We contend that it is important to make a distinction between extreme climate events and extreme impact events. Extreme climate events are defined purely by their statistical relationship to past climate conditions and may or may not have direct ties to impacts. Extreme impact events, in contrast, are defined directly by having nonlinear impacts on social and biological systems. For example, isolated days with temperatures above the 95th percentile for that calendar day do not necessarily cause significant impacts to natural and human systems. Some extreme impact events are sensitive to the timing of extreme climate conditions, such as temperature extremes during the flowering of rice plants reducing yield (Jagadish et al. 2007). Moreover, extreme impact events can be caused by a combination of non-extreme physical conditions, such as damage due to flooding caused by a combination of non-extreme precipitation and frozen ground.

Extreme climate conditions affect components, or sectors, of biological and social systems. While one can imagine a large number of sectors being affected by extreme events, several California sectors have been identified in the literature as particularly susceptible to extreme climate events. Table 1 lists sectors that are sensitive to negative impacts from extreme climate events. We also provide examples of metrics (quantifiable units) that can be used as indicators of the magnitude of extreme impact events in these sectors.

Table 1 California sectors most sensitive to negative impacts from extreme climate events

Sector	Sector description	Example metrics
Air quality	Addresses U.S. Environmental Protection Agency (EPA) criteria pollutants and other air pollution.	Concentration of ozone and particulate matter.
Agriculture	Production of crops, animal products, produce.	Farmland loss. Crop, meat, dairy, egg production and yields. Animal deaths.
Ecosystems	Natural biomes. Forests and marine/coastal ecosystems treated separately.	Loss of habitat, biodiversity, pollination services, etc.
Forestry	Includes forest, woodland, scrubland, and grassland ecosystems. Also includes lumber industry, recreation, and other human uses.	Loss of habitat and natural resources due to wildfire. Financial losses to lumber and other businesses. Loss of homes and infrastructure.
Energy	Electrical generation and transmission capacity.	Peak and baseload electrical demand. Frequency, duration, and spatial extent of blackouts and brownouts.
Marine/coastal	Includes marine, estuary, river delta, brackish wetland, beach, and coastal bluff ecosystems. Also includes fishing and tourism, recreation, and other human uses.	Beach and coastal bluff erosion. Sewage overflow events. Damage to coastal infrastructure, interruption of coastal economic activities.
Public health	Includes issues relating to human existence and welfare.	Incidences of death, illness, and interruption of activities due to illness or injury.
Water	Includes the quality, transportation, and management of freshwater supply. Includes flood control and sewage processing.	Flood frequency and consequent financial losses. Level of needed water conservation or rationing.

It is important to understand the quantitative relationships between changes in climatic conditions and particular metrics for sectoral impacts, such as how changes in heat wave frequency and intensity impact heat-related mortality. Multiple sectors are affected by the same extreme climatic conditions; for example, drought conditions affect water supply for urban and agricultural uses as well as for hydropower. Impacts in one sector can affect other sectors and thus, it is also important to quantify those inter-sector relationships.

Ideally, the definition of extreme impact events and these quantitative relationships would be arrived at by a "bottom-up approach," characterizing climate conditions that generate extreme impact events through an assessment involving each sector's stakeholders and analysts to identify thresholds of particular concern for on-the-ground planning. A study of such bottom-up definitions, however, does not exist, to our knowledge. We see this study and the associated report, Mastrandrea et al. (2009), as setting the stage for more formal research in this direction. Mastrandrea et al. (2009) synthesized existing research to characterize the current understanding of the direct impacts of extreme events across sectors and the interactions between sectors as they are affected by extreme events.

2 Projections of extreme events

Given the lack of a consolidated set of bottom-up definitions of extreme impact events, we chose to take a more traditional approach for this projection analysis. This analysis employs downscaled climate projections from six Global Climate Models (GCMs), two emissions scenarios and two downscaling methods, produced for the 2008 Public Interest Energy Research (PIER) Scenarios Report (Cayan et al. 2008). Previous assessments of climate change impact projections for California have used a smaller subset of climate projections (one of the two downscaling methods, fewer GCMs) and fewer categories of extreme events than are analyzed here. By using this larger suite of projections, we are able to more fully characterize the range of uncertainty in projected changes, and address issues of consistency or spread among projections.

First, we compute projections for a set of ten indicators that have been extensively studied in the recent literature (Tebaldi et al. 2006), and whose projected behavior from GCM experiments provides a ready benchmark for comparison.

Second, we rely on standard definitions of extremes from Extreme Value Theory (EVT) statistics (e.g., Coles 2001), focusing on the characterization of the tail behavior of daily precipitation and maximum and minimum temperatures.

We perform these analyses at 58 separate grid points extracted from the full set of downscaled climate projections. Each grid point is representative of a county in California (closest to the county geographical centroid), and the locations are representative of areas of importance from an economic, social, and ecological perspective (e.g., major urban areas, Central Valley, Wine Country, higher elevation regions, coastal areas).

2.1 Downscaled climate projections for California

Two statistical downscaling techniques have been applied to climate projections for California, the Bias Correction and Spatial Downscaling (BCSD) method (Wood et al. 2004) and the Constructed Analog downscaling (CAD) method (Hidalgo et al. 2008). These methods have produced a suite of projections for daily temperature and precipitation at a 1/8° grid scale (about 12 km) for the entire state under a higher and lower emissions scenario, *Special Report on Emissions Scenarios* (SRES) A2 and B1, respectively (Nakicenovic et al. 2000). With BCSD, this has been done for six GCMs (NCAR PCM1, GFDL CM2.1, CNRM CM3, Max Planck Institute ECHAM5, NCAR CCSM3 and MIROC 3.2 at medium resolution), and for CAD this has been done for the first three of the same GCMs (Cayan et al. 2008). Maurer and Hidalgo (2008) compare the ability of these methods to reproduce observed temperature and precipitation extremes. They conclude that the CAD method has greater skill for fall and winter low-temperature extremes and summer high-temperature extremes. For daily precipitation extremes, both methods are limited in their ability to reproduce observed wet and dry extremes. The authors note that this is a reflection of the general low skill of GCMs in reproducing daily precipitation variability.

As part of our analysis, we further compare the BCSD and CAD methods' ability to reproduce characteristics and trends in observed temperature and precipitation extremes in the context of the analyses outlined above. We also compare analyses of projections produced using the two downscaling methods.

2.2 Indicators for extreme (impact) events

Constructed "climatic indicators" are often developed to represent the climate conditions that can cause extreme impact events. We use a set of indicators that have become well known, at least in the climate science literature (Frich et al. 2002; Tebaldi et al. 2006). These indicators are defined as annual indices, such as the longest run of consecutive dry days in the year, or of consecutive days with maximum temperatures above a certain threshold (usually defined with respect to the climatological distribution). These indices have been designed with some practical concern in mind. For example, they are arguably mild definitions of extremes and we can expect GCM output to be appropriate for constructing them. Because they represent mild definitions, we expect the observational record to contain enough cases to allow a robust characterization of each indicator's climatological distribution and trends. They are designed in consideration of societal impacts rather than ecosystem vulnerability. Nonetheless, they offer a multi-dimensional picture of changes beyond average conditions, particularly when computed from downscaled data that provides insight into potential regional changes.

In the following tables, we list the climate indicators that we use to assess projected changes (Frich et al. 2002; Tebaldi et al. 2006). Admittedly, these are still far from an optimal definition of extremes in light of the needs of impact assessment studies. In fact, some criticize these indicators for their fixed thresholds, questionable impact relevance, problematic statistical properties, and biases at the boundaries of the historical period (Zhang et al. 2005; Tebaldi et al. 2006; Alexander et al. 2006). We view these indices, however, as a first step toward bridging the gap between an abstract definition of extremes and definitions that may be dictated by particular concerns. We hope to see such an approach complemented by future research reversing this process, i.e., defining relevant extreme-impact thresholds in consultation with stakeholders, and then defining indicators accordingly.

3 Indicator results

3.1 Comparison to observations

We derived time series for each of the 12 indicators described in Tables 2 and 3 for each county (each county-representative grid point, as described above), based on daily observed

Table 2 Climatic indicators of temperature extremes and associated impact events

Climatic indicator	Impact event
Frost days: Number of days in a year with an absolute minimum temperature below 0°C	Crop and plant death, Insect infestation.
Growing season length: Length of time between the first and last five consecutive days with mean temperature above 5°C (41°F).	Crop and plant growth
Warm nights (and warm summer nights): Percentage of days in the year (and in the period May through September) when the minimum temperature is above the 90th percentile of the climatological distribution for that calendar day.	Heat-related Illness
Warmest three nights: Warmest spell of three consecutive nights	Heat-related Illness
Heat wave duration: Length of the maximum period of at least five consecutive days with a maximum temperature higher by at least 5°C (9°F) than the climatological norm for that calendar day. Computed only over the warm season (May through September)	Heat-related Illness, Fire frequency and intensity, Electricity demand, Crop and plant growth
Hot spell duration: Length of the maximum period of at least five consecutive days with a maximum temperature higher by at least 5°C (9°F) than the climatological norm for that calendar day. Computed over the entire calendar year.	Heat-related Illness, Fire frequency and intensity, Electricity demand, Crop and plant growth

data and downscaled GCM simulations. We computed climatological means for both observed data and modeled datasets over the period 1950–2000. In addition, we computed trends for each of the indicators, both within the observational period (1950–2000) and over the entire length of the simulated datasets (1950–2100). We checked for autocorrelation in the time series of the indicators' annual values and failed to detect it at any significant level. Therefore, we compute linear trends by ordinary least square and consider the value and statistical significance (through the common t-test at a 5% level) of the year coefficient so derived. We compare these results across the suite of GCMs, the two downscaling methods, and the two SRES scenarios. Note that because we compare across downscaling methods, we limit our analysis to the common subset of GCM simulations available under both methods.

Before examining future projections, it is important to investigate whether the downscaled model simulations reproduce observed characteristics of these indicators in

Table 3 Climatic indicators of precipitation extremes and associated impact events

Climatic indicator	Impact event
Precipitation intensity: Annual total precipitation divided by the number of wet days.	Floods, Erosion
Consecutive dry days: Maximum number of consecutive dry days.	Wildfires, Water availability
10 mm days: Number of days with precipitation greater than 10 millimeters (mm).	Floods, Crop yields
Heavy precipitation fraction: Fraction of total precipitation from events exceeding the 95th percentile of the distribution of wet day amounts. How much of precipitation comes in heavy events?	Floods
Five-day precipitation: Maximum five-day precipitation total.	Floods

terms of geographical distribution and magnitude. Figures 1, 2, 3, 4 and 5 show maps for an illustrative set of five of these indicators: Frost Days, Heat Wave Duration, Warmest Three Nights, Consecutive Dry Days, and Precipitation Intensity. Similar results for the rest of the indicators are available as Supplementary Online Material (SOM). In the first row of each figure, we compare climatological means computed for 1950–2000 observed, the BCSD ensemble mean, and the CAD ensemble mean (e.g., the average of the number of Frost Days for each year in the observed dataset, compared to that simulated by the BCSD and CAD ensembles). In the second row, we compare trends across the same three datasets over the same time period, with solid circles indicating statistically significant trends at the 5% level.

Both downscaling approaches closely reproduce the mean number of Frost Days, Consecutive Dry Days, and the temperature of the Warmest Three Nights per year around the state. In general, both downscaling approaches somewhat underestimate Heat Wave Duration and the fraction of years with at least one Heat Wave, with CAD closer to observed magnitudes, and BCSD closer to observed frequencies. Both generally reproduce the pattern of relative regional differences in Heat Wave Duration. CAD likewise underestimates Precipitation Intensity in some parts of the state, while BCSD slightly

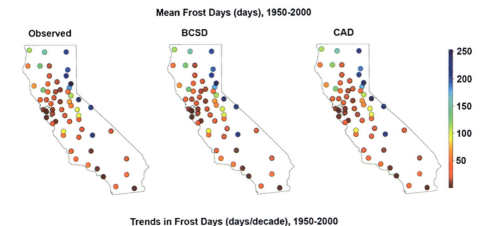

Fig. 1 Climatological means and trends for frost days, 1950–1999, by county. *Top row*, mean values for observed (1950–1999) and downscaled GCM simulations (1950–1999) ensemble means (BCSD, center and CAD, right panels). *Bottom row* trend values computed from the same datasets. *Filled circles* indicate significant trends at the 5% level

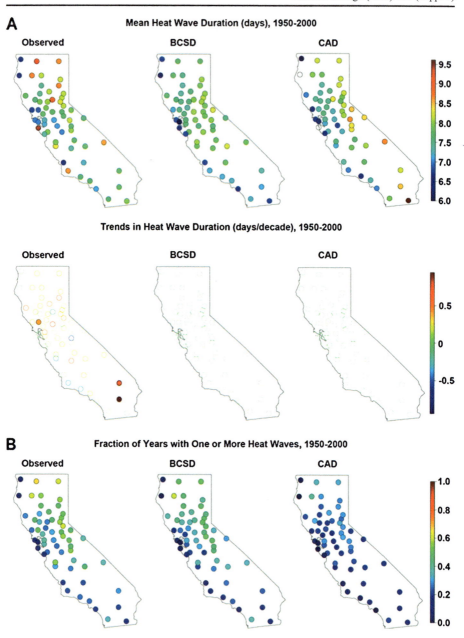

Fig. 2 Climatological values and trends for heat wave duration, 1950–1999, by county. **a** Like Fig. 1, for Heat Wave Duration. **b** Frequency of years with at least one Heat Wave

overestimates in limited regions, but generally reproduces the observed pattern. For the other indicators the observed climatological mean for each indicator is reproduced quite accurately, both in terms of geographical differences and actual values by the ensemble mean of the downscaled GCM simulations, with only one exception. Both methods overestimate the mean value of Heavy Precipitation Fraction (the percentage of total

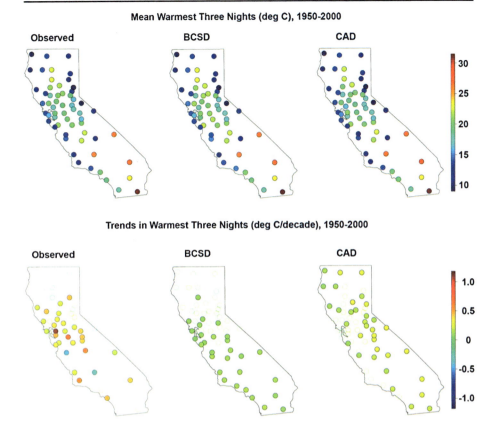

Fig. 3 Climatological values and trends for warmest three nights, 1950–1999, by county. Like Fig. 1, for Warmest Three Nights

precipitation falling in very wet days, defined as days above the 95th percentile of the 1961–1990 climatological distribution), but they do maintain the geographically differentiated features observed.

When we consider the simulation of trends, the results are more diverse across indicators and methods. We summarize in Table 4 the agreement on the sign of the trends and their statistical significance. Here we summarize the general findings. When observed indicators show widespread significant trends, the ensemble mean of the downscaled simulations correctly estimates the direction. The downscaled simulations often, however, tend to "spread" the values and significance of the trends smoothly across space, while the observations generally show more heterogeneous spatial patterns. This is not surprising, given the statistically downscaled nature of these datasets, which may tend to conserve smooth spatial patterns from their parent coarse-grid GCM simulations. When the observed indicators do not show significant trends, the simulated values do not either. However, the underlying trends for precipitation indicators are at times in opposite directions when comparing the two downscaling methods' output. In general, BCSD displays drying tendencies across the state over the observed period while CAD suggests wetter conditions. The observed precipitation data do not help in resolving this issue, showing a mixed pattern of increasing and decreasing trends indicating large natural variability.

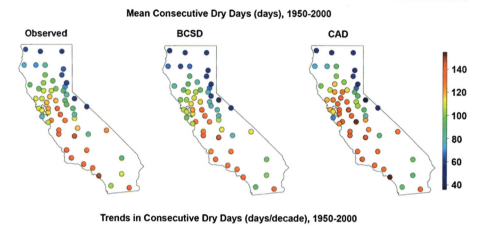

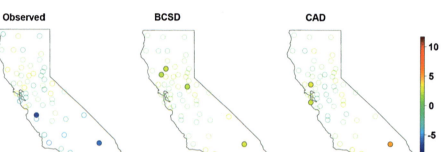

Fig. 4 Climatological values and trends for consecutive dry days, 1950–1999, by county. Like Fig. 1, for Consecutive Dry Days

Some of the indicators related to temperature extremes already show the widespread significant trends in the direction expected under a warming climate (e.g., fewer Frost Days, more Warm Nights, and warmer Warmest Three Nights). The simulation ensemble means reproduce these overall tendencies and their significance, although they generally underestimate the slope of the trends. Overall, simulations of extreme behavior related to temperature should elicit greater confidence due to the agreement between downscaling methods and between simulations and observations. Precipitation, as to be expected, poses a greater challenge. From the observational viewpoint, no significant trends are consistently detected in the California region. Modeled data are in agreement only on the absence of a consistent observed signal, but the underlying tendencies towards increasing or decreasing precipitation intensity are not consistent.

3.2 Future projections of indicators

Figures 6, 7, 8, 9 and 10 show future trends under the two scenarios (along the rows) and the two downscaling methods (along the columns) for the five illustrative indicators. Similar results for the rest of the indicators are available in the SOM. Indicators of temperature extremes tell a very consistent story: the trends to be expected in a warming climate are produced almost identically by the two methods' ensemble means: fewer Frost

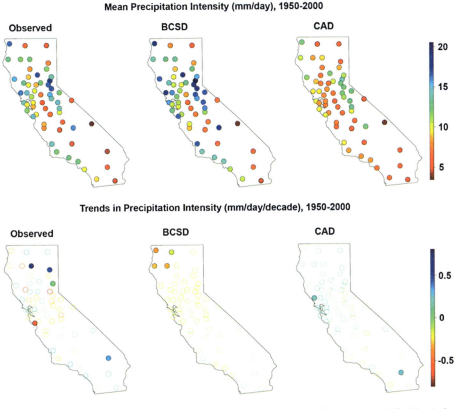

Fig. 5 Climatological values and trends for precipitation intensity, 1950–1999, by county. Like Fig. 1, for Precipitation Intensity

Days, longer Heat Wave Duration and more frequent Heat Waves, and warmer Warmest Three Nights. Growing Season and Hot Spell Duration also lengthen, and Warm Nights and Warm Summer Nights increase (not shown). The trends are all significant across the region, and they generally show an intensifying gradient from West to East (from the coast to the interior) with the exception of Warm Nights/Warm Summer Nights, where the gradient appears to be more North to South. There are significant differences in the magnitude of the trends when comparing SRES B1 and A2, with larger changes under the latter, higher emissions scenario. These findings are in perfect agreement with a multi-GCM study of the same indices (Tebaldi et al. 2006). That study took a global and continental perspective, but the strong agreement of temperature indices allows us to find commonalities even at these very different regional scales.

The picture for precipitation extremes indicators is much less coherent. The two methods produce different projections with regard to trends and their statistical significance. Consecutive Dry Days is the only indicator that does not relate to changes in precipitation intensity. The BCSD simulations indicate significant increases in the length of dry spells, confined to the northern part of the state under B1 and more widespread under A2. CAD simulations do not attribute any significance to changes under B1, but agree with lengthening of dry spells under A2, although with less widespread significance than the other method.

The remaining four indicators look at various aspects of change in precipitation intensity. Precipitation Intensity, representing the average amount of precipitation on a wet day in a

Table 4 Observed 1950–2000 trends in indicators of temperature and precipitation extremes and agreement with downscaled model simulations

Index	Observed	20C3M BCSD	20C3M CAD
Frost days	Decreasing trends with scattered significant values across state, and geographically heterogeneous values.	Decreasing trends, wider coverage of significant values than observed, more homogeneous values, and generally smaller in absolute value than observed	Same as BCSD
Growing season length	Increasing trends with scattered significant values, same patterns as Frost Days, fewer overall significant values.	Similar patterns as observed, with scattered significant values.	Same as BCSD
Hot spells duration	Very few and isolated significant trends, both increasing and decreasing. Likely due to chance.	More homogeneously increasing trends, but very few significant, likely only by chance, like obs.	Same as BCSD
Heat wave duration	Similar to Hot Spells, with both increasing and decreasing trends, even more obviously due to chance.	Few small increasing trends, likely to be significant only by chance.	Same as BCSD
Warm (summer) nights	Strong pattern of increasing significant trends mainly interior and South.	Fairly uniform coverage of increasing significant trends all over the state with values lower than observed.	Same as BCSD
Warmest three nights	Same as Warm Nights	Increasing and significant trends, uniformly located all over the state, with smaller and more geographically homogeneous values than observed.	Same as BCSD with generally larger values of trends.
Consecutive dry days	Mix of non-significant increasing and decreasing trends over the state.	Same as obs.	Same as obs.
Precipitation intensity	Mix of non-significant increasing and decreasing trends over the state.	No significant trends. Values are mostly negative.	No significant trends. Values are mostly positive.
Days w/Precip >10 mm	Generally increasing trends but very few scattered significant values. Likely only by chance occurring.	No significant trends. Values are mostly negative.	No significant trends. Values are mix of positive and negative.
% Precip in very wet days	Mix of non-significant increasing and decreasing trends over the state.	Same as obs.	No significant trends. Values are mostly positive
Maximum 5-day total Precip	Mix of non-significant increasing and decreasing trends over the state.	No significant trends. Values are mostly negative.	No significant trends. Values are mostly negative.

given year, is the most direct indicator. The two methods here disagree in the ensemble mean result. BCSD does not produce significant trends, and hints at diminishing intensity, while CAD shows significant increasing trends, especially in the northern half of the state. Similar behavior is shown for Heavy Precipitation Fraction. Total Five-Day Precipitation sees better agreement in the patterns across the methods, with positive but not statistically significant trends decreasing from north to south. Puzzlingly, 10 mm Days, the number of days with precipitation amounts exceeding 10 mm, shows consistently decreasing trends,

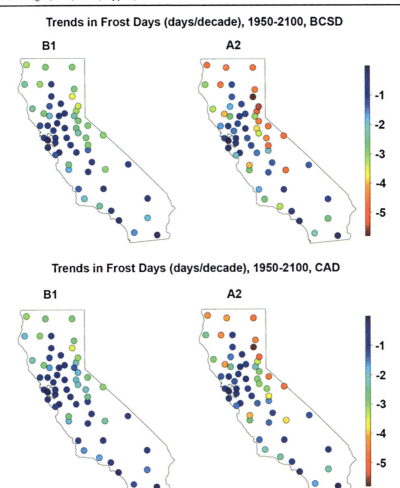

Fig. 6 Trends in frost days, 1950–2100, by county. Trends under SRES B1 (*left*) or SRES A2 (*right*) computed from downscaled data from BCSD (*top*) or CAD (*bottom*). *Filled circles* indicate significant trends at the 5% level

relatively more so for BCSD than CAD. In general, BCSD simulations produce trends towards drier conditions and less intensifying precipitation than do CAD simulations.

We are setting the bar high in attempting to detect changes at point locations, but results at this scale are arguably more meaningful for impact analysis and local decision-making. These indicators would likely show a stronger signal if we averaged the grid point results into a regional average. Averaging of indicators' time series across a large region is expected to cancel out low frequency "wiggles" and bring out the signal that may exist at a regional level, overcoming spatial patchiness and revealing either a decreasing or increasing trend in most cases. In fact, an analysis (not shown) of GCM-derived indicators (without downscaling) in the Western region of the United States (California and Nevada) for the same five precipitation indicators showed significant increasing trends for all but 10 mm Days, suggesting that the behavior is not an artifact of the downscaling process. Table 5 lists direction and significance of projected trends for each index in detail.

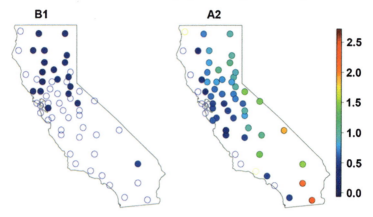

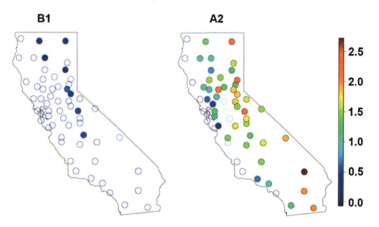

Fig. 7 Trends in heat wave duration, 1950–2100, by county. **a** Like Fig. 6, for Heat Wave Duration. **b** Frequency of years with at least one heat wave

4 Return level analysis

An intuitive way of characterizing extremes under current and future conditions is through the concepts of return level or return period. Such analyses are conducted by applying Extreme Value Theory (EVT) to time series of observed data or model simulations. To fit a statistical distribution to the extreme values of a climate variable, EVT uses either the maximum values over a predefined period (year, season, month) or all the values over a given threshold (to be optimally estimated). A member of a family of parametric distributions (the Generalized Extreme Value distribution, or GEV, in the case of maximum values, or the generalized Pareto distribution in the case of excess over threshold data) is fitted to these extreme observations. Thus, the statistical characterization applies specifically to the large values of the climate variable, rather than to its entire climatological distribution (Coles 2001).

Because of the need to perform this analysis over many locations, variables, and models and for both downscaling methods, we chose to extract three-day average maxima and fit GEV

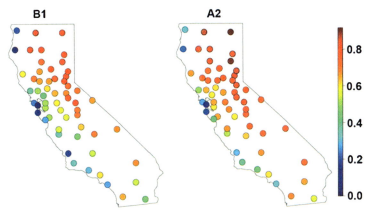
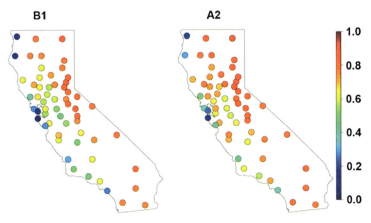

Fig. 7 (continued)

distributions, rather than using threshold exceedances that would require a case-by-case estimation of thresholds.

After fitting a GEV distribution to a climate variable (e.g., observed annual highest maximum temperature at a given location over the period 1950–2000), the estimated parameters of the GEV distribution determine a functional relation between values of that variable and their expected return period. Thus, for a given fit of the GEV, a value 'x' (say 125°F) of the annual maximum at a given location may be associated with a return period of 'y' years (say 100 years), meaning that under the climate conditions represented by the record used to fit the distribution one would expect to experience that value 'x' of annual maximum only once every 'y' years (e.g., one would expect that the annual maximum would reach 125°F only once in a hundred years on average, or with probability 0.01 in any given year).

Alternatively, the concept of return level is characterized by fixing a return period, say 100 years, and determining the value of the extreme that can be expected to recur on average once in that period. The use of parametric statistical distributions also provides a means of characterizing confidence intervals around the estimated return values and periods

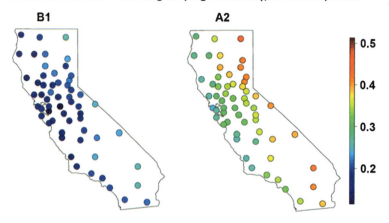

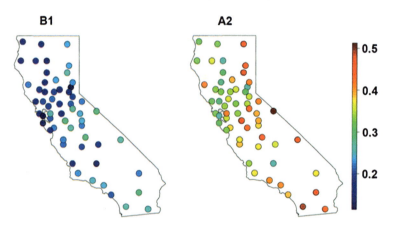

Fig. 8 Trends in warmest three nights, 1950–2100, by county. Like Fig. 6, for Warmest Three Nights

and thus determining the significance of the changes projected. Note that a 100-year return period is far more "extreme" than the indices discussed in the previous section of this paper, and 100-year return values are thus much rarer. In turn, the associated impacts would be expected to be more severe as well.

Of course, any statistical analysis is conditional on the data at hand. Extreme Value Theory handles the uncertainty in the fit itself by providing confidence intervals around each individual return level curve that is fitted to a given set of records (or model data). Structural uncertainties characteristic of climate models' approximations are addressed here by fitting return level curves to an ensemble of climate model simulations, from different GCMs and two alternative downscaling methods. These models and methods span a range of structural assumptions, and thus our results should bracket a relevant set of alternative future outcomes.

By comparing return level curves derived from the observed record to those derived from model simulations of the same period, we also address the basic necessity of validating these models' ability to reproduce observed extremes behavior.

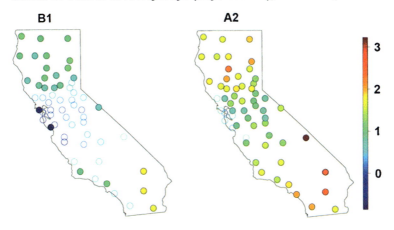

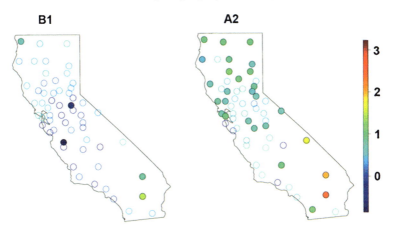

Fig. 9 Trends in consecutive dry days, 1950–2100, by county. Like Fig. 6, for Consecutive Dry Days

4.1 Return level analysis results

In the results that follow, we focus on multiple-day extreme events (e.g., annual highest three-day average maximum/minimum temperature) to address the persistency of extreme conditions that are more likely to represent extreme impact events. Results based on daily extremes were found to be qualitatively similar (not shown). We also focus on temperature rather than precipitation extremes. The less coherent signal in precipitation trends found in the previous section is consistent with what was found (but not shown here) when applying EVT to daily precipitation series and comparing return levels and periods between present and future climate conditions.

Time series of daily minimum and maximum temperature from downscaled model simulations were analyzed for the 58 locations representative of the 58 counties of California in two separate segments: one representative of current climate (1950–2000), and one of future conditions (2050–2100). The latter is available under two emission scenarios: SRES A2 and B1. Observed time

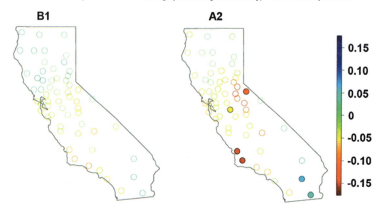

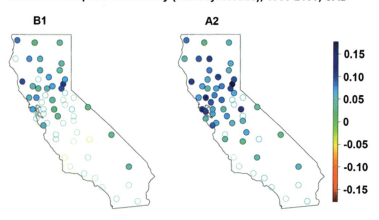

Fig. 10 Trends in precipitation intensity 1950–2100, by county. Like Fig. 6, for Precipitation Intensity

series at the same 58 locations are also available and can be used for validation. As we aim to compare return levels associated with long return periods (100 years), we chose to use 50-year time series to enhance the precision of our estimates, and we fit GEV distributions to these time series by maximum likelihood. We note, however, that GEV distributions are generally fitted to stationary distributions of random variables, while there are trends for temperature over both of these periods, and thus the GEV parameters will more closely reflect the values toward the end of each time period. Therefore, we emphasize that the results should be interpreted "qualitatively," as we have low confidence in the absolute magnitudes of calculated return values and focus instead on the significance of changes between the two time periods.

Keeping these caveats in mind, we can ask several questions about the behavior of annual extremes:

- How does the return period of a given return value (extreme temperature) change, from current to future conditions?
- How does the return value change, for a given return period?
- Are there geographically differentiated trends?

Table 5 Direction and significance of projected trends in Indicators of temperature and precipitation extremes

Index	1950–2100 BCSD (SRES B1 and A2)	1950–2100 CAD (SRES B1 and A2)
Frost days	Decreasing trends, All significant. A2 projections exhibit steeper trends than B1 patterns. West to east positive gradient in the values.	Same as BCSD
Growing season length	Similar geographical pattern as Frost Days, with all significant positive values.	Same as BCSD
Hot spells duration	Increasing trends, more significant under A2, all significant, west to east gradient.	Same as BCSD, with generally larger values.
Heat wave duration	Increasing trends, more significant under A2, west to east gradient.	Same as BCSD, with generally larger values.
Warm (summer) nights	Increasing significant trends all over the state.	Same as BCSD but with lower positive values
Warmest three nights	Increasing and significant trends, uniformly located all over the state.	Same as BCSD.
Consecutive dry days	Increasing trends, but becoming homogeneously significant only under A2.	Increasing trends, but only a scatter of significant values and only under A2. Values are lower than BCSD.
Precipitation intensity	Non significant, decreasing trends.	Significant increasing trends (limited to Northern California under B1, all over under A2).
Days w/Precip. >10 mm	Decreasing trends all over the state, significant under A2.	Decreasing trends, with scattered significance only under A2.
Percent Precip. in very wet days	No significant trends for either scenarios.	Increasing and significant for the northern part of the state under B1 and more widespread under A2.
Maximum 5-day total Precip.	No significant trends.	No significant trends.

4.1.1 Current climate simulations and observed records

We found good agreement between the return level curves derived from observations and the envelope spanned by the two methods' downscaled simulations. This is true for all counties and both of the climate variables that we analyzed: highest annual three-day-average maximum temperature and highest annual three-day-average minimum temperature.

Figure 11 shows results for two counties that we chose as representative of two different broad California climate zones (Northern vs. Southern California): Shasta and Imperial.

The two panels of each column show return level curves for one of two climate variables (highest annual three-day-average maximum temperature and highest annual three-day-average minimum temperature) in one county, based on the period 1950–2000. Blue curves are return level curves (solid) and confidence intervals (dashed) from the six GCM twentieth-century simulations using BCSD. Green curves are from the subset of three GCM twentieth-century simulations using CAD. The black lines are estimated from observations. The general message is that for all indices and counties the black solid lines are bracketed by the colored lines or very close to their envelope. As a whole, therefore, the climate simulations represent the return levels (or periods) of the observed climate for these climate variables realistically, and we are justified

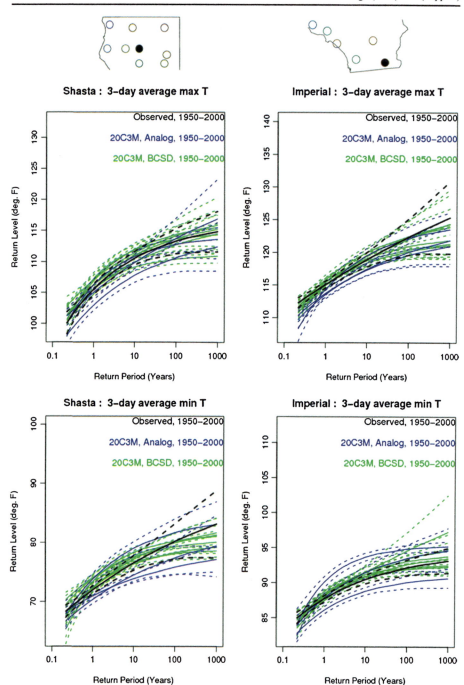

Fig. 11 Return level curves for observed and model simulated highest annual three-day maximum and minimum temperature in example counties, 1950–1999. Return level curves for maximum (*upper row*) and minimum (*lower row*) three-day-average temperature, estimated on the basis of annual maxima from the period 1950–1999 for Shasta (*left column*) and Imperial (*right column*) counties. The locations of the counties are indicated in the maps at the top. *Black solid line* is curve estimated from observed dataset. *Green lines* are curves estimated using BCSD, *blue lines* are estimated using CAD. *Dashed lines* are corresponding 95% confidence intervals

in taking a multi-model, multi-method approach to the projection of future changes. However, the range of uncertainty across models and methods is substantial, and therefore, rather than on specific return level or return period magnitudes we focus on the significance of the changes, represented by the changes in the ensemble mean projections and in their uncertainty range under future conditions.

4.1.2 Future changes in return period and return levels

We examine shifts in the curves between current (1950–2000) and future (2050–2100) periods, under two different emissions scenarios. First, we present an overview of projected changes in return period for each county (Figs. 12 and 13), under B1 or A2 for

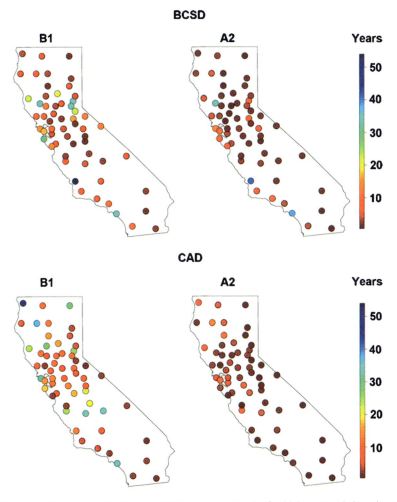

Fig. 12 Projected return periods of current 100-year return levels, for highest annual three-day-average maximum temperature, by county. Ensemble mean estimates of projected 2050–2100 return periods for current 100-year return levels, under the B1 emissions scenario (*left column*), and the A2 emissions scenario (*right column*). Panels on top row are ensemble means from BCSD downscaled GCM simulations, panels in bottom row from CAD. Values of the color scale correspond to the new return period of a current 100-year event (e.g., less than 10 years for a *dark red circle*)

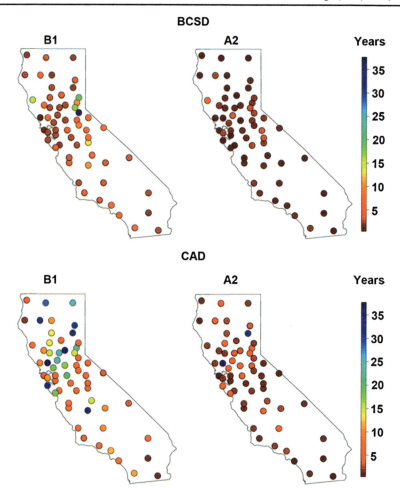

Fig. 13 Projected return periods of current 100-year return levels, for highest annual three-day-average minimum temperature, by county. Ensemble mean estimates of projected 2050–2100 return periods for current 100-year return levels, under the B1 emissions scenario (*left column*), and the A2 emissions scenario (*right column*). Panels on top row are ensemble means from BCSD downscaled GCM simulations, panels in bottom row from CAD. Values of the color scale correspond to the new return period of a current 100-year event (e.g., less than 5 years for a *dark red circle*)

both BCSD and CAD simulations. Colored circles represent the new return periods of temperatures that, under current climate, are expected to return only once every 100 years (smaller numbers of years associated with warmer colors). The smaller the number of years (compared to the current 100-yr period), the greater the likelihood that such temperatures will be experienced more frequently, and the larger the change toward more extreme conditions.

There is some degree of spatial heterogeneity to the projections, mainly under SRES B1, and one can find areas where the frequency of current 100-yr extreme values "only" doubles or triples (green to blue dots). Most of the milder changes are located along the coast. However there is a large prevalence of deep red colors across the maps for both variables, indicating return periods of 10 years or less, and thus at least a ten-fold increase in the frequency of such conditions. The agreement of models (not shown here) is stronger under A2 than B1, as is the agreement between the two methods. Note that in these figures we are computing return periods only from the three GCM simulations that are common to the two downscaling methods, but including the additional 3 GCMs available for BCSD does not change the corresponding maps (i.e., the ensemble averages) appreciably.

An alternative way of characterizing the patterns of change is to ask what temperature will replace that which is associated with the current 100-year return period. Figures 14 and 15 display these results, similar maps with circles colored according to temperature values (in degrees F) correspond to the 100-yr return level as it is currently estimated and as estimated under future conditions during the second half of the twenty-first century under the two scenarios. There is very good agreement between the two downscaling methods' ensemble mean projections, both in terms of values and in terms of their regionally diversified distribution. Changes are significantly stronger under the A2 scenario, with increases in the return level of double digits of degrees F across the entire region.

To provide a measure of the inter-model spread, in Fig. 16 we show sets of curves for the same two counties: Shasta and Imperial (one in each column). The figure shows three-day-average maximum temperature as estimated on the basis of the downscaled simulations by BCSD (top

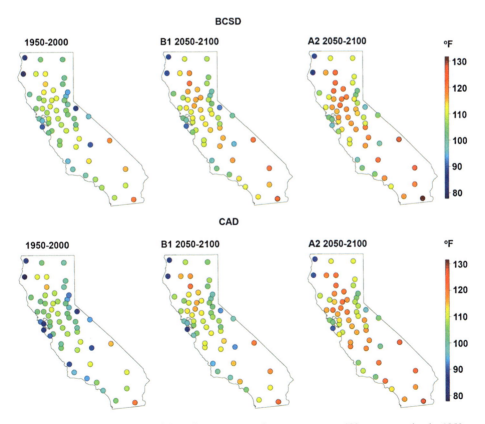

Fig. 14 Simulated highest annual three-day-average maximum temperature 100-year return levels, 1950–1999 and 2050–2099, by county. 100–year return levels of highest annual three-day-average maximum temperature (ensemble averages of three GCMs as in Figs. 12 and 13), for current simulations (1950–1999; *left column*) and future projections (2050–2099) under SRES B1 (*middle column*) and A2 (*right column*). BCSD results along the top row, CAD results along bottom row

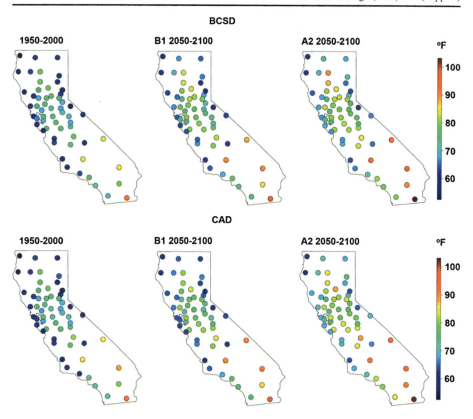

Fig. 15 Simulated highest annual three-day-average minimum temperature 100-year return levels, 1950–1999 and 2050–2099, by county. 100-year return levels of highest annual three-day-average minimum temperature (ensemble averages of three GCMs as in Figs. 12 and 13), for current simulations (1950–1999; *left column*) and future projections (2050–2099) under SRES B1 (*middle column*) and A2 (*right column*). BCSD results along the top row, CAD results along bottom row

panels) and CAD (bottom panels). The interpretation would be the same using return level curves derived for extremes of minimum temperatures. We call attention to two sets of comparisons.

First, consider a vertical line intersecting the x-axis at 100 years. For most indices/counties the set of orange/red curves corresponding to the downscaled future simulations under respectively B1 and A2 are situated above the envelope formed by the black curves (estimated from the twentieth century segment of the simulation). The lack of overlap between the two sets of colored (future return levels) vs. black (current return levels) solid curves indicates a significant shift of the return levels associated with the 100-yr return period.

The figure also displays the spread of values across the methods/GCMs simulation. Again, because of this spread we do not suggest focusing on the specific magnitudes calculated, but

Fig. 16 Individual model simulations of three-day annual maximum temperature in example counties, 1950–1999 and 2050–2099. Return level curves for highest annual three-day-average maximum temperatures for Shasta (*left column*) and Imperial (*right column*) counties. The locations of the counties are indicated in the maps at the top. Top row projections for BCSD, bottom row projections for CAD. Each panel compares three sets of curves. *Black*: current climate simulations (1950–1999). *Orange*: B1 (2050–2099). *Red*: A2 (2050–2099). Shading captures the range of all 95% confidence intervals around the individual curves

rather on the significance of the projected shift between current and future conditions. In addition, some of the fits produced by the GEV maximum likelihood calculations produce curves that appear to belong to the Frechet family of unbounded extreme distributions. This

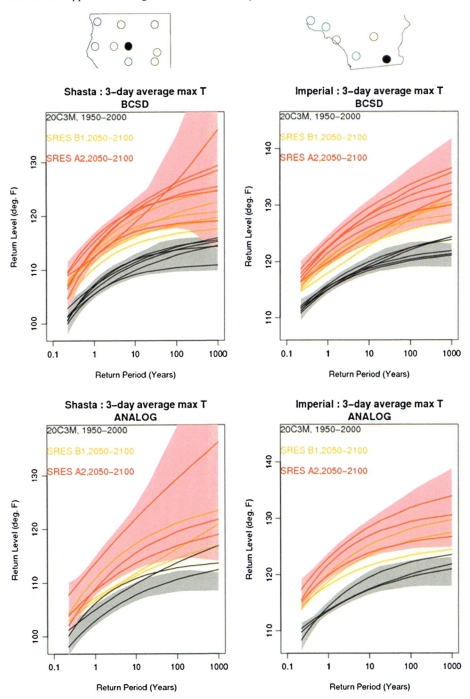

result is most likely an artifact of the fit, and further supports our emphasis on qualitative comparison of the significance of the shifts between time periods, rather than focusing on the specific numerical results themselves. Moreover, the highest temperatures produced through these statistical calculations under the A2 scenario may not be physically realistic.

Second, consider a horizontal line that intersects the black curves at their 100-yr return level. The corresponding periods along the x-axis corresponding to where the line intersects the orange and red curves are the new return periods under future conditions of what is now the 100-yr return level. In this case too, the colored curves lie to the left of the black curves (with no overlap), signaling a significant shift (decrease) in the return period. Here as well the ensemble mean curve does not tell the entire story, and the uncertain range of new return periods is substantial. There is nonetheless a clear separation between the envelope of black curves and that of the colored curves, meaning that what is now a range of extreme conditions associated with a given return period will recur considerably more often in the future.

5 Conclusions

This paper distinguishes between extreme climate events and extreme impact events. Extreme climate conditions (e.g., temperatures above a given threshold of the climatological distribution) do not necessarily induce extreme impacts. Ideally, the definition of extreme impact events and subsequent transfer functions linking them to climate conditions would be arrived at by an interactive process involving scientists and relevant stakeholders and analysts. Such a process would involve analyzing historical events and their impacts to better quantify interactions and to identify thresholds of particular concern for on-the-ground planning. We see this study as setting the stage for more formal research in this direction, building from the extensive body of literature on the impacts of climate change in California. A next step, for example, is to couple the projections presented here with research on specific interconnections between extreme events and sectoral impacts described above, in order to better understand the importance of the impacts from interactions between sectors.

We produce new projections of changes in the frequency and intensity of extreme events in the future across climate models, emissions scenarios, and downscaling methods, and for each county in California. In general, model simulations using either of the two downscaling approaches reproduce most of the characteristics of observed patterns of extreme climate conditions in California. Some exceptions, and the range of individual model results, are outlined in the discussion. As a whole, however, we conclude that employing this multi-model, multi-method ensemble of climate simulations to project future changes in extreme events is justified, based on this comparative analysis.

Changes in extreme events related to high temperatures are found to be very consistent across simulations downscaled by both methods. Coherent changes over California suggest significant increases in the severity of hot spells (both in length and intensity) and decreases in frost days and—more generally—cold spells. Larger warming is projected inland compared to coastal areas. A significant difference in the magnitude of these changes in temperature extremes is found when comparing the two scenarios, A2 and B1, suggesting that mitigation would limit the severity of these changes.

For indicators and EVT analyses of precipitation, our inquiries failed to detect a significant signal of change, with inconsistent behavior when comparing simulations across different GCMs and different downscaling methods. Simulations using BCSD, for example, show a tendency toward drier conditions (longer dry spells), while simulations using CAD do not exhibit this behavior. Similarly, simulations using CAD indicate widespread significant increasing trends in precipitation intensity, which are not found in the BCSD simulations. Parallel studies that use larger regional averages suggest a more consistent picture of a general lengthening of dry spells and increasing precipitation intensity for this region as a whole, but we cannot support the same findings at the local scale represented through our grid-point level analyses.

We see this research as a first step toward informing more formal vulnerability assessment in the context of extreme events in California. Vulnerability is often considered as related to exposure to a stress, sensitivity to that stress, and adaptive capacity to cope with that stress. The projections presented in this paper provide information regarding the exposure to extreme events in different regions of California. The synthesis of impacts and interactions identified in the literature in Mastrandrea et al. (2009) is a first step toward characterizing the sensitivity of specific sectors to extreme events. Moving forward, further refinements of information regarding exposure and sensitivity must be integrated with assessment of the adaptive capacity of specific sectors, regions, and populations, to identify specific vulnerabilities and inform strategies to reduce those vulnerabilities (see, e.g., Mastrandrea et al. 2010 for further discussion of this process). Since patterns of extreme events can change considerably even under lower emissions scenarios, vulnerability assessment in the context of extremes is particularly important for informing adaptation strategies.

Acknowledgements We thank Scripps Institute of Oceanography (in particular Mary Tyree) and the Public Interest Energy Research Program for making data and model simulations available for these analyses. We deeply thank John Pfefferle for his invaluable help in preparing this paper, and David Lobell and Gregor Horstmeyer for useful discussions and input. We also thank all the researchers we have consulted for information on their research, which has formed the basis for our summary of current understanding of the impacts of extreme events on California.

References

Alexander LV, Zhang X, Peterson TC, Caesar J, Gleason B, Tank AK, Haylock M, Collins D, Trewin B, Rahimzadeh F, Tagipour A, Kumar KR, Revadekar J, Griffiths G, Vincent L, Stephenson DB, Burn J, Aguilar E, Brunet M, Taylor M, New M, Zhai P, Rusticucci M, Vazquez-Aguirre JL (2006) Global observed changes in daily climate extremes of temperature and precipitation. J Geophys Res D-Atm 111:D05109

Cayan D, Tyree M, Dettinger M, Hidalgo H, Das T, Maurer E, Bromirski P, Graham N, Flick R (2008) Climate change scenarios and sea level rise estimates for California 2008 Climate Change Scenarios Assessment. PIER Technical Report CEC-500-2009-014-F

Coles S (2001) An introduction to statistical models of extremes. Springer, New York

Easterling DR, Meehl GA, Parmesan C, Changnon SA, Karl TR, Mearns LO (2000) Climate extremes: observations, modeling, and impacts. Science 289:2068–2074

Frich P, Alexander LV, Della-Marta P, Gleason B, Haylock M, Tank AK, Peterson T (2002) Observed coherent changes in climatic extremes during the second half of the twentieth century. Clim Res 19:193–212

Hallegatte S, Hourcade JC, Dumas P (2007) Why economic dynamics matter in assessing climate change damages: illustration on extreme events. Ecol Econ 62:330–340

Hidalgo H, Dettinger M, Cayan D (2008) Downscaling with constructed analogues: daily precipitation and temperature fields over the United States. PIER Technical Report CEC-500-2007-123

Jagadish SVK, Craufurd PO, Wheeler TR (2007) High temperature stress and spikelet fertility in rice (*Oryza sativa* L). J Exp Bot. doi:101093/jxb/erm003

Katz RW, Brown BG (1994) Sensitivity of extreme events to climate change: the case of autocorrelated time-series. Environmetrics 5:451–462

Kunkel KE, Pielke RA, Changnon SA (1999) Temporal fluctuations in weather and climate extremes that cause economic and human health impacts: a review. Bull Am Meteorol Soc 80:1077–1098

Mastrandrea MD, Tebaldi C, Snyder CP, Schneider SH (2009) Current and future impacts of extreme events in California. PIER Technical Report CEC-500-2009-026-D

Mastrandrea MD, Heller NE, Root TL, Schneider SH (2010) Bridging the gap: linking climate-impacts research with adaptation planning and management. Clim Chang 100:87–101

Maurer EP, Hidalgo HG (2008) Utility of daily vs. monthly large-scale climate data: an intercomparison of two statistical downscaling methods. Hydrol Earth Syst Sci 12:551–563

Mearns LO, Katz RW, Schneider SH (1984) Changes in the probabilities of extreme high temperature events with changes in global mean temperature. J Clim Appl Meteorol 23:1601–1613

Meehl GA, Karl T, Easterling DR, Changnon S, Pielke R, Changnon D, Evans J, Groisman PY, Knutson TR, Kunkel KE, Mearns LO, Parmesan C, Pulwarty R, Root T, Sylves RT, Whetton P, Zwiers F (2000) An introduction to trends in extreme weather and climate events: observations, socioeconomic impacts, terrestrial ecological impacts, and model projections. Bull Am Meteorol Soc 81:413–416

Meehl GA, Stocker TF, Collins WD, Friedlingstein P, Gaye AT, Gregory J, Kitoh A, Knutti R, Murphy JM, Noda A, Raper SCB, Watterson IG, Weaver AJ, Zhao ZC (2007) Global climate projections. In: Solomon S, Qin D, Manning M, Marquis M, Averyt K, Tignor MMB, Miller HL Jr, Chen Z (eds) Climate Change 2007: the scientific basis, contribution of Working Group I to the Fourth Assessment Report of the IPCC. Cambridge University Press, Cambridge, pp 747–845

Nakicenovic N, Alcamo J, Davis G, de Vries B, Fenhann J, Gaffin S, Gregory K, Grubler A, Jung TY, Kram T et al (2000) Intergovernmental Panel on Climate Change Special report on emissions scenarios. Cambridge University Press, Cambridge

Parry M, Canziani O, Palutikof J, van der Linden P, Hanson C (2007) Climate Change 2007: impacts, adaptation, and vulnerability, contribution of Working Group II to the Fourth Assessment Report of the IPCC. Cambridge University Press, Cambridge

Solomon S, Qin D, Manning M, Marquis M, Averyt K, Tignor MMB, Miller HL Jr, Chen Z (2007) Climate Change 2007: the scientific basis, contribution of Working Group I to the Fourth Assessment Report of the IPCC. Cambridge University Press, Cambridge

Tebaldi C, Hayhoe K, Arblaster JM (2006) Going to the extremes: an intercomparison of model-simulated historical and future changes in extreme events. Clim Chang 79:185–211

Trenberth KE, Jones PD, Ambenje P, Bojariu R, Easterling DR, Tank AK, Parker D, Rahimzadeh F, Renwick JA, Rusticucci M, Soden BJ, Zhai P (2007) Observations: surface and atmospheric climate change. In: Solomon S, Qin D, Manning M, Marquis M, Averyt K, Tignor MMB, Miller HL Jr, Chen Z (eds) Climate Change 2007: the scientific basis, Contribution of Working Group I to the Fourth Assessment Report of the IPCC. Cambridge University Press, Cambridge, pp 747–845

Wood AW, Leung LR, Sridhar V, Lettenmaier DP (2004) Hydrologic implications of dynamical and statistical approaches to downscaling climate model outputs. Clim Chang 62:189–216

Zhang XB, Hegerl G, Zwiers FW, Kenyon J (2005) Avoiding inhomogeneity in percentile-based indices of temperature extremes. J Clim 18:1641–1651

Potential increase in floods in California's Sierra Nevada under future climate projections

Tapash Das · Michael D. Dettinger · Daniel R. Cayan · Hugo G. Hidalgo

Received: 9 March 2010 / Accepted: 17 June 2011 / Published online: 24 November 2011
© Springer Science+Business Media B.V. 2011

Abstract California's mountainous topography, exposure to occasional heavily moisture-laden storm systems, and varied communities and infrastructures in low lying areas make it highly vulnerable to floods. An important question facing the state—in terms of protecting the public and formulating water management responses to climate change—is "how might future climate changes affect flood characteristics in California?" To help address this, we simulate floods on the western slopes of the Sierra Nevada Mountains, the state's primary catchment, based on downscaled daily precipitation and temperature projections from three General Circulation Models (GCMs). These climate projections are fed into the Variable Infiltration Capacity (VIC) hydrologic model, and the VIC-simulated streamflows and hydrologic conditions, from historical and from projected climate change runs, allow us to evaluate possible changes in annual maximum 3-day flood magnitudes and frequencies of floods. By the end of the 21st Century, all projections yield larger-than-historical floods, for both the Northern Sierra Nevada (NSN) and for the Southern Sierra Nevada (SSN). The increases in flood magnitude are statistically significant (at $p<=0.01$) for all the three GCMs in the period 2051–2099. The frequency of flood events above selected historical thresholds also increases under projections from CNRM CM3 and NCAR PCM1 climate models, while under the third scenario, GFDL CM2.1, frequencies remain constant or decline slightly, owing to an overall drying trend.

T. Das (✉) · M. D. Dettinger · D. R. Cayan
Division of Climate, Atmospheric Sciences, and Physical Oceanography, Scripps Institution of Oceanography, University of California San Diego, Mail Code 0224, La Jolla, CA 92093-0224, USA
e-mail: tapash.das@ch2m.com

M. D. Dettinger · D. R. Cayan
United States Geological Survey, La Jolla, CA, USA

H. G. Hidalgo
School of Physics and Center for Geophysical Research, University of Costa Rica, San Jose, Costa Rica

T. Das
CH2MHILL, Inc., San Diego, CA 92101, USA

These increases appear to derive jointly from increases in heavy precipitation amount, storm frequencies, and days with more precipitation falling as rain and less as snow. Increases in antecedent winter soil moisture also play a role in some areas. Thus, a complex, as-yet unpredictable interplay of several different climatic influences threatens to cause increased flood hazards in California's complex western Sierra landscapes.

1 Introduction

Floods cause immense damage to human societies globally and are thought to be the costliest type of natural disaster in economic and human terms. For example, in United States (US) the total estimated fiscal year 2003 flood damage was almost $2.5 billion (Pielke et al. 2002). Flood damage fluctuates from year to year, but Pielke and Downton (2000) present an increasing trend in US flood damage over the past century.

Floods are particularly dangerous in California, where topography, exposure to heavy moisture-laden storm systems, and extensive human development and infrastructure in low lying areas add to the risks. In California, most of the large historical floods have arisen from two general mechanisms: (i) winter general floods covering large areas and (ii) spring and early summer snowmelt floods, mostly from the higher elevations of the central and southern Sierra Nevada (Roos 1997). The largest winter floods have historically been most catastrophic and have caused billions of dollars in damages (Roos 1997). Floods in California are linked to winter-spring atmospheric circulations (Cayan and Riddle 1992), with the largest floods most often associated with atmospheric rivers (Ralph et al. 2006; Neiman et al. 2008). These influences, and hence the floods they produce, may be amplified in coming decades as climate becomes warmer, as projected by almost all global climate models.

California is currently faced by many urgent issues concerning floods and will need to make decisions in the context of possible climate changes. For example, the State's long history of major floods, together with ongoing tightening of water-supply options in the State, have led the California Water Plan Update in 2005 to identify integrated flood management actions as a crucial part of improved water supply management DWR (2005a). Such integration must address inherent and longstanding conflicts between management of California's reservoirs for flood protection versus their management for dry-season water supplies. But even as flood flows are accepted as a component of future water-supply solutions for the state, the recognized potential for flood-induced failures of aging levee systems, for widespread flood damages, and for disruptions of freshwater conveyances throughout the Central Valley, and especially in the Sacramento-San Joaquin Delta, threatens the State's communities and infrastructures, as well as its largest scale water-supply conveyances. Furthermore, there is a growing understanding that floods and floodplains play fundamental roles in the sustainability of Delta and Central Valley ecosystems and water quality (Healey et al. 2008). In addressing these pressing challenges, informed decisions now require greater understanding and, eventually, reliable projections of future flood responses to likely 21st Century climate changes. That understanding and those projections are still very much matters of research, of which the present study is an early example.

Long-term changes in hydro-meteorological variables have already been documented across the western US, including large increases in winter and spring temperatures (e.g., Dettinger and Cayan 1995; Bonfils et al. 2008), substantial declines in the volumes of snow pack in low and middle latitudes (Lettenmaier and Gan 1990; Knowles and Cayan 2004), significant declines in April 1st snow water equivalent (SWE) (Mote et al. 2005; Pierce et al. 2008), shifts toward more rainfall and less snowfall (Knowles et al. 2006), earlier streamflow in snowfed rivers (Dettinger and Cayan 1995; Cayan et al. 2001; Stewart et al. 2005), and sizeable increases in winter streamflows as fractions of water-year totals (Dettinger and Cayan 1995; Stewart et al. 2005). Recently Barnett et al. (2008) performed a formal multivariate detection and attribution study and showed that observed changes in the mountainous regions of the western US of minimum temperature, SWE as a fraction of precipitation, and center timing of streamflow co-vary during the period 1950–1999. They concluded, with a high statistical confidence, that up to 60% of the observed trends in those variables have been driven by increasing greenhouse effects and attendant temperature increases. In response to projected continuing increases in greenhouse-gas emissions, current climate change projections by the end of the 21st century California uniformly yield warming by at least a couple of degrees Celsius, and, although great uncertainties exist about future changes in long-term average precipitation rates in California (e.g., Dettinger 2005; Cayan et al. 2008a,b), it is generally expected that extreme precipitation episodes will become more extreme as the climate changes (Easterling et al. 2000).

Many studies have been performed to investigate the potential impact of these kinds of climate changes on California's hydrology at seasonal and longer time scales (for example, see Lettenmaier and Gan 1990; Miller et al. 2003; Knowles and Cayan 2004; Dettinger et al. 2004; Maurer 2007; Cayan et al. 2008a; Cayan et al. 2010), finding changes in streamflow timing, warm season vs. cool season runoff, low flow season discharges, and overall flow totals. Given all of the observed and predicted hydrologic changes across the western US, it is reasonable to wonder whether there may be changes in the frequencies and magnitudes of floods as well. Projected climate changes may affect California's flood magnitudes and frequencies, and even the mechanisms by which floods are unleashed, in a variety of ways. Tendencies towards larger storm precipitation totals or more frequent large storms, or the reverse, could directly change the opportunities for floods. Projected warming trends would continue or increase recent trends toward having more precipitation fall as rain and less as snow (Knowles et al. 2006; Das et al. 2009), which could affect both magnitudes and frequencies of floods by increasing the catchment areas from which precipitation (rain) runs off most rapidly or by increasing the frequency of storms with large rainfall runoff totals. Warming may also change catchment water balances to alter antecedent hydrologic conditions that set the stage for floods, e.g., by increasing rates and opportunities for evapotranspiration to sap soil-moisture reservoirs.

In California, the magnitude of 100-year floods estimated from the early historical period was smaller than recently observed floods, suggesting either that the records used to estimate 100-year floods was too short or subject to recent changes in climate (Anderson et al. 2006; DWR 2005a, b; 2007). Dettinger et al. (2004) evaluated future flood risks in specific California basins and found increased risks from warming alone. Miller et al. (2003) found increased likelihood of more floods in California under climate change. Hamlet and Lettenmaier (2007) identified increased flood risk tendencies in warmer basins along the coasts of Washington, Oregon and California using 20th century precipitation and temperature in a hydrological model. Milly et al. (2008) argued "now is an opportune moment to update the analytic strategies used for planning such grand investments under an uncertain and changing climate." More recently, Dettinger et al. (2009) and Raff et al. (2009) predicted general increases in flood magnitude with climate change.

Although uncertainties abound, planning and investments for flood management in California must somehow accommodate potentially changing flood challenges due to climate changes. The projected climate changes may affect California's flood regimes in several ways, such as (Dettinger et al. 2009):

- Potential to intensify or ameliorate of flood magnitudes
- Potential for increase, or decrease, frequencies of floods
- Potential for changing seasonalities and mechanisms of floods

In each case, such changes will remain uncertain. For example, arguments can be made that, with more water vapor and heat in the atmosphere, storms and thus floods should become more intense (e.g., Trenberth 1999) or larger overall; however, as the polar regions warm more quickly than the lower latitudes, the equator-to-pole temperature differences are expected to decrease (Jain et al. 1999) which generally would be expected—from basic geophysical fluid dynamics considerations—to weaken mid latitude storm tracks. Which change will dominate California's future flood regimes? Because the humid tropics have the strongest greenhouse effect, more heat will need to be transported poleward as the world warms and thus mid latitude storms (an important mechanism for poleward heat transport) may become more frequent. Alternatively, declining equator-to-pole temperature differences might tend to weaken storm tracks and reduce storm frequencies. Even with weakening storm tracks, however, trends towards warmer storms may result in more precipitation as rain rather than snow (Knowles et al. 2006; Das et al. 2009), changing the mechanisms and opportunities for flood generation in California's higher and wetter catchments to increase the percentage of storms that generate floods. In such a case, would weakening storm tracks or increasing rain fractions dominate to decrease or increase the numbers of overall floods? In a warming region, with presumed increases in the demands and opportunities for evaporation and transpiration, soils are likely to generally be drier (unless total precipitation increases considerably), which could reduce risks of floods. Alternatively, though, as long as some snow is still being deposited in the state, it is likely that warmer winters—at least—may induce more mid-winter snowmelt and thus wetter soils in winter, which might make flooding more likely. Thus future flood regimes are likely to depend on delicate and difficult balancing between several different kinds of climate-change effects and hydrologic responses. An integrated, hydrologic modeling approach is thus required to assess the net or overall flood-regime changes that might result.

The present study simulates possible changes in annual maximum 3-day flood magnitudes and frequencies for floods exceeding selected historical thresholds under a range of projected climate changes. Potential flood conditions were simulated by forcing the Variable Infiltration Capacity (VIC) hydrologic model with downscaled daily precipitation and temperature projections from three General Circulation Models (GCMs) (CNRM CM3, GFDL CM2.1 and NCAR PCM1). Further analysis was performed to examine possible mechanisms with the associated flood regime change.

This article is organized as follows. Section 2 presents data sets and models used in our study. A description of the method is given in Section 3. Section 4 presents model results and possible mechanisms for the projected flood regime changes. A summary and conclusions are given in Section 5.

2 Datasets and models

2.1 Observed meteorology

Daily observations of precipitation (P), maximum daily temperature (Tmax), minimum temperature (Tmin), and wind speed gridded to $1/8°$ spatial resolution across California were obtained from the Surface Water Modeling Group at the University of Washington (http://www.hydro.washington.edu; Hamlet and Lettenmaier 2005). The data are based on the National Weather Service co-operative network of weather observations stations, augmented by information from the higher quality Global Historical Climatology Network (GHCN) stations. The dataset also relies on monthly PRISM data fields (Daly et al. 1994) to adjust for elevation effects on precipitation and temperature. Such corrections are necessary because topography in the study region strongly determines not only spatial patterns of precipitation (and temperature) but also basin-scale to regional totals of precipitation. The Hamlet and Lettenmaier (2005) dataset is available for the period 1915 through 2003, but hydrologic simulations in this study were of the period 1950–1999.

2.2 Global climate models and downscaling

Because global-scale climate models operate on very coarse spatial grids (with grid cell centers separated by 1 to 2° of latitude and longitude or about 100-km to 200-km), the representation of topography and thus of orographic enhancements of precipitation are almost entirely muted (Dettinger et al. 2004; Dettinger 2005; Maurer 2007; Cayan et al. 2008a,b). Downscaling methods are used to obtain local-scale representations of surface weather from the regional-scale atmospheric variables provided by GCMs. Daily P, Tmax and Tmin from three GCMs (CNRM CM3 model, GFDL CM2.1 and NCAR PCM1) were downscaled to 1/8 ° resolution (about 12-km) using the Constructed Analogues (CA; Hidalgo et al. 2008, 2009; Maurer and Hidalgo 2008) statistical downscaling method. These three GCMs were selected from among the six GCMs used by the California Climate Action Team in its recent California Climate Change Assessment (Cayan et al. 2009; Franco et al. 2011, in this issue). The selection of the global models was guided by several different factors, including having a good representation of typical patterns of El Nino-Southern Oscillation (ENSO) and Pacific Decadal Oscillation (PDO) and realistic representation of mean climate and its decadal variability over California. Another selection criterion was the availability of daily precipitation and temperature data required for the CA downscaling procedure, and to set up a hydrological model to simulate future floods. Flood responses to historical climates, as simulated by each GCM under 20c3m estimates of historical emissions, and to future climates under the medium-high emission scenario (SRES A2; IPCC 2007) are simulated and evaluated here. In the A2 emission scenario, emissions accelerate throughout the 21st Century, and, by the end of the 21st century, atmospheric CO_2 concentrations reach more than triple their pre-industrial levels. This emission scenario yields significant climate change; thus, if climate change could alter flood conditions in California, we expect that evidence will emerge from simulations using this scenario. However, because a broader range of scenarios is not evaluated here, the present results cannot be interpreted as prediction of flood changes but rather as examples of levels of flood change that could plausibly develop in the 21st Century.

In the constructed analogues (CA) method, downscaled climate fields at 1/8° are obtained by constructing linear combinations of previously observed weather patterns, including adjustments for model biases. A library of gridded meteorological observations, based on Maurer et al. (2002) from the period 1950 to 1999, coarsened to match the GCM grid is searched at each GCM time step to locate 30 best analogues of the GCM output for a given day. The 30 previously observed daily patterns most similar (at GCM resolution) to the GCM pattern to be downscaled are then joined by a linear regression to obtain a linear combination that best matches, on the GCM grid, the GCM pattern to be downscaled. Once the best combination of past patterns and the regression-based linear combination of them are obtained, the same linear combination is applied to the original high-resolution Maurer et al. (2002) fields from the same historical days to obtain a high-resolution version of the meteorological conditions corresponding to the daily GCM pattern in question. Downscaling by CA and by the bias correction and spatial downscaling (BCSD), another statistical downscaling methodology, yield results that are quantitatively similar (Maurer and Hidalgo 2008). An advantage of the CA method over the BCSD method is that CA can capture changes in the diurnal cycles of temperatures and changing weather extremes in the GCM outputs. The original version of the CA method (Hidalgo et al. 2008) does not include as extensive a bias-correction of the GCM data as does BCSD, instead including only the bias-correction that was obtained through working with anomalies of the GCM data. But in this work we add a simple post-downscaled bias-correction for precipitation, as follows. As a first step, for every 12-km VIC gridcell, we compute ratios of long term monthly observed precipitation (Maurer et al. 2002) and long term monthly precipitation of GCM historical simulation (so called 20c3m simulation) for the period 1950–1999, for 12 months. In the second step, these ratios are multiplied by the daily values of downscaled precipitation for the period 1950–2099 (thus using the multiplier factor to the downscaled future precipitation). By doing this, no day-to-day correspondence between the model historical simulation and observations could be expected, however the downscaled mean seasonal cycle of precipitation, at 12-km grid scale, is expected to be much the same as the 50 year observed climatology for downscaled model historical simulations. Interested readers are referred to an investigation of advantages of bias-correction procedures (Maurer et al. 2010a).

2.3 Hydrological model

To simulate daily streamflow during the period 1951–1999 and under 21st Century climate change conditions, we used the VIC distributed macro-scale hydrological model (Liang et al. 1994; Cherkauer et al. 2003). Defining characteristics of VIC are the probabilistic treatment of sub-grid soil moisture capacity distribution, the parameterization of baseflow as a nonlinear recession from the lower soil layer, and the unsaturated hydraulic conductivity at each particular time step is treated as a function of the degree of soil saturation (Liang et al. 1994; Maurer et al. 2002). Surface runoff uses an infiltration formulation based on the Xinanjiang model (Wood et al. 1992), while baseflow follows the ARNO model (Todini 1996). It uses a tiled representation of the land surface within each model grid cell, allowing sub-grid variability in topography, infiltration, and land surface vegetation classes (Liang et al. 1994; Maurer et al. 2002). Derived variables such as radiation, humidity and pressure are estimated within the model based on the input P, Tmax and Tmin values using the algorithms of Kimball et al. (1997) and Thornton and Running (1999). The calibrated soil parameters for VIC are the same as those used in Barnett et al. (2008) and Hidalgo et al. (2009). The vegetation cover was obtained from the North American Land Data Assimilation System (NLDAS) and was held static through the entire simulation period. The VIC model was run at a daily time step, with a 1-hour snow model time step in water balance mode, and using a 1/8 by 1/8° grid resolution. Using the gridded meteorological forcing, along with the physiographic characteristics of the catchment (for example, soil and vegetation properties), VIC calculates a suite of hydrologic variables, including runoff, baseflow, soil moisture, actual evapotranspiration and Snow Water Equivalent (SWE) in the snowpack.

VIC has been used extensively in a variety of water resources applications from studies of climate variability, hydrologic forecasting and climate change studies (for example, Nijssen et al. 2001; Maurer et al. 2002; Sheffield and Wood 2007; Barnett et al. 2008; Wood and Lettenmaier 2006; Dettinger et al. 2011; Cayan et al. 2010). Mote et al. (2005) found reasonable agreement between the spatial pattern of observed SWE and the VIC simulated values. The ability of the model to simulate soil moisture at some of the points is found satisfactory when compared with the measurements (Maurer et al. 2002). VIC simulated streamflow validates well with observations, for relatively larger catchments at monthly temporal scale when the model has been calibrated using streamflow data (Maurer et al. 2002; Hidalgo et al. 2009). Hamlet and Lettenmaier (2007) performed a study using VIC simulated extreme streamflows to examine flood characteristics across the West (including California), and argued that VIC can be applied to investigate relative changes in flood risks. In the present study, the main focus is to investigate the changes of floods risk under future climate projections.

2.4 Study areas

In the present study, flood regimes are analyzed for Northern and Southern parts of the western slopes of the Sierra Nevada (Fig. 1). Drainage areas for the basins simulated here in the Northern Sierra Nevada (NSN) and Southern Sierra Nevada (SSN) are approximately 46,080 km^2 and 15,120 km^2 respectively. Topographic elevation for NSN varies from 150 m (m ASL) to 2480 m, with an average elevation of 1360 m. Annual precipitation spatially varies with topographic structure, from 290 mm to 2050 mm, with an average value of 940 mm. Annual temperature spatially varies from -0.3^0 C to 16.4^0 C, averaging 8.1^0 C. Topographic elevation for SSN spatially varies from 150 m (m ASL) to 3520 m, with an average elevation of 1840 m. Annual precipitation varies from 345 mm to 1430 mm; with an average value of 990 mm. Annual temperature spatially varies from -4.6^0 C to 16.2^0 C, averaging 6.6^0 C. Because the NSN area is at lower elevation and warmer, the majority of the streamflow comes directly from winter precipitation and is mixed rain-snow dominated, whereas SSN rivers are dominated by springtime snowmelt runoff (Dettinger et al. 2004).

3 Methods

The VIC simulated runoff and baseflow were combined and routed, as in Lohmann et al. (1996), to obtain daily streamflows for locations near the bottom of seven California river basins (Fig. 1). NSN streamflow was computed by adding the routed streamflow from the Sacramento River at Bend Bridge, the Feather River at Oroville and the Yuba River at Smartville. SSN streamflow was obtained by summing daily streamflows from the four main tributaries of the San Joaquin River: the Stanislaus at New Melones Dam, the Tuolumne River at New Don Pedro, the Merced at Lake McClure, and the San Joaquin at Millerton Lake. As stated previously, the main aim is to study possible changes in flood occurrences and flood magnitudes. A historical 99th percentile value of daily streamflow (from November through July) for the period 1951–1999 was obtained from routed flows as simulated with historical meteorology and with the historical simulations from each GCM. These flow values were used as thresholds to define floods for frequency-of-flooding evaluations.

To examine changes in flood magnitudes, we use 3-day maximum annual streamflows simulated by VIC as driven by the downscaled climate model meteorology for the period 1951–2099. The 3-day flood, a measure commonly used for flood planning purposes in California (DWR 2006; Chung et al. 2009), is important because flood-control reservoir space in the Sierra is large enough so that discharge totals over 3 or more days are more representative of uncontrolled flood risks than are shorter term totals. The 3-day peak flow has thus become a standard for climate change impact studies in California (e.g., Brekke et al. 2009; Maurer et al. 2010a, b).

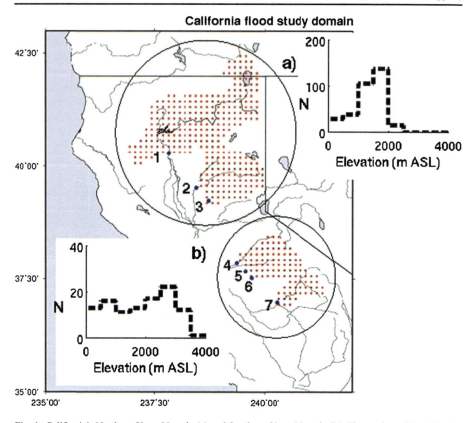

Fig. 1 California's Northern Sierra Nevada (**a**) and Southern Sierra Nevada (**b**). The northern Sierra Nevada consists of the drainage areas from the Sacramento River at Bend Bridge (1), the Feather River at Oroville (2) and the Yuba River at Smartville (3). The southern Sierra Nevada consists of the drainage areas from the four main tributaries of the San Joaquin river: the Stanislaus at New Melones Dam (4), the Tuolumne River at New Don Pedro (5), the Merced at Lake McClure (6) and the San Joaquin at Millerton Lake (7). Histograms show the distribution of the elevations of VIC grid cells

Flood magnitudes of several return periods (*T*), in years, computed from model simulated historical period (1951–1999) were compared with the flood magnitudes at the corresponding return periods based on flows simulated for 2001–2049 and 2051–2099. The 3-day maximum floods for each year were calculated along with their probabilities of exceedance (assigned using the Weibull plotting position) from the three time periods 1951–1999, 2001–2049, and 2051–2099. Then the inverse of the probability (frequency factor, *K*) (Chow et al. 1988; Shabri 2002) was calculated assuming a log-Pearson type III distribution. The *K* values were plotted against the base10 logarithm of the 3-day streamflow maximum. A least-squares linear regression was computed from the data along with 99% regression confidence intervals. From these fits and confidence intervals, flood magnitudes with different return periods ranging from 2 years through 50 years were computed from the fitted flood frequency curves from each of the time epochs. A statistical method described in the appendix was used to determine if the change of the flood magnitude under future climate projections with compared to model historical period is significant at a high statistical confidence level (at $p=0.01$ level).

4 Results and discussions

4.1 Climate change projections

Water year precipitation and annual temperature values, as deviations from model simulated historical period (1951–1999) normal, from the three GCMs over the NSN and SSN are depicted in Table 1. The changes in annual precipitation and temperature were computed using the area-averaged precipitation and temperature across the drainage areas of NSN and SSN. GCMs used in this study cover a range of climate change sensitivities, so that the projections yield moderate increases of annual temperature between 0.5°C and 1.0°C by 2049, with increases between about 2°C to 3°C by 2099 (Table 1). The GFDL model is warmest, followed by CNRM. Because of naturally occurring decadal and multi-decadal precipitation variations as well as differences among the GCMs, trends in precipitation projections are less steady or unanimous. The GFDL model projects overall drying, whereas the CNRM model projects about a 10% increase in precipitation, by 2049. Precipitation ranges from near historical normal to about 8% reduction by 2099. PCM shows moderate warming and moderate increase of precipitation (Table 1).

The number of winter days with precipitation occurring as snow divided by the total number of winter days with precipitation is calculated as follows. A given wet day (day with precipitation above 0.1 mm), in the period November through March, was classified as a snowy day if snowfall (S) was greater than 0.1 mm. As in the VIC (and also in Das et al. 2009), S was calculated using the equation below:

$$S = \begin{cases} 0 & \text{for } T \geq T_{rain} \\ P \cdot \left(\frac{T - T_{rain}}{T_{snow} - T_{rain}} \right) & \text{for } T_{snow} < T < T_{rain} \\ P & \text{for } T \leq T_{snow} \end{cases}$$

where, T is the daily average temperature, T_{snow} is the maximum temperature at which snow falls and T_{rain} is the minimum temperature at which rain falls. To be consistent with the VIC model simulations, the values of T_{snow} and T_{rain} were set to -0.5°C and 0.5°C respectively.

Table 1 Annual temperature change (°C), annual precipitation change (%), and annual snow days as a percentage of wet days change (%) from historical period (1951–1999) in the Northern Sierra Nevada and Southern Sierra Nevada for the 3 GCMs A2 greenhouse-gas emission scenarios

	ΔT (°C)		ΔP (%)		ΔSnow days as percentage of wet days (%)	
	2001–2049	2051–2099	2001–2049	2051–2099	2001–2049	2051–2099
Northern Sierra Nevada						
CNRM CM3	0.9	3.0	10.7	0.5	−13.7	−36.0
GFDL CM2.1	1.2	3.3	−0.5	−10.1	−16.6	−47.8
NCAR PCM1	0.6	2.1	3.6	3.0	−10.5	−39.3
Southern Sierra Nevada						
CNRM CM3	1.0	3.1	10.5	−8.9	−4.9	−19.7
GFDL CM2.1	1.2	3.5	−2.2	−14.5	−7.0	−25.4
NCAR PCM1	0.7	2.1	4.7	5.4	−4.6	−16.6

The projections yield reductions of winter-total snowy days as a fraction of winter-total days with precipitation (indicating a decrease in days with snowfall); larger changes are projected for NSN and the reductions are amplified by the late 21st century (Table 1).

4.2 Changes in extreme precipitation

There is an indication of increased magnitudes in the 3-day precipitation among the highest values, (longest return intervals), except in NCAR PCM1 (Fig. 2). With respect to the climate projections, in general, there is an increased frequency of the extreme precipitation events for the future projected scenarios, except for the GFDL for SSN which shows extreme precipitation events staying either same or decreasing slightly (Table 2). In terms of the frequency of occurrence of extreme precipitation events, precipitation is considered an extreme event when the daily area-averaged precipitation over the NSN and SSN is larger than 99 percentile values computed from the GCM simulated historical period (1951–

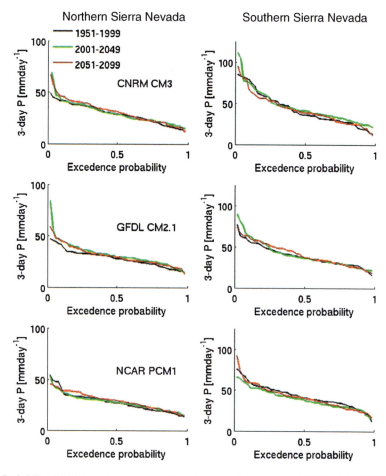

Fig. 2 Probability distributions for 3-day annual maximum precipitation from CNRM CM3, GFDL CM2.1 and NCAR PCM1 GCMs. Climate change period (2000–2099) from SRES A2 scenarios. 1951–1999 period is from GCM simulated historical period. (*Left*) Northern Sierra Nevada, (*Right*) Southern Sierra Nevada

Climatic Change (2011) 109 (Suppl 1)

Table 2 Frequency (number of events per year on average) of extreme precipitation in the Northern Sierra Nevada and Southern Sierra Nevada from three coupled ocean–atmosphere general circulation models, each forced by historical and 21st century greenhouse-gas A2 emission scenarios

	1951–1999	2001–2049	2051–2099
Northern Sierra Nevada			
CNRM CM3	2.7	3.8	4.2
GFDL CM2.1	2.7	3.0	3.2
NCAR PCM1	2.7	3.5	3.5
Southern Sierra Nevada			
CNRM CM3	2.7	4.0	3.2
GFDL CM2.1	2.7	2.6	2.7
NCAR PCM1	2.7	3.0	3.3

1999). The climate model simulations yield increases in extreme precipitation events by about 11% to 40% by 2049 for NSN, and between 18% through 55% by 2099. For SSN, the changes of extreme precipitation frequency are about −4% through 48% by 2049, and 0% through 22% by 2099 (Table 2). Furthermore, when aggregating the extreme precipitation events for the winter period (DJF), climate models suggest an increase in the frequency of extreme precipitation events for the projected period (2000–2099) in both the NSN and SSN. Thus, overall, the projections tend towards larger storm totals and more frequent extreme-precipitation events.

4.3 Flood magnitudes

We begin our evaluations by comparing probability distributions for 3-day annual streamflow from VIC simulations as driven by downscaled climate model meteorology from CNRM CM3, GFDL CM2.1 and NCAR PCM1 GCMs for NSN and SSN. Annual 3-day maximum flows increase for the longest return periods (largest floods), except for NCAR PCM1 for the period 2051–2099 which shows a slight decrease (Fig. 3).

A flood frequency analysis of the 3-day floods as simulated by VIC in response to the historical observed meteorology is shown in Fig. 4 (*top panels*). The climate change impact on the flood discharge magnitude was evaluated by comparing flood frequency curves estimated from the future simulations (Fig. 4) to those estimated from historical simulations. Table 3 presents flood discharges for various return periods, ranging from 2-years to 50-years, in absolute and relative terms. The hypothesis that the best prediction of flood discharge for a certain return period T in years (Q_T) from the historical period (1951–1999) was statistically different from the best prediction of Q_T under the future climate was tested using a statistical method described in the appendix. As can be seen, for the NSN there is a general increase in the Q_T in the future climate. These increases are statistically significant (at $p=0.01$ level) for the CNRM, GFDL simulations of the first half of the 21st century (2001–2049) while the increases of Q_T for all return periods are statistically significant for the late period of the 21st century (2051–2099) (Table 3). For SSN, the increase of flood discharges for all the return periods for CNRM, and flood discharges with more frequent return periods (up to 5-years) for GFDL are statistically significant for the first half of 21st century (2001–2049). PCM1 simulated flood discharges for the 2001–2049 epoch declined compared to model historical. However, by 2051–2099, there is an increase in the 3-day flows for the all 3-GCMs, including the GCM models that project 8–15% declines in overall precipitation, and the increases are statistically significant with high confidence.

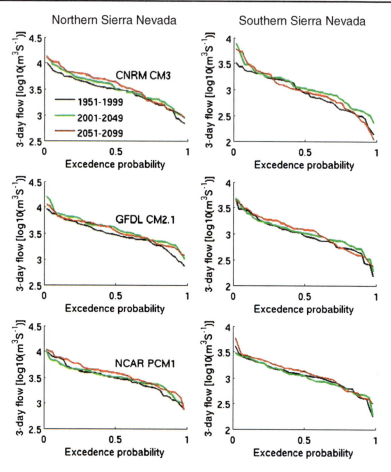

Fig. 3 Same as Fig. 2, except using 3-days maximum annual streamflows as simulated by downscaled meteorology from CNRM CM3, GFDL CM2.1 and NCAR PCM1 GCMs

4.4 Flood occurrences

Historically, most of the floods in California's NSN have occurred from November to March (Fig. 5, *left top panel*). As in the observations, the largest simulated floods occurred during winter (DJF) and they correspond to so-called "pineapple express" events—in which strong low-level Pacific jets transport copious water vapor and warm temperatures to the West Coast (Cayan and Peterson 1989; 1993; Pandey et al. 1999; Cayan and Riddle 1992; Dettinger 2004; Mo et al. 2005; Ralph et al. 2006; Dettinger 2011). During spring (MAM), smaller floods are common and usually are driven by snowmelt on sunny days. These two kinds of floods in the SSN were evaluated in terms of whether it rained or not during the days of the flood (Fig. 5, *right top panel*). There are "wet" or precipitation-dominated floods, mostly in winter, and "dry" or snowmelt-dominated floods, mostly in summer (see also Roos 2006). As in the NSN, the largest floods in the SSN occur during the winter (DJF) corresponding to precipitation-dominated floods, associated with midlatitude, winter storms. Snowmelt-dominated floods occur mainly during the spring-summer (MJJ) and are associated with high-

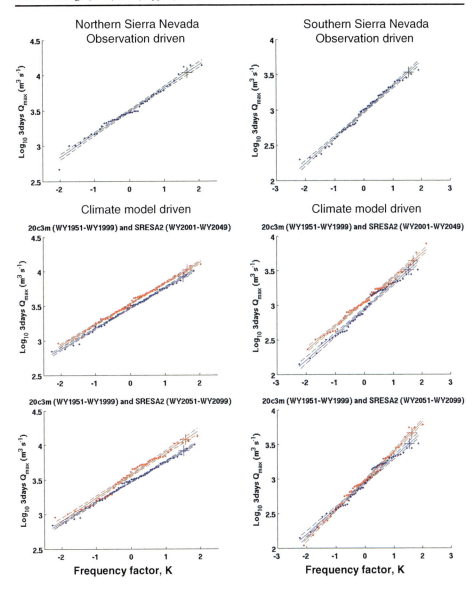

Fig. 4 Frequency of the three-day annual maximum streamflows. Log Pearson III distributions have been fitted to each flood series. Dotted curves: confidence limit for regression line (at 0.01 confidence level). Panels on first row shows flood frequency curves developed using observed meteorology driven VIC simulated streamflows from the period 1951–1999. Second and third row panels compares flood Frequency curves developed using downscaled CNRM CM3 driven VIC simulated streamflows from period 1951–1999 (blue curves) with time periods 2001–2049 and 2051–2099. In the figures, "+" symbol mark flood magnitude with 20-years return period. (*Left*) Northern Sierra Nevada, (*Right*) Southern Sierra Nevada

pressure patterns over the western US that bring higher than normal temperatures and cloud-free days through atmospheric subsidence.

Figure 5 (second to fourth row panels) illustrates floods in Californian NSN and SSN using VIC simulated streamflows as driven by the downscaled CNRM meteorology for the periods 1951–1999, 2001–2049 and 2051–2099. A few of the projected floods have

Table 3 Three-day maximum floods from flood frequency analyses using log-Pearson III distributions; boldface where statistically different from floods in 1951–1999 period (at p=0.01). Numbers in italics indicate flood discharges are decreased compared to 1951–1999 period

T (In Years)	3-days flood discharge			Percentage difference	
	Q20c3m	Qearly	Qlate	(Qearly-Q20c3m)/ Q20c3m	(Qlate-Q20c3m)/ Q20c3m
	(WY1951– WY1999)	(WY2001– WY2049)	(WY2051– WY2099)	(%)	(%)
Northern Sierra Nevada					
CNRM CM3					
2	3050	**3410**	**3800**	12	25
5	5220	**6060**	**7020**	16	34
10	6750	**8170**	**9500**	21	41
20	8240	**10440**	**12080**	27	47
50	10190	**13730**	**15680**	35	54
GFDL CM2.1					
2	3110	**3800**	**3590**	22	15
5	5220	**6430**	**5830**	23	12
10	6730	**8550**	**7510**	27	12
20	8230	**10880**	**9260**	32	13
50	10220	**14350**	**11730**	40	15
NCAR PCM1					
2	3070	**3250**	**3750**	6	22
5	5200	*4980*	**6150**	−4	18
10	6770	*6120*	**7850**	−10	16
20	8380	*7190*	**9520**	−14	14
50	10590	*8550*	**11740**	−19	11
Southern Sierra Nevada					
CNRM CM3					
2	910	**1080**	**960**	19	5
5	1790	**2170**	**2150**	21	20
10	2480	**3160**	**3280**	27	32
20	3220	**4360**	**4620**	35	43
50	4260	**6320**	**6780**	48	59
GFDL CM2.1					
2	970	**1070**	**1120**	10	15
5	1820	1880	**2130**	3	17
10	2500	2510	**2940**	0	18
20	3230	*3190*	**3810**	−1	18
50	4290	*4170*	**5070**	−3	18
NCAR PCM1					
2	1140	*1080*	**1230**	−5	8
5	1910	*1800*	**2190**	−6	15
10	2430	*2320*	**2930**	−5	21
20	2910	*2830*	**3720**	−3	28
50	3510	3510	**4860**	0	38

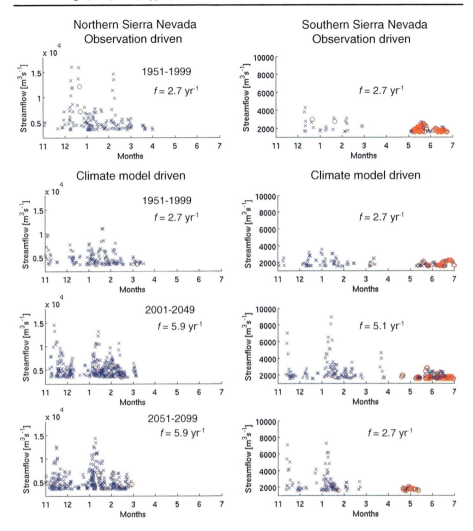

Fig. 5 Floods in Californian Northern Sierra Nevada (*Left*) and Southern Sierra Nevada (*Right*). Panels on first row shows floods using observed meteorology driven VIC simulated streamflows from the period 1951–1999. Second and third row panels show floods using downscaled CNRM CM3 driven VIC simulated streamflows from periods 1951–1999, 2001–2049 and 2051–2099. Numbers (*f*) in each panel indicate frequency of the floods per year on average. In the figure, "X" symbols are precipitation-driven floods and red circles are snowmelt driven floods. Flood is defined if streamflow in a given day for the period November through July is larger than 99 percentile value computed from the time period 1951–1999

unprecedentedly large magnitudes compared to the historical simulations. Table 4 presents the frequency (number of events per year on average) of flood events in the NSN and SSN. In general, the frequency of floods increases in the future, except under the GFDL climate in SSN where numbers of flood events remain constant or decline slightly; recall that GFDL is the GCM (among those considered here) that dries most in the 21st century. There is no clear indication of a change in the seasonality of the floods; however the simulations suggest declining numbers of snow-melt driven floods for SSN under projected climate, particularly in the late 21st century (2051–2099).

Table 4 Frequency (number of events per year on average) of the floods in the Northern Sierra Nevada and Southern Sierra Nevada from three coupled ocean–atmosphere general circulation models, each forced by historical and 21st century greenhouse-gas A2 emission scenarios

	1951–1999	2001–2049	2051–2099
Northern Sierra Nevada			
CNRM CM3	2.7	5.9	5.9
GFDL CM2.1	2.7	2.6	3.0
NCAR PCM1	2.7	3.6	6.4
Southern Sierra Nevada			
CNRM CM3	2.7	5.1	2.7
GFDL CM2.1	2.7	2.5	2.4
NCAR PCM1	2.7	2.8	4.0

4.5 Mechanisms of change in flood risk

Although changes in flood magnitude and occurrence ultimately result from changes in storm sizes (Dettinger 2011), temperatures, and frequencies, there are other factors, including rainfall-receiving area, areas where rain falls on snow, and antecedent soil moisture conditions that also influence flood characteristics and changes under climate change. In this section, several possible mechanisms for each flood characteristic change are discussed.

Relationships between precipitation and areas receiving rainfall are illustrated by plotting the conditions during all storm events historically and in first and second halves of the 21st Century for NSN (Fig. 6) and SSN (Fig. 7). To make these plots, for each flood event, precipitation is summed over 3 consecutive days and the fraction of catchment receiving rain (rather than snow) is averaged, where the 3-day windows constitute the identified flood day, the day before, and the day before that. Notably, comparison of panels along the rows in Figs. 6 and 7 shows that some projected storms have precipitation totals with unprecedentedly large magnitudes compared to the historical simulations. However, more generally, the slopes of the clouds of dot in Figs. 6 and 7 indicate that larger floods (represented by larger circle sizes in the figures) have typically been associated with storms with large precipitation totals. Nonetheless, there is considerable scatter and some large floods have occurred during smaller storms. Also a few storms with larger precipitation totals produce small floods. This kind of scatter appears across the almost the full range of rain-receiving areas from little rainy area to entirely rainy. Consequently we infer that differences in the saturation of soils and land surfaces when the various storms arrived also probably play important roles in determining whether many storms are flood-generating versus non-flood-generating. In particular, in the moderately warmer climates projected for the 21st Century, whether or not the climate is drier or wetter overall, snowpacks typically continue to be formed each winter, although they may be less persistent and thinner. The snowpacks presumably are wetter and experience more frequent midwinter melt cycles, so that soils in the wintertime Sierra Nevada may be wetter than historical as a result of more frequent percolation of winter snowmelts into the soils (Dettinger et al. 2009).

To further explore the role of areas that receive rainfall in the flood-generating storms, we computed basin area receiving rainfall as a percentage of total basin area for each rainfed flood (X's in Fig. 5) for NSN and SSN. Table 5 lists the average areas that receive

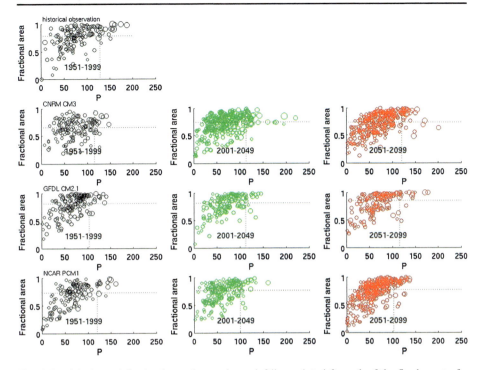

Fig. 6 Precipitation and fractional area that receives rainfall are plotted for each of the flood events, for Northern Sierra Nevada. Panel in first row shows result from observational driven VIC simulation. Second, third and fourth row panels show results using three climate models. For each of the flood events, precipitation (fractional area) is summed (averaged) for 3-consecutive days: the identified flood day, the day before, and the day before that. In the figure, symbol size varies depending on the flood day streamflow. Dotted vertical lines show 90 percentile values of precipitation from each flood population time period. Dotted horizontal lines show 50 percentile values of fractional area from each flood population time period

rainfall, averaged over the rainfed floods in the model historical period (1951–1999) and from climate projections epochs (2001–2099). The extent to which a given amount of warming increases the rainfall receiving areas differs from basin to basin depending on the topography of the basins. Altitudes in SSN rise much higher than in NSN. As a result, the NSN is more dominated by catchment areas at middle to low altitudes, and thus receives rain on larger fractions of the basin in many storms, than in SSN. As in Table 5, except for one exception, there are increases in areas that receive rainfall during the flood days under climate change projections in all three GCMs in both NSN and SSN, by 3% to 15% for NSN and 7% to 30% for SSN by 2051–2099.

McCabe et al. (2007) have shown that, historically, with warming in the western US, the occurrence of rain-on-snow events has declined at low sites and increased at high sites. The decreasing numbers of low-altitude rain-on-snow events reflect declining periods when snow is on the ground available to receive rain. The average percentages of basin area receiving rainfall when there is an antecedent snowpack on the ground, computed from the VIC simulations, for the precipitation driven flood days, as shown in Fig. 5, indicate considerable decreases of the rain-on-snow areas for NSN, especially for the late 21st century. Surprisingly, there is not much change for SSN, perhaps due to countervailing influences of low areas with less snow to be rained upon and high areas with snow that are reached more frequently by rain under the projected warming trends. These countervailing

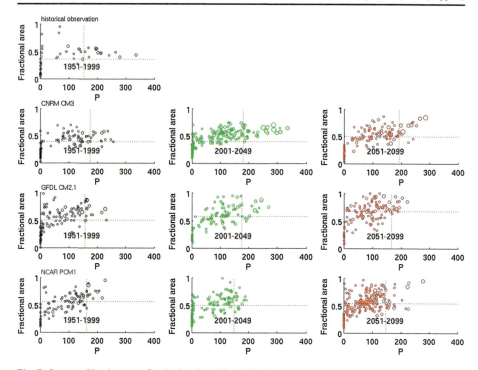

Fig. 7 Same as Fig. 6, except for the Southern Sierra Nevada

influences may be more important in the SSN than in the NSN because of the larger range of elevations in the SSN.

As long as there is generally some snow on the ground in the catchment, warming is likely to increase the number of occasions when the snow will melt even in mid-winter. To investigate the role of antecedent soil moisture in the generated floods, we compared cumulative distribution functions of antecedent soil moisture 2 days prior to the flood events simulated by VIC. These antecedent soil moisture contents increased in the NSN only under the CNRM projections (soil moisture stayed the same or decreased in the other two GCMs). In the SSN, antecedent soil moisture increased in two of the three GCMs (*not shown*).

5 Summary and conclusions

This study investigated how future climate changes might affect flood risk in the Sierra Nevada Mountains of California. The study uses downscaled daily precipitation and temperature simulations from three global climate models (GCMs) under an accelerating (A2) greenhouse-gas emissions scenario, to drive the VIC hydrologic model for catchments on the west slopes of the Sierra Nevada Mountains. In making VIC runs from historical and from projected climate change GCM simulations, we explore possible changes in annual maximum 3-day flood magnitudes and frequencies of floods greater than selected historical thresholds.

Analyses of future projections of flooding reveal a general tendency toward increases in the magnitudes of 3-day flood events. By the end of the 21st Century, all three projections yield larger floods for both the moderate elevation Northern Sierra Nevada (NSN) watershed and for the high elevation Southern Sierra Nevada (SSN) watershed, even for

Climatic Change (2011) 109 (Suppl 1)

Table 5 River basin area that receives rainfall (area with average temperature larger than 0.5°C), as a percentage of total River basin area. Table also contains percentage of basin area that receives Rainfall on Snow. The values are averaged from all precipitation-driven floods, in the Northern Sierra Nevada and Southern Sierra Nevada, under model historical simulation and climate change projections from three coupled ocean–atmosphere general circulation models. In parentheses, changes are computed from model simulated historical period (1951–1999)

	1951–1999	2001–2049	2051–2099
Percentage of basin area that receives Rainfall			
Northern Sierra Nevada			
CNRM CM3	55.2	64.0 (+15.9)	60.8 (+10.1)
GFDL CM2.1	66.0	66.2 (+0.3)	68.3 (+3.5)
NCAR PCM1	54.8	59.7 (+8.9)	64.0 (+16.8)
Southern Sierra Nevada			
CNRM CM3	34.6	38.7 (+11.8)	45.1 (+30.3)
GFDL CM2.1	44.7	51.0 (+14.1)	55.3 (+23.7)
NCAR PCM1	45.6	43.4 (−4.8)	48.9 (+7.2)
Percentage of basin area that receives Rainfall on Snow			
Northern Sierra Nevada			
CNRM CM3	23.9	24.9 (+4.2)	12.7 (−46.9)
GFDL CM2.1	15.9	11.3 (−28.9)	8.9 (−44.0)
NCAR PCM1	17.2	12.5 (−27.3)	10.2 (−40.7)
Southern Sierra Nevada			
CNRM CM3	12.3	15.7 (+27.6)	11.6 (−5.7)
GFDL CM2.1	16.9	11.5 (−32.0)	15.4 (−8.9)
NCAR PCM1	13.7	15.6 (+13.9)	14.4 (+5.1)

GCM simulations with 8–15% declines in overall precipitation. The increases in flood magnitude are statistically significant (at $p=0.01$ level) for all three GCMs for the period 2051–2099 (Fig. 4 and Table 3). By the end of the 21st Century, the magnitudes of the largest floods increase to 110% to 150% of historical magnitudes. Realistically, the half-century windows used to estimate changing 50-year flood magnitudes are short. However, we wanted to present here indications of the potential for changes in magnitudes of the largest floods under medium-high greenhouse-gas emissions with attendant medium-high climate changes over California. As part of the analysis described here, we also computed changes in flood magnitudes with return periods 2-years, 5-years, 10 years, and 20-years (as shown in Table 3), return periods more confidently estimated with the half-century windows analyzed. These more confidently estimated results (for smaller flood levels) corroborate the 50-year flood changes presented in the body of this paper.

Large winter (DJF) floods have occurred during the historical period in both the NSN and SSN (Fig. 5, *top panels*). These winter floods are most often outcomes from Pacific storm-generated immediate runoff from the warmer, rainfall affected parts of the Sierra watersheds. Spring floods in the NSN often arise from storms and conditions similar to the DJF floods, but with lower magnitudes. In contrast, in the SSN, spring (MJJ) floods do not often occur during storm events and instead are more commonly snowmelt fed. These MJJ floods rise to magnitudes comparable to the DJF floods. In the projections, the frequency of floods (above historical 99th percentile thresholds) increases under the CNRM and PCM1 climate changes, while under the GFDL climate, which has the strongest drying trend over the 21st Century, the frequency remained constant or declined slightly. The VIC simulations have a tendency to yield fewer snowmelt driven floods for SSN, particularly under the late 21st century (2051–2099). In general there is a projected tendency for more frequent winter rainfall generated floods and fewer snowmelt floods as climate warms.

The increased flood frequencies appear to derive from three factors: increases in the sizes of the largest storms, increased storm frequencies, and days with more precipitation falling as rain and less as snow (Figs. 6 and 7; Table 5). Increases in antecedent winter soil moisture also play a role in some areas, particularly in the SSN. Thus, several mechanisms may end up contributing simultaneously to increased flood hazards in California's western Sierra Nevada catchments. The particular mixes of influences on flood magnitudes and frequencies at the multi basin scales considered here are unlikely to apply everywhere at smaller scales, from basin to basin, and do not even appear to be consistent from climate-change scenario to scenario. Thus more detailed simulations and analyses, with broader ensembles of climate-change projections, will be needed to develop much confidence in real-world, actionable predictions about flood-regime changes.

The varied results here emphasize the uncertainties related to the present generation of GCM's, manifested in different magnitudes of warming and quite strongly different changes (neutral to negative) in precipitation over the 21st Century. However, the current results present us with three separate versions of the future of floods (one from each of the GCMs). The differences between flood responses to the various scenarios considered here are not unexpected, as flood is a result of combination from precipitation, temperature and antecedent catchment conditions. Future research involving GCMs (Dettinger 2005; Maurer 2007; Pierce et al. 2009), downscaling (Maurer et al. 2010a) and hydrological model simulations (Maurer et al. 2010b) will be needed to objectively merge results from several GCMs into a single master estimate of changing flood frequencies and their associated uncertainties. (*see* e.g. Raff et al. 2009). Nonetheless, the size of changes simulated here, under the different projections, are evidence that plausible levels of 21st Century climate change are likely to affect future flood magnitudes and frequencies in significant ways. The strong tendencies for the several scenarios investigated here to agree in certain qualitative and even quantitative ways also suggests that some strong consensus regarding future flood probabilities may be possible outcomes from such expansions of the research agenda illustrated here.

VIC is a macroscale hydrologic model and its use in the present study is only to assess, through a rather coarse lens, flood tendencies that might occur under climate changes. We would not suggest that the quantitative flood discharges computed using VIC be used for structural designs or details of flood management. From these analyses, we conclude (as others have, for example Dettinger et al. 2004; Miller et al. 2003; Anderson et al. 2006; Dettinger et al. 2009; Raff et al. 2009) that flood regimes in Californian Sierra Nevada are vulnerable to change with future climate change. With the models used here, flood magnitudes and frequencies tend overall (but not universally) to increase, even under scenarios with significant drying. A common feature among the projections is warming; however, tendencies towards intensification of the most severe storms also are common (Dettinger 2011). These broad qualitative mechanisms and flood responses may prove to be robust, but precision in the changing flood frequency estimates is not yet available. Flood frequency estimation procedures that more directly incorporate non-stationarity will need to be developed to permit observations-based frequency estimation for detailed planning of flood management under future climate change (see also Griffis and Stedinger 2007; Sivapalan and Samuel 2009; Raff et al. 2009). Precision simulation-based flood frequency estimation will have to await more precise climate-change projections and hydrologic models of flood responses.

Acknowledgments We thank the Program for Climate Model Diagnosis and Intercomparison (PCMDI) and the WCRP's Working Group on Coupled Modelling (WGCM) for the WCRP CMIP3 multi-model dataset. Support of this dataset is provided by the Office of Science, U.S. Department of Energy. We acknowledge particularly the GCM modeling groups at CNRM, NCAR and GFDL for GCM output. Thanks to Michael Anderson and John T. Andrew at California Department of Water Resources for their valuable discussions. We thank three anonymous reviewers and editors for useful comments. The study was supported by both the CALFED Bay-Delta Program-funded postdoctoral fellowship grant provided to TD and the California Energy Commission-funded California Climate Change Center. Partial salary support for TD from Environment and Sustainability Initiative at UC San Diego (now Sustainability Solutions Institute) through a seed funding grant is also acknowledged. The California Energy Commission PIER Program through the California Climate Change Center, the NOAA RISA Program via the CNAP RISA, and DOE through grant DE-SC0002000, provided partial salary support for DC. Much of the contribution of HH was done while he was a Project Scientist at SIO. He is now partially funded through research projects VI-805-A9-224 and VI-808-A9-180 of the University of Costa Rica.

Appendix: Test of significance for flood magnitude

For two regression-based estimates of log-transformed flood magnitudes with recurrence interval of r years, $f_1 \sim N(m_1, s_1)$ (e.g., from simulated historical annual-flood series) and $f_2 \sim N(m_2, s_2)$ (from simulated future annual-flood series), the probability that $f_2 > f_1$ is

$$\int_{-\infty}^{+\infty} \frac{1}{\sqrt{2\pi}s_2} e^{-\frac{1}{2}\left(\frac{x-m_2}{s_2}\right)^2} \int_{-\infty}^{x} \frac{1}{\sqrt{2\pi}s_1} e^{-\frac{1}{2}\left(\frac{y-m_1}{s_1}\right)^2} dy\, dx$$

or, after rearranging the problem in a more readily generalized form,

$$\int_{-\infty}^{+\infty} \frac{s_1}{\sqrt{2\pi}s_2} e^{-\frac{1}{2}(x-(m_2-m_1)/s_1)^2 s_1^2/s_2^2} \left[\frac{1}{\sqrt{2\pi}} \int_{-\infty}^{x} e^{-\frac{y^2}{2}} dy\right] dx$$

Upon numerical integration of this distribution for various values of (s_2/s_1) and $(m_2-m_1)/s_1$, and then identification of the values $\mu_{s2/s1}$ of $(m_2-m_1)/s_1$ corresponding to $\mathrm{Prob}(f_2>f_1)=0.95$ or 0.99 as a function of s_2/s_1 (Fig. 8), the null hypothesis that $f_2 <= f_1$ can be rejected (at $p=0.05$ and $p=0.01$ levels) when $(m_2-m_1)/s_1 > \mu_{s2/s1}$.

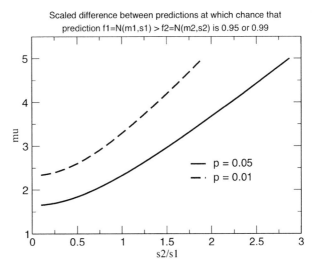

Fig. 8 Values $\mu=(m_2-m_1)/s_1$ as a function of s_2/s_1 where the probability that $f_1 \sim N(m_1,s_1)$ is greater than $f_2 \sim N(m_2,s_2)$ is 0.95 or 0.99

References

Anderson M, Miller NL, Heiland B, King J, Lek B, Nemeth S, Pranger T, Roos M (2006) Climate change impacts on flood management. Chapter 6, Progress on incorporating climate change into management of California's water resources. California Department of Water Resources Progress Report. Governor's Climate Initiative Report

Barnett TP, Pierce DW, Hidalgo HG, Bonfils C, Santer BD, Das T, Bala G, Wood A, Nazawa T, Mirin A, Cayan D, Dettinger M (2008) Human-induced changes in the hydrology of the western US. Science. doi:10.1126/science.1152538

Bonfils C, Duffy PB, Santer BD, Wigley TML, Lobell DB, Phillips TJ, Doutriaux C (2008) Identification of external influences on temperatures in California. Clim Chang 87(Suppl 1):S43–S55. doi:10.1007/s10584-007-9374-9

Brekke LD, Maurer EP, Anderson JD, Dettinger MD, Townsley ES, Harrison A, Pruitt T (2009) Assessing reservoir operations risk under climate change. Water Resour Res 45:W04411. doi:10.1029/2008WR006941

Cayan DR, Peterson DH (1989) The influence of North Pacific atmospheric circulation on streamflow in the West. PACLIM. AGU monograph, No 55, American Geophysical Union. 375–397

Cayan DR, Peterson DH (1993) Spring climate and salinity in the San Francisco Bay Estuary. Water Resour Res 2:293–303

Cayan DR, Riddle L (1992) Atmospheric circulation and precipitation in the Sierra Nevada: Proceedings, International Symposium on Managing Water Resources During Global Change. American Water Resources Association, Reno, Nevada, Nov 1–5

Cayan DR, Dettinger MD, Hanson R, Brown T, Westerling A (2001) Investigation of climate change impacts on water resources in the California region, Department of Energy Accelerated Climate Prediction Initiative (ACPI) Progress Report, 1/19/01, Scripps Institution of Oceanography, U.S. Geological Survey, Desert Research Institute, p 26

Cayan DR, Maurer EP, Dettinger MD, Tyree M, Hayhoe K (2008a) Climate change scenarios for the California region. Clim Chang 87(suppl 1):21–42. doi:10.1007/s10584-007-9377-6

Cayan DR, Lures AL, Franco G, Hanemann M, Croes B, Vine E (2008b) Overview of the California climate change scenarios project. Clim Chang 87(suppl 1):S1–S6. doi:10.1007/s10584-007-9352-2

Cayan D, Tyree M, Dettinger M, Hidalgo H, Das T, Maurer E, Bromirski P, Graham N, Flick R (2009) Climate change scenarios and sea level rise estimates for the California 2009 climate change scenarios assessment. California Climate Change Center. CEC-500-2009-014-F, p 64. Available online: http://www.energy.ca.gov/2009publications/CEC-500-2009-014/CEC-500-2009-014-F.PDF

Cayan DR, Das T, Pierce DW, Barnett TP, Tyree M, Gershunov A (2010) Future dryness in the southwest US and the hydrology of the early 21st century drought. Proc Natl Acad Sci 107(50):21271–21276. doi:10.1073/pnas.0912391107

Cherkauer KA, Bowling LC, Lettenmaier DP (2003) Variable infiltration capacity cold land process model updates, Global Plan. Change 38:151–159

Chow VT, Maidment DR, Mays LW (1988) Applied hydrology. Mcgraw-Hill International Editions, Civil Engineering Series

Chung F, Anderson J, Arora S, Ejeta M, Galef J, Kadir T, Kao K, Olson A, Quan C, Reyes E, Roos M, Seneviratne S, Wang J, Yin H, Blomquist N (2009) Using future climate projections to support water resources decision making in California, California Energy Commission Technical Report CEC-500-2009-052-F, August 2009

Daly C, Neilson RP, Phillips DL (1994) A statistical-topographic model for mapping climatological precipitation over mountainous terrain. J Appl Meteorol 33:140–158

Das T, Hugo G, Hidalgo MD, Dettinger DR, Cayan DW, Pierce CB, Barnett TP, Bala G, Mirin A (2009) Structure and Detectability of trends in hydrological measures over the Western US. J Hydrometeorol 10:871–892. doi:10.1175/2009JHM1095.1

Dettinger MD (2004) Fifty-two years of "pineapple-express" storms across the West Coast of North America. US Geological Survey, Scripps Institution of Oceanography for the California Energy Commission. PIER project report. CEC-500-2005-004, 20 Pp. Available online at http://www.energy.ca.gov/2005publications/CEC-500-2005-004/CEC-500-2005-004.PDF

Dettinger MD (2005) From climate change spaghetti to climate change distributions for 21st Century. San Francisco Estuary and Watershed Science 3(1)

Dettinger MD (2011) Climate change, atmospheric rivers and floods in California—A multimodel analysis of storm frequency and magnitude changes: Journal of American Water Resources Association 47:514–523

Dettinger MD, Cayan DR (1995) Large-scale atmospheric forcing of recent trends toward early snowmelt runoff in California. J Climate 8:606–623

Dettinger MD, Cayan DR, Meyer MK, Jeton AE (2004) Simulated hydrologic responses to climate variations and change in the Merced, Carson, and American River basins, Sierra Nevada, California, 1900–2099. Clim Chang 62:283–317

Dettinger M, Hidalgo H, Das T, Cayan D, Knowles N (2009) Projections of potential flood regime changes in California. Public Interest Energy Research, California Energy Commission, Sacramento, CA. Available online at http://www.energy.ca.gov/2009publications/CEC-500-2009-050/CEC-500-2009-050-D.PDF

Dettinger MD, Ralph FM, Hughes M, Das T, Neiman P, Cox D, Estes G, Reynolds D, Hartman R, Cayan D, Jones L (2011) Requirements and designs for a winter storm scenario for emergency preparedness and planning exercises in California. Natural Hazards. doi:10.1007/s11069-011-9894-5

DWR (2005a) Bulletin 160–05 California Water Plan Update

DWR (2005b) Flood warning: Responding to California's Flood Crisis. State of California, The Resources Agency, Department of Water Resources, January 2005

DWR (2006) Progress on incorporating climate change into planning and management of California's Water Resources, California Department of Water Resources, Technical Memorandum Report, July 2006

DWR (2007) A California challenge - flooding in the Central Valley. A report to the Department of Water Resources, State of California, October 2007

Easterling DR, Meehl GA, Parmesan C, Changnon SA, Karl TR, Mearns LO (2000) Climate extremes: observations, modeling, and impacts. Science 2889(5487):2068–2074

Franco G, Cayan D, Moser S, Hanemann M, Jones M-A (2011) Second California Assessment: Integrated Climate Change Impacts Assessment of Natural and Managed Systems. Climatic Change. In this Issue

Griffis VW, Stedinger JR (2007) Incorporating climate change and variability into bulletin 17B LP3 Model, World Environmental and Water Resources Congress 2007: Restoring Our Natural Habitat, American Society of Civil Engineers, 8

Hamlet AF, Lettenmaier DP (2005) Production of temporally consistent gridded precipitation and temperature fields for the continental U.S. J Hydrometeorol 6:330–336

Hamlet AF, Lettenmaier DP (2007) Effects of 20th Century warming and climate variability on flood risk in the Western U.S. Water Resour Res 43:W06427. doi:10.1029/2006WR005099

Healey M, Dettinger M, Norgaard R (eds) (2008) The state of Bay-Delta science, 2008: CALFED Science Program, 174 p., http://science.calwater.ca.gov/publications/sbds.html.

Hidalgo HG, Dettinger MD, Cayan DR (2008) Downscaling with constructed analogues: daily precipitation and temperature fields over the United States. California Energy Commission, PIER Energy-Related Environmental Research. CEC-500-2007-123. p. 48. Available online: www.energy. ca.gov/2007publications/CEC-500-2007-123/CEC-500-2007-123.PDF

Hidalgo HG, Das T, Dettinger MD, Cayan DR, Pierce DW, Barnett TP, Bala G, Mirin A, Wood AW, Bonfils C, Santer BD, Nozawa T (2009) Detection and attribution of climate change in streamflow timing of the Western United States. J Clim 22(13):3838–3855

Intergovernmental Panel on Climate Change (2007) Climate Change 2007: The physical science basis. Contribution of working group I to the fourth assessment report of the intergovernmental panel on climate change. In: Solomon, S., D. Qin, M. Manning, Z. Chen, M. Marquis, K.B. Averyt, M. Tignor and H.L. Miller (eds.)]. Cambridge University Press, Cambridge, United Kingdom and New York, NY, USA, p 996

Jain S, Lall U, Mann ME (1999) Seasonality and interannual variations of Northern Hemisphere temperature: Equator-to-pole gradient and ocean–land contrast. J Clim 12:1086–1100

Kimball JS, Running SW, Nemani R (1997) An improved method for estimating surface humidity from daily minimum temperature. Agric For Meteorol 85:87–98

Knowles N, Cayan DR (2004) Elevational dependence of projected hydrologic changes in the SanFrancisco Estuary and watershed. Clim Chang 62(1):319–336

Knowles N, Dettinger M, Cayan D (2006) Trends in snowfall versus rainfall for the Western United States. J Clim 19(18):4545–4559

Lettenmaier DP, Gan TY (1990) Hydrologic sensitivities of the Sacramento–San Joaquin River Basin, California, to global warming. Water Resour Res 26:69–86

Liang X, Lettenmaier DP, Wood EF, Burges SJ (1994) A simple hydrologically based model of land surface water and energy fluxes for GSMs. J Geophys Res 99(D7):14,415–14,428

Lohmann D, Nolte-Holube R, Raschke E (1996) A large scale horizontal routing model to be coupled to land surface parameterization schemes. Tellus 48A:708–721

Maurer EP (2007) Uncertainty in hydrologic impacts of climate change in the Sierra Nevada, California under two emissions scenarios. Clim Chang 82(3–4):309–325. doi:10.1007/s10584-006-9180-9

Maurer EP, Hidalgo HG (2008) Utility of daily vs. monthly large-scale climate data: an intercomparison of two statistical downscaling methods. Hydrol Earth Syst Sci 12:551–563

Maurer EP, Wood AW, Adam JC, Lettenmaier DP, Nijssen B (2002) A long-term hydrologically-based data set of land surface fluxes and states for the conterminous United States. J Clim 15:3237–3251

Maurer EP, Hidalgo HG, Das T, Dettinger MD, Cayan DR (2010a) The utility of daily large-scale climate data in the assessment of climate change impacts on daily streamflow in California. Hydrol Earth Syst Sci 14:1125–1138, 113

Maurer EP, Brekke LD, Pruitt T (2010b) Contrasting lumped and distributed hydrology models for estimating climate change impacts on California watersheds. J Am Water Resour Assoc 46(5):1024–1035. doi:10.1111/j.1752-1688.2010.00473

McCabe GJ, Clark MP, Hay LE (2007) Rain-on-snow events in the Western United States. BAMS, 1–10

Miller NL, Bashford KE, Strem E (2003) Potential impacts of climate change on California hydrology. J Amer Water Resoures Assoc 771–784

Milly PCD, Betancourt J, Falkenmark M, Hirsch RM, Kundzewicz ZW, Lettenmaier DP, Stouffer RJ (2008) Stationarity is dead: whither water management? Science 319(5863):573–574

Mo KC, Chelliah M, Carrera ML, Higgins RW, Ebisuzaki W (2005) Atmospheric moisture transport over the United States and Mexico as evaluated in the NCEP regional reanalysis. J Hydrometeorol 6:710–728

Mote P, Hamlet AF, Clark MP, Lettenmaier DP (2005) Declining mountain snowpack in western North America. Bull Am Meteorol Soc 86:39–49

Neiman PJ, Ralph FM, Wick GA, Lundquist JD, Dettinger MD (2008) Meteorological characteristics and overland precipitation impacts of atmospheric rivers affecting the West Coast of North America based on 8 years of SSM/I. J Hydrometeor 9:22–47

Nijssen B, O'Donnell GM, Lettenmaier DP, Lohmann D, Wood EF (2001) Predicting the discharge of global rivers. J Clim 14:3307–3323

Pandey GR, Cayan DR, Georgakakos KP (1999) Precipitation structure in the Sierra Nevada of California during winter. J Geophys Res 104:12019–12030

Pielke RA Jr, Downton MW (2000) Precipitation and damaging floods: trends in the United States, 1932–97. J Clim 13(20):3625–3637

Pielke RA Jr, Downton MW, Barnard Miller JZ (2002) Flood damage in the United States, 1926–2000: a reanalysis of national weather service estimates. UCAR, Boulder

Pierce DW, Barnett TP, Bala G, Mirin A, Wood AW, Bonfils C, Santer BD, Nozawa T (2008) Detection and attribution of streamflow timing changes to climate change in the Western United States. J Clim 22:3838–3855

Pierce DW, Barnett TP, Santer BD, Gleckler PJ (2009) Selecting global climate models for regional climate change studies. Proc Natl Acad Sci. doi:10.1073/pnas.0900094106

Raff DA, Pruitt T, Brekke LD (2009) A framework for assessing flood frequency based on climate projection information. Hydrol Earth Syst Sci 13(1–18)

Ralph FM, Neiman PJ, Wick GA, Gutman SI, Dettinger MD, Cayan DR, White AB (2006) Flooding on California's Russian river: Role of atmospheric rivers. Geophys Res Lett 33:L13801. doi:10.1029/2006GL026689

Roos M (1997) The top ten California floods of the 20th Century. In: CM Isaacs, Tharp VL (eds.) Proceedings of the Thirteenth Annual Pacific Climate 742 (PACLIM) Workshop. April 15–18, 1996. Interagency Ecological Program. Technical Report 53. 743 California Department of Water Resources, pp. 9-18

Roos M (2006) Flood management practice in northern California. Irrig Drain 55:S93–S99

Shabri A (2002) A comparison of plotting formulas for the Pearson Type III distributions. J Technol 36 (C):61–74

Sheffield J, Wood EF (2007) Characteristics of global and regional drought, 1950–2000: analysis of soil moisture data from off-line simulation of the terrestrial hydrologic cycle. J Geophys Res 112:D17115. doi:10.1029/2006JD008288

Sivapalan M, Samuel JM (2009) Transcending limitations of stationarity and the return period: process-based approach to flood estimation and risk assessment. Hydrol Process 23:1671–1675. doi:10.1002/hyp.7292

Stewart IT, Cayan DR, Dettinger MD (2005) Changes toward earlier streamflow timing across Western North America. J Clim 18:1136–1155

Thornton PE, Running SW (1999) An improved algorithm for estimating incident daily solar radiation from measurements of temperature, humidity, and precipitation. Agric For Meteorol 93:211–228

Todini E (1996) The ARNO rainfall-runoff model. J Hydrol 175:339–382

Trenberth KE (1999) Conceptual framework for changes of extremes of the hydrological cycle with climate change. Clim Chang 42:327–339

Wood AW, Lettenmaier DP (2006) A testbed for new seasonal hydrologic forecasting approaches in the western U.S. Bull Am Meteorol Soc 87:12. doi:10.1175/BAMS-87-12-1699,1699-1712

Wood EF, Lettenmaier DP, Zartarian VG (1992) A land-surface hydrology parameterization with subgrid variability for general circulation models. J Geophys Res 97:2717–2728

Simulating cold season snowpack: Impacts of snow albedo and multi-layer snow physics

D. Waliser · J. Kim · Y. Xue · Y. Chao · A. Eldering ·
R. Fovell · A. Hall · Q. Li · K. N. Liou · J. McWilliams ·
S. Kapnick · R. Vasic · F. De Sale · Y. Yu

Received: 12 February 2010 / Accepted: 21 October 2011 / Published online: 24 November 2011
© Springer Science+Business Media B.V. 2011

Abstract This study used numerical experiments to investigate two important concerns in simulating the cold season snowpack: the impact of the alterations of snow albedo due to anthropogenic aerosol deposition on snowpack and the treatment of snow physics using a multi-layer snow model. The snow albedo component considered qualitatively future changes in anthropogenic emissions and the subsequent increase or decrease of black carbon deposition on the Sierra Nevada snowpack by altering the prescribed snow albedo values. The alterations in the snow albedo primarily affect the snowpack via surface energy budget with little impact on precipitation. It was found that a decrease in snow albedo (by as little as 5–10% of the reference values) due to an increase in local emissions enhances snowmelt and runoff (by as much as 30–50%) in the early part of a cold season, resulting in reduced snowmelt-driven runoff (by as much as 30–50%) in the later part of the cold season, with the greatest impacts at higher elevations. An increase in snow albedo associated with reduced anthropogenic emissions results in the opposite effects. Thus, the most notable impact of the decrease in snow albedo is to enhance early-season snowmelt and to reduce late-season snowmelt, resulting in an adverse impact on warm season water resources in California. The timing of the sensitivity of snow water equivalent (SWE), snowmelt, and runoff vary systematically according to terrain elevation; as terrain elevation increases, the peak response of these fields occurs later in the cold season. The response of SWE and surface energy budget to the alterations in snow albedo found in this study shows that the effects of snow albedo on snowpack are further enhanced via local snow-albedo

D. Waliser · Y. Chao · A. Eldering
JPL/CALTECH, Pasadena, CA, USA

J. Kim (✉) · Y. Xue · R. Fovell · A. Hall · Q. Li · K. N. Liou · J. McWilliams · S. Kapnick · R. Vasic ·
F. De Sale · Y. Yu
UCLA, Los Angeles, CA, USA
e-mail: jkim@atmos.ucla.edu

R. Vasic
National Centers for Environmental Prediction, Silver Springs, MD, USA

feedback. Results from this experiment suggest that a reduction in local emissions, which would increase snow albedo, could alleviate the early snowmelt and reduced runoff in late winter and early spring caused by global climate change, at least partially. The most serious uncertainties associated with this part of the study are a quantification of the relationship between the amount of black carbon deposition and snow albedo—a subject of future study. The comparison of the spring snowpack simulated with a single- and multi-layer snow model during the spring of 1998 shows that a more realistic treatment of snow physics in a multi-layer snow model could improve snowpack simulations, especially during spring when snow ablation is significant, or in conjunction with climate change projections.

1 Introduction

The snowpack in the Sierra Nevada region is important to California's water resources. The high elevation snowpack serves as a natural reservoir that stores fresh water during the wet, cold season and releases it gradually during the dry, warm season. About 60% of the water supply for Southern California comes from melting Sierra Nevada snowpack. Snowmelt also affects hydropower generation in California (Vicuna et al. 2008). The impact of global warming on the Sierra Nevada snowpack has become one of the leading topics in regional climate studies for California (Leung and Ghan 1999; Kim 2001, 2005; Kim et al. 2002). However, snowpack projections under global warming scenarios suffer from large uncertainties. Figure 1 shows the simulated southern/central California surface air temperature and Sierra Nevada snowpack changes for the middle-of-the-road (SRES-A1B) greenhouse gas concentration scenario from 16 global climate models (GCMs) that contributed to the Intergovernmental Panel on Climate Change's (IPCC's) 4th Assessment Report (2007). All models agree on the general increase of the low-level temperature (Fig. 1a) and the decrease in the Sierra Nevada SWE with the model ensemble mean projection (black line) indicating 80% loss by the year 2050 (Fig. 1b). However, the inter-GCM spread is substantial; the simulated temperature changes in 2050 range between 1 K and 2 K among the GCMs. The uncertainty in the projected SWE is even larger with losses

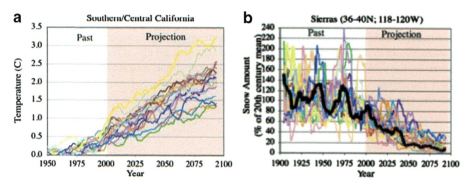

Fig. 1 The time series of the annual-mean (**a**) surface air temperature and (**b**) SWE normalized by the 20th century ensemble mean value (i.e.,% of the 20th century mean SWE) over the regions corresponding to the southern/central California and the Sierra Nevada, respectively, during the 20th and 21st centuries simulated by the 16 GCMs including BCC, BCCR, CGCM, CNRM, CSIRO, GFDL, GISS, FGOALS, INM, IPSL, MIROC3.2, ECHAM5, MRI, CCSM3, HadCM3, and HadGEM1, contributing to the IPCC Assessment Report 4. The *thick black line* in (**b**) represents the average over all GCMs

ranging from 20% to 95%. Reducing the level of uncertainty is crucial for establishing reliable adaptation and mitigation plans. The large inter-model spread in the SWE projections largely result from the differences in model physics, but also due to differences in the models' resolution and their ability to capture detailed orographic effects (Giorgi et al. 1997; Kim 2001). Thus, fine-scale modeling with improved snow physics is crucial for reducing uncertainties for the snowpack projections in mountainous California and the western United States.

The snow budget in the Sierra Nevada is affected by a number of factors, such as insolation, air temperature, and orography. Previous studies on the impact of climate change on the Sierra Nevada snowpack have focused solely on the low tropospheric warming (e.g., Leung and Ghan 1999; Kim 2001, 2005; Kim et al. 2002; Cayan et al. 2008) since this affects two key factors determining the snow budget: rainfall-snowfall partitioning and snow ablation. For a more comprehensive understanding and projection of the Sierra Nevada snowpack in future climate, it is necessary to investigate the role of other factors that also affect the snow budget.

Snow albedo is among the most important local parameters in shaping the spatiotemporal variations in snowpack. Surface insolation, and more specifically the portion of insolation absorbed by the snowpack, is the leading energy source in the evolution of snowpack, especially during the melting period (Marks and Dozier 1992). Thus variations in snow albedo can exert significant impact on snowpack during the course of accumulation and ablation. The surface albedo of sufficiently deep snowpack, and in turn the amount of the insolation absorbed by the snowpack, depends largely on the ice grain size and impurities within or at the surface of ice grains (e.g., Wiscombe and Warren 1980; Warren and Wiscombe 1980; Yang et al. 1997; Mölders et al. 2008). Previously, the impact of snow grain size has been incorporated into snow albedo formulation in terms of snow age or surface temperature or a combination of both, and has been examined in a number of evaluation studies (e.g., Yang et al. 1997; Sun et al. 1999; Molotch and Bales 2006; Mölders et al. 2008). There exist, however, only a limited number of studies (e.g., Hadley et al. 2007; Painter et al. 2007, 2010; Flanner et al. 2009; Steltzer et al. 2009) on the alteration in snow albedo and its impact on surface hydrology due to dust and black carbon (BC) particles deposited on snowpack. This is an important concern because the amount of BC deposition on snowpack is closely related with anthropogenic emissions. Thus, anthropogenic emissions of black carbon that have bearing on the causes and characteristics of global climate change include an influence on local snowpack by altering snow albedo.

The potential importance of BC deposition on snow albedo in the Sierra Nevada region can be inferred from previous studies. In a series of theoretical studies, Wiscombe and Warren (1980) and Warren and Wiscombe (1980) showed that impurities in snowpack such as dust and BC reduce snow albedo primarily in the spectral range shorter than 1 micrometer (mm) where most of the solar energy resides. For an ice grain radius of 1,000 mm, for example, their calculations show that the average snow albedo for the wavelengths between 0.4 mm and 1 mm varies from near unity (i.e., almost total reflection of insolation) for pure snow to about 0.4 with a presence of a moderate amount of soot within the snow layer. They also showed that the impact of soot on snow albedo tends to decrease for smaller snow grain sizes; however, for the snow grain size of 100 mm, soot concentration of 1 ppmw can still reduce snow albedo from near unity to below 0.9. A reduction of surface albedo by 0.1 can increase the absorbed insolation by over 60 Wm^{-2} during midday under clear conditions and 20 Wm^{-2} for daily averages. Thus their theoretical study shows that the snow albedo and the absorbed insolation at the surface of the snowpack can be altered by a significant amount by the deposition of black carbon.

Recently, observational studies (Husar et al. 2001; VanCuren et al. 2005) reveal that depositions of dusts and BC of local and Asian origins can alter snow albedo and snow ablation in the western U.S. region. A series of observational studies (Lawrence et al. 2010; Painter et al. 2007, 2010; Flanner et al. 2009; Steltzer et al. 2009) examine the origins of dust particles and their impacts on the snowpack in the Colorado Rockies. Similar impacts of dust and BC depositions on snow have also been observed in the polar region (McConnell et al. 2007). Significant anthropogenic emissions in California, in conjunction with prevailing westerly winds that transport fine particulates into the Sierra Nevada region, can alter the snow albedo in the region. Thus, the sensitivity of the Sierra Nevada snowpack to the deposition of BC needs investigation.

Another challenge in simulating snowpack is the complexity in the physical processes in the interior of the snowpack. Snow models that have been used in climate simulations range from a relatively simple single layer model that considers only limited physical processes within the snowpack to state-of-the-art multi-layer models that can resolve a number of important physical processes (e.g., Yang et al. 1997; Sun et al. 1999; Slater et al. 2001; Ek et al. 2003; Xue et al. 2003). Many regional climate models including the newly developed community model, the Weather Research and Forecast (WRF) community model (Skamarock et al. 2005), use a single-layer snow model. In general, a single-layer snow model overly simplifies important physical processes such as the heat transfer, snow compaction, phase changes in energy balance, and refreezing of snowmelt water. One of the shortcomings of many single-layer snow models in simulating long-term snow variations, especially during the ablation period, is that for snowmelt to occur, the temperature of the entire snow layer must rise above the freezing point. Note that this is not a universal method. For example, the Noah land surface model computes snowmelt as a part of the surface energy balance calculation, and does not require the entire snow layer to be above the freezing temperature (Chang et al. 1999; Koren et al. 1999). In reality, the near-surface snow layer can readily warm up relative to deeper levels to begin the melting process. Incorporating this realism into a model would alter snowpack loss rate significantly, not only for the spring snow ablation period but also for the winter snow accumulation period. Recognizing the shortcomings of single-layer snow models, several multi-layer snow models have been developed for various land-surface schemes (e.g., Sun et al. 1999; Dai et al. 2003). A snow model intercomparison study showed that multi-layer snow models generally outperform single-layer models in simulating long-term snowpack variations (Slater et al. 2001; Bowling et al. 2002; Nissen et al. 2002; Luo et al. 2003; Xue et al. 2003; Rutter et al. 2008). In particular, the multi-layer treatment of snowpack tends to result in faster snowmelt during spring, representing an alleviation of systematic errors in many single-layer model results (Slater et al. 2001). Considering the importance of the long-term snow budget in California's water resources, snowpack simulations on the basis of single- and multi-layer snow physics representations need a close examination.

This study examines qualitatively the impact of the alteration of snow albedo by BC deposition and the multi-layer treatment of snow physics on simulating the cold season snowpack in two regional climate model (RCM) experiments. The results from this study will be used to design more comprehensive future experiments for quantifying these effects in regional climate study with an emphasis on California and the western U.S. region using the Regional Earth System Model (RESM). Experimental designs for examining the impact of snow albedo alterations and the multi-layer snow scheme are presented in Section 2. Sections 3 and 4 present the results obtained in the snow albedo and the multi-layer snow scheme experiments, respectively. Conclusions and discussions based on these experiments are presented in Section 4.

2 Experimental design

The results presented in the following sections are generated in two separate numerical experiments: (1) a fine-resolution simulation for the California region in which the impact of snow albedo changes on the Sierra Nevada snowpack is investigated, and (2) a coarse resolution simulation over a North American region for investigating the snowpack simulation based on single- and multi-layer snow models. Both experiments used the WRF model version 2.2.1 (Skamarock et al. 2005) with 28 atmospheric layers but with different horizontal resolutions and land-surface schemes. The WRF model parameterizations for atmospheric physical processes used in these experiments include the rapid radiative transfer model (RRTM) longwave radiation scheme (Mlawer et al. 1997), Dudhia (1989) shortwave radiation, and the WRF single-moment with simple ice cloud microphysics scheme (WSM-3). Details of WRF and the parameterized physics can be found on the WRF model website (http://wrf-model.org) and will not be presented here. Specifics for the two experiments are presented below.

2.1 The impact of snow albedo on the Sierra Nevada snowpack

In the snow albedo sensitivity study, the model domain covers California with a 12-kilometer (km) horizontal resolution (Fig. 2a). Note that parameterized cumulus convection is inactive in this fine-resolution simulation. The Noah land-surface scheme (Chang et al. 1999; LeMone et al. 2008) with four soil layers and a single-layer snow scheme is used in this experiment. Details on the Noah and the snow models are presented in Mahrt and Pan (1984), Pan and Mahrt (1987), Kim and Ek (1995), Chang et al. (1999), Koren et al. (1999), and LeMone et al. (2008).

The simulation period covers the 7 month period October 2050–April 2051 using the initial and lateral boundary forcing obtained from the results in a 36-km resolution WRF model simulation that in turn is driven by a climate scenario from the National Center for Atmospheric Research Community Climate System Model, version 3.0 (NCAR-CCSM3) corresponding to the Special Report on Emissions Scenarios (SRES) A1B emission scenario. Details of the 36-km regional climate run are presented in Kim et al. (2009, 2010). This allows us to zoom in the fields simulated in a coarse resolution run over a region of

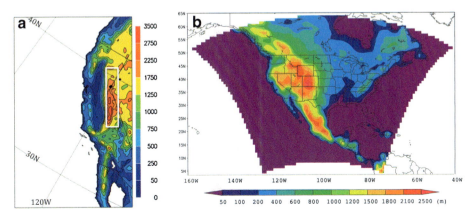

Fig. 2 The model domains and terrain representations in (**a**) the snow albedo sensitivity study (12 km) and (**b**) multi-layer snow model sensitivity study (80 km). The units are in meters. The area marked by a *rectangle* in (**a**) indicates the Sierra Nevada region

interests with manageable computational resources while avoiding an excessive spectral gap between the large-scale forcing data and the regional simulation (Kim et al. 2009, 2010).

2.2 Snowpack differences between single- and multi-layer snow models

The domain for the multi-layer snow model study covers the conterminous U.S. region at an 80 km horizontal resolution (Fig. 2b). The WRF model physics selected for the experiment are the same as in the snow albedo sensitivity study except that the Kain-Fritsch cumulus parameterization scheme (Kain and Fritsch 1993, 1998) is activated due to this experiment's coarse horizontal resolution, and that the Noah land-surface scheme is replaced with the Simplified Simple Biosphere (SSiB) model with three soil layers. In addition, the SSiB/Monin-Obukhov scheme (Xue et al. 1991, 2003) is used for computing surface turbulent fluxes. The simulation is performed for the three-month period April–June 1998 in which snow ablation is important and the differences among snow schemes are largest in general (Slater et al. 2001).

The differences in snowpack simulations in single and multi-layer snow schemes are examined using the WRF-SSiB model. For the single- and multi-layer snow model simulations, the SSiB-1 and SSiB-3 models that use a single- and three-layer snow models respectively, are separately coupled with the WRF model. The physics in the SSiB-1 and SSiB-3 models are identical except the snow scheme. For more details of SSiB and the snow models used in lieu of SSiB-1 and SSiB-3, readers are referred to Sun et al. (1999) and Xue et al. (2003).

3 Results

3.1 The impact of snow albedo on the Sierra Nevada snowpack

The impact of the alterations in snow albedo on the cold season Sierra Nevada snowpack is investigated in a sensitivity study in which five cold season simulations are performed using five different snow albedo specifications. The control run utilizes the default snow albedo values provided with the Noah model as a part of the WRF package on the basis of satellite observations (Robinson and Kukla 1985), but without considering the physical processes that can alter snow albedo such as aging, aerosol deposition, compaction. The prescribed snow albedo values are used to calculate the grid-mean surface albedo value by area-weighted averaging the snow albedo and the landuse-dependent snow-free albedo in conjunction with the fractional snowcover area calculated in terms of SWE within each model grid. More details of the calculation of the fractional snow cover and the corresponding surface albedo value are presented in Chang et al. (1999). The default area-mean snow albedo values used in the control run are 0.6 in the three Sierra Nevada regions described below. In the four sensitivity runs, the default snow albedo values are modified to be 90%, 95%, 105%, and 110% of the control run. The corresponding snow albedo values used in the sensitivity simulations range from 0.54 for the 90% run to 0.66 for the 110% run. Considering Warren and Wiscombe (1980) showed that snow albedo could be reduced from near unity to below 0.94 with 0.1 ppmw BC concentration for 100 mm-size snow grains and even by a larger amount for larger snow grains and/or higher BC concentration, the range of snow albedo variations in these sensitivity studies are realistic possibilities. The two smaller snow albedo runs (0.9 and 0.95 times the default values) represent the cases in which BC emissions in California, thus the BC deposition on the Sierra Nevada snowpack, will continue to increase in the future. The other two runs with larger snow albedo values (1.05 and 1.1 times the default values) represent the cases in which anthropogenic emissions will be reduced by successful implementation of recent mandates by California's governor (Steiner et al. 2006). The sensitivity runs based on the snow albedo values prescribed in this way can be used for qualitative examinations; the qualitative approach is inevitable because the amount of aerosol deposits on the Sierra Nevada snowpack and the quantitative relationship between aerosol deposition and snow albedo remain poorly understood. The model data for elevations roughly above the 1,750 m level within the Sierra Nevada region (marked by the white box in Fig. 2a) are analyzed according to elevation ranges defined at 500 m intervals.

The snowfall and SWE in the control run (Fig. 3) vary significantly according to terrain elevation as well as month. Note that all simulations are driven by the CCSM3 climate scenario representing 2050. Thus, some features (e.g., the disappearance of snow by end of March) are due to the projected warming, not by shortcomings in the RCM. The timing of the maximum snowfall is similar in all elevations, but the amount increases with increasing elevation (Fig. 3a). This is because higher elevation regions remain above the freezing level for longer periods during the cold season and because precipitation generally increases with increasing elevation in the region (Soong and Kim 1996; Kim 1997, 2001). The variation according to elevation range is significantly amplified in the SWE field relative to the snowfall amount (Fig. 3b). In December, the SWE in the highest elevation range (above 3,000 m) is about five times the value in the lowest elevation range (2,000–2,250 m), even though the snowfall is only twice as large. The large variations in the SWE according to terrain elevation are due to larger snowmelt in the lower elevation ranges. Below 2,500 m, snowpack is almost completely depleted in February; above that level, snow depletion occurs 1 month later. The simulated occurrence of the peak snowfall and SWE in November and December, respectively, does not correspond well with the historical data that show the peak snowfall in California climatologically occurs in the period January–February (Kim and Lee 2003). This may be due to shortcomings of WRF, interannual variations in precipitation, and/or due to the warmer climate of the mid-twenty-first century. Note that the large-scale forcing in this study is obtained from a GCM projection that contains its own biases, not from more accurate present-day reanalysis data.

The area-mean surface albedo varies systematically according to terrain elevations in response to the changes in snow albedo (Fig. 4). The maximum surface albedo occurs in December when the SWE is also largest, especially in the regions above 2,250 m where snow cover is significant. The differences in the surface albedo among the sensitivity runs

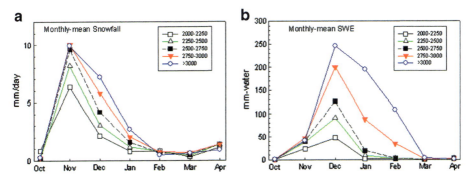

Fig. 3 The monthly-mean snowfall and SWE in the Sierra Nevada region above the 2,000 m level simulated in the 12-km resolution control run

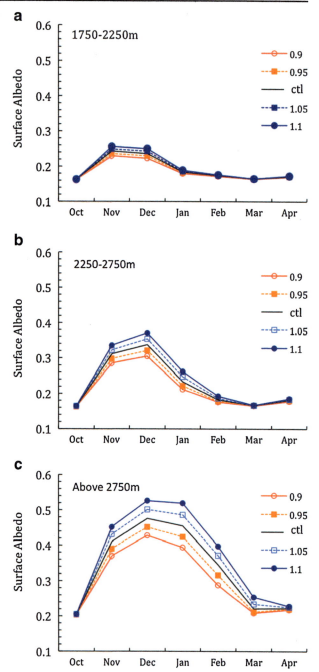

Fig. 4 The monthly-mean surface albedo values in the five simulations in the three elevation ranges: **a** 1,750–2,250 m, **b** 2,250–2,750 m, and **c** above the 2,500 m level. The numbers in the legends indicate the ratio between the snow albedo values used in the corresponding sensitivity study and in the control run (ctl). For example, 0.95 represents the simulation in which the snow albedo values are 95% of the control run

increase with increasing elevation. In the lowest elevation range (Fig. 4a), the maximum difference in the monthly-mean surface albedo between the ±10% snow albedo change runs is 0.04 in November and December; it becomes 0.11, almost as large as the snow albedo differences between the ±10% runs, in the highest elevation region in January (Fig. 4c). This suggests that the effects of snow albedo changes on the area-mean surface albedo are amplified via snow-albedo feedback. The period of maximum differences in the surface albedo among the sensitivity runs tend to occur later as terrain elevation increases. In the lowest elevation range, the maximum differences appear in the November–December period (Fig. 4a), but the differences are largest during January in the highest elevation range (Fig. 4c). This is because snow lasts longer in higher elevation regions due to colder temperatures.

The surface energy budget component that is most directly affected by the alterations in snow albedo is the amount of absorbed insolation at the surface. The differences in the absorbed surface insolation among the five runs range from less than 2 Wm^{-2} at the lowest elevation (Fig. 5a) to over 10 Wm^{-2} at the highest elevation (Fig. 5c). These variations in

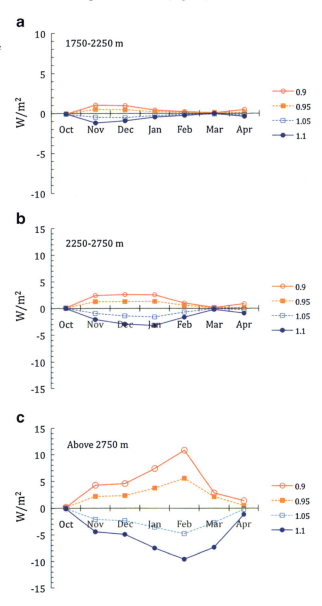

Fig. 5 The changes in the absorbed surface insolation due to the snow albedo changes in the three elevation ranges: **a** 1,750–2,250 m, **b** 2,250–2,750 m, and **c** above the 2,750 m level

the absorbed surface solar energy according to terrain elevation are a direct consequence of the elevation dependence in the surface albedo sensitivity shown in Fig. 4 because the surface insolation remains similar for all elevations (not shown). Considering that the changes in the surface radiative forcing between the late 20th century and the late 19th century by the combined effects of the increases in greenhouse gases and sulfate aerosols is about 2 Wm^{-2} (Meehl et al. 2003), the result shows that anthropogenic alterations in snow albedo via BC deposition can exert substantial climate forcing in regions of significant snow cover.

The most notable feature in the sensitivity of the Sierra Nevada snowpack, represented here in terms of SWE, to snow albedo is that the snowpack sensitivity increases with increasing terrain elevation (Fig. 6). In the lowest region (1,750–2,250 m), the SWE varies according to the prescribed snow albedo values, but the magnitude of the SWE sensitivity is small (Fig. 6a). The SWE sensitivity to snow albedo becomes larger in higher regions. With a 10% increase in snow albedo from the control run (or an increase in snow albedo by 0.06 in physical units), the February SWE is doubled in the mid-elevation range, 2,250–2,750 m, (Fig. 6b) and the March SWE becomes nearly five times as large as in the control run above the 2,750 m (Fig. 6c). The impact of the decrease in snow albedo on SWE is also larger in higher elevations. Another noticeable result is that the timing of the peak sensitivity varies according to both the sign of the snow albedo changes and terrain elevations. The peak percentage reduction in SWE due to a decrease in snow albedo occurs earlier than the peak percentage increase in SWE due to an increase snow albedo. This is most evident in the mid-elevation range (2,250–2,750 m), where the largest reduction in the SWE due to decreased snow albedo (red lines) occurs in January, and the largest impact of the increased snow albedo (blue lines) occurs in February. The actual time lag varies according to the interval of the terrain elevation range over which the average is taken; however, the timing of the peak responses presented above remains consistent. These differences in the timing of the peak SWE response (Fig. 6) to snow albedo are due to the fact that, relative to the control run, decreases (increases) in snow albedo accelerate (decelerate) snow loss, and in turn further decrease (increase) albedo. This reveals that the impact of the alterations in snow albedo on the temporal evolution of SWE is further amplified through snow-albedo feedback.

The response of the 2-m air temperature to the prescribed snow albedo changes also varies according to terrain elevation and the magnitude of the snow albedo change (Fig. 7). The prescribed snow albedo changes have minimal impacts on the surface air temperature in the lowest elevation range; with snow albedo changes by ±10% resulting in temperature changes of less than 0.04 K. The surface air temperature sensitivity to the albedo changes increases in higher altitudes; by nearly 0.1 K in the mid-elevation range (Fig. 7b) and by over 0.2 K in the highest elevation range (Fig. 7c). Not only the magnitude but also the timing of the peak sensitivity varies with both elevation and the sign of the snow albedo change in a similar way as other variables. In general, the peak warming due to the decrease in snow albedo occurs earlier than the peak cooling by the increase in snow albedo. The skin temperature responds to the snow albedo changes similarly, but with almost twice the magnitudes (not shown).

The sensible heat flux varies according to the changes in snow albedo (Fig. 8) in a similar way as the absorbed insolation. The decrease in snow albedo results in a larger increase in the skin temperature than in the surface air temperature as discussed above, and in turn, results in the increase in the sensible heat flux. The increase in snow albedo causes the opposite effect on the sensible heat flux. The sensible heat flux sensitivity also varies according to elevation and month. The sensible heat flux sensitivity increases with increasing terrain height; about 1 Wm^{-2} in the lowest elevation range (Fig. 8a) and around

Fig. 6 The ratio of the SWE in the four sensitivity simulations to that in the control simulation in the three elevation ranges: **a** 1,750–2,250 m, **b** 2,250–2,750 m, and **c** above the 2,750 m level

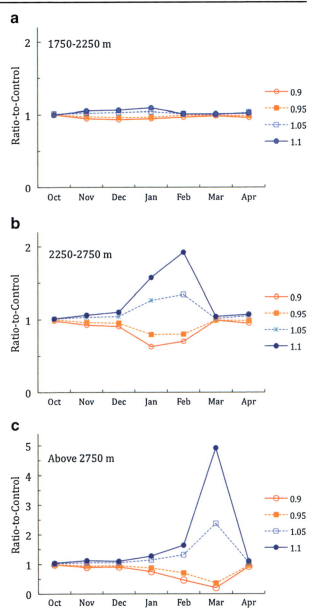

10 Wm^{-2} in the highest elevation range in response to ±10% changes in the snow albedo. Also, the peak response of the sensible heat to the decreased snow albedo occurs earlier than that to the increase in snow albedo.

With the decrease in snow albedo, snowmelt increases in earlier months of the cold season. The timing of the increased snowmelt also appears in later months as elevation increases (Fig. 9). The snowmelt changes in the lowest elevation range (Fig. 9a) are negligible because snowfall in the low elevation region melts quickly due to higher air temperatures. Thus the impact of snow albedo changes on the snowmelt in the low-elevation region is small. In the mid-elevation region (Fig. 9b), the snowmelt increases by

Fig. 7 The sensitivity of the 2-m air temperature to the prescribed snow albedo changes

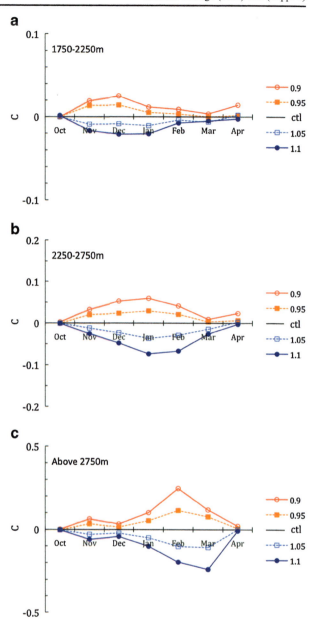

5% in November, followed by a 10% decrease in January in response to the decrease in snow albedo by 10% (red solid line with open circles) because the decrease in snow albedo depletes SWE by melting in the early part of the season and reduces SWE for late-season melt. Peak snowmelt changes (i.e., early-season increase and late-season decrease) in the highest region (Fig. 9c) due to smaller snow albedo values occur one-month later than in the mid-elevation region and with much larger magnitudes. Thus, the most notable impact of the decrease in snow albedo is to enhance early-season snowmelt and to reduce late-season snowmelt, resulting in an adverse impact on warm season water resources in California. The two experiments with larger snow albedo values (the blue solid and dashed

Fig. 8 Same as Fig. 6, but the sensible heat flux

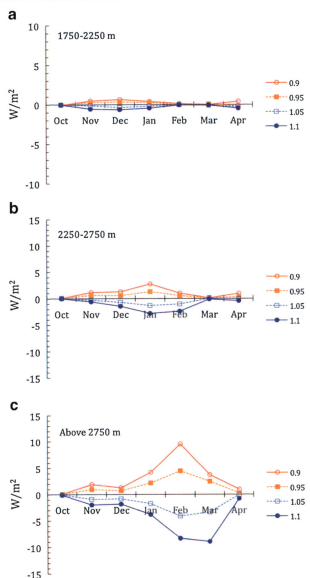

lines in Fig. 9) show that an increase in snow albedo will suppress snowmelt in the early part of the cold season and will enhance it in the later part of the season. Such a change might help to partially alleviate the adverse impact of global warming on California water resources. The timing of peak impact of the altered snow albedo on the snowmelt also varies with elevation similarly as for SWE—that is, the peak response appears later in higher regions than in lower regions, especially in the cases of increased snow albedo.

The changes in snowmelt due to the alterations in snow albedo result in notable changes in runoff in the early and late part of the cold season (Fig. 10). Decreases in snow albedo result in runoff increases in the early part of the cold season and decreased runoff in the late part of the cold season. Increases in snow albedo result in opposite effects; a decrease in runoff during the early cold season and an increase in runoff in the late cold season. This

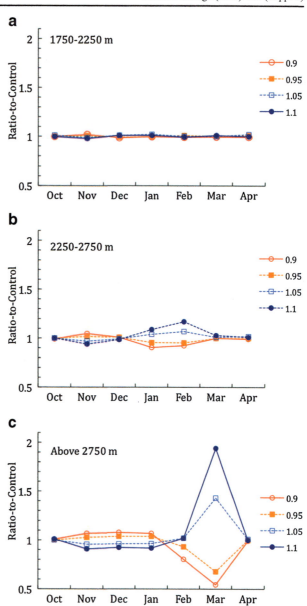

Fig. 9 The sensitivity of the snowmelt to the snow albedo changes. The sensitivity is presented in terms of the ratio of snowmelt in each sensitivity study to that in the control run

runoff response to increased snow albedo is qualitatively consistent with the corresponding responses of SWE and snowmelt. Similar to the SWE and snowmelt, the timing of peak response of the simulated runoff to snow albedo occurs later in the cold season as terrain elevation increases.

3.2 Snowpack differences between single- and multi-layer snow models

An additional uncertainty in simulating snowpack, especially snowmelt processes, derives from the formulation of physical processes within snowpack such as snow compaction, heat

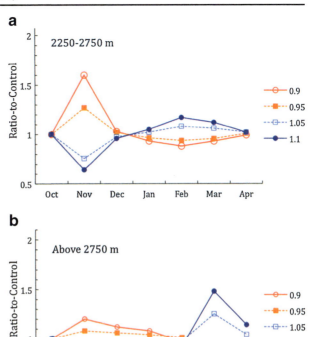

Fig. 10 The runoff sensitivity to the snow albedo changes in the mid- and high-elevation ranges

conduction, snow grain growth, and the retention of liquid water within snow, among others (Yang et al. 1997; Sun et al. 1999). Accommodating these processes in simulating snowpack requires an efficient snow cover layering system and has led to the development of several multi-layer snow models (e.g., Sun et al. 1999; Dai et al. 2003). Single-layer snow models that are widely used in many broadly-applied climate models often lack important physical processes such as the development of the vertical temperature gradient within snowpack that plays a crucial role in snowmelt. This is an important concern in assessing the impact of anthropogenic climate change on California's water resources. In this study, we compare two snowpack simulations simulated using a single- and three-layer snow models within the context of the WRF-SSiB model in order to examine the snowpack treatment in regional climate simulations.

The SSiB-3 uses the three-layer version of the Simple Atmosphere-Snow Transfer (SAST) model of Sun et al. (1999) that was developed on the basis of up-to-date comprehensive and complex snow schemes (Anderson 1976; Jordan 1991) with a number of simplifications and improvements. The SAST model includes three prognostic variables: specific enthalpy, SWE, and snow depth. Using enthalpy instead of temperature in the energy conservation equation greatly simplifies the computational procedure for calculating phase change within snowpack. The model also retains important physical processes such as snow compaction, heat conduction, snow grain growth, and snow melting. Details of the SAST model and the coupling of the SAST and SSiB-3 are presented in Sun et al. (1999), Sun and Xue (2001), and Xue et al. (2003). The SSiB-1 model is an early version of SSiB-3 in which a single-layer snow scheme is utilized. Details of SSiB-1 can be found in Xue et al. (1991).

Two seasonal simulations in which the WRF model coupled with SSiB-1 and SSiB-3 are used for the period April–June 1998 over the North American domain (Fig. 2b). Both runs are initialized from the National Centers for Environmental Prediction (NCEP)/NCAR reanalysis (Kalnay et al. 1996). The large-scale forcing along the lateral boundaries is obtained from the reanalysis data as well. The SWE fields during May 1998 in the two simulations are compared with the observed data over the western U.S. region for evaluation. For more details of the observed snow data, the readers are referred to Mote (2003) and Mote et al. (2005a, b). The differences in the simulated SWE between the single- and three-layer snow models are presented for two sub-regions in the western United States (Fig. 11): Pacific coastal region (P) that includes the Coastal Range, the Sierra Nevada and the Cascades, and the Rocky Mountains region (R). The orography in the P region varies with characteristic zonal length scales between 50 and 100 km and is not well represented by the 80 km resolution. The zonal length scale of the terrain in R (Rocky Mountains) is of hundreds of kilometers and is represented better than in P.

The SWE fields simulated with SSiB-1 (Fig. 11b) and SSiB-3 (Fig. 11c) are compared against the observational data (Fig. 11a). Both snow models significantly underestimate SWE in P, perhaps due to the lack of representation of the orography in P. The snow-covered area in Oregon and California compares well with the observed data, but the significant SWE in northern California and in northern Washington is absent in both runs. Both snow models represent the SWE field in R better than in P. The major SWE centers along the Rocky Mountains in R appear both simulations. However, the spatial details of the observed SWE distribution such as the SWE maximum in the northeastern Arizona region are missing in both simulations, showing the problem with coarse spatial resolution.

Figure 12 presents the model errors in terms of the difference between the simulated and observed monthly-mean SWE and the corresponding root-mean-square error (RMSE). In P, both the single- and three-layer model underestimates SWE, especially in the northern California and the central Sierra Nevada regions. The three-layer model results in slightly larger bias (−4.46 mm) than the single layer model (−4.27 mm) in the area-mean SWE values, but the RMSE in the three-layer simulation is slightly smaller than the single layer model (11.16 mm versus 11.33 mm). Overall, the difference in the area-mean bias and the RMSE between the two snow model simulations is small for the P region. The SWE bias in R shows that the three-layer snow model could improve the snowpack simulation over the single-layer snow model. The significant SWE bias of +5.38 mm in the single-layer snow simulation is reduced to −0.22 mm in the three-layer model results. The three-layer snow model simulation also improved the RMSE from 17.4 mm to 3 mm in the R region. Evaluations of the results for Canada and northeastern United States (not shown) also reveal that a significant improvement in simulating snowpack could be achieved for those regions by the use of a multi-layer snow model.

These results show that the lack of representation of physical processes within the snowpack, in this case most likely the vertical transfer of solar and thermal energy and the associated development of temperature gradient within the snowpack can be a source of significant errors in simulating snowpack during the snowmelt period. The results in this study are highly qualitative, mainly due to the coarse horizontal resolution; however, the improvement in simulating snowpack during major ablation periods by the use of a multi-layer snow model is well demonstrated. Moreover, for some time, climate models (especially those used for global climate projection studies) will be applied with even lower resolution so these results are quite relevant to understanding the uncertainties

Fig. 11 The monthly mean SWE in millimeters (mm): **a** Observation, and simulated with, **b** a single layer snow model (SSiB-1) and **c** a 3-layer snow model (SSiB-3)

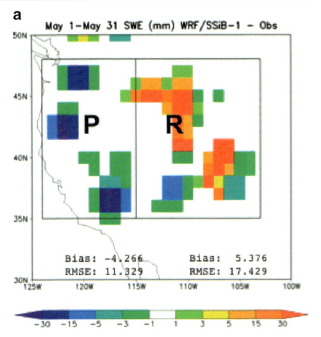

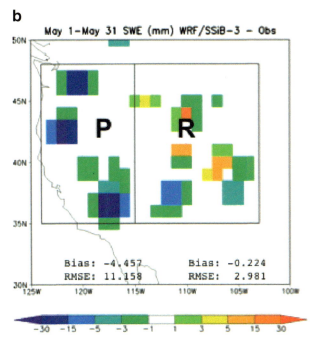

associated with these projections. A high-resolution experiment in conjunction with more detailed observed data e.g., the fine-resolution snow analysis by the National Operational Hydrologic Remote Sensing Center (www.nohrsc.nws.gov/) is necessary for a more quantitative analysis of the differences between the single- and multi-layer snow model results. This is a subject of our follow-up experiment.

Fig. 12 The monthly mean SWE (mm) simulation errors against observation: **a** a single layer snow model (SSiB-1) and **b** a 3-layer snow model (SSiB-3)

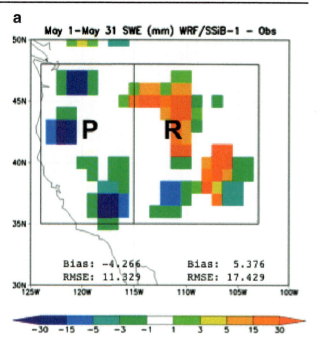

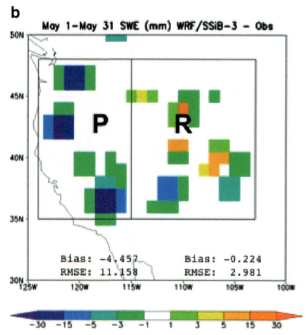

4 Conclusions and discussions

Cold season snowpack plays a crucial role in determining the warm-season water supply in California and the western U.S. region. Apart from water supply, snowpack in the Sierra Nevada Mountains plays an important role in tourism and in establishing and maintaining the region's ecosystems. Uncertainties in snowpack projections can be caused by a number of reasons. To better understand the uncertainties in projecting the impact of anthropogenic climate change on California hydrology, specifically snowpack, due to the alterations in snow albedo due to possible changes in the deposition of anthropogenic BC and the representation of snow physics in a numerical model, we have carried out and analyzed two regional climate model (RCM) experiments focused on these two issues.

The possible effects of the deposition of anthropogenic BC on snow albedo and subsequently on the evolution of the Sierra Nevada snowpack have been investigated using the WRF model by varying the default snow albedo field provided with the Noah LSM. To represent the impact of changes in aerosol deposition (e.g., changes in local emission or transport) on snow albedo in a qualitative way, the snow albedo is decreased or increased by 5% and 10% of the default value in four sensitivity simulations. The ±10% alteration in the default snow albedo corresponds to ±0.06 in physical units that can occur with a moderate amount of BC deposition on snowpack. The decrease and increase of snow albedo can, for example, qualitatively represent the increase and decrease of local anthropogenic BC emissions, respectively. The simulations are performed for the cold season October 2050–April 2051 using the initial and lateral boundary forcing data from an NCAR-CCSM3 climate change projection generated with the SRES-A1B emission scenario.

The control simulation—using the default observation-based snow albedo values provided with the Noah LSM—shows that the snowfall amount and SWE in the Sierra Nevada region vary significantly according to terrain elevation. The elevation dependence is caused mainly by the large low-level temperature variations associated with significant variations in terrain heights, the tendency of increasing precipitation with increasing terrain elevation (e.g., Kim 1997; Kim and Lee 2003), and the location of freezing level that appears between the 2,000 m and 2,250 m levels in the region during the winter. Previous studies over complex terrain (e.g., Giorgi et al. 1997; Kim 2001) also found significant differences in precipitation and snow accumulation across the seasonal mean freezing level altitude.

Decreased snow albedo (increased aerosol deposition) promotes early season snowmelt and runoff, resulting in smaller SWE in the early part of the cold season. The reduced early season SWE in turn results in reduced snowmelt and runoff in the later part of the season. Thus, reduced snow albedo that can occur by the increase in local emissions and subsequent BC deposition on the Sierra Nevada snowpack can enhance earlier snowpack depletion. This will further reduce warm season water resources and add to the effects of the low-level warming induced by increased greenhouse gases. An increase in snow albedo, possibly from the reduction in anthropogenic BC emissions, reduces snowmelt and runoff, and thus increases SWE, in the early part of the cold season. The resulting increase in early season SWE enhances late season snowmelt and runoff. This can partially alleviate one of the most important adverse impacts, the loss of snowpack earlier in the cold season to deplete the snowmelt runoff in spring and early summer, of the anthropogenic global warming on the Sierra Nevada snowpack and California's warm season water resources.

Examinations of the response of the major components in surface energy balance, the grid-mean surface albedo, the absorbed surface insolation and the sensible heat flux, show that the impact of the snow albedo changes is most pronounced in high elevation regions where snow cover is more significant. The grid-mean surface albedo that is calculated by an area-weighted average of the snow albedo and snow-free albedo, varies with snow albedo, most noticeably in higher elevation range. Decreases (increases) in snow albedo result in an increase (decrease) in the absorbed surface insolation, surface and 2-m temperatures, and sensible heat fluxes as well. The variations in SWE, surface energy fluxes, snowmelt, and the grid-mean surface albedo in response to the alterations in snow albedo support that the alterations in snow albedo is further amplified through local snow-albedo feedback.

This study's examination of the sensitivity of snowpack hydrology to changes in snow albedo is qualitative—in the sense that hypothetical values of albedo are utilized; however, it suggests that changes in anthropogenic emissions can influence the warm-season water supply in California via the snow albedo and, subsequently, the cold-season snowpack. The largest uncertainties in the current study are due to the lack of quantitative knowledge on two important factors: the amount of aerosol deposition on the Sierra Nevada snowpack and the relationship that links the amount and type of aerosol deposition and snow albedo. This study's results reveal that the modification of snow albedo, possibly due to anthropogenic aerosol deposition, can result in significant alterations in the Sierra snowpack, and as a result, changes in the warm season water supply in California. Quantifying the amount of anthropogenic aerosol deposition on the Sierra snowpack and the associated snow albedo changes is a topic of future research. Combined with the large impact of snow albedo on the snowpack, uncertainties in future emissions and transports of aerosols into the region can be an important factor in assessing future snowpack in the region.

A comparison of the SWE fields simulated using the WRF-SSiB model with the single-layer and multi-layer snow models indicates that systematic errors in the SWE during a major snow ablation period simulated using a single-layer snow model could be alleviated significantly by using a multi-layer snow model. The latter typically includes more comprehensive treatment of vertical variations in the absorption of insolation and, subsequently, the temperature gradient within the snowpack, in addition to other important physical processes such as snow compaction, heat conduction, and snow grain growth. The results in the Pacific coastal regions are inconclusive, possibly due to the use of a horizontal resolution still too coarse to represent the orographic variations in the region. However, the improvement in simulating snowpack is clearly shown in the Rocky Mountains region where the terrain is represented better than in the Pacific coastal region. Despite shortcomings due to the relatively coarse spatial resolution, this study shows that it is important to include a comprehensive snow model in assessing the climate change impact on water resources in California. Future works on this subject will focus on the Sierra Nevada region with a finer spatial resolution and a more detailed observational dataset in order to examine the model errors associated with the numerical structure and the representation of the underlying physical processes.

Development of the formulations to account for the impact of aerosol depositions on snow albedo and the treatment of physical processes, as well as overall snow modeling, has been difficult due to the lack of reliable observational data. Thus, comprehensive observational studies–including high resolution remote sensing data that characterize the optical, thermal and physical properties of the snow pack, particularly in conjunction with process/modeling studies, are necessary to improve snow modeling.

Acknowledgements The research described in this paper was performed as an activity of the Joint Institute for Regional Earth System Science and Engineering (JIFRESSE) via a Memorandum of Understanding between the UCLA and the JPL/CALTECH. This work was supported by the NASA Energy and Water Cycle Study (NEWS), NA07OAR4310226, the Ministry of Environment, Korea (Grant No. 1600-1637-301-210-13), and University of California Lab Fees Research Program (Award No. 09-LR-09-116849-SORS). Jet Propulsion Laboratory's Supercomputing and Visualization Facility and the NASA Advanced Supercomputing (NAS) Division provided the data storage and computational resources for this study.

References

Anderson E (1976) A point energy and mass balance model of a snow cover, NOAA Tech Rep., NWS, 19, Office of Hydrology, National Weather Service, Silver Spring, Maryland

Bowling L et al (2002) Simulation of high latitude hydrological processes in the Torne_Kalix basin: PILPS Phase 2(e): 1. Experiment description and summary inter-comparison. Global Planet Change 38:1–30

Cayan D, Maurer E, Dettinger M, Tyree M, Hayhoe K (2008) Climate change scenarios for the California region. Clim Chang 87:S21–S42

Chang S, Hahn D, Yang C, Norquist D, Ek M (1999) Validation study of the CAPS model and land surface scheme using the 1987 Cabauw/PILPS dataset. J Appl Meteor 38:405–422

Dai Y, Zeng Z, Dickinson RE, Baker I, Bonan GB, Bosilovich MG, Denning AS, Dirmeyer A, Houser PR, Niu G, Oleson KW, Schlosser CA, Yang Z (2003) The common land model. Bull Am Meteorol Soc 84:1013–1023

Dudhia J (1989) Numerical study of convection observed during the winter monsoon experiment using a mesoscale two-dimensional model. J Atmos Sci 46:3077–3107

Ek M, Mitchell K, Lin Y, Rogers E, Grunmann P, Koren V, Gayno G, Tarpley J (2003) Implementation of Noah land surface model advances in the National Centers for Environmental Prediction operational mesoscale Eta model. J Geophys Res 108(D22):8851. doi:10.1029/2002JD003296

Flanner MG, Zender CS, Hess PG, Mahowald NM, Painter TH, Ramanathan V, Rasch PJ (2009) Springtime warming and reduced snow cover from carbonaceous particles. Atmos Chem Phys 9:2481–2497

Giorgi F, Hurrell JW, Marinucci MR, Beniston M (1997) Elevation dependency of the surface climate signal: a model study. J Clim 10:288–296

Hadley O, Ramanathan V, Carmichael G, Tang Y, Corrigan C, Roberts G, Mauger G (2007) Transpacific transport of black carbon and find aerosols (D<2.5 mm) into North America. J Geophys Res 112: D05309. doi:10.1029/2006JD007632

Husar R, Tratt D, Schichtel B, Falke S, Li F, Jaffe D, Gasso S, Gill T, Laulainen N, Lu F, Reheis M, Chun Y, Westphal D, Holben B, Gueymard C, McKendry I, Kuring N, Feldman G, McClain C, Frouin R, Merrill J, DuBois D, Vignola F, Murayama T, Nickovic S, Wilson W, Sassen K, Sugimoto N, Malm W (2001) Asian dust events of April 1998. J Geophys Res 106:18317–18330

Jordan R (1991) One-dimensional temperature model for a snow cover, Special Report. 91-1b, Cold Regions Res. and Eng. Lab., Hanover, New Hampshire

Kain J, J Fritsch (1993) 'Convective parameterization for mesoscale models: The Kain-Fritsch scheme', The representation of cumulus convection in numerical models. Meteor. Monogr. No. 24. Amer. Meteor. Soc. 165–170

Kain J, Fritsch J (1998) Multi-scale convective overturning in mesoscale convective systems: reconciling observations, simulations, and theory. Mon Weather Rev 126:2254–2273

Kalnay E, Kanamitsu M, Kistler R, Collins W, Deaven D, Gandin L, Iredell M, Saha S, White G, Woolen J, Zhu Y, Leetmaa A, Reynolds R, Chelliah M, Ebisuzaki W, Higgins W, Janowiak J, Mo K, Ropelewski C, Wang J, Jenne R, Joseph D (1996) The NCEP/NCAR 40-year reanalysis project. Bull Am Meteorol Soc 77:437–471

Kim J (1997) Precipitation and snow budget over the southwestern United States during the 1994–1995 winter season in a mesoscale model simulation. Water Res Res 33:2831–2839

Kim J (2001) A nested modeling study of elevation-dependent climate change signals in California induced by increased atmospheric CO_2. Geophys Res Lett 28:2951–2954

Kim J (2005) A projection of the effects of the climatic change induced by increased CO_2 on extreme hydrologic events in the Western U. S. Clim Chang 68:153–168

Kim J, Ek M (1995) A simulation of surface energy budget and soil water content over the HAPEX/ MOBILHY forest site. J Geophys Res 100(D10):20,845–20,854

Kim J, Lee J-E (2003) A multi-year regional climate hindcast for the western United States using the mesoscale atmospheric simulation model. J Hydrometeor 4:878–890

Kim J, Kim T, Arritt RW, Miller N (2002) Impacts of increased atmospheric CO_2 on the hydroclimate of the Western United States. J Clim 15:1926–1942

Kim J, Chao Y, Eldering A, Fovell R, Hall A, Li Q, Liou K, McWilliams J, Waliser D, Xue Y, Kapnick S (2009) 'A projection of the cold season hydroclimate in California in mid-twenty-first century under the SRES-A1B emission scenario', California Climate Change Report, CEC-500-2009-029-D, California Energy Commission

Kim J, Fovell R, Hall A, Liou K, Xue Y, Kapnick S, Qu X, McWilliams J, Li Q, Waliser DE, Eldering A, Chao Y, Friedel R (2010) 'A projection of the cold season hydroclimate in California in mid-twenty-first century under the SRES-A1B emissions scenario'. Climatic Change, submitted

Koren V, Schaake J, Mitchell K, Duan Q-Y, Baker J (1999) A parameterization of nowpack and frozen ground intended for NCEP weather and climate models. J Geophys Res 104(D16):19569–19585. doi:10.1029/1999JD900232

Lawrence CR, Painter TH, Landry CC, Neff J (2010) The contemporary physical and chemical composition of aeolian dust deposition to terrestrial ecosystems of the San Juan Mountains, Colorado, USA. J Geophys Res. doi:10.1029/2009JG001077

LeMone MA, Tewari M, Chen F, Alfieri JG, Niyogi D (2008) Evaluation of the Noah land surface model using data from a fair-weather IHOP_2002 day with heterogeneous surface fluxes. Mon Weather Rev 136:4915–4941

Leung R, Ghan S (1999) Pacific Northwest climate sensitivity simulated by a regional climate model driven by a GCM: part II: 2XCO$_2$ simulations. J Clim 12:2031–2053

Luo L, Robock A, Vinnikov K, Schlosser CA, Slater A, Boone A, Etchevers P, Habets F, Noilhan J, Braden H, Cox P, de Rosnay P, Dickinson R, Dai Y, Zeng Q, Duan Q, Schaake J, Henderson-Sellers A, Gedney N, Gusev Y, Nasonova O, Kim J, Kowalczyk E, Mitchell K, Pitman A, Shmakin A, Smirnova T, Wetzel P, Xue Y, Yang Z (2003) Effects of frozen soil in soil temperature, spring infiltration, and runoff: results from the PILPS 2(d) experiment at Vaidai, Russia. J Hydrometeorol 4:334–351

Mahrt L, Pan H (1984) A two-layer model of soil hydrology. Bound Lay Meteor 29:1–20

Marks D, Dozier J (1992) Climate and energy exchange at the snow surface in the alpine region of the Sierra Nevada, 2, snow cover energy balance. Water Resour Res 28(11):3043–3054. doi:10.1029/92WR01483

McConnell J et al (2007) 20th-Century doubling in dust archived in an Antarctic Peninsula ice core parallels climate change and desertification in South America. Proc Natl Acad Sci U S A 104:5743–5748

Meehl G, Washington W, Wigley T, Arblaster J, Dai A (2003) Solar and greenhouse gas forcing and climate response in the twentieth century. J Clim 16:426–444

Mlawer E, Taubman S, Brown P, Iacono M, Clough S (1997) Radiative transfer for inhomogeneous atmosphere: RRTM, a validated correlated-k model for the longwave. J Geophys Res 102:16663–16682

Mölders N, Luijting H, Sassen K (2008) Use of atmospheric radiation measurement program data from Barrow, Alaska, for evaluation and development of snow-albedo parameterizations. Meteorol Atmos Phys 99:199–219

Molotch N, Bales R (2006) Comparison of ground-based and airborne snow surface albedo parameterizations in an alpine watershed: Impact on snowpack mass balance. Water Resour Res 42:W05410. doi:10.1029/2005WR004522

Mote P (2003) Trends in snow water equivalent in the Pacific Northwest and their climatic causes. Geophys Res Lett 30:1601. doi:10.1029/2003GL017258

Mote P, Dyer J, Grundstein A, Robinson D, Leathers D (2005a) 'Evaluation of new snow depth and mass data sets for North America', Proceedings 15th Conf. on Applied Climatology. Savannah, Georgia. American Meteorol. Soc., JPL. 10

Mote P, Hamlet A, Clark M, Lettenmaier D (2005b) Declining mountain snowpack in western North America. BAMS 86:39–49

Nissen B et al (2002) Simulation of high latitude hydrological processes in the Torne_Kalix basin: PILPS phase 2(e): 2. Comparison of model results with observations. Global Planet Change 38:31–53

Painter T, Barrett A, Landry C, Neff J, Cassidy M, Lawrence C, McBride K, Farmer G (2007) Impact of disturbed desert soils on duration of mountain snow cover. Geophys Res Lett 34:L12502. doi:10.1029/2007GL030284

Painter TH, Deems J, Belnap J, Hamlet A, Landry CC, Udall B (2010) Response of Colorado River runoff to dust radiative forcing in snow. Proc Natl Acad Sci. doi:10.1073/pnas.0913139107

Pan H, Mahrt L (1987) Interaction between soil hydrology and boundary-layer development. Bound Lay Meteor 38:185–202

Robinson D, Kukla G (1985) Maximum surface albedo of seasonally snow-covered lands in the northern hemisphere. J Climate Appl Meteor 24:402–411

Rutter N, Essery R, Pomeroy J, Altimir N, Andreadis K, Baker I, Barr A, Bartlett P, Deng H, Elder K, Ellis C, Feng X, Gelfan A, Goodbody G, Gusev Y, Gustafsson D, Hellström R, Hirota T, Jonas T, Koren V, Li W, Luce C, Martin E, Nasonova O, Pumpanen J, Pyles D, Samuelsson P, Sandells M, Schädler G, Shmakin A, Smirnova T, Stähli M, Stöckly R, Strasser U, Su H, Suzuki K, Takata K, Tanaka K, Thompson E, Vesala T, Viterbo P, Wiltshire A, Xue Y, Yamazaki T (2008) 'Evaluation of forest snow processes models (SnowMIP2)'. J Geophys Res, Submitted

Skamarock W, Klemp J, Dudhia J, Gill D, Baker D, Wang W, Powers J (2005) A description of the advanced research WRF version 2. NCAR/TN-468+STR, 88 pp

Slater A, Schlosser CA, Desborough C, Pitman A, Henderson-Sellers A, Robock A, Vinnikov K, Mitchell K, Boone A, Braden H, Chen F, Cox P, Rosnay P, Dickinson R, Dai Y, Duan Q, Entin J, Etchevers P, Gedney N, Gusev Y, Habets F, Kim J, Koren V, Kowalczyk E, Nasonova O, Noilhan J, Schaake S,

Shmakin A, Smirnova T, Verseghy D, Wetzel P, Xue Y, Yang Y, Zeng Q (2001) The representation of snow in land-surface schemes; results from PILPS 2(d). J Hydrometeorol 2:7–25

Soong S, Kim J (1996) Simulation of a heavy precipitation event in California. Clim Chang 32:55–77

Steiner A, Tonse S, Cohen R, Goldstein A, Harley R (2006) Influence of future climate and emissions on regional air quality in California. J Geophys Res 111:D18303. doi:10.1029/2005JD006935

Steltzer H, Landry CC, Painter TH, Anderson J, Ayres E (2009) Biological consequences of earlier snowmelt from desert dust deposition in alpine landscapes. Proc Natl Acad Sci. doi:10.1073/pnas.0900758106

Sun S, Xue Y (2001) Implementing a new snow scheme in Simplified Simple Biosphere Model (SSiB). Adv Atmos Sci 18:335–354

Sun S, Jin J, Xue Y (1999) A simplified layer snow model for global and regional studies. J Geophys Res 104:19,587–19,597

VanCuren R, Cliff S, Perry K, Jimenez-Cruz M (2005) Asian continental aerosol persistence above the marine boundary layer over the eastern North Pacific: continuous aerosol measurements from Intercontinental Transport and Chemical Transformation 2002 (ITCT 2 K2). J Geophys Res 110: DD095S90. doi:10.1029/2004JD004973

Vicuna S, Leonardson R, Hanemann M, Dale L, Dracup J (2008) Climate change impact on high elevation hydropower generation in California's Sierra Nevada: a case study in the upper American River. Clim Chang 87:S123–S137

Warren S, Wiscombe W (1980) A mode for the spectral albedo of snow. II. Snow containing atmospheric aerosols. J Atmos Sci 37:2734–2745

Wiscombe W, Warren S (1980) A mode for the spectral albedo of snow. I. Pure snow. J Atmos Sci 37:2712–2733

Xue Y, Sellers P, Kinter J III, Shukla J (1991) A simplified biosphere model for global climate studies. J Clim 4:345–364

Xue Y, Sun S, Kahan D, Jiao Y (2003) The impact of parameterizations in snow physics and interface processes on the simulation of snow cover and runoff at several cold region sites. J Geophys Res 108. doi:10.1029/2002JD003174

Yang Z, Dickinson R, Robock A, Vinnikov K (1997) Validation of snow submodel of the biosphere-atmosphere transfer scheme with Russian snow cover and meteorological observational data. J Clim 10:353–373

Climatic Change (2011) 109 (Suppl 1)
DOI 10.1007/s10584-011-0300-9

Human-induced changes in wind, temperature and relative humidity during Santa Ana events

Mimi Hughes · Alex Hall · Jinwon Kim

Received: 9 March 2010 / Accepted: 21 September 2011 / Published online: 24 November 2011
© Springer Science+Business Media B.V. 2011

Abstract The frequency and character of Southern California's Santa Ana wind events are investigated within a 12-km-resolution downscaling of late-20th and mid-21st century time periods of the National Center for Atmospheric Research Community Climate System Model global climate change scenario run. The number of Santa Ana days per winter season is approximately 20% fewer in the mid 21st century compared to the late 20th century. Since the only systematic and sustained difference between these two periods is the level of anthropogenic forcing, this effect is anthropogenic in origin. In both time periods, Santa Ana winds are partly katabatically-driven by a temperature difference between the cold wintertime air pooling in the desert against coastal mountains and the adjacent warm air over the ocean. However, this katabatic mechanism is significantly weaker during the mid 21st century time period. This occurs because of the well-documented differential warming associated with transient climate change, with more warming in the desert interior than over the ocean. Thus the mechanism responsible for the decrease in Santa Ana frequency originates from a well-known aspect of the climate response to increasing greenhouse gases, but cannot be understood or simulated without mesoscale atmospheric dynamics. In addition to the change in Santa Ana frequency, we investigate changes during Santa Anas in two other meteorological variables known to be relevant to fire weather conditions—relative humidity and temperature. We find a decrease in the relative humidity and an increase in temperature. Both these changes would favor fire. A fire behavior model accounting for changes in wind, temperature, and relative humidity simultaneously is necessary to draw firm conclusions about future fire risk and growth associated with Santa Ana events. While our results are somewhat limited by a relatively small sample size, they illustrate an observed and explainable regional change in climate due to plausible mesoscale processes.

M. Hughes (✉)
National Oceanic and Atmospheric Administration, Earth System Research Laboratory, Physical Sciences Division, 325 Broadway, Boulder, CO 80305, USA
e-mail: Mimi.hughes@noaa.gov

A. Hall · J. Kim
Department of Atmospheric and Oceanic Sciences, University of California, Los Angeles, Box 951565, Los Angeles, CA 90095, USA

Keywords regional climate · climate change · downslope winds · fire weather

1 Introduction

The cool, relatively moist fall and winter climate of Southern California is often disrupted by dry, hot days with strong winds, known as Santa Anas, blowing out of the desert. The Santa Ana winds are a dominant feature of the cool season climate of Southern California (Conil and Hall 2006), and they have important ecological impacts. The most familiar is their influence on wildfires: Following the hot, dry Southern California summer, the extremely low relative humidities and strong, gusty winds associated with Santa Anas introduce extreme fire risk, often culminating in wildfires with large economic loss (Westerling et al. 2004). Less widely-known but just as important is their impact on coastal-ocean ecosystems: The strong winds induce cold filaments in sea-surface temperature (SST) with an associated increase in biological activity (Castro et al. 2006; Trasvina et al. 2003). This decrease in SST and increase in biological activity is likely due in part to increased mixing in the oceanic boundary layer, although an increase in dust deposition on the ocean surface during these events could also increase biological activity (Hu and Liu 2003; Jickells et al. 2005).

A recent investigation of the dynamics of Santa Ana winds (Hughes and Hall 2010) found that both local and synoptic conditions control their formation. When strong synoptically-forced offshore flow impinges on Southern California's topography, offshore momentum can be transported to the surface, causing Santa Ana conditions. However, Hughes and Hall (2010) found that there are many days with Santa Ana conditions that are not associated with this type of strong synoptic forcing. Rather, for a large fraction of the Santa Ana days, offshore winds are forced by a local temperature gradient between the cold desert and warmer air over the ocean at the same altitude. The temperature gradient induces a hydrostatic pressure gradient pointing from the desert to the ocean, which is reinforced by the negative buoyancy of the cold air as it flows down the sloped surface of the major topographical gaps.

This study investigates the response of the frequency and intensity of Santa Ana events and associated meteorological conditions to anthropogenic forcing. Santa Ana wind events cannot be simulated without resolving the coastal mountain ranges separating the Mojave desert from the Southern California Bight. This requires resolution higher than roughly 10–15 km. Hughes and Hall (2010) showed that even the relatively high-resolution North American Regional Reanalysis (NARR, 32 km horizontal resolution) has unrealistically weak Santa Ana winds. Because its coarse resolution does not adequately resolve the coastal topography, it cannot develop the tight desert-ocean temperature gradient often driving Santa Anas. However, when the Eta analysis data (a data set very similar to NARR) are dynamically downscaled with a much higher resolution regional atmospheric model, Santa Ana events are well-reproduced (Hughes and Hall 2010). This indicates that conditions leading to Santa Anas are implicit in the coarser resolution Eta data set, even if the reanalysis model itself is incapable of generating them realistically because of its inability to resolve the mesoscale effects that lead to strong surface winds (i.e. blocking, channeling, density flow). This dynamical downscaling technique has also been proven numerous times to give a more realistic view of local climate conditions in other contexts (e.g., Diffenbaugh et al. 2005; Leung and Ghan 1999; Kim and Lee 2003).

It follows that global models even coarser in resolution than the NARR product, such as those used for global climate change simulations, likely have almost no signature of these strong offshore winds, even if they do contain information highly relevant for Santa Ana formation. In our case, we are examining changes in Santa Anas implicit in the National

Center for Atmospheric Research (NCAR) Community Climate System Model (CCSM) global climate change scenario run. This model has a grid-point equivalent resolution of roughly 1.4°, obviously much too coarse to resolve Southern California's coastal ranges. To resolve the topography and draw out the simulation's implications for Santa Anas, we therefore must downscale it with a regional atmospheric model.

As we demonstrate, the regional simulation and the dynamical framework describing Santa Ana wind development of Hughes and Hall (2010) together allow us to identify how large-scale changes in the simulated future climate of the CCSM simulation affect the frequency and intensity of Santa Ana events. Our results indicate that as the climate adjusts to anthropogenic forcing, the frequency of Santa Ana days is reduced. Although the reduction in Santa Ana events could suggest a reduction in fire occurrence as the climate responds to anthropogenic forcing, we further investigate two other meteorological variables known to affect fire in the region, temperature and relative humidity (Moritz et al. 2010). In both cases, these meteorological variables change in ways favorable for fire occurrence, with relative humidity decreasing on days with Santa Ana conditions, and temperature increasing. Thus the implications of anthropogenic climate change for fire conditions in the region are ambiguous.

2 Weather Research and Forecast (WRF) simulation

The climate change experiment is carried out by downscaling late 20th century and mid 21st century time slices from a climate change scenario simulation done with the NCAR Community Climate System Model 3 (CCSM3). This dynamical downscaling was performed with the Weather Research and Forecast (WRF) model, version 2.2.1 (Skamarock et al. 2005). The model solves a non-hydrostatic momentum equation in conjunction with the thermodynamic energy equation. The model features multiple options for advection and parameterized atmospheric physical processes. The physics options selected in this experiment include the NOAH land-surface scheme (Chang et al. 1999), the simplified Arakawa Schubert (SAS) convection scheme (Hong and Pan 1998), the Rapid Radiative Transfer Model (RRTM) longwave radiation scheme (Mlawer et al. 1997), Dudhia (1989) shortwave radiation, and the WRF Single-Moment (WSM) 3-class with simple ice cloud microphysics scheme. For more details on the physics options, readers are referred to the website http://wrf-model.org. The model domain covers the western United States at a 36-km horizontal resolution, with the inner 12-km nest spanning the entire state of California and adjacent coastal zone. Both domains have 28 atmospheric and 4 soil layers in the vertical.

WRF was driven by the global climate data generated when the Special Report on Emissions Scenarios (SRES) A1B emission scenario is imposed on CCSM3 (Nakicenovic and Swart 2000). The emission scenario assumes balanced energy generation between fossil and non-fossil fuel; the resulting carbon dioxide (CO_2) emissions are located near the averages of all SRES emission scenarios. The CO_2 concentrations in the WRF simulations were fixed at 330 parts per million, volume (ppmv) and 430 ppmv during the late 20th century and mid 21st century periods, respectively.

Regional climate for the late 20th century and mid-21st century periods is calculated from a total of 20 cool season (October–March) WRF simulations spanning two time periods: 1971–1981 and 2045–2055. Individual WRF runs were initialized at 00UTC October 1 of the corresponding years using the CCSM3 output data. All simulations continued for the remaining 6-month period without re-initialization by updating the large-scale forcing along the lateral boundaries at three-hour intervals. The focus on the cool season simulations is appropriate in this case because Santa Ana winds have a very strong seasonality, with peak occurrence in December, and very few strong offshore winds from April to September.

3 Santa Ana response to a changing climate

The first step in quantifying anthropogenic changes in Santa Ana (SA) wind occurrence and characteristics is to create a SA index (SAt). As in Hughes and Hall (2010), our SA index is simply the offshore (i.e., southwestward) daily-mean wind strength at the exit of the largest gap in southern California (blue box, Fig. 1). The advantage of this index, in contrast to previously-defined SA indices (e.g. Miller and Schlegel 2006; Raphael 2003; Sommers 1978; D. Danielson, personal communication), is that it provides information about SA occurrence and intensity, but does not introduce assumptions about mechanisms possibly causing Santa Anas, whether local or synoptic. Figure 1 shows the composite surface winds for days with SAt greater than 10 ms^{-1} for the 10 cool season WRF simulations corresponding to the late 20th century time slice (1971–1981). The composite SA wind field exhibits characteristics we expect for SA events: strong offshore (that is, roughly northeasterly) winds throughout most of Southern California, with the strongest winds on the leeward slopes of the mountains and through the gaps in the topography, most notably across the Santa Monica mountains.

Is there any change in the number of SA days due to anthropogenic forcing? To answer this question, we quantify the number of SA days per season in the WRF downscaling of CCSM. To the extent that there is a difference in climate between the two WRF simulations, we know it is likely due to the effect of anthropogenic forcing, since that is the

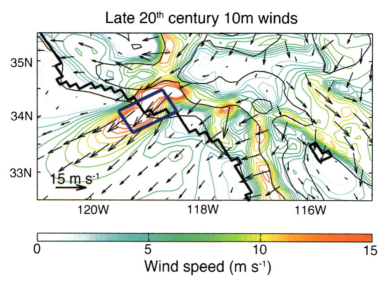

Fig. 1 Average winds for days with Santa Ana time series greater than 10 ms^{-1} for late 20th century simulation. *Arrows* show total wind; color contours show wind speed contoured every 0.25 ms^{-1} from 0 to 15 ms^{-1}. Reference *arrow* of length 15 ms^{-1} is shown on bottom *left*. Only every third grid point is plotted for clarity. *Black* contours show model terrain, plotted every 800 m (m) starting at 100 m. The *thick black* contour shows coastline at 12-km resolution

only sustained and systematic difference between the two simulations. Figure 2 shows the average number of days with SAt greater than 10 ms^{-1} for the late 20th and mid 21st century WRF simulations. There is nearly a 20% reduction in the total number of SA days per year in the mid 21st century run. The mid 21st century mean is less than the late 20th century mean with 90% confidence using a directional t-test. This result is insensitive to the threshold used in SA identification (for thresholds ranging from 8 ms^{-1} to 14 ms^{-1}), although the statistical significance level varies slightly with threshold. In the following sections, we explore the mechanism by which the frequency of SA wind events is reduced to lend more credibility to this result.

4 Understanding reduced Santa Ana frequency

The atmospheric dynamics associated with SA winds were recently investigated by Hughes and Hall (2010). They validated and analyzed SA events in a 6-km resolution Southern California climate reconstruction. The reconstruction was accomplished by downscaling reanalysis data corresponding to the years 1995–2006 using WRF's predecessor, the NCAR/Pennsylvania State University Mesoscale Model, Version 5 (MM5, Grell et al. 1994). They found that SAs arise from a combination of two mechanisms – one with a synoptic extent covering much of the western U.S., and another more local process confined to Southern California. Here we briefly review important results from this study, as they are relevant to our explanation of reduced SA frequency resulting from anthropogenic forcing.

Previous studies identified large-scale mid-tropospheric conditions as the driver of SAs (e.g., Sommers 1978). If there is a large synoptic-scale pressure gradient causing strong offshore winds over Southern California at mountain-top level, this causes surface flow as the offshore momentum is transferred to the surface in a stably stratified atmosphere. This often occurs when a high surface pressure anomaly is located over the Great Basin, with a corresponding high geopotential height anomaly at 700 hPa centered over Oregon. Though this synoptic mechanism contributes to offshore flow in Southern California, Hughes and Hall (2010) also found that if a large temperature gradient exists between the cold desert

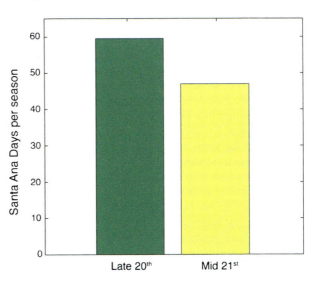

Fig. 2 Number of days per season with SAt greater than 10 m s^{-1} in the (*green bar*) late 20th and (*yellow bar*) mid 21st century simulations

surface and the warm ocean air at the same altitude (approximately 1.2 km), it causes a localized offshore pressure gradient near the surface. This generates katabatic offshore flow in a thin layer near the surface, as the negatively buoyant, cold desert air flows down the sloped surface of the gaps. The negative buoyancy can be written as a pressure gradient force (Parish and Cassano 2003):

$$\beta = \frac{g\theta'}{\theta_0} \sin(\alpha) \tag{1}$$

where g=9.8 ms^{-1} is gravitational acceleration, θ' is the temperature deficit of the cold layer, θ_0 is the average temperature in the cold layer, and α is the slope of the topography. To calculate β, Hughes and Hall (2010) used the average desert surface temperature for θ_0, the average slope of the topography through the largest gap for α (approximately 1°, or 1 km drop over 50 km; see Fig. 1) and the temperature difference between the air over the cold desert surface (i.e., the cold layer) and air over the ocean at the same altitude (representative of the ambient atmosphere) for θ'.

These two mechanisms can act independently (and often do), causing mild SA winds, or combine to force the largest magnitude offshore winds. Their joint contribution to SAt can be modeled statistically by a bivariate regression model to predict SAt based on two parameters representative of each mechanism

$$\widehat{SAt}(u,\beta) = A^*u + B^*\beta + C \tag{2}$$

where u, the southwestward (and mainly geostrophic) wind speed at 2 km above mean sea level (AMSL), represents the synoptic forcing and β represents the local thermodynamic forcing. In the MM5 reconstruction, this regression model captured almost all variability in SAt (correlation between SAt and $\widehat{SAt}$ was 0.93), with about 1/8 of the variability accounted for by the synoptic mechanism, more than half by the local mechanism, and the remainder by in-phase variability of the two mechanisms impossible to unambiguously ascribe to one or the other.

Figure 3 shows the regression model's representation of SAt, $\widehat{SAt}$, plotted against SAt for the WRF late 20th century and mid 21st century cool season simulations on days with positive u. The high degree of correspondence (correlation coefficient=0.88, and 0.86 for the late 20th and mid 21st simulations, respectively) confirms that, as with the MM5 reconstruction, the two mechanisms are primarily responsible for determining SAt in the WRF simulations forced by CCSM3 data. The regression model coefficients are shown in Table 1 for both decades. Parameters A and B are within 10% of one another, and C is close to zero for both decades. Table 1 also shows the variance explained by each of the terms of the regression model. Similar to Hughes and Hall (2010), more variance is explained by the local mechanism than by the synoptic mechanism, although in the case of the current WRF downscaling of CCSM data, the variance is more equally partitioned.

Because the terms within it correspond to distinct physical processes, the regression model allows us to identify the changes in forcing responsible for the reduction in SAt. To understand which term of the regression model is causing reduced $\widehat{SAt}$, we calculate the total contribution of β and u to $\widehat{SAt}$ separately and then sum over days with $\widehat{SAt}$ greater than 10 ms^{-1} for the two time periods (Fig. 4).

Turning our attention first to the synoptic forcing represented by u in Eq. 1 and the second column grouping in Fig. 4, we see that the WRF simulation shows almost no change between the present and future simulations. Thus the reduction in SAt between present and future simulations is not primarily due to a systematic change in the low-to-

Fig. 3 Actual SAt plotted against that predicted by the bivariate regression model (SÂt) for (**a**) late 20th century simulation, and (**b**) mid 21st century simulation. *Red dashed line* shows SAt=SÂt

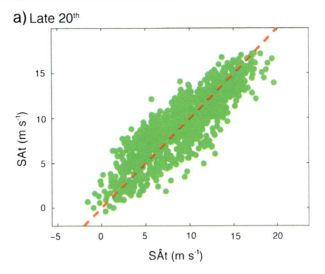

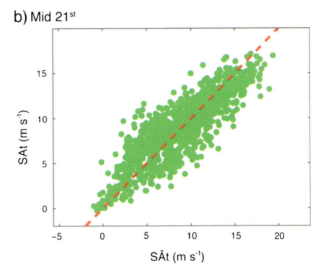

Table 1 Parameters of the bivariate regression model Eq. 2, and variance explained by its terms. Units are shown in the header row

		Late 20th	Mid 21st
Regression model parameters	A (dimensionless)	0.33	0.36
	B (s^{-1})	2704	2443
	C (m s^{-1})	0.08	0.15
Percent variance explained by terms of regression model	u	23	27
	β	40	33
	covariance	14	14
	error	23	26

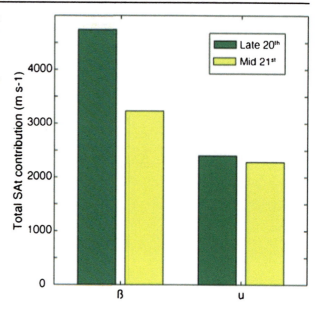

Fig. 4 Total contribution of (left grouping) β and (right grouping) u to SAt. Contributions were calculated by summing the product of the regression model parameters and (*left column*) β or (*right column*) u for days with SAt greater than 10 ms^{-1}

mid-troposphere geostrophic wind and associated synoptic-scale pressure gradients when offshore winds blow. Focusing instead on the katabatic forcing of β in Eq. 1 and the first column grouping in Fig. 4, we see that the 21st century time period shows over 1/3 less contribution to SÂt from β than the present day scenario. This is likely because land masses warm more quickly in response to increased radiative forcing than the oceans (e.g. Trenberth et al. 2007), reducing the temperature gradient between the cold desert and warm ocean.

A signature of the differential warming of land and ocean is illustrated in Fig. 5, which shows changes in seasonal mean desert air temperature at 1.2 km AMSL (near the desert surface) as well as the air temperature over the ocean at the same altitude—the two components of θ' used to calculate β — between the late 20th and mid 21st century simulations. Both locations exhibit warming, but the desert near-surface air temperature

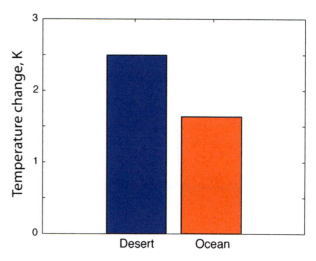

Fig. 5 Change in mean temperature between the late 20th and mid 21st century simulations 1.2 km AMSL at the desert surface (*blue bar*) and over the ocean surface (*red bar*). Mean changes are significant at beyond the 99% level using a two-tailed *t* test. 'Desert' and 'Ocean' locations were chosen to correspond to the points used in the regression model (i.e., desert point is 34.99 N, 117.83 W, and ocean point is 34.19 N, 119.55 W)

warms about a degree Celsius more than the air over the ocean at the same altitude, consistent with previous regional climate change projection studies (e.g., Kim et al. 2002; Christensen et al. 2007). The magnitude of the katabatic pressure gradient force, β, is directly proportional to the difference between these two temperatures. As the desert warms faster than the air over the ocean, a large temperature gradient between the two areas becomes less likely in wintertime, and large β becomes less frequent. Because the only sustained difference between the two periods is the signature of the anthropogenic increase in radiative forcing, this decrease in the desert-ocean temperature gradient is likely anthropogenically forced. It follows that the reduction in β and the resulting decrease in SA frequency also likely arises from anthropogenic forcing.

5 Santa Ana annual cycle, temperature, relative humidity, and implications for fire

In the previous section, we showed that, because the air close to the desert surface warms more quickly in response to anthropogenic greenhouse gas forcing than the air 1.2 km above the ocean, the number of strong offshore wind events per year will likely decline over the following decades. Wildfire activity in this region is strongly tied to the occurrence of SA winds, primarily during fall after the hot, dry summer (e.g., Moritz et al. 2010; Westerling et al. 2004; Keeley et al. 2004). A comparison of the annual cycle of SA occurrence in the two time slices (Fig. 6) reveals that all months with the exception of January experience reduced SA conditions in the mid-21st century compared with the late 20th century; however, the only months where the change is statistically significant are October and February. In fact, the confidence level for the decrease in October events is 92.5%, higher than for the annual mean change.

It is tempting to conclude that this reduction in SA events could lead to a reduction in wildfire activity. However, this may be a misleading conclusion because the reduction in SAs is accompanied by changes in other meteorological variables also determining fire risk. For example, fire severity tends to increase as temperature increases (Moritz et al. 2010). Therefore, the local increases in simulated temperature (Fig. 5) should favor more severe wildfire outbreaks.

Another meteorological variable critical to determining fire conditions is the near-surface relative humidity (Moritz et al. 2010). To examine whether this variable exhibits any changes in our regional climate change experiment, we calculate average relative humidity

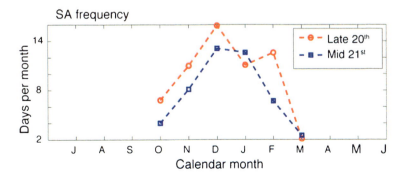

Fig. 6 Annual cycle of number of days per month with $SAt > 10\ ms^{-1}$ for each calendar month from October (O) through March (M) for the late 20th (*red dashed line* and *circle markers*) and mid 21st (*blue dashed line* and *square markers*) century time slice

2 m above the surface (2mRH) at locations surrounding the Los Angeles basin on days with SA conditions (Fig. 7a; criteria for location selection is noted in caption). The WRF

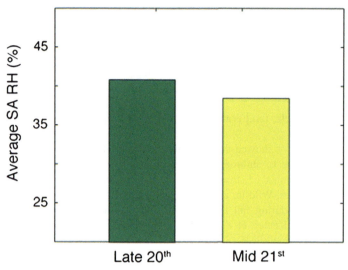

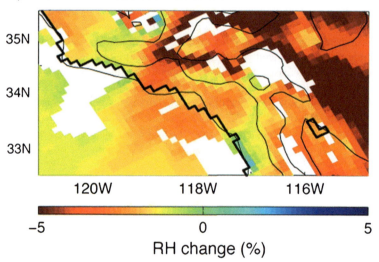

Fig. 7 **a** Average relative humidity (RH,%) 2 m above the surface (hereafter, 2mRH) on days with SAt>10 ms^{-1}, at locations surrounding the Los Angeles basin, for the late 20th century (*green bar*) and mid 21st century scenarios (*yellow bar*). Locations included in the average had to satisfy four criteria: 1) On land, 2) Latitude between 33.5 N and 34.7 N, 3) Longitude between 116.5 W and 120.5 W, and 4) 2mRH change had to be statistically significant. **b** Difference between the average 2mRH on days with SAt>10 ms^{-1} in the mid 21st century simulation and the late 20th century simulation. Terrain and coastline are shown as in Fig. 1. Negative values indicate the mid 21st century simulation values are less than the late 20th century values. Grid points with a change that had less than a 90% confidence interval with a two-tailed *t*-test are *colored white*

simulations show a reduction in average 2mRH during SA days. The future scenario is nearly 2.5 percentage points drier surrounding the Los Angeles basin during SA events than the present-day scenario.

Why does the 2mRH decrease under climate change when simulated Santa Ana events occur in Southern California? Although global RH is not expected to change significantly under climate change conditions (Held and Soden 2006), over dry land surfaces this may not be the case due to an absence of a moisture source to maintain constant RH levels in warmer air. In fact, if we look at the distribution of 2mRH change in the WRF scenario (Fig. 7b) we see that the largest reductions in 2mRH on SA days are located in the Mojave desert interior. The desert surface responds to positive radiative forcing by warming strongly (Fig. 5), and because the desert soil has almost no moisture to release to the atmosphere, 2mRH must decrease. Because much of the air in the coastal areas during SA conditions has its origin in the desert interior (Fig. 1), 2mRH in most of the domain is also reduced, albeit less so than in the desert interior. Notably, the largest non-desert reductions in 2mRH tend to be located in the areas with strongest winds on SA days (cf. Figs. 1 and 7b).

The resulting implications for fire incidence of the combination of these three meteorological variables (wind, RH, and temperature) are ambiguous, since strong offshore wind frequency decreases while RH decreases and temperature increases. To get quantitative estimates on the meteorological implications for fire incidence, a fire behavior model would be necessary to untangle the respective roles these three variables will play in the future. Westerling et al. (2006) recently showed that wildfire activity has increased in Southern California over the past three decades. If the RH and temperature effects dominate the wind speed effects, wildfire activity may well continue to increase as climate change accelerates.

6 Conclusions

This study investigates the frequency of Southern California's SA wind events within a high-resolution dynamical downscaling of two time slices of the NCAR CCSM3 climate change scenario simulation. One time slice corresponds to the late 20th century, the other to the mid 21st century. This particular CCSM simulation imposes the SRES-A1B emission scenario. In the high resolution simulation the SA events per year are reduced approximately 20% in the mid 21st compared with the late 20th century. Because the only forced difference between the two time slices is the level of anthropogenic forcing, the change in SA events is likely caused by anthropogenic forcing. The reduction in the frequency of SAs is also associated with a reduction in their mean intensity.

We use a bivariate regression model to reproduce the Santa Ana time series, where two known forcing mechanisms are used as independent (predictor) variables: (1) synoptically-forced strong offshore winds at the mountain tops whose momentum is transported to the surface, and (2) local thermodynamically forced winds caused by a katabatic pressure gradient that arises from the thermal contrast between cold desert air and warmer air over the adjacent ocean. The regression model reproduces approximately 80% of the variability in the Santa Ana time index and also reproduces the significant reduction in SA events between the two time slices. We use the regression model to partition the anthropogenic change in SA events into contributions from synoptic and katabatic components. We find a large reduction in katabatic forcing. This is caused by the larger transient warming of the desert surface than the air over the ocean at the same altitude. This reduces the likelihood of a large temperature deficit developing in the desert in wintertime and therefore reduces the likelihood of large katabatic forcing. We see no change in the likelihood of synoptically forced SAs.

These results are not inconsistent with previous results: Miller and Schlegel (2006) identified an anthropogenically-forced seasonality change in large-scale pressure patterns associated with SAs, with a reduction in SA events in all cool season months except December comparing the late 20th century to mid 21st century (their Fig. 3). When the statistical significance of the month-by-month changes are taken into consideration, their results agree with ours, despite the difference in GCM, time periods considered, and different SA identification criteria. Moreover, these investigators did not examine the katabatic forcing mechanism, which we found to be of primary importance in creating anthropogenic change in SAs.

The role SA winds play in spreading wildfire in the region (e.g., Moritz et al. 2010; Westerling et al. 2004; Keeley et al. 2004) suggests their reduction in frequency could lead to reduced wildfire in Southern California. However, we find that two other meteorological variables important to fire risk and growth, RH and temperature, also change significantly and systematically in the region during SA events. RH on SA days is reduced, and temperature increases; both of these changes would favor fire development. Also, this study does not address changes in other parameters critical to fire behavior, such as available fuel. Moreover, ignition events (i.e., humans lighting matches in the coastal chaparral shrubland), another major factor affecting fire frequency, will probably increase with population (Syphard et al. 2007). Therefore, a fire behavior model is needed to predict how fire incidence will change under anthropogenic climate change conditions, and no conclusions can be drawn from this study alone about future wildfire occurrence associated with SA events.

There are other societal implications of this anthropogenic reduction of SA wind events that could be significant and should be explored further. A reduction in the frequency and intensity of SA winds has implications for coastal marine ecosystems, which respond favorably to SA conditions, and to the air quality in the Los Angeles basin, which is better during SA events. The ecological effects of nutrient loss for the Southern California Bight and the decline in air quality during winter could be quantified with regional simulations of oceanic biogeochemistry and atmospheric chemistry.

This study portrays one possible change in the meteorological conditions associated with SA wind events, and – more importantly – identifies the mechanism causing the change. However, it has two caveats that arise ultimately from the computational costs of high-resolution dynamical downscaling. First, each time-slice is only 10-years long, and longer time-slices are needed to produce results with greater statistical significance to account for the impact of internal climate variability on SA wind occurrence. Second, this study represents only one emission scenario/general circulation model/regional climate model combination, making it impossible to provide quantitative information about the uncertainty of its projected changes. In addition, to the extent that the smaller temperature increase in the atmosphere over the ocean is due to the larger oceanic heat capacity, the reduction in thermodynamic forcing of SAs might be a feature of the transient climate change that will return to pre-industrial levels once the climate equilibrates. Nevertheless, the reduction in SA wind events due to anthropogenic climate change is significant because it illustrates an observed and explainable regional change in climate due to plausible mesoscale processes. Further, despite the fact that surface temperature changes are due to a well-known response of the climate system to an increase in greenhouse gases, the resultant change in SA wind events cannot be simulated or understood without mesoscale atmospheric processes, thus requiring a high spatial resolution atmospheric model for its detection.

Acknowledgements Mimi Hughes is supported by a National Research Council Postdoctoral Associateship and National Science Foundation ATM-0735056, which also supports Alex Hall. Part of this work was performed using the National Center for Atmospheric Research supercomputer allocation 35681070. The research described in this paper was performed as an activity of the Joint Institute for Regional Earth System Science and Engineering, through an agreement between the University of California, Los Angeles, and the Jet Propulsion Laboratory, California Institute of Technology, and was sponsored by the National Aeronautics and Space Administration. Preprocessing of the Community Climate System Model data was also partially funded by the "National Comprehensive Measures against Climate Change" Program by Ministry of Environment, Korea (Grant No. 1600-1637-301-210-13) National Institute of Environmental Research, Korea. Computational resources for this study have been provided by Jet Propulsion Laboratory's Supercomputing and Visualization Facility and the National Aeronautics and Space Administration's Advanced Supercomputing Division.

References

Castro R, Mascarenhas A, Martinez-Diaz-de Leon A, Durazo R, Gil-Silva E (2006) Spatial influence and oceanic thermal response to Santa Ana events along the Baja California peninsula. Atmosfera 19 (3):195–211

Chang S, Hahn D, Yang C, Norquist D, Ek M (1999) Validation study of the CAPS model and land surface scheme using the 1987 Cabauw/PILPS dataset. J Appl Meteorol 38:405–422

Christensen JH, Hewitson B, Busuioc A, Chen A, Gao X, Held I, Jones R, Kolli RK, Kwon W-T, Laprise R, Magaña Rueda V, Mearns L, Menéndez CG, Räisänen J, Rinke A, Sarr A, Whetton P (2007) Regional climate projections. In: Solomon S, Qin D, Manning M, Chen Z, Marquis M, Averyt KB, Tignor M, Miller HL (eds) Climate Change 2007: The Physical Science Basis. Contribution of Working Group I to the Fourth Assessment Report of the Intergovernmental Panel on Climate Change. Cambridge University Press, Cambridge, United Kingdom and New York, NY, USA

Conil S, Hall A (2006) Local regimes of atmospheric variability: a case study of Southern California. J Clim 19(17):4308–4325

Diffenbaugh N, Pal J, Trapp R, Giorgi F (2005) Fine-scale processes regulate the response of extreme events to global climate change. PNAS 102(44):15 774–15 778

Dudhia J (1989) Numerical study of convection observed during the winter monsoon experiment using a mesoscale two-dimensional model. J Atmos Sci 46(20):3077–3107

Grell G, Dudhia J, Stauffer D (1994) A description of the fifth-generation Penn State/NCAR Mesoscale Model (MM5). Tech. rep., NCAR Tech. Note NCAR/TN-398+STR.

Held IM, Soden BJ (2006) Robust responses of the hydrological cycle to global warming. J Clim 19:5686–5699

Hong S, Pan H (1998) Convective trigger function for a mass flux cumulus parameterization scheme. Mon Weather Rev 126:2599–2620

Hu H, Liu W (2003) Oceanic thermal and biological responses in Santa Ana winds. Geophys Res Lett 30 (11):1596. doi:10.1029/2003GL017208

Hughes M, Hall A (2010) Local and synoptic mechanisms causing Southern California's Santa Ana winds. Clim Dyn 34(6):847–857. doi:10.1007/s00382-009-0650-4

Jickells T, An ZS, Andersen KK, Baker AR, Bergametti G, Brooks N, Cao JJ, Boyd PW, Duce RA, Hunter KA, Kawahata H, Kubilay N, laRoche J, Liss PS, Mahowald N, Prospero JM, Ridgwell AJ, Tegen I, Torres R (2005) Global iron connections between desert dust, ocean biogeochemistry, and climate. Science 308:67–71. doi:10.1126/science.1105959

Keeley JE, Fotheringham CJ, Moritz MA (2004) Lessons from the October 2003 wildfires in southern California. J Forest 102:26–31

Kim J, Lee J-E (2003) A multi-year regional climate hindcast for the western United States using the mesoscale atmospheric simulation model. J Hydrometeorol 4:878–890

Kim J, Kim T-K, Arritt RW, Miller N (2002) Impacts of increased atmospheric CO2 on the hydroclimate of the western United States. J Clim 15:1926–1942

Leung L, Ghan S (1999) Pacific Northwest climate sensitivity simulated by a regional climate model driven by a GCM. Part I: control simulations. J Clim 12:2010–2030

Miller NL, Schlegel NJ (2006) Climate change projected fire weather sensitivity: California Santa Ana wind occurrence. Geophys Res Lett 33. doi:10.1029/2006GL025808

Mlawer E, Taubman S, Brown P, Iacono M, Clough S (1997) Radiative transfer for inhomogeneous atmosphere: RRTM, a validated correlated-k model for the longwave. J Geophys Res 102(D14):16663–16682

Moritz M, Moody TJ, Krawchuk MA, Hughes M, Hall A (2010) Spatial variation in extreme winds predicts large wildfire locations in chaparral ecosystems. GRL. Submitted

Nakicenovic N, Swart R (2000) Special report on emissions scenarios. Edited by Nebojsa Nakicenovic and Robert Swart, pp. 612. ISBN 0521804930. Cambridge, UK: Cambridge University Press, July 2000.

Parish T, Cassano J (2003) Diagnosis of the katabatic wind influence on the wintertime Antarctic surface wind field from numerical simulations. Mon Weather Rev 131:1128–1139

Raphael MN (2003) The Santa Ana winds of California. Earth Interact 7:1–13

Skamarock W, Klemp J, Dudhia J, Gill D, Baker D, Wang W, Powers J (2005) A description of the advanced research WRF version 2. Tech. rep., NCAR Tech. Note, NCAR/TN468+STR.

Sommers WT (1978) LFM forecast variables related to Santa Ana wind occurences. Mon Weather Rev 106:1307–1316

Syphard A, Radeloff V, Keeley J, Hawbaker T, Clayton M, Stewart S, Hammer R (2007) Human influence on California fire regimes. Ecol Appl 17:1388–1402

Trasvina A, Ortiz-Figueroa M, Herrera H, Coso MA, Gonzlez E (2003) Santa Ana winds and upwelling filaments off northern Baja California. Dyn Atmos Oceans 37(2):113–129

Trenberth K, Jones P, Ambenje P, Bojariu R, Easterling D, Klein Tank A, Parker D, Rahimzadeh F, Renwick J, Rusticucci M, Soden B, Zhai P (2007) Observations: surface and atmospheric climate change. Climate Change 2007: The Physical Science Basis. Contribution of Working Group I to the Fourth Assessment Report of the Intergovernmental Panel on Climate Change. Solomon S, Qin D, Manning M, Chen Z, Marquis M, Averyt K, Tignor M, Miller H (eds) Cambridge, United Kingdom and New York, NY, USA: Cambridge University Press.

Westerling AL, Cayan DR, Brown TJ, Hall BL, Riddle LG (2004) Climate, Santa Ana winds and autumn wildfires in Southern California. EOS 85(31):289–296

Westerling AL, Hidalgo HG, Cayan DR, Swetnam TW (2006) Warming and earlier spring increases Western U.S. forest wildfire activity. Science 313:940–943. doi:10.1126/science.1128834

Adapting California's water system to warm vs. dry climates

Christina R. Connell-Buck · Josué Medellín-Azuara · Jay R. Lund · Kaveh Madani

Received: 9 March 2010 / Accepted: 21 September 2011 / Published online: 24 November 2011
© Springer Science+Business Media B.V. 2011

Abstract This paper explores the independent and combined effects of changes in temperature and runoff volume on California's water supply and potential water management adaptations. Least-cost water supply system adaptation is explored for two climate scenarios: 1) warmer-drier conditions, and 2) warmer conditions without change in total runoff, using the CALVIN economic-engineering optimization model of California's intertied water supply system for 2050 water demands. The warm-dry hydrology was developed from downscaled effects of the GFDL CM2.1 (A2 emissions scenario) global climate model for a 30-year period centered at 2085. The warm-only scenario was developed from the warm-dry hydrology, preserving its seasonal runoff shift while maintaining mean annual flows from the historical hydrology. This separates the runoff volume and temperature effects of climate change on water availability and management adaptations. A warmer climate alone reduces water deliveries and increases costs, but much less than a warmer-drier climate, if the water supply system is well managed. Climate changes result in major changes in reservoir operations, cyclic storage of groundwater, and hydropower operations.

1 Introduction

Climate change can have many forms and impacts, including rising sea level, melting permafrost, increased flooding, decreased snowpack, and more frequent wildfires. Yet

---

C. R. Connell-Buck (✉)
Hydrologic Sciences Graduate Group, University of California-Davis, One Shields Ave, Davis, CA 95616, USA
e-mail: crconnell@ucdavis.edu

J. Medellín-Azuara · J. R. Lund
Department of Civil and Environmental Engineering, University of California-Davis, Davis, CA 95616, USA

K. Madani
Department of Civil, Environmental, and Construction Engineering, University of Central Florida, Orlando, FL 32816, USA

rising temperatures and changes in precipitation (and related runoff) drive most such changes. These two major drivers can have independent and combined effects on California's water resources and economy. Characterized by a Mediterranean climate (wet winters and warm, dry summers), California's urban and agricultural water supply depends heavily on water storage in snowpack, reservoirs, and aquifers to meet demands through the dry growing season. Warming in the western United States is reducing snow water equivalents (depth of water in the snowpack if melted) (Hamlet et al. 2005) and has affected deliveries and reservoir storage levels for the State Water Project and Central Valley Project (Anderson et al. 2008).

Downscaled global climate models applied to California and the western United States have been applied to explore possible changes in streamflow peaks, timing and volume, snowmelt, snow water equivalent, and evapotranspiration (Cayan et al. 2008b; Hamlet et al. 2007; Miller et al. 2003). Field data as well as modeling indicate a shift in spring runoff since the 1940s as warming temperatures shift runoff to earlier in the year (Dettinger and Cayan 1995; Maurer 2007; Stewart et al. 2004). This shift in streamflows may influence water management, the extent and character of ecosystems, and changes in estuarine inflows and salinity in the Sacramento-San Joaquin Delta (Cayan et al. 2001, 2008a; Knowles and Cayan 2004).

Climate change studies largely assess hydrologic response of surface water characteristics. However, groundwater response to climate change also has been investigated by linking global climate models to regional groundwater models, to estimate climate influences on conjunctive management of surface and groundwater resources (Hanson and Dettinger 2005; Scibek and Allen 2006). Changes in both surface and groundwater components of the hydrologic system will affect water management challenges and opportunities.

The IPCC Fourth Assessment Report 2007 describes emission scenarios used in General Circulation Models (GCMs) and summarizes regional climate change projections for temperature and precipitation (Christensen et al. 2007). Cayan et al. (2008b) describe some of these models and emission scenarios as they pertain to California. The Parallel Climate Model (PCM1) and NOAA's GFDL CM2.1 model provide simulations for California suggesting warming temperatures ranging from 1.5 to 4.5°C by the end of the century, depending on the emissions scenario. Warming is projected to decrease the share of precipitation falling as snow and to increase the portion falling as rain (Bedsworth and Hanak 2008; Cayan et al. 2008b; Hanak and Lund 2008). Unlike the warming trend for temperatures, climate models present less consensus regarding changes in precipitation. Projections of annual precipitation for northern and southern California range widely from a decrease of 26% to an increase of 18%, though most global climate models indicate moderate changes (Cayan et al. 2008b).

Changing hydrologic conditions have important implications for California water management. Previous studies have assessed economic impacts and water management adaptation to combined warmer, wetter, and drier climates (Medellin-Azuara et al. 2008; O'Hara and Georgakakos 2008; Tanaka et al. 2006). Since precipitation projections include both drier and wetter conditions, Tanaka et al. (2006) applied the CALVIN model to explore integrated management adaptations to a warmer-drier climate and a warmer-wetter climate for 2100 demands. Changing runoff patterns may necessitate changing reservoir operations, thus Medellin-Azuara et al. (2008) explored optimized multi-reservoir operation adaptations to a warm-dry climate with year 2050 demands. Other work assesses the economic impacts of climate change and population growth in semi-arid urban environments and considers the effectiveness of adaptation strategies including storage capacity expansions (O'Hara and Georgakakos 2008).

The two primary drivers affecting water supply, temperature and precipitation, have not before been parsed out to analyze their independent and combined effects on California water management adaptation. This paper uses CALVIN model results to compare a warmer climate scenario and a warmer-drier scenario centered on 2085 with updated 2050 water demand estimates. This results in effects on California's water supply and water management adaptations for a year 2050 level of development using the more extreme climate change model results (temperature and precipitation effects centered on 2085). This is meant to test the system with the greatest amount of temperature increase and reduction in precipitation to see how it performs and adapts. Results indicate that temperature rise alone does not tend to increase water shortages greatly if system operations adapt. Modified surface water operations, conjunctive use of ground and surface waters, and hydropower generation are explored as strategies to mitigate economic costs from warmer and warmer-drier climates.

2 Methods

This study's approach to perturbing the hydrology for climate change conditions involves mapping streamflows to corresponding index basins and then applying perturbation ratios (Connell 2009). Perturbed hydrology is then input to CALVIN to assess economic impacts and water management adaptations to changed water supply volume and timing. The approach used is similar to that employed in Tanaka et al. (2006) and Medellin-Azuara et al. (2008). Two climate change scenarios were developed to explore the independent and combined effects of temperature and precipitation. A warm-dry hydrology was developed from downscaled results of the GFDL CM2.1 (A2 emissions scenario) global climate model for a 30-year period centered at 2085 (Connell 2009; Zhu et al. 2003). A warm-only climate was developed by adjusting warm-dry and historical hydrologies to maintain historical average annual runoff while capturing the shift in runoff timing expected from warming temperatures (Connell 2009). Global climate models simulating a warm-only scenario for California were not available.

2.1 The CALVIN model

CALVIN (CALifornia Value Integrated Network) is a hydro-economic optimization model of California's statewide water supply system (http://cee.engr.ucdavis.edu/calvin/). The CALVIN model includes 44 reservoirs, 28 groundwater basins, and 54 economically-represented urban and agricultural demand areas (Fig. 1). Using HEC-PRM, a network flow reservoir optimization solver developed by the U.S. Army Corps of Engineers, CALVIN operates surface and groundwater resources and allocates water over a historical (1921–1993) hydrologic record. Economically driven, CALVIN allocates water to minimize total statewide water scarcity of agricultural and urban water use within physical and environmental constraints (Draper et al. 2003). Water scarcity occurs whenever a user's economic target for water is not met. This results in a corresponding scarcity cost for each demand. Historical flow data from existing surface and integrated surface-groundwater models are typically used (Draper et al. 2003; Zhu et al. 2003).

2.1.1 Agricultural and urban water demands

Agricultural and urban water demands are estimated for year 2050 level of development, and account for population growth, urban water conservation, agricultural land use

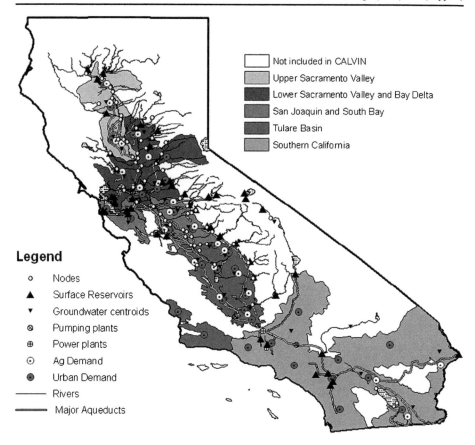

Fig. 1 Hydrologic basins, demand areas, major inflows and infrastructure represented in CALVIN (adapted from Lund et al. 2007)

conversion, likely crop price conditions, and technology-related crop farming yield improvements. Values for agricultural water use are derived using the Statewide Agricultural Production model (SWAP http://swap.ucdavis.edu), an ancillary optimization model that maximizes farm profit for each agricultural demand area using positive mathematical programming (Howitt 1995). The version of SWAP used for this paper includes 21 Central Valley Project Model (or CVPM after USBR 1997) regions in the Central Valley plus Coachella, Imperial, and Palo Verde irrigation districts in Southern California. A detailed description of the marginal value of water in agriculture for these regions appears in Howitt et al. (2009). For each agricultural production region, a loss function is derived for the economic losses from water scarcity, obtained by numerical integration of the marginal value of water in agriculture, for every month. Data from georeferenced land use surveys from the California Department of Water Resources (DWR) are employed. Average applied water by CVPM and/or hydrological regions and crop groups is employed in the base case from DWR estimates of applied water per crop group.[1] Land conversion from agricultural to urban uses follows Landis and Reilly (2002) 2050

---

[1] Available at: http://www.water.ca.gov/landwateruse/anaglwu.cfm#

urbanization trends. Technological change is assumed to increase crop yields by 29% by 2050 (Howitt et al. 2009). For the regions covered in CALVIN (representing 89% of statewide applied water use for 2005), 2050 agricultural water use totals 29.7 BCM/yr (compared to total water availability of about 43 BCM/year, as detailed in Table 1). The effect of climate change on the economic value of water for agriculture in SWAP is represented as changes in yields from changing temperature, precipitation, and CO_2 concentrations (Howitt et al. 2009). Thus cropping patterns under historical, warm-only, and warm-dry scenarios allow adaptation such as shifts to higher value and more climate tolerant crops.

Urban water shortage loss functions follow methods described in Jenkins et al. (2003) with population growth projections and urban water demands for year 2050 (Jenkins et al. 2007). Constant price-elasticity demand functions for water are estimated for every economically-represented urban area in CALVIN. Total base urban water use in CALVIN by 2050 is projected at 16.4 BCM/yr, representing about 92% of estimated

Table 1 Changes in California's water supply for warm-dry and warm-only climate scenarios (average annual totals)

	Statewide	Sacramento Valley	San Joaquin Valley	Tulare Basin	Southern California
Warm-dry Change in Precipitation					
% Change[a]	−27%	−24%	−30%	−33%	−
cm/year	−5.8	−5.8	−5.8	−6.1	−
Rim Inflows (MCM/yr)					
Historical/Warm-only	34,824	23,577	7,078	3,485	684
Warm-dry	25,031	18,253	4,372	1,953	453
% Change	−28%	−23%	−38%	−44%	−34%
Net Reservoir Evaporation (m/yr)					
Historical	1.6	1.1	1.9	2.0	1.6
Warm-only	1.8	1.3	2.1	2.5	1.9
% Change	15%	17%	9%	23%	19%
Warm-dry	2.2	1.6	2.7	2.7	2.2
% Change	37%	43%	37%	32%	36%
Groundwater Inflows (MCM/yr)					
Historical/Warm-only	8,359	2,748	1,444	4,168	−
Warm-dry	7,525	2,368	1,277	3,880	−
% Change	−10%	−14%	−12%	−7%	−
Local Accretion (MCM/yr)					
Historical/Warm-only	5,449	4,377	577	495	−
Warm-dry	3,812	3,226	336	250	−
% Change	−30%	−26%	−42%	−49%	−
Local Depletions (MCM/yr)					
Historical/Warm-only	1,786	629	66	1,090	−
Warm-dry	3,966	1,370	442	2,154	−
% Change	122%	118%	566%	98%	−

[a] Percent change relative to historical in all cases

Adapted from (Connell 2009)

statewide urban demands in 2050. Water conservation includes per capita use reduction from 908 to 837 l/per capita/per day in 2050 with an estimate of 54 million[2] people in CALVIN covered urban areas (Jenkins et al. 2007).

2.1.2 Perturbed hydrology

CALVIN uses 72 years of monthly hydrology (1921–1993) to represent hydrologic variability. Hydrologic processes perturbed for climate change include flows into Central Valley reservoirs (rim inflows), net evaporation rates at reservoirs, groundwater inflows, and net local accretions. The GFDL CM2.1 model with a higher emissions scenario (A2 scenario) was selected for this study. Outputs from the global climate model simulated a warm-dry scenario with 4.5°C increases in annual temperature by the end of the century and variable amounts of decreased precipitation for watersheds across the state (Cayan et al. 2008b). Downscaling the A2 scenario, using bias correction and spatial downscaling (BCSD) (Maurer and Hidalgo 2008), yields estimated temperature and precipitation effects on streamflow for 18 index basins and 21 groundwater basins for a 30-year period centered on 2085. This unpublished data follows methods presented in Maurer (2007) and Maurer and Duffy (2005), and was made available for this study.

A two step process was employed to perturb CALVIN rim inflows, 1) mapping index basin streamflows from a downscaled climate model to CALVIN rim inflows, 2) applied permutation ratios from the index basins to the corresponding historical CALVIN streamflows. Streamflows for select rivers in California are referred to as index basins. Of the 18 available index basins, 13 were used to map to CALVIN rim inflows. From north to south, these include: Smith River at Jedediah Smith State Park, Trinity River at Trinity Reservoir, Sacramento River at Shasta Dam, Feather River at Oroville, American River at Folsom Dam, Cosumnes River at McConnell, Mokelumne River at Pardee, Calaveras River at New Hogan, Stanislaus River at New Melones Dam, Merced River at Pohono Bridge, Tuolumne River at New Don Pedro, San Joaquin River at Millerton Lake, and Kings River at Pine Flat Dam. Statistical analysis, geographic location, and knowledge of hydrological processes characterizing each basin guided assignment of appropriate matches (Connell 2009).

Using GCM-based streamflows for these 18 index basins, permutation ratios capturing the effects of magnitude and timing shifts in streamflows were used to perturb all CALVIN rim inflows for the warm-dry scenario. This method maps hydrologic changes in index basin streamflows to CALVIN's 37 rim inflows producing a new climate change time series with historical hydrologic variability for each rim inflow (Connell 2009; Zhu et al. 2003). The permutation ratio approach captures the hydrologic variability present in the historical record, but does not allow for increasing variability driven by climate change. It also assumes stationarity of mapping such that a single index basin represents the flows at the rim inflows overtime.

For the warm-only hydrology, as with the warm-dry series, permutation ratios were applied to the historical time series to capture the effect of warming, reflecting a shift in the timing of peak flow. The warm-dry time series was then multiplied by the ratio of average historical flows to average warm-dry flows. As a result, the warm-only time series mirrors the timing of the warm-dry climate but preserves historical average annual streamflow volume.

---

[2] Department of Finance 2050 population estimate available at http://www.dof.ca.gov/research/demographic/reports/projections/p-1/

2.1.3 Other climate perturbed hydrologic processes

In addition to rim flows, climate-adjusted hydrologic processes include net reservoir evaporation, groundwater inflows, and net local accretions. Changes in reservoir evaporation were based on an empirical linear relationship derived between historical monthly average net reservoir evaporation rates and monthly average air temperature and precipitation (Zhu et al. 2003). For this study the main drivers for net evaporation rates are temperature and precipitation. For the warm-only scenario, annual volume of precipitation is assumed unchanged; therefore change in precipitation was set to zero and only changes in temperature increased net reservoir evaporation.

Changes in groundwater storage are calculated as changes in deep percolation, obtained from an empirical cubic relationship between precipitation and recharge from the Central Valley Groundwater-Surface Water Model or CVGSM (USBR 1997). This relationship was used to perturb groundwater inflows for the warm-dry scenario. Since estimates of deep percolation depend solely on precipitation in the present study, historical time series of groundwater inflows were not perturbed for the warm-only scenario. As a result, effects from reduced snowpack and earlier melting on groundwater recharge are not represented in the warm-only scenario. However, timing and magnitude of historical and warm-dry scenario time series of groundwater storage were close, suggesting this simplifying approach is appropriate.

Rim inflows (over 70% of valley inflows) enter the Central Valley from the mountain regions outside the major water demand areas, whereas net local accretions enter the valley floor within the major demand areas. Net local accretions combine local accretions and local depletions. Changes in local surface water accretion are affected by changes in deep percolation and precipitation. Changes in these factors from the downscaled global climate model for groundwater basins were used to perturb net local accretions for the warm-dry scenario. Since precipitation was assumed unchanged from historical hydrology, the historical time series for local accretions and depletions were used in the warm-only scenario.

Permutation ratios for the warm-dry scenario generally decrease flow and shift flow timing with earlier snowmelt. Under the warm-dry climate, precipitation decreases across the Central Valley by 27%, a total of 4.7 BCM, (Table 1). This amounts to 5.8 cm/yr less precipitation in the Sacramento and San Joaquin valleys, and 6.1 cm/yr less in the Tulare Basin. Drier conditions also affect rim inflows, net evaporation rates from reservoirs, groundwater inflow, and net local accretions. In all regions, rim inflows and groundwater inflows decrease while evaporation from reservoirs increases. Average rim inflow volumes aggregated across the state decrease 28% from historical levels (Table 1).

3 Results and discussion

3.1 Water supply results

Optimized water deliveries are compared to delivery targets for each urban and agricultural demand area to estimate water scarcity and its costs. The difference between water allocated by CALVIN at each demand site and the target water demand defines the region's water scarcity. Economic loss functions assign a scarcity cost to each region's scarcity. Table 2 shows statewide scarcity volumes and costs in agriculture and urban centers for each economically optimized scenario. The second column also indicates each sector's

Table 2 Statewide average annual water scarcity, scarcity cost, willingness to pay and percent of water deliveries by 2050 (in $2008)

Scenario	Willingness to Pay ($/TCM[1])	Scarcity Cost ($M/yr)	Scarcity (MCM[2]/yr)	Delivery (% of Target)
Historical				
Agriculture	188	201.1	1,075	96.4
Urban	457	46.8	39	99.8
Total		247.9	1,114	
Warm-only				
Agriculture	188	207.0	1,104	96.3
Urban	457	48.5	40	99.7
Total		255.5	1,144	
Warm-dry				
Agriculture	215	1423.0	9,456	68.2
Urban	1043	108.5	134	99.2
Total		1531.5	9,590	

1. TCM is Thousand Cubic Meters

2. MCM is Million Cubic Meters

willingness to pay for additional water, an indication of how water is economically allocated in the system. For this reason, urban demands (with a high willingness to pay) incur little scarcity and the brunt of water scarcity falls on agricultural production, where more senior water rights holders are paid to forego use. This pattern of water scarcity under optimized operations is common in previous CALVIN studies (Draper et al. 2003; Medellin-Azuara et al. 2008; Tanaka et al. 2006). Low willingness to pay in some urban areas indicates that some level of urban water scarcity is economically optimal, likely a result of low water retail prices or high supply costs.

Statewide water scarcity increases by 2.7%, with warm-only conditions compared to the historical climate. In contrast, scarcity under warm-dry conditions is over eight times higher than scarcity with historical conditions (Table 2). Climate warming decreases water deliveries and increases water scarcity, however, drier conditions combined with climate warming proves far more costly. Increases in scarcity costs for warm-only conditions are only 3.1% higher than historical costs, whereas warm-dry scarcity costs are over six times higher than historical costs. Relatively small additional scarcity from the warm-only climate arises due to the ability of large storage reservoirs to adapt to the seasonal shift of runoff. This is in line with classical reservoir operations theory that reservoirs with over-year storage capability are affected much less by seasonal changes in flows (Hazen 1914). Many approaches are available to quantify the over-year storage character of a reservoir (Hoshi et al. 1978; Lettenmaier and Burges 1977; Vogel et al. 1999). All are imperfect. Here we present both the simple ratio of storage capacity to mean annual flow, and the Coefficient of Variation (C_v) as presented in Vogel et al. (1999). C_v is the standard deviation of annual inflow divided by the mean of annual inflows. Seasonal storage characterizes reservoirs with C_v less than .3 (Vogel et al. 1999). This indicates that all reservoirs presented in Table 3 have some over-year storage capability. However, only eight of the presented reservoirs have storage capacities greater than their historical mean annual inflow (MAI). Therefore, most large reservoirs in California have both seasonal and over-year (drought) storage which serves well to buffer the changes in water supply related to warmer conditions.

Climatic Change (2011) 109 (Suppl 1)

Table 3 Percent of years filled and ratio of storage capacity to mean annual inflow (MAI) under each climate scenario, and the Coefficient of Variation for selected surface water facilities

Facility	% Years filled			Storage : Hist/WO MAI	Storage : WD MAI	C_v
	Hist.	WO	WD			
Clair Engle Lake	53	46	22	1.68	1.97	0.46
Shasta Lake	97	92	49	0.80	0.95	0.36
Whiskeytown Lake	100	100	53	0.25	0.34	0.51
Black Butte Lake	99	97	67	0.35	0.46	0.74
Lake Oroville	100	99	88	0.90	1.18	0.48
New Bullards Bar Reservoir	100	99	76	0.56	0.81	0.51
Englebright Lake	100	100	100	0.01	0.02	0.46
Camp Far West Reservoir	92	92	79	0.24	0.36	0.51
Clear Lake & Indian Valley Res	42	38	17	1.14	1.53	0.79
Lake Berryessa	14	8	4	4.15	4.89	0.73
Folsom Lake	100	99	50	0.33	0.45	0.49
Pardee Reservoir	81	88	35	0.29	0.45	0.50
New Hogan Lake	44	40	22	1.94	2.67	0.76
New Melones Reservoir	86	83	3	2.19	3.55	0.50
Turlock Reservoir	75	75	3	0.07	0.10	0.44
Lake Lloyd/Lake Eleanor	40	31	1	0.62	1.04	0.41
Hetch Hetchy Reservoir	51	54	8	0.43	0.88	0.40
New Don Pedro Reservoir	76	83	6	1.29	2.56	0.49
Lake McClure	75	81	7	0.99	1.53	0.57
Eastman Lake	19	19	1	2.02	2.85	0.97
Hensley Lake	39	44	17	1.02	1.55	0.86
Millerton Lake	49	76	29	0.24	0.38	0.50
Pine Flat Reservoir	99	100	13	0.57	1.03	0.54
Lake Kaweah	100	100	54	0.31	0.58	0.63
Lake Success	89	90	74	0.60	0.87	0.79
Lake Isabella	29	44	15	0.79	1.32	0.64
Grant Lake	24	3	0	0.36	0.52	0.36
Long Valley Reservoir	4	0	0	0.90	1.55	0.41

[1] Hist/WO MAI is Mean Annual Inflow for the historical and warm-only scenarios

[2] WD MAI is Mean Annual Inflow for the warm-dry scenario

3.2 Changes in storage operations

A changing climate affects average levels of aggregated statewide surface water storage. Both warm-only and warm-dry hydrologies reduce the average annual peak from historical levels (Fig. 2). Fig. 2 compares average monthly volumes of aggregated statewide surface storage for warm-only and warm-dry climates to the maximum, average, and minimum storage levels with historical hydrology. For example, the historical minimum storage series in Fig. 2 is comprised of the minimum statewide aggregated storage volume for each month in the 72 year modeling period. Both warmer and warmer-drier hydrologies shift the peak average storage earlier in the year. Other studies (Anderson et al. 2008; Madani and Lund

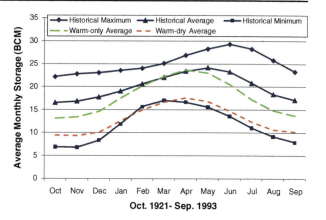

Fig. 2 Monthly aggregated statewide surface water storage (over period of record, 1921–1993) for historical, warm-only, and warm-dry climates

2009) also suggest about a month earlier shift in peak reservoir storage. Storage with warm-only hydrology shadows historical storage levels during February through April, but has lower levels the rest of the year. This suggests a changing pattern of drawdown and refill for reservoirs statewide. This trend is present in the Columbia River Basin as well (Lee et al. 2009). Surface storage with warm-dry conditions is close to the minimum monthly storage conditions from historical hydrology (Fig. 2). Recall that in CALVIN, all reservoirs and other infrastructure exist and are operated for the entire 72 years of hydrology.

Reservoirs in California typically have a regular drawdown-refill cycle, as reservoirs fill in winter and spring (wet season) and are drawn down in the dry season. The amplitude of this cycle represents seasonal reservoir storage. Annual amplitudes of aggregated statewide surface water storage increase for warm-only and warm-dry conditions compared to historical hydrologic conditions (Fig. 2). This amounts to a 28% and 2% increase in average amplitudes compared to historical amplitudes for warm-only and warm-dry, respectively. This greater swing in drawdown-refill storage within the water year (October-September) reflects the value of capturing winter and spring flows for use in the dry season.

For individual reservoirs, larger amplitudes for surface water storage indicate greater activity in reservoir operations to store and release water to meet water demands. This is especially true for warm-only conditions such that most of the storage amplitude ratios of warm-only to historical exceed one (Fig. 3). This reflects the case that reservoirs are refilling in the wet season

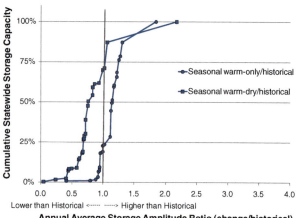

Fig. 3 Comparison of seasonal storage amplitudes for warm-only and warm-dry hydrologies for 26 reservoirs

to higher levels than with historical hydrology, but are still drawn down to meet water demands in the dry season. Figure 3 shows that 75% of the statewide storage capacity under warm-only conditions has larger seasonal storage amplitudes than occurs with historical hydrology. Amplitudes with warm-dry hydrology decrease since the reservoirs do not refill to the extent they do with historical hydrology, even though they are drawn down further. About 70% of statewide storage has smaller seasonal amplitudes with a warm-dry climate (Fig. 3).

3.2.1 Frequency of filling surface water storage facilities

Linear optimization model outputs include shadow prices or Lagrange multipliers for infrastructure capacity constraints and environmental and policy constraints. These values estimate the marginal benefit to the objective function of small changes in each constraint. For example, when a reservoir reaches capacity (an upper-bound constraint), the shadow price is the amount by which the objective function value would improve if the storage capacity was increased by one unit.

Additionally, capacity constraint shadow prices can identify months in which reservoir capacity (or constraint for the conservation pool) is reached. Table 3 shows a selection of surface reservoirs in geographic order north to south and the percent of years they fill based on the number of years (of 72) for which any given month has a negative shadow value. Warm-only hydrology usually increases the frequency of filling and almost always increases the value of increased storage. Pardee Reservoir, Millerton Lake, Lake McClure, Hensley Lake, New Don Pedro Reservoir, Hetch Hetchy Reservoir, Lake Isabella, Lake Success, and Pine Flat fill more frequently with warm-only hydrology than historical, due to earlier and higher peak spring and winter flows. Otherwise, reservoirs often fill the same number of years or slightly less frequently. In contrast to the warm-only response, the frequency of reservoirs reaching their capacity is much less with warm-dry conditions. Many reservoirs will not fill in most years if California's climate tends toward warmer-drier conditions.

The nature of climate change is crucial as to whether additional storage relieves water scarcity and adds flexibility to operating the system or goes unused if the reservoirs are rarely filled. The warm-dry climate decreases the frequency of filling as the reservoirs are now on streams with reduced flow volumes.

3.2.2 Cyclic groundwater storage and conjunctive use

Optimized statewide groundwater storage over the period of record is shown in Fig. 4. Its annual and inter-annual oscillations indicate periods of drawdown and refill seasonally and

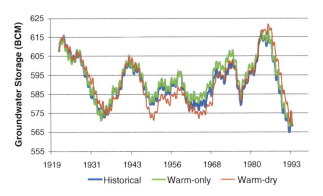

Fig. 4 Monthly Central Valley aggregated groundwater storage over the 72-year period

between drought and wet years, the latter referred to as cyclic storage. Using groundwater in dry periods when surface water is scarce, and banking peak flows to groundwater storage during multi-year wet periods is what cyclic storage is all about. Results reflect economically optimal groundwater storage but do not explicitly suggest artificial groundwater recharge facilities are in operation. CALVIN includes a limited number of artificial recharge facilities: Owens Valley, Mojave, Coachella, MWD, and Santa Clara Valley (Ritzema et al. 2001).

Warm-only storage generally traces historical levels or are slightly higher. The warm-dry scenario drives operational changes that make greater use of groundwater storage than in the historical and warm-only scenarios. Higher highs and lower lows in groundwater storage volume for warm-dry conditions on a decadal time scale reflect this (Fig. 4).

The proportion of deliveries from groundwater are shown in Fig. 5. More deliveries come from groundwater during dry years. Consistent with other analyses, the warm-only scenario shadows results of the historical hydrology, for over-year cyclic groundwater storage. The steeper slope and larger range in percent groundwater use with warm-dry conditions suggests greater coordination of ground and surface waters.

Conjunctive use is defined as the coordinated management of surface and groundwater resources. The CALVIN model economically-optimizes use of groundwater and surface water resources conjunctively to meet urban and agricultural demands. The role of conjunctive use for southern California water supply was previously explored in which the value of conjunctive use programs along the Colorado River Aqueduct, in Coachella Valley, and north of the Tehachapi Mountains were examined, showing that conjunctive use programs, in coordination with water transfers, can add operational flexibility to the system and decrease reliance on water imports (Pulido-Velazquez et al. 2004). Here, conjunctive use within the Central Valley is assessed as a management adaptation to a warm-only and warm-dry climate.

As with scarcity, percent of groundwater use for each region's supply portfolio is comparable between historical and warm-only climate scenarios (Table 4). In general, a larger portion of Tulare's water supply comes from groundwater pumping. In contrast the Sacramento Valley relies mostly on surface water, especially in non-drought years. Only the Sacramento Valley incurs scarcity (about 1%) under historical and warm-only conditions. This occurs because the willingness to pay for water is greater in the San Joaquin Valley and the Tulare Basin. Therefore, to minimize system-wide economic costs, available water preferentially goes to these higher paying demands first and shorts demands in the north (through water transfers). Likewise, under warm-dry conditions when surface water resources are less available, the Sacramento Valley pumps additional groundwater,

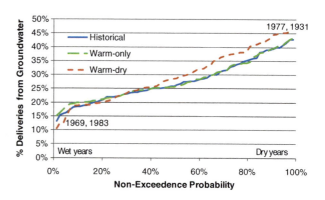

Fig. 5 Annual variability in statewide use of groundwater

Table 4 Water portfolios for Sacramento Valley, San Joaquin Valley, and the Tulare basin for each scenario for drought and non-drought periods

Supply source per region	Drought period			Non-drought period		
	Historical	Warm-only	Warm-dry	Historical	Warm-only	Warm-dry
Sacramento Valley						
Groundwater Pumping	54%	54%	46%	28%	28%	22%
Surface Water	38%	38%	19%	63%	63%	48%
Agricultural Re-Use	7%	7%	5%	7%	7%	5%
Recycling & Desalination	0%	0%	0%	0%	0%	0%
Scarcity	1%	1%	30%	1%	1%	25%
San Joaquin Valley						
Groundwater Pumping	49%	48%	48%	32%	33%	26%
Surface Water	46%	46%	21%	62%	62%	45%
Agricultural Re-Use	5%	5%	3%	5%	5%	3%
Recycling & Desalination	0%	0%	0%	0%	0%	0%
Scarcity	0%	0%	27%	0%	0%	26%
Tulare Basin						
Groundwater Pumping	67%	67%	58%	41%	43%	39%
Surface Water	30%	30%	16%	56%	54%	37%
Agricultural Re-Use	2%	2%	2%	2%	2%	2%
Recycling & Desalination	0%	0%	0%	1%	1%	1%
Scarcity	0%	0%	23%	0%	0%	22%

decreases its surface water use, and incurs a greater percentage of scarcity than the San Joaquin Valley or Tulare Basin (Table 4). In all cases, deliveries from groundwater increase during drought periods when surface water is less available. Groundwater pumping is a much larger contributor to supply in all regions for times of drought compared to non-drought years. This highlights the economic value of using surface supplies during wet periods and increasing use of groundwater supplies during dry periods over the variable hydrologic record.

3.3 Hydropower

In-state hydropower is supplied by two major resources, namely low elevation (lower than 305 m) and high elevation systems. CALVIN includes large, low elevation multi-purpose reservoirs in California which produce 26% of in-state hydropower. For historical hydrology, low-elevation hydropower benefits for an average year are $445 million/yr. With the warm-only climate, hydropower benefits are virtually unchanged (less than 2% decrease). This is mainly due to the fairly large size of most low-elevation reservoirs in California which provide significant operational flexibility for seasonal shifts in streamflow (Madani and Lund 2009). However, this system is directly affected by changes in total annual inflow reduction. Therefore, hydropower generation and revenue losses are much greater when the climate gets drier or streamflows are reduced from increases in evapotranspiration, reducing the "fuel" available to hydropower plants. This amounts to a 4.5% decrease (20 $M/yr decrease) in hydropower benefit for warm-dry conditions in an average year and 6.5% decrease (388 $M/yr from 415 $M/yr for historical) for a dry year.

Climate change effects on the high elevation hydropower system, which includes more than 150 hydropower plants located below low-storage, high-head, single-purpose hydropower reservoirs that produce the rest of the in-state hydropower, have been explored in another study by Madani and Lund (2010), using CALVIN's complementary hydropower model, EBHOM (Energy-Based Hydropower Optimization Model) (Madani and Lund 2009). Their results also indicate 2% revenue reduction from the current 1,791 $M/yr for the warm-only climate. The average high-elevation hydropower revenue reduction for warm-dry conditions was found to be 14%, mainly due to the changes in the runoff pattern, lower snowpack volumes, earlier peaks, and the limited flexibility in operations due to small storage capacity.

4 Limitations

Like any large-scale hydro-economic optimization model, CALVIN has limitations (Harou et al. 2009). With a model of this size and extent, data availability and quality present challenges. It has simple representations of environmental constraints, groundwater storage and flow, and hydropower. There is no inter-annual variability on groundwater pumping costs, by year type (i.e. wet, dry, or normal years), or groundwater in storage (reflection of groundwater elevation). As an optimization model, there are shortcomings of perfect foresight (which can reduce the economic value of optimally-operated storage by a factor of 2–5) (Draper et al. 2003) and it optimistically combines management alternatives. Considerations of water quality (except as it relates to water treatment costs) are not represented. Limitations of CALVIN are more comprehensibly discussed elsewhere (Jenkins et al. 2001; Jenkins et al. 2004).

For this particular study three specific limitations should be mentioned. First, urban water use and scarcity costs are assumed constant for all three climate conditions. This approach does not account for use of additional conservation measures if the climate becomes warmer and drier. Second, urban footprint projections from Landis and Reilly (2002) do not take into account shocks in the housing market that may slowdown conversion from agricultural to urban land uses. Thus agricultural water scarcity may be lower for all three modeled scenarios. Third, a warm-dry hydrology may reduce yields for some crops in California (Adams et al. 2003; Lobell et al. 2007). Similar estimates of changes to urban water use in response to warmer-drier conditions are not available. Thus, water demands for these three scenarios are a static projection towards year 2050; the bias introduced will depend on whether warmer climate increases per capita use, or whether reductions in supply lead to additional urban water conservation. Also, since CALVIN economically optimizes water deliveries based on scarcity cost curves, water allocations are driven by the water demand targets and willingness to pay assigned to agricultural and urban regions. Uncertainty in estimates for these target levels for 2050 introduces uncertainty into the overall model results.

Another limitation relates to the bias implicit in the estimated warm-only hydrology. The permutation ratio's dual representation of warming and reduced runoff may introduce a bias for flows in months where the permutation ratio mostly represents effects of warming. This could overestimate streamflows during these times. Additionally, construction of warm-only stream-flows neglects increased evapotranspiration and decreased soil moisture effects on annual runoff volumes due to increased temperatures. Initial results from WEAP (Yates et al. 2005), a rainfall run-off model, suggest mean annual runoff may decrease as much as 11% with increased climate warming even if historical precipitation is maintained (Null et al. 2009). Warm-only streamflows in this study do not capture these effects. This limitation can be addressed either by using mean annual streamflow ratios by year type or by using a downscaled simulation of hydrology that follows a warm-only pattern, when available. In addition, future work on the response of evaporation rates to changes in precipitation and temperature would be useful in refining the approach used to perturb net reservoir evaporation.

The index mapping approach assumes the paired CALVIN/index basin matches are static. Additionally, perturbation of CALVIN's 72 years of historical hydrology limits climate changed hydrology to the inter-annual variation present in the historical record. Therefore, this study does not explore effects of increased climate variability.

There is also uncertainty in how groundwater will be affected by a changing climate and the warm-only scenario in this study assumes historical conditions for groundwater. Losses in groundwater storage and variable pumping costs could increase variability in groundwater versus surface water use. Despite these limitations, CALVIN provides some insights on promising management alternatives, relative costs, and the system's response to a wide range of hydrologic and other system-wide conditions.

5 Conclusions

California has many management options for adapting and mitigating costs of climate induced changes in water supply volume and timing. Combining these options and employing many of them, water scarcity and its cost as well as storage volumes appear to be more sensitive to reductions in runoff than to temperature increases alone. Temperature rise alone does not tend to increase water shortages greatly if system operations adapt to respond to the timing shift of peak runoff and storage. However, agriculture remains the most vulnerable user to water shortages under all climate scenarios. Total agricultural scarcity statewide increases less than 3% for warm-only compared to historical conditions with annual losses just over $206 million. In contrast, water shortages exceeding 20% of agricultural target deliveries occur for warm-dry conditions, reducing annual agricultural production by over $800 million, as determined by scarcity costs. This amounts to about 2% of California's annual agricultural value.[3]

With recurring wet and dry periods in the hydrologic record, groundwater resources are important in helping meet demands during droughts when surface water is unavailable. Cyclic groundwater storage has a larger role in a warmer and drier climate compared to just a warmer climate. Reoperation adaptations are aided by shifting some drought storage from surface reservoirs to groundwater during multi-year wet periods.

These climate scenarios also affect surface water storage and operations. Surface water operations confirm findings of other studies that reservoir storage levels peak earlier in the year under warmer climates. Surface water storage volumes are lower during summer and statewide storage is exercised more (having greater amplitudes) under warmer-drier conditions. For warm-only conditions, increased storage capacity in wet months may be valuable to help capture increased peak flows in winter months, though recharging peak flows to groundwater is an alternative to surface storage. Under a warmer-drier climate, increasing the system's surface storage capacity may not alleviate climate induced water scarcity. Additional surface storage may not be utilized in most years simply because less water is available to store. Under either scenario, changing reservoir operations in conjunction with a suite of management adaptations

---

[3] Available at http://www.cdfa.ca.gov/statistics/.

(i.e. conjunctive use, water markets, adapted hydropower generation) serves well to reduce water scarcity and economic cost from climate change.

Acknowledgements This work was supported by the California Energy Commission Public Interest Energy Research (PIER) program.

References

Adams RM, Wu J, Houston LL (2003) Climate change and California, appendix IX: the effects of climate change on yields and water use of major California crops. California Energy Commission, Public Interest Energy Research (PIER), Sacramento

Anderson J, Chung F, Anderson M, Brekke L, Easton D, Ejeta M, Peterson R, Snyder R (2008) Progress on incorporating climate change into management of California's water resources. Clim Chang 87:S91–S108

Bedsworth L, Hanak E (2008) Preparing California for a changing climate. Public Policy Institute of California, San Francisco

Cayan DR, Kammerdiener SA, Dettinger MD, Caprio JM, Peterson DH (2001) Changes in the onset of spring in the western United States. Bull Am Meteorol Soc 82(3):399–415

Cayan DR, Luers AL, Franco G, Hanemann M, Croes B, Vine E (2008a) Overview of the California climate change scenarios project. Clim Chang 87:S1–S6

Cayan DR, Maurer EP, Dettinger MD, Tyree M, Hayhoe K (2008b) Climate change scenarios for the California region. Clim Chang 87:S21–S42

Christensen JH, Hewitson B, Busuioc A, Chen A, Gao X, Held I, Jones R, Kolli RK, Kwon W-T, Laprise R, Magaña Rueda V, Mearns L, Menéndez CG, Räisänen J, Rinke A, Sarr A, Whetton P (2007) Regional climate projections. Climate change 2007: the physical science basis. Contribution of Working Group I to the Fourth Assessment Report of the Intergovernmental Panel on Climate Change

Connell CR (2009) Bring the heat but hope for rain: adapting to climate warming in California. Masters Thesis, University of California, Davis

Dettinger MD, Cayan DR (1995) Large-scale atmospheric forcing of recent trends toward early snowmelt runoff in California. J Clim 8(3):606–623

Draper AJ, Jenkins MW, Kirby KW, Lund JR, Howitt RE (2003) Economic-engineering optimization for California water management. J Water Resour Plann Manag 129(3):155–164

Hamlet AF, Mote PW, Clark MP, Lettenmaier DP (2005) Effects of temperature and precipitation variability on snowpack trends in the western United States. J Clim 18(21):4545–4561

Hamlet AF, Mote PW, Clark MP, Lettenmaier DP (2007) Twentieth-century trends in runoff, evapotranspiration, and soil moisture in the western United States. J Clim 20(8):1468–1486

Hanak E, Lund JR (2008) Adapting California's water management to climate change. Public Policy Institute of California

Hanson RT, Dettinger MD (2005) Ground water/surface water responses to global climate simulations, Santa Clara-Calleguas Basin, Ventura, California. J Am Water Resour Assoc 41(3):517–536

Harou JJ, Pulido-Velazquez M, Rosenberg DE, Medellin-Azuara J, Lund JR, Howitt RE (2009) Hydro-economic models: concepts, design, applications, and future prospects. J Hydrol 375:627–643

Hazen A (1914) Storage to be provided in impounding reservoirs for municpal water supply. Trans Am Soc Civ Eng 77:1539–1640

Hoshi K, Burges SJ, Yamaoka I (1978) Reservoir design capacities for various seasonal operational hydrology models. Proc JSCE 273:121–134

Howitt RE (1995) Positive mathematical-programming. Am J Agr Econ 77(2):329–342

Howitt RE, Medellin-Azuara J, MacEwan D (2009) Estimating economic impacts of agricultural yield related changes. California Energy Commission, Sacramento, CA

Jenkins MW, Draper AJ, Lund JR, Howitt RE, Tanaka SK, Ritzema R, Marques GF, Msangi SM, Newlin BD, Van Lienden BJ, Davis MD, Ward a KB (2001) Improving California water management: optimizing value and flexibility. University of California Davis, Davis

Jenkins MW, Lund JR, Howitt RE (2003) Using economic loss functions to value urban water scarcity in California. J Am Water Works Assoc 95(2):58-+

Jenkins MW, Lund JR, Howitt RE, Draper AJ, Msangi SM, Tanaka SK, Ritzema RS, Marques GF (2004) Optimization of California's water supply system: results and insights. J Water Resour Plann Manag-Asce 130(4):271–280

Jenkins MW, Medellin-Azuara J, Lund JR (2007) California urban water demands for year 2050. California Energy Commission, PIER Program. CEC-500-2005-195, Sacramento, California

Knowles N, Cayan DR (2004) Elevational dependence of projected hydrologic changes in the San Francisco Estuary and Watershed. Clim Chang 62:319–336

Landis JD, Reilly M (2002) How we will grow: baseline projections of California's urban footprint through the year 2100. Project Completion Report, Department of City and Regional Planning, Institute of Urban and Regional Development, University of California, Berkeley., Berkeley, CA

Lee SY, Hamlet AF, Fitzgerald CJ, Burges SJ (2009) Optimized flood control in the Columbia River basin for a global warming scenario. J Water Resour Plann Manag-Asce 135(6):440–450

Lettenmaier DP, Burges SJ (1977) Operational approach to preserving skew in hydrologic-models of long-term persistence. Water Resour Res 13(2):281–290

Lobell DB, Cahill KN, Field CB (2007) Historical effects of temperature and precipitation on California crop yields. Clim Chang 81(2):187–203

Lund JR, Hanak E, Fleenor W, Howitt R, Mount J, Moyle P (2007) Envisioning futures for the Sacramento-San Joaquin River Delta. Public Policy Institute of California, San Francisco

Madani K, Lund J (2010) Estimated impacts of climate warming on California's high-elevation hydropower. Clim Chang 102(3):521–538

Madani K, Lund JR (2009) Modeling California's high-elevation hydropower systems in energy units. Water Resour Res 45:W09413. doi:10.1029/2008WR007206

Maurer EP (2007) Uncertainty in hydrologic impacts of climate change in the Sierra Nevada, California, under two emissions scenarios. Clim Chang 82(3–4):309–325

Maurer EP, Duffy PB (2005) Uncertainty in projections of streamflow changes due to climate change in California. Geophys Res Lett 32(3):5

Maurer EP, Hidalgo HG (2008) Utility of daily vs. monthly large-scale climate data: an intercomparison of two statistical downscaling methods. Hydrol Earth Syst Sci 12(2):551–563

Medellin-Azuara J, Harou JJ, Olivares MA, Madani K, Lund JR, Howitt RE, Tanaka SK, Jenkins MW, Zhu T (2008) Adaptability and adaptations of California's water supply system to dry climate warming. Clim Chang 87:S75–S90

Miller NL, Bashford KE, Strem E (2003) Potential impacts of climate change on California hydrology. J Water Resour Plann Manag 39(4):771–784

Null SE, Viers JH, Mount JF (2009) The naked Sierra Nevada: anticipated changes to unimpaired hydrology under climate warming. Unpublished data

O'Hara JK, Georgakakos KR (2008) Quantifying the urban water supply impacts of climate change. Water Resour Manag 22(10):1477–1497

Pulido-Velazquez M, Jenkins MW, Lund JR (2004) Economic values for conjunctive use and water banking in southern California. Water Resour Res 40(3):15

Ritzema RS, Newlin BD, Van Lienden BJ (2001) Appendix H: Infrastructure. CALFED Report: Improving California Water Management: Optimizing Value and Flexibility.

Scibek J, Allen DM (2006) Modeled impacts of predicted climate change on recharge and groundwater levels. *Water Resources Research*, 42(W11405).

Stewart IT, Cayan DR, Dettinger MD (2004) Changes in snowmelt runoff timing in Western North America under a 'Business as Usual' climate change scenario. Clim Chang 62:217–232

Tanaka SK, Zhu TJ, Lund JR, Howitt RE, Jenkins MW, Pulido MA, Tauber M, Ritzema RS, Ferreira IC (2006) Climate warming and water management adaptation for California. Clim Chang 76(3–4):361–387

USBR (1997) Central Valley project improvement act programmatic environmental impact statement. United States Bureau of Reclamation, Sacramento

Vogel RM, Lane M, Ravindiran S, Kirshen P (1999) Storage reservoir behavior in the United States. J Water Resour Plann Manag 125(4):245–254

Yates D, Sieber J, Purkey D, Huber-Lee A (2005) WEAP21—a demand-, priority-, and preference-driven water planning model: Part 1, model characteristics. Water Int 30(4):487–500

Zhu TJ, Jenkins MW, Lund JR (2003) Appendix A: climate change surface and groundwater hydrologies for modeling water supply management. Available at <http://cee.engr.ucdavis.edu/faculty/lund/CALVIN/ReportCEC/AppendixA.pdf>, Department of Civil and Environmental Engineering, University of California- Davis, Davis, California.

Climate change impacts on two high-elevation hydropower systems in California

Sebastian Vicuña · John A. Dracup · Larry Dale

Received: 9 March 2010 / Accepted: 21 September 2011 / Published online: 24 November 2011
© Springer Science+Business Media B.V. 2011

Abstract This paper describes research to estimate the effects of climate change on two high-elevation hydropower systems in California: the Upper American River Project, operated by the Sacramento Municipal Utility District, and the Big Creek system, operated by Southern California Edison. The study builds on a previous model of the Upper American River Project, which is here modified and extended for use to simulate two hydropower systems under various conditions. Future operations of the two high-elevation systems are simulated using climate change scenarios provided for the Second California Assessment. These scenarios suggest reduced precipitation and reduced runoff for both systems, and a shift toward runoff earlier in the year. The change in the hydrograph is somewhat greater for the Upper American River Project system, because its basins lie at a lower elevation. Reduced runoff directly reduces energy generation and revenues from both systems. Because the Upper American River Project system is projected to have greater spills with warmer climate conditions, it also has greater reduction in energy generation and revenues. Both systems continue to meet peak historical power demands in summer under most climate projections. However, if the number of heat waves increases in the late summer (September), reservoir operating strategies may need to be modified.

1 Introduction

As climate change becomes more evident, there is an increasing interest in determining its potential effects on various sectors of the economy. Hydropower

---

S. Vicuña (✉)
Centro Interdisciplinario de Cambio Global, Pontificia Universidad Catolica de Chile, Santiago, Chile
e-mail: svicuna@uc.cl

J. A. Dracup
Department of Civil and Environmental Engineering, University of California, Berkeley, CA, USA

L. Dale
Energy and Technologies Division, Lawrence Berkeley National Laboratory, Berkeley, CA, USA

generation is one of the most critical sectors, because of its effect on global greenhouse gas emissions; it is also especially vulnerable to changes in climate and hydrologic conditions. These issues are of particular importance in California where on average, hydropower constitutes about 18% of in-state energy generation (Aspen Environmental and M-Cubed 2005; http://energyalmanac.ca.gov/electricity/index.html), much of it associated with providing energy during peak demand periods and to help stabilize the grid. Half the energy generation occurs at a high elevation (above 1,000 ft [304.8 m]) in systems that have less storage capacity but higher natural head than their lower-elevation counterparts.

Evaluating the effects of climate change on hydropower energy resources in California has produced a series of research papers. Tanaka et al. (2006) used the CALVIN model hydropower (among others) to estimate the effects of climate change across the state. Vicuña et al. (2008) developed a deterministic linear programming model and used it to estimate the impacts of climate change on the high-elevation Upper American River Project (UARP), located within the Rubicon and Upper American River basins of California and operated by Sacramento Municipal Utility District (SMUD). Madani et al. (2007) and later Madani and Lund (2010) developed an energy-based linear programming model, which was used to project the effects of climate change on a large set of hydropower units in the state.

In this paper we present a modified version of the model described in Vicuña et al. (2008) to estimate the impact of climate change on operations at two locations, the Big Creek system operated by Southern California Edison (SCE) and the UARP. Whereas the UARP system is located in the Northern Sierra Nevada, the Big Creek system is located in the upper San Joaquin river basin in the Southern Sierra Nevada. The Big Creek system has a different configuration and is located at a higher elevation than the UARP. Including both systems in this study provides a more complete picture of the effects of climate change on high-elevation hydropower in California.

Section 2 of this paper introduces the method used to model historical operations of the two high-elevation hydropower systems and to estimate the effects of climate change. Section 3 presents the hydropower systems covered in this study—the UARP and Big Creek systems. Section 4 describes the model used to estimate climate impacts on hydropower, including the calibration of the model. Section 5 covers the estimated effects of climate change on the two systems; the paper is concluded in Section 6.

2 Approach

The major steps involved in our method include, first, calibrating a simulation model of historical operations of hydropower systems, and then using that calibrated model to study the effects of climate change on UARP and Big Creek system operations.

2.1 Simulation of historical operations

To simulate operations of the high-elevation hydropower systems, we constructed an optimization model that uses as input a daily time series of hydrologic conditions at various inflow locations and that gives as output detailed daily operations of each system, including reservoir storage, reservoir release through turbines, and reservoir release though spillways or directly to a river to satisfy streamflow requirements.

The simulation model is based on a sequential multistep linear programming (LP) optimization routine that determines daily operations for the various components of a hydropower system. The model is developed as an "energy-and-storage-driven" model. The objective function of the model includes both a value for electricity generation and a value for reservoir storage as needed to meet unscheduled peak demand, and spinning operations. This differs from the model developed in Vicuña et al. (2008) and tested in Madani et al. (2008) in which energy generation was the only driver in the model. This new addition improves the representation of system operations, as will be explained later.

Using an approach explained in Madani and Lund (2009) and Olivares (2008), we estimate energy revenues using a continuous piecewise linear representation of energy prices derived from a distribution of hourly California Independent System Operator (ISO) prices specific to each month. The electricity price data used in this model is consistent with prices used in other modeling efforts in the 2008 Climate Change Impacts Assessment Project (e.g., Medellín et al. 2008).

Operations of a high-elevation hydropower system are affected by the overall demand for electricity generation, including operational releases for peaking and following real-time loads (SMUD 2001). Based on discussions with UARP system operators, we included a value in the objective function to reflect the benefit of holding reserve storage to meet unscheduled demand peaks and spinning operations. This option value of storage in the model is estimated from monthly historical operations, as explained in greater detail below.

A common characteristic of high-elevation hydropower systems in California and in other western states is that a large amount of their water supplies is derived from melting snowpacks, or snowmelt. Accordingly, in the spring, when inflows are determined by snowmelt quantities, reservoir operators can anticipate reservoir inflows some months ahead based on current assessments of the accumulated winter snowpack. In the late fall, when inflows depend more on current precipitation levels, reservoir operators have difficulty anticipating inflows even a few weeks ahead. At that time, operators can do little more than assume future inflows will equal the historical average.

To represent in the LP model this level of foresight of future inflows and to replicate an operator's level of information, we introduced a series of modifications to the model developed in Vicuña et al. (2008). The optimization routine is performed as a series of overlapping, moving horizons of 12 months. This approach is similar to the one applied by Hooper et al. (1991). Within the 12 month time horizon, the model sets limits on available hydrologic information to approximate the level of uncertainty experienced by reservoir operators. In the short and medium time frames we replicated this level of knowledge by introducing a variable factor that weighs actual and average inflows to estimate information about inflow conditions available to the LP model at given times. The ability of reservoir operators to forecast inflows varies from one month to the other according to assigned weighting factors. For example, in October (fall) operators have little or no ability to forecast inflow in future months. Accordingly, in the model the weighting factor for that month is equal to 0 for future months, reflecting complete uncertainty about future inflows beyond what is known about average historical inflows. Alternatively in May the weighting factor is close to 1 (perfect forecast), because operators have a very high degree of certainty about inflows during the subsequent months of the snowmelt season (June, July, August, September), but 0 afterward.

In order to avoid the need of a storage value at the end of the modeling horizon, we performed the simulation of a given time period as a series of overlapping windows of 12 months, each one starting 5 days apart from each other. Only the first day in each window of optimization is kept as a valid result, discarding the following days (and months) of the window.

2.2 Development of future climate change scenarios

The model described in the previous section was calibrated using historical data from our two case studies to better represent historical reservoir operations, as explained in section 4. The calibrated model is used to estimate the effects of climate change on selected critical system operations variables such as reservoir storage, energy generation and revenues, and system spills, among others. The term spills is used in this work to include both intentional releases by system operators to meet minimum streamflow requirements for the preservation of aquatic life downstream of the turbines and unplanned releases in response to overcapacity issues. In other words spills means all releases not flowing through penstocks upstream of power houses. An important output from the model is the final outflow from the hydropower system downstream into the valley floor. This model output can help identify possible adaptation measures, such as improved flood control, that could be needed under future climate scenarios. With this approach, we can obtain the effects of climate change not only on energy generation and revenues, but also on how the operation of these types of systems may need to be modified under these scenarios.

In order to develop the climate change scenarios, we use direct time series of hydrologic conditions projected by the variable infiltration capacity (VIC) model run for a given climate change scenario. The VIC model is a macroscale, distributed, physically based hydrologic model that balances both surface energy and water over a grid mesh. It has been applied successfully at resolutions ranging from a fraction of a degree to several degrees latitude by longitude. A description of the hydrologic VIC model can be found in Nijssen et al. (1997). The inflows at different locations in the system are developed using a weighting scheme of the runoff output from the VIC model at the surrounding grids over the given hydropower system. This approach, unlike the perturbation ratio approach that is used in Vicuña et al. (2007, 2008) and Madani and Lund (2010), allows the representation of potential changes in climate variability that could be associated with climate change scenarios. However, the approach doesn't allow the explicit inclusion of uncertainty among climate change scenarios, as is treated in Vicuna et al. (2010), and is thus deterministic in terms of representing future climate conditions.

3 Case studies

We applied the optimization model to evaluate high-elevation hydropower operations under historical and climate change conditions. We selected two high-elevation systems in California as case studies: the Upper American River Project (UARP) and the Big Creek system. The UARP, operated by SMUD, is located in the Upper American and Rubicon basins [headwaters at 9,900 ft (3,018 m)]; the Big Creek system, operated by SCE, is located in the upper San Joaquin basin [headwaters at 14,000 ft (4,267 m)]. Figure 1 shows the locations of both systems.

The UARP system was constructed between 1957 and 1985. It includes 11 reservoirs that can impound more than 425 TAF [524 million m^3] of water, eight powerhouses that can generate as much as 688 megawatts (MW) of electricity, and about 28 miles [45 kilometers (km)] of power tunnels and penstocks. The system is located above the CVP's Folsom Dam. Annual runoff to the UARP system is about 1,000 thousand acre-feet (TAF) [1,233 million m^3]. The project currently is undergoing relicensing by the Federal Energy Regulatory Commission (FERC), a process that produces sufficient data to conduct our

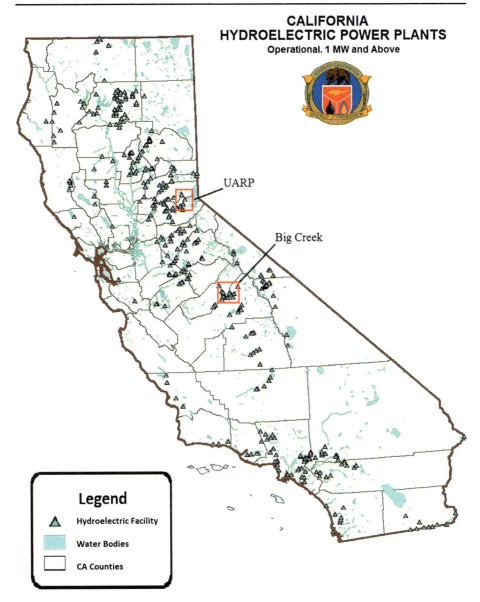

Fig. 1 Locations of Upper American River Project and Big Creek hydropower systems within rectangles. Background map provides the location of all hydropower systems in California (above 1 MW). http://www.energy.ca.gov/hydroelectric/hydro_power_plants.html

case study. Table 1 summarizes the major characteristics of the seven components of the UARP system.

The Big Creek system, one of the largest and oldest high-elevation hydropower systems in California, consists of nine power plants that provide a total generation capacity of approximately 1,000 MW, along with about 54.3 miles [87.4 km] of power tunnels and penstocks. Annual runoff to the Big Creek system is about 1,800 TAF [2,220 million m^3].

Table 1 Components of UARP system

Parameter		Component						
		Loon Lake	Robbs Peak	Union Valley	Jones Fork	Jaybird	Camino	White Rock
Elevation	(feet)	6,410	5,231	4,870	5,450	4,450	2,915	1,850
	(meters)	1,953.8	1,594.4	1,484.4	1,661.2	1,356.4	888.5	563.9
Head	(feet)	1,099	361	420	581	1,535	1,066	856
	(meters)	335.0	110.0	128.0	177.1	467.9	324.9	260.0
Reservoir capacity	(TAF)	79	1.3	277	46	3.3	0.82	17
	(cubic meters)	97,407,000	1,602,900	341,541,000	56,718,000	4,068,900	1,011,060	20,961,000
Reservoir depth	(feet)	165	21	360	52	141	76	186
	(meters)	50.3	6.4	109.7	15.8	43.0	23.2	56.7
Depth/head	(%)	15	6	86	9	9	7	22
Penstock capacity	(cubic feet per second)	999	1,249	1,576	291	1,344	2,099	3,948
	(cubic meters per second)	28.3	35.4	44.6	8.2	38.1	59.4	111.8
Generating capacity	(MW)	82	29	46.7	11.5	144	150	224

The system has a storage capacity of 560 TAF [690 million m^3], distributed among six major reservoirs in the upper San Joaquin River watersheds (see Fig. 1). The Big Creek system provides 90% of SCE's hydropower resources. The system feeds into the Pacific Gas & Electric Company's Kerckhoff Reservoir and hydropower station and CVP's Millerton Lake. Like the UARP system, Big Creek is undergoing FERC relicensing. Table 2 summarizes the major characteristics of the nine components of the system.

The UARP and Big Creek systems are typical high-elevation hydropower systems (Aspen Environmental, M-Cubed 2005). In both cases the elevations of all components exceeds 1,000 ft [304.8 m](higher in the case of the Big Creek system). In both cases the ratio of storage to runoff is also relatively small: 0.4 (425 TAF/1,000 TAF) [524 million m^3/1,233 million m^3] for the UARP and 0.3 (560 TAF/1,800 TAF)[690 million m^3/2,200 million m^3] for the Big Creek system. Finally, except for Union Valley Reservoir in the UARP system, reservoir depth is a small fraction of the head in these systems, a feature characteristic of many high-elevation hydropower systems in the Sierra.

4 Calibration of model using historical operations

This section presents and examines simulation model results for two high-elevation hydropower systems under historical conditions. To simulate historical operations of the UARP and Big Creek systems, we used as input the time series of daily inflows into the reservoirs that comprise each system. We used historical data for 1985–2000 for the UARP system, and 1982–2002 for the Big Creek system; these data sets were provided by SMUD and SCE, respectively. Figure 2 shows the interannual variability of hydrologic conditions for both systems. Both systems show similar patterns of annual runoff, but the runoff for the Big Creek system is higher.

We used the LP optimization model described in Section 2 to simulate historical operations of both systems. Figure 3 illustrates the impact of including the electricity price and the storage option value in the model to simulate historical operations at the UARP system. (For the sake of brevity we present results for the UARP system and note that the results for the Big Creek system are similar.) The figure compares monthly energy generation as a percent of total generation for four cases:

- actual historical operations,
- operations simulated by a version of the model including only the monthly energy price (energy-driven model)
- operations simulated by the energy-driven model developed by Madani et al. (2007)
- operations simulated by the version of the model including the energy price and storage option value (energy and storage driven model)

It is apparent that the model driven purely by energy prices tends to overpredict generation in the winter months and underpredict generation in the spring months. Not surprisingly, this generation pattern closely follows the monthly pattern of ISO energy prices, ignoring among other things the value of holding storage to cover unscheduled spikes in late summer electricity demand. (Figure 3a) (Section 2.1 describes the energy price data set used). On the other hand, the version of the model driven by a mix of energy and storage-option values simulates historical operations quite accurately. The monthly pattern of the calibrated storage value— high in months prior to the peak electricity demand period and low at other times—is consistent with reservoir operating criteria as described to us by electric utility staff.

Springer

Table 2 Components of the Big Creek System

Parameter		Component								
		Portal PH[a]	Big Creek No. 1	Big Creek No. 2	Big Creek No. 2A	Big Creek No. 3	Big Creek No. 4	Big Creek No. 8	East-wood PS[b]	Mammoth Pool PH
Elevation	(feet)	7,643/7,328	6,950	4,810	5,370	2,230	1,403	2,943	6,670	3,330
	(meters)	2,330/2,234	2,118	1,466	1,637	680	428	897	2,033	1,015
Head	(feet)	230	2,131	1,858	2,418	827	418	713	1,338	1,100
	(meters)	70.1	649.5	566.3	737.0	252.1	127.4	217.3	407.8	335.3
Reservoir capacity	(TAF)	125/64	89	0.06	136	0.99	26	0.05	1.5	120
	(cubic meters)	154,125,000/78,912,000	109,737,000	73,980	167,688,000	1,220,670	32,058,000	61,650	1,849,500	147,960,000
Reservoir depth	(feet)	18	95	15	64	46	55	5	38	205
	(meters)	5.5	29.0	4.6	19.5	14.0	16.8	1.5	11.6	62.5
Depth/head	(%)	8	4	1	3	6	13	1	3	19
Penstock capacity	(cfs)	715	678	467	654	3,265	3,437	1,305	2,303	2,542
	(cms)	20.2	19.2	13.2	18.5	92.5	97.3	37.0	65.2	72.0
Generating capacity	(MW)	10.8	88.2	66.5	46.7	174.5	100	75	200	190

[a] The Portal PH (PH=powerhouse) system comprises two reservoirs: Lake Thomas A. Edison and Florence Lake, resulting in two elevations and two capacities for Portal PH.

[b] PS=Power station

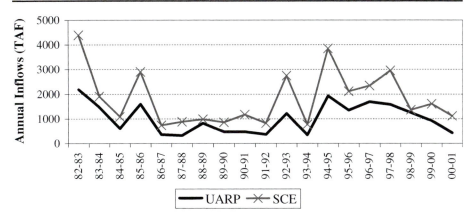

Fig. 2 Annual runoff into the UARP and Big Creek systems

A. Energy Price, ISO spot market ($/MWh)

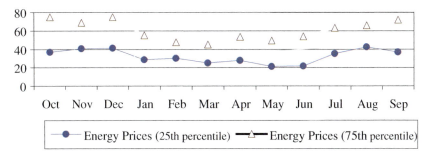

B. Monthly Percent of Annual Generation

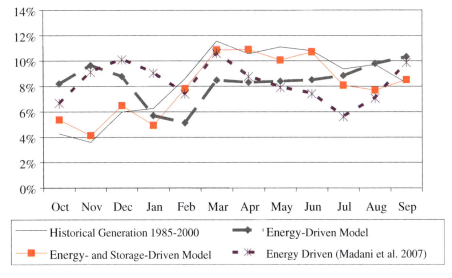

Fig. 3 a Monthly average ISO spot energy prices throughout the year. **b** Historical monthly percent energy generation in the UARP system compared with various model results: energy-driven, energy- and storage-driven

Including the calibration values and running the model configuration that includes the energy- and storage-driven objective function tends to result in larger spills, increased reservoir storage, reduced energy generation, and generation that is shifted to months (especially spring months) that are subject to lower energy prices

5 Results for climate change scenarios

5.1 Forecasts of future hydrologic conditions

The Second California Assessment provides the climate change scenarios to be used in studies performed for the Second Biennial Science Report to the California Climate Action Team. In this particular assessment we used bias correction and spatial (BCSD) downscaled data provided for a set of scenarios presented in Table 3 (Maurer and Hidalgo 2008).

Cayan et al. (2009) provided VIC-simulated runoff under all historic and all future climate scenarios. Inflows to the different components of these hydropower systems were derived by weighting the runoff of this output at different grids in order to represent historic streamflow conditions. Table 4 summarizes the primary characteristics of the hydrologic conditions for four periods: (1) historical (1974–2010), as represented by the scenario based on the VIC model; (2) early twenty-first century (2011–2040); (3) mid twenty-first century (2041–2070); and (4) late twenty-first century (2071–2100). For each variable, we show the actual values for the historical period and averages of all the climate change scenarios listed in Table 3. For future periods we show the average change as a percent change from historical conditions.

As shown in Table 4 both systems show a slight decrease in annual inflow early in the century but the downward trend grows over time. By midcentury, this reduction in runoff is greater. The projections show larger reductions in the Big Creek system than in the UARP. This result is consistent with the projected changes in climatic conditions in Northern and Southern California (California Climate Change Center website at http://meteora.ucsd.edu/cap/).

Table 3 Climate change scenarios used in the analysis

GCM[a]	Emission Scenario[b]	
CNRMCM3[c]	B1	A2
GFDL CM21	B1	A2
NCAR PCM1[d]	B1	A2
MIROC32MED	B1	A2
MPIECHAM5[d]	B1	A2
NCARCCSM3	B1	A2

[a] GCM=General circulation model.

[b] A2 and B1 are two of the future carbon emissions scenarios developed by the Intergovernmental Panel on Climate Change in its *Special Report on Emissions Scenarios*. A2 reflects a future involving relatively high carbon dioxide emissions; B1 reflects a future involving lower carbon emissions. The scenarios considered in this study are intended to bracket the uncertainty among models that predict climatic conditions for California.

A description of the scenarios and GCMs can be found on the California Climate Change Center website at http://meteora.ucsd.edu/cap/. Accessed January 24th 2010.

Table 4 Average hydrologic conditions for various periods for the UARP and Big Creek hydropower systems. Information for future periods is given as a percent over historic conditions

Variable	Period	UARP System	Big Creek System
Annual runoff in million m^3	Historical	1,238	2,209
	Early 21st C.	−0.3%	−1%
	Mid-21st C.	−9.3%	−14.1%
	Late 21st C.	−10.1%	−17.8%
Percent runoff during snowmelt season	Historical	68.9%	87.8%
	Early 21st C.	−8%	−3.4%
	Mid-21st C.	−9.5%	−6.4%
	Late 21st C.	−21.3%	−13.8%
Average 90th percentile flow during winter months in m^3/s	Historical	34	20
	Early 21st C.	12.4%	24.1%
	Mid-21st C.	7.5%	25.9%
	Late 21st C.	24.1%	70.7%

These projections also indicate a shortening of the snowmelt season with decreased summer runoff, consistent with the notion that increasing temperatures will shift streamflow to earlier in the water year. This trend is seen in all scenarios and for all periods (except for one UARP system scenario in the early twenty-first century). Although the direction of these trends is the same for both systems, the change is greater in the UARP system than in the Big Creek (Fig. 4).

Averaging projections for the UARP system under historical hydrologic conditions reveals that almost 70% of total runoff occurs during the typical snowmelt season. This is reduced by an average of more than 20% for all future projections. Under future hydrologic conditions in the basins feeding the UARP system, about 55% of runoff is predicted to occur from March through September. In contrast, under historical conditions for the Big Creek system, snowmelt season runoff is almost 90% of annual runoff, a percentage that falls to slightly more than 75% under future (end-of-century) projections. This difference in hydrologic pattern is attributed to the difference in the elevations of the systems, with the southern Sierra Nevada being higher than the northern Sierra Nevada.

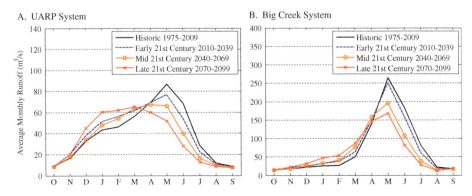

Fig. 4 Changes in monthly hydrologic conditions in the UARP (a) and Big Creek (b) systems

Consistent with the shorter snowmelt season, the projections show a trend toward increasing flows during the winter months. This result is expected given projected increases in temperature, although the effect is somewhat counterbalanced by reduced precipitation and runoff. Even given the largest reductions in flows projected by the end of the century, however, the increase in temperature is enough to increase extreme runoff under all scenarios.

5.2 Impacts of climate change scenarios on hydropower system operations

We ran the calibrated energy- and storage-driven model to evaluate the impacts of the climate change scenarios on operations of the UARP and Big Creek systems. The impact of climate change on these systems is summarized by changes in generation, revenue, potential generation, reservoir spills, and downstream releases (to Folsom and Millerton reservoirs). For each variable, we first show the average value for the historical period. For each future period, we show the average change (the percent change from historical conditions) based on results for all scenarios (Table 5).

On average both systems experience a reduction in energy generated and revenues as a result of climate change. As observed by Vicuña et al. (2008) and others (e.g. Madani and Lund 2010), the model projections suggest that that the reduction in revenues caused by climate change is smaller than the reduction in energy generated. This result reflects the shifting of reservoir storage between months by the model, in order to minimize the value of the loss of generation revenues as a result of dryer conditions and declining inflows.

The impact of dryer hydrologic conditions on hydropower operations is illustrated by comparing changes in runoff to changes in energy generation and revenues across climate scenarios for the end-of-the-century period (Fig. 5).

Table 5 Comparison of outputs for the two hydropower systems for various periods

Variable	Period	UARP System	Big Creek System
Energy generation in GWh[a]/year	Historical	1,976	3,580
	Early 21st C.	−2.1%	−0.6%
	Mid-21st C.	−8.2%	−8%
	Late 21st C.	−12.2%	−10.4%
Energy generation revenues in million $/year	Historical	130	212
	Early 21st C.	−1.3%	0.7%
	Mid-21st C.	−5.8%	−4.7%
	Late 21st C.	−8.5%	−7.8%
Average August power capacity in MW	Historical	654	1,034
	Early 21st C.	−0.2%	−0.1%
	Mid-21st C.	−0.2%	−0.1%
	Late 21st C.	−0.1%	−0.2%
Average Spills[b] in m^3/s	Historical	8	98
	Early 21st C.	19.2%	−0.5%
	Mid-21st C.	−1.8%	−17.3%
	Late 21st C.	10.8%	−21.8%

[a] GWH=gigawatt-hours.

[b] Includes downstream river release plus releases not generating releases

Figure 5 includes the best-fit linear relation for all four data sets, based on the primary properties of the relationships presented in Table 6.

Figure 5 corroborates the positive relationship between changes in energy generation and revenues and changes in system runoff. In addition, the slope of the lines relating energy generation and runoff for both systems is greater than the slope of the lines relating revenues and runoff. This result reflects the flexibility that system operators have to shift electricity generation to periods when energy commandsa higher price. Coefficients for the UARP system are smaller than those for the Big Creek system for both types of relations. Thus, for a given change in annual runoff, the UARP system has a larger reduction in annual generation and revenues than does the Big Creek system. This result is attributed to the hydrologic changes in the UARP system being larger than those in the Big Creek system, due to its lower elevation. The projected hydrology in the UARP system is more "inconvenient," with more water flowing in months when it is not needed (winter months) and less when it is needed (late spring and summer months).

Also, as can be seen in Fig. 5, this change in runoff timing affects UARP system operations by reducing the relative increase in revenues associated with increases in runoff. This can be checked by looking for those markers representing scenarios that have an increase in runoff. It is apparent that the markers representing increased outputs for those scenarios are all situated below the 45° line, indicating that changes in output are smaller than changes in runoff. We see also that the linear relation between revenues and runoff in this system is weaker than the relation between energy generation and runoff in this system. The difference between the coefficients of those relations (see Table 6) is larger than the

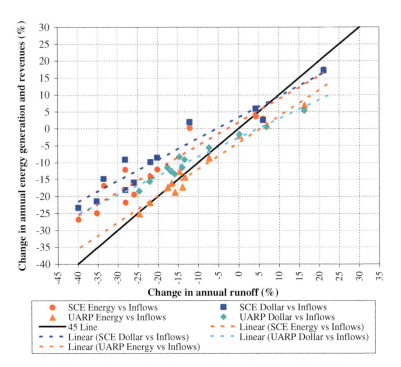

Fig. 5 Comparison between changes in hydrologic conditions and UARP and Big Creek system outputs for the late twenty-first century

Table 6 Summary of linear correlations between changes in annual runoff and system outputs

Variable	UARP System		Big Creek System	
	Energy vs. Runoff	Revenues vs. Runoff	Energy vs. Runoff	Revenues vs. Runoff
R^2	0.97	0.96	0.93	0.92
Coefficient	0.78	0.56	0.69	0.62

difference in the case of the Big Creek system. For the Big Creek system, for most scenarios (except for the driest), the increase in revenues exceeds the increase in runoff.

The increasing inconvenience of the projected hydrology is evidenced by the changes in system operations illustrated in Fig. 6. For three periods (historical, early twenty-first century, and late twenty-first century), the charts in Fig. 6 compare three primary variables in system operations: monthly reservoir releases via turbines, spills, and storage.

Figure 6 shows that both systems have a similar timing of reservoir release through turbines, except the releases tend to be larger in early spring in the UARP system. The two

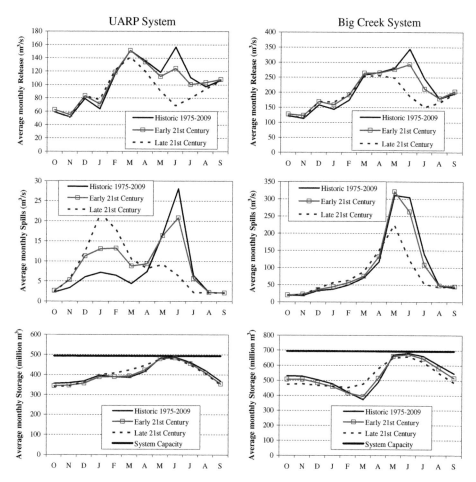

Fig. 6 Summary of simulated operations for UARP and Big Creek systems during three periods

systems' timing of spills differs, however. Under historical conditions, spills in both systems are concentrated in the spring, with the UARP system experiencing relatively more winter spills than Big Creek. By the late twenty-first century the Big Creek system spills are still concentrated in the spring and summer months while the UARP spills have largely shifted to the early months of the year.

On the other hand, climate change has relatively little impact on reservoir storage in both locations. Storage goes up somewhat early in the year (late winter and early spring) and is reduced slightly in the late spring and summer but these changes are small particularly for the UARP system reflecting the smaller reduction in runoff projected for that system. These results reflect the flexibility generally accorded to high-elevation hydropower systems. Without flood control restrictions operators of these systems can effectively respond to changing inflow conditions by filling and emptying reservoirs.

As a result, climate change is not projected to have much impact on the capacity of the two hydropower systems to generate energy when demand is at its peak (summer months). The average system power capacity in August is reduced by a maximum of 0.6%. One reason for this insensitivity is that much of the available head is fixed for these high-elevation systems.

These results apply to the month of August, which in California typically has the largest demand for energy. This demand is determined primarily by the occurrence of heat waves (Miller et al. 2008). As temperatures increase overall due to climate change, energy demands may also rise (Miller et al. 2008; Hayhoe et al. 2004) and extend the time during which reserve storage has a high value. Figure 7 shows the effect that temperature-related changes could have on both the UARP and Big Creek systems. The figure compares the frequency of heat waves (percent days where temperatures exceed the historical 95th percentile), against the frequency of peak generation episodes (when system power capacity generation exceeds 99% of total capacity at a daily timestep) between May through September. This information is presented for the historical and late twenty-first century for

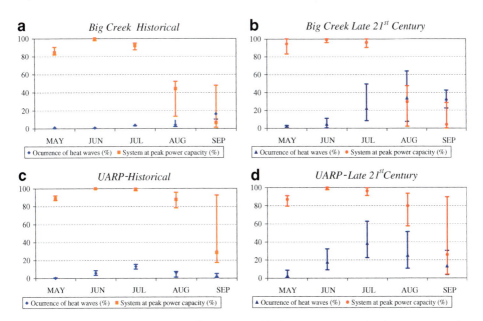

Fig. 7 Comparison between occurrence of heat waves and hydropower system performing at peak capacity

both systems, averaged across all 12 scenarios. Upper and lower bound impacts are presented as well.

It is apparent that historically, both systems have had sufficient capacity and storage to supply peak energy demands (Fig. 7). However over time, increasing heat waves will lead to a point where peak demand periods will exceed peak capacity days on the Big Creek system. The UARP system is expected to meet future August peak demands more easily but September peak demands will exceed system capacity there as well. It is important to mention that a 99% level in system capacity is a stringent constraint. The simulation of system operations is shown to be robust at smaller commitment levels.

In our final analysis, we apply the model to examine the impacts of climate change on flows and potential flooding downstream of the hydropower systems. Releases from the UARP system become inflows to Folsom Reservoir, a key component of the CVP. Folsom Reservoir has a low storage capacity compared to its inflow conditions suggesting that climate change could potentially cause floods affecting the city of Sacramento, which lies downstream of the reservoir. A similar (but less critical) situation occurs for the Big Creek system, because it is upstream of Millerton Dam—another important component of the CVP that supplies water to agricultural users in the Southern San Joaquin Valley.

Climate projections for the UARP system suggests a decrease in average daily outflow and an increase in maximum daily outflow by the end of the Century (Table 7). The finding of a decrease in median daily outflow for most scenarios is consistent with the slight reduction in runoff associated with most climate projections for this region. Of greater concern is the increase in maximum flows projected in our study for all climate scenarios. The increase is sometimes dramatic; on average for all scenarios maximum flows are predicted to increase about 75%, which potentially could lead to flooding in the city of Sacramento.

6 Conclusions

This paper describes the results of our work to estimate the impacts of climate change on two high-elevation hydropower systems in California: SMUD's Upper American River Project (UARP) and SCE's Big Creek System.

Under all the climate change scenarios we examined, both systems would experience a reduction in runoff and an earlier runoff associated with a reduction in precipitation and an increase in temperatures. There are crucial differences in hydrologic impacts between the two systems, however. Because precipitation reduction is projected to be greater in Southern California, the reduction in annual runoff will be greater in the Big Creek System than in the UARP system. On the other hand, the projections show that the change in the hydrograph pattern is larger in the UARP than in the Big Creek system, because the Big Creek System is at a lower elevation than the UARP.

Associated with the projected changes in hydrologic conditions, and considering the average results for future scenarios, both systems should experience a reduction in both energy generation and associated revenues. The greater change in the hydrograph pattern for the UARP system produces a larger impact for a given change in annual runoff than in the Big Creek system.

Finally, although these systems could experience a reduction in both energy generation and associated revenues, the results show that both hydropower systems should remain able to supply peak power demand during the spring and early summer days in both Northern and Southern California. Unless some operating policies are modified, however, both

Table 7 Levels of daily outflow from the UARP system under historical conditions and the late twenty-first century scenario

Flow (m³/s)	Period	Scenario											
		CNRMCM3-A2	GFDL CM21-A2	NCAR PCM1-A2	MIROC32 MED-A2	MPIE CHAM5-A2	NCARCC SM3-A2	CNRM CM3-B1	GFDL CM21-B1	NCAR PCM1-B1	MIROC32 MED-B1	MPIEC HAM5-B1	NCARCC SM3-B1
Median	Historical	26	29	29	29	28	28	29	29	29	28	27	28
	Late 21st C.	24	20	26	20	26	27	25	23	25	27	25	22
99th percentile	Historical	165	198	174	167	164	167	153	172	153	150	161	170
	Late 21st C.	152	142	137	133	151	184	145	142	145	194	144	191
Maximum	Historical	941	2,244	759	1,338	1,222	1,434	873	1,919	873	2,065	1,060	1,211
	Late 21st C.	2,892	3,078	1,468	2,118	2,194	2,178	2,401	2,352	2,401	3,236	1,661	1,923

systems could have difficulty meeting an increased power demand in late summer associated with an increase in the occurrence of heat waves.

Results from this work are similar to previous assessments of climate change impacts on hydropower systems in California. However the level of detail provided by the method used provides more useful information in terms of the likely changes in system operations that would need to be incorporated within climate change adaptation strategies.

Acknowledgments We would like to acknowledge Kevin Cini and Tom Watson from Southern California Edison, Scott Flake and Dudley McFadden from Sacramento Municipal Utility District, and Guido Franco from the California Energy Commission. Funding for this project came from the Public Interest Energy Research (PIER) Program of the California Energy Commission (Project No. MR-07-03A).

This paper was prepared as the result of work sponsored by the California Energy Commission (Energy Commission) and the California Environmental Protection Agency (Cal/EPA). It does not necessarily represent the views of the Energy Commission, Cal/EPA, their employees, or the State of California. The Energy Commission, Cal/EPA, the State of California, their employees, contractors, and subcontractors make no warrant, express or implied, and assume no legal liability for the information in this paper; nor does any party represent that the uses of this information will not infringe upon privately owned rights. This report has not been approved or disapproved by the Energy Commission or Cal/EPA; nor has the Energy Commission or Cal/EPA passed upon the accuracy or adequacy of the information in this paper.

References

Aspen Environmental, M-Cubed (2005) Potential changes in hydropower production from global climate change in California and the western United States. Prepared in support of the 2005 Integrated Energy Policy Report Proceeding. California Energy Commission, Sacramento

Cayan D, Tyree M, Dettinger M, Hidalgo M, Das T, Maurer E, Bromirski P, Graham N, Flick R (2009) Climate change scenarios and sea level rise estimates for the California 2009 climate change scenarios Assessment. California Energy Commission, Sacramento, CEC-500-2009-014-F

Hayhoe K, Cayan DR, Field C, Frumhoff P, Maurer E, Miller N, Moser S, Schneider S, Cahill K, Cleland E, Dale L, Drapek R, Hanemann WM, Kalkstein L, Lenihan J, Lunch C, Neilson R, Sheridan S, Verville J (2004) Emissions pathways, climate change, and impacts on California. Proc Natl Acad Sci 101 (34):12422–12427

Hooper ER, Georgakakos AP, Lettenmaier DP (1991) Optimal stochastic operation of Salt River Project Arizona. ASCE J Water Resour Plann Manag 117(5):566–587

Madani K, Lund JR (2009) Modeling California's high-elevation hydropower systems in energy units. Water Resour Res 45:W09413. doi:10.1029/2008WR007206

Madani K, Lund JR (2010) Estimated impacts of climate warming on California's high/elevation hydropower. Climatic Change. doi:10.1007/s10584-009-9750-8

Madani K, Lund JR, Jenkins MW (2007) Sierra's high elevation hydropower and climate change. Project report for the California Hydropower Reform Coalition. Center for Watershed Sciences, University of California, Davis. August

Madani K, Vicuña S, Lund J, Dracup J, Dale L (2008) Different approaches to study the adaptability of high-elevation hydropower systems to climate change: the case of SMUD's Upper American River Project. World Water and Environmental Resources Congress 2008. May 12–16, 2008, Honolulu, Hawaii

Maurer EP, Hidalgo HG (2008) Utility of daily vs. monthly large-scale climate data: an intercomparison of two statistical downscaling methods. Hydrol Earth Syst Sc 12:551–56

Medellín J, Connell CR, Madani K, Lund J, Howitt, RE (2008) Water management adaptation with climate change. UC Davis.

Miller NL, Hayhoe K, Jin J, Auffhammer M (2008) Climate, extreme heat, and electricity demand in California. J Appl Meteorol Clim 47(6):1834–1844

Nijssen B, Lettenmaier DP, Liang X, Wetzel SW, Wood E (1997) Streamflow simulation for continental-scale basins. Water Resour Res 33(4):711–724

Olivares MA (2008) Optimal hydropower reservoir operation with environmental requirements. PhD dissertation, Department of Civil and Environmental Engineering, University of California – Davis, Davis, CA

Sacramento Municipal Utility District (SMUD) (2001) The Upper American River Project initial information package (IIP). FERC Project No. 2101, SMUD, Sacramento, California. Available at: http://hydrorelicensing. smud.org/docs/docs_iip.htm. Accessed 22 January 2010

Tanaka ST, Zhu T, Lund JR, Howitt RE, Jenkins MW, Pulido MA, Tauber M, Ritzema RS, Ferreira IC (2006) Climate warming and water management adaptation for California. Climatic Change 76(3–4):361–387

Vicuna S, Dracup JA, Lund JR, Dale LL, Maurer EP (2010) Basin scale water systems operations under climate change hydrologic conditions: methodology and case studies. Water Resour Res 46:W04505. doi:10.1029/2009WR007838

Vicuña S, Maurer EP, Joyce B, Dracup JA, Purkey D (2007) The sensitivity of California water resources to climate change scenarios. J Am Water Resour Assoc 43(2):482–498. doi:10.1111/j.1752-1688.2007.00038

Vicuña S, Leonardson R, Dale L, Hanemann M, Dracup J (2008) Climate change impacts on high elevation hydropower generation in California's Sierra Nevada: a case study in the Upper American River. Climatic Change. doi:10.1007/s10584-007-93

The impact of price on residential demand for electricity and natural gas

Felipe Vásquez Lavín · Larry Dale ·
Michael Hanemann · Mithra Moezzi

Received: 9 March 2010 / Accepted: 21 September 2011 / Published online: 24 November 2011
© Springer Science+Business Media B.V. 2011

Abstract Climate change will affect the demand of many resources that households consume, including electricity and natural gas. Although price is considered an effective tool for controlling demand for many resources that households consume, including electricity and natural gas, there is disagreement about the exact magnitude of the price elasticity. Part of the problem is that demand is confounded by block pricing and the interrelated consumption of electricity and natural gas, which prevent easy estimation of price impacts. Block pricing suggests that the purchaser controls the marginal price of a commodity by the quantity purchased, turning price into an endogenous variable. Interrelated consumption indicates that demand for one resource is affected by the price of another. These complications have made difficult the estimation of the price elasticity of demand for resources and consequently the household-level impact of climate change, which will affect resource supplies. This paper evaluates statistical tools for estimating the joint demand for natural gas and electricity when both resources face a block price setting and develops estimates of own and cross price elasticity. We use data from the Federal Residential Energy Consumption Survey, along with utility price data, to estimate the household demand for electricity and natural gas in California as separate commodities. We then use a joint estimation procedure to evaluate the household demand for natural gas and

F. V. Lavín (✉)
Departamento de Economía and Núcleo Científico en Economía Ambiental y de Recursos Naturales, Universidad de Concepción, Concepción, Chile
e-mail: fvasquez@udec.cl

L. Dale
Lawrence Berkeley National Laboratory, University of California, Berkeley, CA, USA
e-mail: lldale@lbl.gov

M. Hanemann
Economics Department, Arizona State University, Tempe, AZ, USA
e-mail: Hanemann@berkeley.edu

M. Moezzi
Ghoulem Research, Mill Valley, CA, USA
e-mail: mmmoezzi@gmail.com

electricity. Finally, we evaluate the degree to which block pricing and interrelated demand affect the price elasticity of demand for the two resources. The paper ends by noting the continuing uncertainty surrounding the use of price to manage household demand for electricity and natural gas.

1 Introduction

Climate change will affect the demand for many of the resources that households consume, including electricity and natural gas. Although price is considered an effective tool for controlling demand for many resources that households consume, including electricity and natural gas, there is disagreement about the exact magnitude of price elasticity. Part of the problem is that demand for resources is confounded by block pricing and interrelated consumption of electricity and natural gas, which prevent easy estimation of price impacts. Block pricing suggests that the purchaser controls the marginal price of a commodity by the quantity purchased, turning price into an endogenous variable. Interrelated consumption indicates that demand for one commodity is affected by the price of another.

These complications have made difficult the estimation of the price elasticity of demand for the three resources and consequently the household-level impact of climate change, which will affect the price of these resources. This paper evaluates statistical tools for estimating the joint demand for natural gas and electricity when both resources face a block price setting using a discrete continuous choice (DCC) model and develops estimates of own and cross price elasticity.[1] First, we use information from the Federal Residential Energy Consumption Survey, along with utility price data, to estimate the household demand for electricity and natural gas, as separate commodities. We then use a joint estimation procedure to evaluate the household demand for natural gas and electricity.

In Section 2 we describe block pricing and linked demand for electricity and natural gas, and then examine how those factors affect the price elasticity of demand for household utility commodities. In Section 3 we summarize estimates found in the literature for the price elasticity of demand for the two resources and note a wide variability among estimates. Section 4 describes the statistical technique we have developed for estimating demand elasticities. The results of our work are presented in Section 5. Section 6 offers conclusions.

1.1 The residential use of natural gas and electricity

Examining average household expenditures for natural gas, water, and electricity in the Western Census Division in 2006 (Table 1), we see that average natural gas and water expenditures are nearly equal, whereas average electricity expenditures are double the other two. Seventy percent of households in the West use some natural gas. About one-third of natural gas expenditures represent water heating, and the rest space heating [Energy Information Administration (EIA) 2004]. Natural gas expenditures for water heating usually are a fairly small part of the household utility budget, although in certain months and seasons they may appear major.

More California households use natural gas (85%) than do those in most other states and virtually all households with access to natural gas use it for water heating. Eighty-five percent of California's residential water heaters are gas-fired; only 11% are electric (EIA

---

[1] The endogeneity problem can also be solved by means of a natural experiment (Ito 2010).

Climatic Change (2011) 109 (Suppl 1)

Table 1 Average 2006 annual household income and expenditures for commodities for the Western Census Division

Category	Household average (2006 US Dollars)
After-tax income	63,574
Expenditures	57,486
Utilities, fuels, and public services	3,101
Electricity	1,042
Natural gas	423
Telephone	1,081
Water and other public services	485
Gasoline and motor oil	2,382

Consumer Expenditures Survey 2006, Bureau of Labor Statistics

2008).[2] Among California households that have gas water heating, that end use accounts for 41%, on average, of a household's natural gas use. Natural gas prices for the residential sector generally are structured as a two-tier system, with a baseline price and an excess-of-baseline price. Natural gas utilities buy gas on the national market and are allowed to pass purchase costs onto their customers. While prices in both tiers vary from month to month, the quantities designated as baseline stay the same within a season. Consumers do not necessarily know in advance the price per unit they will charged for natural gas.

Electricity consumption generally is subject to block rates as well, as described in the following section.

1.2 Block pricing

As noted above, markets for electricity and natural gas often are characterized by block pricing, which complicates the estimation of their consumer demand functions and therefore the calculation of both price and income elasticities. Block pricing introduces a nonlinearity to the budget constraint (a piecewise-linear budget constraint), which, when developing demand functions, makes the price that people pay for the marginal unit of consumption an endogenous variable. From the point of view of social science, there is debate about the prices consumers actually perceive (Nieswiadomy and Molina 1991).

In a simple uniform price system, people decide how much to purchase given a constant price for the commodity. With a piecewise-linear budget constraint, however, consumer actions determine not only the quantity purchased, but also the price paid at the margin. The fact that consumer actions represent a joint decision regarding price and quantity must be taken into account when estimating demand for such commodities.

Figure 1 illustrates the simplest case of an increasing block pricing system that comprises two tiers. Here, the initial price for the marginal unit is p_1, the price paid for each unit consumed until the quantity reaches W*. Beyond that quantity, the marginal price increases to p_2. A consumer whose demand function resembles D1 in Fig. 1 will consume W^1 and pay p_1 for each unit. In contrast, a consumer whose demand function resembles D2

---

[2] These percentages were based on preliminary tables for the 2005 Residential Energy Consumption Survey (EIA 2008).

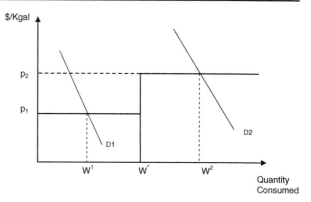

Fig. 1 Demand functions D1 and D2 with block pricing

in Fig. 1 will consume a quantity equal to W^2, paying p1 for each unit below W^*, and paying p2 for each unit above that threshold. Extension to additional tiers is straightforward.

1.3 Linked demand for electricity and natural gas

In the residential sector, the consumption of natural gas, electricity, and water are interrelated. For example, different models of some appliances may use either natural gas or electricity. Home heating, clothes washing, and cooking can be performed using natural gas or electricity—the consumer chooses which, depending on prices and other factors. In the long run, consumers can switch product models to take advantage of changes in fuel prices. Alternatively, natural gas and water generally are considered complementary goods, because consumers often use natural gas to heat household water. The dual nature of water-energy services means that traditional incentives for conservation, including changes in the price of water, or, independently, changes in the price of energy, will have attenuated effects.

Some of the highest embedded energy use occurs in urban water consumption, particularly residential water that is heated and used in dishwashers, clothes washers, and bathing. About 90% of all natural gas consumed directly for interior residential water heating goes to those three end uses (Cohen et al. 2004). Although this finding has important implications for evaluating the price elasticity of demand for water, the sparcity of data regarding the linked use of energy and water limits our current focus to the joint elasticity of demand for natural gas and electricity.

2 Elasticity of demand for natural gas and electricity—the literature

Early analysis of the price elasticity of demand for consumption goods emphasized dynamic elements and habit formation (Houthakker and Taylor 1970). Using Bayesian procedures and aggregated data at a State level, Maddala et al. (1997) estimated long and short run price elasticities of demand for electricity and natural gas. The recent literature more frequently employs a discrete continuous choice (DCC) model to estimate household price elasticity for a single commodity that is subject to increasing block pricing. DCC models long have been used to estimate demand for residential electricity use [e.g., Reiss and White (2005)]. The use of DCC models to estimate the demand for water is more recent and less common. Some examples are Hewitt and Hanemann (1995) and Olmstead et al. (2007). Recently, some papers have focused on identifying whether people are more sensitive to marginal or average prices on the demand of electricity, and its relationship with empirical evidence of bunching in demand (Borenstein 2010; Ito 2010). We found no studies that use a DCC model to estimate the demand for two goods together and cross price elasticities when both are subject to block pricing.

Two key points emerge from the literature review. First, for all three commodities, the average demand elasticity cited is similar, ranging from −0.2 to −0.5. Second, as discussed further below, there is a wide variation in elasticities around the mean, depending on the study's method, geographic area, time period, and data factors.

2.1 Natural gas

Few econometric studies of natural gas utilize DCC modeling. Bohi and Zimmerman (1984) reported consensus price elasticities of −0.2 in the short term and −0.3 in the long term for natural gas. Compared to electricity, natural gas provides fewer and less obvious opportunities for consumers to reduce their demand in response to price. The use of natural gas in the home (for space heating, water heating, and cooking) is a necessity subject to few dichotomous choices (use/do not use). There are more discrete choices and varied options for cognitively and operationally controlling electricity use (e.g., turning off some lights or using fewer electric appliances).Furthermore, consumer-oriented energy conservation programs overall focus more on electricity than on natural gas. During the California energy crisis of 2000–2001, for example, demand response programs and rate options were developed for electricity, but not natural gas. On the other hand, natural gas bills may vary more from month to month than do electricity bills, especially where gas provides space heating.

A recent study sponsored by the American Gas Association (Joutz and Trost 2007) found an elasticity on the price of natural gas of −0.18 on average nationwide, based only on post-2000 data. The price elasticity in the Western Census Division was lower, at −0.12, with California having particularly low price elasticity of demand for natural gas. The study noted, in addition, a 1% "natural" annual decline in residential consumption of natural gas due to turnover toward more efficient appliances, which is occurring even in the absence of changes in the price of natural gas.

2.2 Electricity

Most recent studies of electricity demand use DCC modeling. A meta-analysis of price elasticity for residential electricity cites 36 studies that provide estimates of long- or short-run income and price elasticity (Espey and Espey 2004). The study results, based on data from various geographic locations and times, encompassed a broad range of elasticities. The mean of short-run price elasticity estimates was −0.35, ranging from −2.01 to −0.004; the mean of long-run price elasticity estimates was −0.85, ranging from −2.25 to −0.04.

Bohi and Zimmerman's (1984) comprehensive review of studies on energy demand found the consensus estimates for residential electricity price elasticities to be −0.2 in the short run and −0.7 in the long run. Garcia-Cerrutti (2000) estimated price elasticities for residential electricity and natural gas demand at the county level in California. For residential electricity, the estimate of the mean was −0.17, with a minimum of −0.79 and a maximum of 0.01. Finally, in a recent study using a similar DCC model and data set, Reiss and White (2005) estimated the price elasticity for residential electricity to be −0.39.

3 Estimating price elasticity of demand for natural gas and electricity—statistical techniques

Complex rate structures and linked demands complicate the estimation of household demand for electricity, natural gas, and water. Below we discuss the problem of estimating household demand in cases of increasing or decreasing block rates. We then present an approach to estimating the joint demand for two goods that are both subject to tiered rates.

3.1 Block pricing

A demand model for estimating the price elasticity of a commodity that is subject to block pricing must take into account the endogeneity of price and it must provide accurate estimates of price elasticity. We believe that, given enough data, the preferred model to account for block pricing is the discrete continuous choice (DCC) model. This model incorporates two decision points: a price decision, at which people choose their rate/consumption block, and a quantity decision, at which people decide how much to consume in that block. Because there are a discrete number of blocks, the first decision is discrete. The second decision, about consumption within a block, is continuous.

For a given block pricing system, DCC considers the probability that an individual chooses a block and consumes a given quantity within that block. Additionally, DCC considers the probability that an individual decides to consume at any of the thresholds (termed *kinks*) that separate two blocks. In this way, the statistical model explains the discrete and continuous decisions that underlie individuals' behavior.

For a single demand model we use a log-log demand function based on Hewitt and Hanemann (1995):

$$\ln w = Z\delta + \beta \ln p + \lambda \ln m + \eta + \varepsilon \tag{1}$$

where w is the observed level of consumption, Z is a vector of individual and weather-related characteristics, p is the marginal price, m is the individual's income, and δ, β and λ are parameters to be estimated. This model has a two-error structure, whereby η represents the unexplained heterogeneity of preferences among consumers, and ε represents additional heterogeneity that is unexplained to the econometrician. If we call w^* the optimal level of consumption and assume that the two error terms are independent and normally distributed with mean zero and variance of σ_η and σ_ε, the contribution to the likelihood function of an individual consuming at a given level in a uniform price system would be:

$$l_i = \ln \left(\frac{1}{\sqrt{2\pi}(\sigma_\eta + \sigma_\varepsilon)} \exp \left(-\frac{1}{2} \left(\frac{\ln w - \ln w^*}{\sigma_\eta + \sigma_\varepsilon} \right)^2 \right) \right) \tag{2}$$

The likelihood function in the case of increasing block prices is more complicated. If there are K blocks, then there exist K - 1 kinks in the budget set; let's call them w_k. For each block there exists a virtual income $\tilde{y}_k = y + d_k$ that enables a consumer to reach a level of consumption inside that block. In order to define d_k, and therefore the virtual income, we must consider that people pay different marginal prices for various units of consumption. Referring to Fig. 1 again, people pay only p_1 for units below W^*. There is an implicit benefit of $p_2 - p_1$ for each unit consumed in this range. Higher levels of consumption are

Climatic Change (2011) 109 (Suppl 1)

associated with higher marginal prices, in essence providing a subsidy for the consumption of the initial, lower-priced quantities. Then,

$$d_k = \begin{cases} 0 & \text{if } k = 1 \\ \sum_{j=1}^{k-1} (p_{j+1} - p_j) w_k & \text{if } k > 1 \end{cases} \tag{3}$$

To build the likelihood function, we must consider the probability that consumption lies in each of the ranges defined by the K blocks and $K - 1$ kinks. Following Olmstead et al. (2007), the likelihood function is:

$$\ln \left[\sum_{k=1}^{K} \left(\frac{1}{\sqrt{2\pi}} \frac{\exp(-s_k^2/2)}{\sigma_v} \right) (\Phi(r_k) - \Phi(n_k)) + \sum_{k=1}^{K-1} \left(\frac{1}{\sqrt{2\pi}} \frac{\exp(-u_k^2/2)}{\sigma_\varepsilon} \right) (\Phi(m_k) - \Phi(t_k)) \right] \tag{4}$$

with:

$$v = \eta + \varepsilon \qquad t_k = \left(\ln w_k - \ln w_k^* \right) / \sigma_\eta$$

$$\rho = corr(v, \rho), \qquad r_k = (t_k - \rho s_k) / \sqrt{1 - \rho^2}$$

$$s_k = \left(\ln w_i - \ln w_k^* \right) / \sigma_v \qquad m_k = \left(\ln w_k - \ln w_{k+1}^* \right) / \sigma_\eta$$

$$u_k = (\ln w_i - \ln w_k) / \sigma_\varepsilon \qquad n_k = (m_{k-1} - \rho s_k) / \sqrt{1 - \rho^2}$$

where w_k^* is the optimal consumption in block K, and w_k is consumption at kink point K.

3.2 Estimating price elasticity

Price elasticity is defined as the percent change in the quantity of a good or service consumed, divided by the percent change in its price. For the simple case of a uniform price, the demand model provides a direct way to estimate price and income elasticity. For example, given the demand function in (1) the price elasticity is $\partial \ln w / \partial \ln p = \beta$.

Estimating elasticity is more complicated for a DCC model, however, because the calculation must consider a change in the entire price structure, e.g., a 1% increase in all prices in an increasing block system. Using the DCC model to calculate elasticity starts with estimating expected consumption based on known values for price, income, and other explanatory variables. If expected consumption is $W = E(w)$, price elasticity is the percent change in W following a 1% rise in all prices in the tiered system. That is:

$$\omega = \frac{\partial W}{\partial \theta} \frac{\theta}{W} \tag{5}$$

where θ represents a 1% change in the price structure. The formula for calculating elasticity in a DCC model is a function of all model parameters, including stochastic components (Appendix A).

3.3 Addressing linked consumption

In the case of two or more commodities that both are subject to increasing block pricing, Lee and Pitt (1987) demonstrate the use of dual theory and virtual prices to estimate price elasticity of demand. Virtual prices are those that would cause people to consume optimally at a predefined level the commodities under analysis. Virtual prices can be determined using the first-order conditions of the maximization of a utility function. In this paper we extend the approach of Olmstead et al. to the case of two goods with two error terms each. Figure 2 illustrates the situation involving two goods, X_1 and X_2, each of which is subject to a two-tier rate structure.

Several consumption patterns are possible in the situation illustrated in Fig. 2 (Lee and Pitt 1987). For example, consumption may occur below both kink points, as represented by line D1. Alternatively, consumption may occur below the kink point in one market and above the kink point in the other as represented by line D1 in one graph and D2 in the other.

Each consumption pattern is represented by a different set of virtual prices and a different formulation of the demand curve. There are nine different possible combinations or patterns of consumption, each of them characterized by a distribution function that can be derived from a general multivariate normal distribution. The statistical technique the authors used to estimate linked consumption is described in Appendices B and C.

4 Price elasticity of demand for natural gas and electricity—results

After describing the data available on consumption of natural gas and electricity, we provide preliminary estimates of price elasticity of demand for the individual and joint consumption of the two resources.

4.1 Natural gas and electricity

We obtained household-level data on gas and electricity consumption from microdata files of the Residential Energy Consumption Survey (RECS), performed in 1993 and 1997 by the Energy Information Administration (see Table 2). The RECS microdata identify energy-consuming assets and annual energy consumption for a statistically

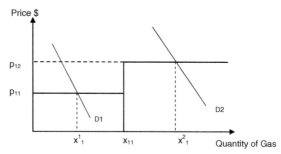
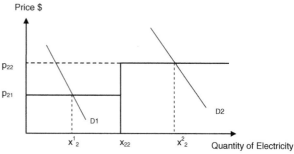

Fig. 2 Linked consumption of two goods subject to increasing block pricing

Table 2 Descriptive statistics for data used in the price elasticity of demand model for natural gas and electricity

Variable	Description	Mean	Standard deviation	Min.	Max.
Cooling	Cooling degree days to base 65	1,063.65	555.99	16	**4,432**
Heating	Heating degree days to base 95.	1,784	977.34	517	**6,271**
Members	Number of members in household	3.16	1.74	1	**12**
Bath	Number of complete bathrooms	1.63	.71	1	**5**
Income	Annual Income: 1=less than $3,000; 25=more than $100,000	17.27	6.54	1	**25**
Own	Categorical home ownership variable. 1=owns, 2=rents, 3=rent is free	1.39	.49	1	**3**
Daily elec. consump.	Daily consumption in kilowatt-hours	16.69	9.30	1.46	**54.08**
Daily gas consump.	Daily consumption in therms	1.45	.78	0.12	**6.93**
Hispanic	Dummy for head of household is of Hispanic origin	0.24	0.43	0	**1**

representative sample of U.S. households, including a subset sample identified as *California households*. The RECS microdata do not reveal tariffs directly. To determine the prices consumers paid for natural gas and electricity in 1993 and 1997, we used the database constructed by Reiss and White (2005), for which each individual in the RECS was matched to a National Oceanic and Atmospheric Administration (NOAA) weather station. Using geographic information system (GIS) techniques, each weather station was assigned a ZIP code and a city, then matched to its associated natural gas and electricity provider(s). For natural gas, data for the areas supplied by providers, as well as prices for 1993 and 1997, were obtained from the websites of the providers and by contacting utility personnel. Data on electricity prices were obtained from the database provided by Reiss and White (2005). Because natural gas prices were obtained for single-family households only, we discarded records of consumers who reported living in condominiums or multi-unit houses, as well as those who did not pay for utilities. Households without natural gas service also were omitted from the analysis. Some of the ranges or definitions for socioeconomic variables differ for the two years of RECS. If possible, we transformed the differences to render the answers comparable; otherwise the data were discarded. Given that RECS reports annual energy consumption, we were able to calculate average natural gas prices for each household (Appendix D).

Tables 3 and 4 present our analytical results using our DCC model for price elasticity of demand for natural gas and electricity. The results are consistent with expectations: price effects are negative and significant; the income effect is positive, although not significant for natural gas. Family size, race, ownership, and house area are all significant explanatory variables in both applications.

Using the results presented in Table 4 and the equations given in Section 4.1, we found a natural gas price elasticity of −0.11 in the log-log model, with a mean expected consumption $(E(X))$ equal to 1.46, which is only 2% higher than the true (observed) mean. For a linear model the elasticity is higher (−0.41) and the prediction of consumption is even more precise. The elasticity for electricity is equal to −0.28 in the log-log model and -.71 in the linear model, and the prediction is accurate in both estimations.

The sign and significance of the coefficients associated with heating and cooling degree days in our regressions indicate that global warming—to the extent it results in

Table 3 Model results for price elasticity of demand for electricity, linear and log-log models

Electricity

Parameters	log-log demand Model			linear demand Model		
	Estimates	Std. err.	Est./s.e.	Estimates	Std. err.	Est./s.e.
Constant	2.4613	0.3597	6.843	10.4922	3.0597	3.429
Cooling ('000)	0.1697	0.0385	4.413	2.5263	0.6831	3.699
Heating ('000)	0.0904	0.0275	3.294	1.0697	0.4885	2.19
Members	0.0721	0.0124	5.793	1.0055	0.2214	4.542
Hispanic	−0.0384	0.0527	−0.728	−0.7944	0.9409	−0.844
Bath	0.1134	0.0393	2.888	2.0036	0.7023	2.853
Own	0.2131	0.0479	4.453	2.9153	0.8483	3.437
Area	12.8712	3.0443	4.228	2.1953	0.5423	4.048
Price	−0.2845	0.1368	−2.079	−1.0605	0.2158	−4.915
Income	0.1402	0.0274	5.113	8.5017	1.6193	5.25
S_N	−0.4166	0.096	−4.339	8.0296	0.2965	27.077
S_E	0.207	0.1846	1.121	1.9895	0.4292	4.635
Mean log-likelihood	−0.6307			−3.44218		
N	590			590		
elasticity	−0.28			−0.72		
mean expected value	17.42972			17.14		
mean observed value	17.1632			17.1632		
% of error	1.55284			−0.11896		

Income has been divided by 10,000 and cooling and heating by 1,000 to help convergence

generally increased temperatures in the state—will increase the demand for electricity and decrease the demand for natural gas. This effect is to be expected, because electricity is used primarily for residential cooling, and natural gas primarily for residential heating.

4.2 Joint demand for natural gas and electricity

We utilized RECS data to estimate the linked demand for natural gas and electricity. The model results are presented separately, using one equation for natural gas demand and one for electricity (Table 5). We were able to estimate only the linear model for this case.

To speed convergence of the model, the income value is divided by 10,000, the cooling and heating degree day values are divided by 1,000, the price units are divided by 10 and the quantity units are also divided by 10.

All the coefficients associated with the variables in Table 5 have the expected signs. The own price coefficients in this model are both negative but only the coefficient for natural gas is significant. The income effect coefficients are also negative in both cases and significant for natural gas but not for electricity. The cross-price effects in both equations are positive, indicating that natural gas and electricity are substitute goods. This result is expected, because in the long run natural gas and electricity are interchangeable for cooking, heating, washing, and other home uses. The cross-price effect of electricity on gas demand (.85) is positive and significant which, following the symmetry

Climatic Change (2011) 109 (Suppl 1)

Table 4 Model results for price elasticity of demand for gas, linear and log-log models

Gas

Parameters	log-log demand Model			linear demand Model		
	Estimates	Std. err.	Est./s.e.	Estimates	Std. err.	Est./s.e.
Constant	−0.4775	0.1066	−4.478	1.1115	0.2842	3.91
Cooling ('000)	−0.0905	0.0377	−2.4	−0.1166	0.0543	−2.148
Heating ('000)	0.0819	0.025	3.283	0.1731	0.0358	4.833
Members	0.0755	0.0117	6.465	0.0863	0.0169	5.118
Hispanic	−0.1557	0.0489	−3.185	−0.1462	0.0702	−2.084
Bath	0.0599	0.0387	1.545	0.1363	0.0562	2.426
Own	0.1882	0.043	4.381	0.2048	0.0632	3.243
Area	12.8188	3.0072	4.263	16.3474	4.2797	3.82
Price	−0.1106	0.049	−2.255	−1.1357	0.3986	−2.849
Income	0.0175	0.0254	0.688	0.1858	0.1231	1.51
S_N	0.4969	0.0139	35.645	−0.6371	0.036	−17.675
S_E	0.0135	0.0061	2.193	−0.2713	0.0661	−4.104
Mean log-likelihood	−0.69233			−0.9461		
N	717			717		
elasticity	−0.11			−0.41		
mean expected value	1.465125			1.433403		
mean observed value	1.433017			1.433017		
% of error	2.240586			0.026933		

condition,[3] implies a cross-price effect of gas on electricity demand that is also positive and significant.

At the means for price and consumption, the own price elasticity of demand implied by these coefficients is −0.007 for electricity and −0.15 for gas. The cross-price elasticity of gas price on electricity is approximately 0.07 and of electricity price on gas is approximately 0.03. These relatively small cross-price elasticity estimates reflect conditions in the natural gas and electricity markets in California, which limit consumer shifting between energy sources. The shifting that occurs is largely limited to purchases of household appliances, e.g., clothes dryers, home heaters, cooking equipment—that use both energy sources and that take place relatively infrequently.

5 Conclusions

The single-commodity price elasticities calculated in this paper for electricity and natural gas are generally consistent with estimates of price elasticity for these commodities reported in the literature. The preliminary estimate of the price elasticity

---

[3] The symmetry condition refers to the statistical interdependence of cross-price elasticities. In this case, the value of the electricity cross-price elasticity on natural gas is constrained by the value of the natural gas price elasticity on electricity.

Table 5 Estimation of linked demand for natural gas and electricity: linear model

	Electricity			Natural gas		
	Estimates	Std. Err.	Est./s.e.	Estimates	Std. Err.	Est./s.e.
Constant	0.6167	0.0292	21.12	0.1542	0.0062	24.87
Cooling ('000)	−1.3714	1.3842	−0.99	−0.011	0.0776	−0.14
Heating ('000)	0.6502	0.618	1.05	0.2412	0.0619	3.90
Electricity price	−0.9741	1.3968	−0.70	0.8509	0.1282	6.64
Natural gas price	-	-	-	−0.326	0.0548	−5.95
Income	−0.7214	0.7317	−0.99	−0.0562	0.034	−1.65
S_N	0.0042	0.0003	14.00	0.0713	0.0013	54.85
S_E	0.9346	0.0298	31.36	0.8077	0.0325	24.85
RHO	0.5733	0.0175	32.76			

of demand for natural gas (−0.11) is somewhat lower than estimates from other studies. Our preliminary estimate of price elasticity of demand for electricity (−.28) is also slightly less than those reported elsewhere. These results suggest that price may be a less effective policy tool for managing these resources than many resource managers may believe.

Our analysis of the joint demand for natural gas and electricity indicates the existence of small, positive and significant cross-price elasticity between these products. This finding is important for several reasons. First, although the cross-price effects are small, they do exist and call into question the accuracy of single-resource demand models. As substitute goods, high natural gas prices during the period of analysis in the early 1990s contributed to a small part of the electricity use that occurred during this period and may have distorted the single-equation estimates of the price elasticity of demand. Although the effect is likely very small in California, in other regions, where shifts between electricity and natural gas may be more common, the effect may be larger.

The importance of joint resource modeling may also be larger when dealing with other, more closely linked resources. For example, natural gas and water are consumed as joint products in California households, with natural gas used to heat water for indoor bathing and washing. In this case, the rise in the price of natural gas that occurred in the last decade increased the price of hot water, and almost certainly affected the quantity of indoor water consumption. This linkage suggests that single-resource models of the demand for water in the past were biased and likely overestimated the price elasticity of demand for water. In future, more accurate estimates of the price elasticity of demand for water may be possible using the econometric technique introduced in this paper.

Finally, it is necessary to remind ourselves of the general uncertainty surrounding estimates of the price elasticity of demand for household commodities such as electricity, natural gas, and water, as was also noted in Bernstein and Griffin (2006). Elasticity is less a "natural law" than a description of the dynamic relationships between price and consumption. This relationship changes depending on time, location, resource, price structure, and model specification. Differences in methodologies and type of data affect the degree to which results from different studies are comparable. In this and most similar estimates of price elasticity, statistical precision is low and uncertainties are high.

Acknowledgements We thank Guido Franco of the California Energy Commission for his generous support of this project.

Appendix A. Price elasticity of demand in discrete continuous choice models

To understand the estimation of price elasticity of demand in a discrete continuous choice model, consider the log-log demand function, which can be written as:

$$\ln w = Z\delta + \beta \ln p + \lambda \ln m + \eta + \varepsilon$$
$$w = w(p_k, y + d_k)e^\eta e^\varepsilon \tag{A-1}$$

Where $w(p_k, y + d_k) = \exp\{Z\delta + \beta \ln p + \lambda \ln m\}$. The expected value of consumption must take into account that the individual might consume in any block or at any kink point. Given the assumption about the distribution of η and ε, the expected consumption level is:

$$W = \sum_{k=1}^{K} w(p_k, y + d_k)e^{\frac{\sigma_\varepsilon^2}{2}}e^{\frac{\sigma_\eta^2}{2}}\pi_k^* + \sum_{k=1}^{K-1} w_k e^{\frac{\sigma_\varepsilon^2}{2}}\lambda_k \tag{A-2}$$

where:

$$\pi_k^* = \Phi\left(\frac{\ln b_k}{\sigma_\eta} - \sigma_\eta\right) - \Phi\left(\frac{\ln a_k}{\sigma_\eta} - \sigma_\eta\right)$$

$$a_k = \frac{w_{k-1}}{w(p_k, y + d_k)}, \quad b_k = \frac{w_k}{w(p_k, y + d_k)}$$

$$\lambda_k = \Phi\left(\frac{\ln c_k}{\sigma_\eta}\right) - \Phi\left(\frac{\ln b_k}{\sigma_\eta}\right)$$

$$c_k = \frac{w_k}{w(p_{k+1}, y + d_{k+1})}$$

We can use this expression to calculate the change in expected consumption given a change in 1% of all prices in the block rate structure. For the log-log demand function given above, it can be shown (see Olmstead et al. 2007) that the elasticity is given by:

$$\frac{1}{W}\frac{\partial W}{\partial \theta} = \frac{\beta\left(\sum_{k=1}^{K}\frac{w(p_k, y+d_k)}{p_k}\psi_k + \sum_{k=1}^{K-1}w_k\chi_k\right) + \lambda\left(\sum_{k=1}^{K}\frac{w(p_k, y+d_k)}{y+d_k}d_k\psi_k + \sum_{k=1}^{K-1}w_k\tau_k\right)}{\Omega} \tag{A-3}$$

with:

$$\psi_k = \left(\pi_k^* - \frac{1}{\sigma_\eta}\left[\phi\left(\frac{\ln b_k}{\sigma_\eta} - \sigma_\eta\right) - \phi\left(\frac{\ln a_k}{\sigma_\eta} - \sigma_\eta\right)\right]\right)$$

$$\chi_k = \frac{1}{\sigma_\eta e^{\sigma_\eta^2/2}}\left(\phi\left(\frac{\ln b_k}{\sigma_\eta}\right) - \phi\left(\frac{\ln a_k}{\sigma_\eta}\right)\right)$$

$$\tau_k = \frac{1}{\sigma_\eta e^{\sigma_\eta^2/2}}\left(\phi\left(\frac{\ln b_k}{\sigma_\eta}\right)\frac{d_k}{y+d_k} - \phi\left(\frac{\ln a_k}{\sigma_\eta}\right)\frac{d_{k+1}}{y+d_{k+1}}\right)$$

$$\Omega = \sum_{k=1}^{K} w(p_k, y + d_k)\pi_k^* + \sum_{k=1}^{K-1} w_k e^{-\frac{\sigma_\eta^2}{2}}\lambda_k.$$

We can see that price elasticity is a complex function of the parameters of the model; it includes a price effect and an income effect produced by the virtual subsidy implicit in the tiered rate structure.

Analogous formulae can be derived for the linear model.

Appendix B. Formulation of linked consumption in discrete continuous choice demand models

Let us consider the following system of demand equations

$$x_{1j} = Z\delta_1 + \beta_{11}p_{1j} + \beta_{12}p_{2j} + \lambda_1 m_j + \varepsilon_1 + \eta_1$$
$$x_{2j} = Z\delta_2 + \beta_{21}p_{1j} + \beta_{22}p_{2j} + \lambda_2 m_j + \varepsilon_2 + \eta_2$$

in which Z is a vector of explanatory variables, p_{1j} and p_{2j} are the prices of the goods and j take the value 1 if the consumption is below the kink point and the value of 2 if consumption of that good is above the kink point. m is the income of the individual and (δ, β, λ) are parameters to be estimated. Finally, $(\varepsilon_1, \eta_1, \varepsilon_2, \eta_1)$ are error terms with the same interpretation given in Hewitt and Hanemann (1995) and Olmstead et al. (2007). The main purpose of our model is to capture correlation between the two goods. Therefore we assume that $(\varepsilon_1, \varepsilon_2)$ distributes as a bivariate normal with correlation coefficient equal to ρ and mean zero. Finally, following the authors mentioned above, we assume that the rest of the error terms are independent among each other and with mean zero as well. Therefore the joint distribution of the error terms is

$$\varepsilon_1 = u_1$$
$$\varepsilon_2 = u_2$$
$$\varepsilon_1 + \eta_1 = v_1$$
$$\varepsilon_2 + \eta_2 = v_2$$

$$f_{\varepsilon_1, \varepsilon_2, \eta_1, \eta_2} \sim MN(0, \Sigma)$$

$$\Sigma = \begin{pmatrix} \sigma_1^2 & \sigma_{12} & 0 & 0 \\ \sigma_{12} & \sigma_2^2 & 0 & 0 \\ 0 & 0 & \sigma_{\eta_1}^2 & 0 \\ 0 & 0 & 0 & \sigma_{\eta_2}^2 \end{pmatrix}$$

where $E(\varepsilon_1^2) = \sigma_1^2$, $E(\varepsilon_2^2) = \sigma_2^2$, $E(\varepsilon_1 + \eta_1)^2 = \sigma_1^2 + \sigma_{\eta_1}^2$, $E(\varepsilon_2 + \eta_2)^2 = \sigma_2^2 + \sigma_{\eta_2}^2$ and $E(\varepsilon_1\varepsilon_2) = \sigma_{12} = \rho\sigma_1\sigma_2$. Furthermore $E(\varepsilon_1(\varepsilon_1 + \eta_1)) = \sigma_1^2$, $E(\varepsilon_2(\varepsilon_2 + \eta_2)) = \sigma_2^2$ $E(\varepsilon_1(\varepsilon_2 + \eta_2))(E(\varepsilon_2(\varepsilon_1 + \eta_1)) = E(\varepsilon_1 + \eta_1)(\varepsilon_2 + \eta_2) = \sigma_{12}$. Then

$$f_{\varepsilon_1, \varepsilon_2, \varepsilon_1 + \eta_1, \varepsilon_2 + \eta_2} \sim MN\left(0, \widehat{\Sigma}\right)$$

$$\widehat{\Sigma} = \begin{pmatrix} \sigma_1^2 & \sigma_{12} & \sigma_1^2 & \sigma_{12} \\ \sigma_{12} & \sigma_2^2 & \sigma_{12} & \sigma_2^2 \\ \sigma_1^2 & \sigma_{12} & \sigma_1^2 + \sigma_{\eta_1}^2 & \sigma_{12} \\ \sigma_{12} & \sigma_2^2 & \sigma_{12} & \sigma_2^2 + \sigma_{\eta_2}^2 \end{pmatrix}$$

Using the marginal distribution and the conditional distribution expression of this distribution, that is $f_{\varepsilon_1, \varepsilon_2, \varepsilon_1+\eta_1, \varepsilon_2+\eta_2} = f_{\varepsilon_1+\eta_1, \varepsilon_2+\eta_2} * f_{\varepsilon_1, \varepsilon_2/(\varepsilon_1+\eta_1),(\varepsilon_2+\eta_2)}$ we can characterize all possible consumption patterns. The marginal distribution is

$$f_{\varepsilon_1+\eta_1, \varepsilon_2+\eta_2} \sim NB\left(0, \overline{\Sigma}\right)$$

$$\overline{\Sigma} = \begin{pmatrix} \sigma_1^2 + \sigma_{\eta_1}^2 & \sigma_{12} \\ \sigma_{12} & \sigma_2^2 + \sigma_{\eta_2}^2 \end{pmatrix}$$

while the conditional distribution is

$$f_{\varepsilon_1, \varepsilon_2/(\varepsilon_1+\eta_1),(\varepsilon_2+\eta_2)}$$

where we use the following property of a partitioned matrix of normal distribution given the mean vector and the matrix of variance and covariance given by

$$\mu = \begin{bmatrix} \mu_1 \\ \mu_2 \end{bmatrix}$$

$$\Sigma = \begin{bmatrix} \Sigma_{11} & \Sigma_{12} \\ \Sigma_{21} & \Sigma_{22} \end{bmatrix}$$

the mean of the variables i conditional on variables j is given by

$$\mu_{i/j} = \mu_i + \Sigma_{ij}\Sigma_{jj}^{-1}\left(x_j - \mu_j\right)$$

and the conditional variance is given by

$$\Sigma_{i/j} = \Sigma_{ii} - \Sigma_{ij}\Sigma_{jj}^{-1}\Sigma_{ji}$$

Using this equation we found that

$$f_{\varepsilon_1, \varepsilon_2/(\varepsilon_1+\eta_1),(\varepsilon_2+\eta_2)} \sim NM\left(\mu_{i/j}, \ \Omega, \ \widehat{\rho}\right)$$

$$\mu_{i/j} = \begin{pmatrix} \frac{1}{1-\rho^2}\left[\rho_1^2 v_1 - \rho^2\left(v_1 - \frac{\sigma_{\eta_1}^2}{\sigma_{12}}v_2\right)\right] \\ \frac{1}{1-\rho^2}\left(\rho_2^2 v_2 - \rho^2\left(v_2 - \frac{\sigma_{\eta_2}^2}{\sigma_{12}}v_1\right)\right) \end{pmatrix}$$

$$\Omega = \begin{pmatrix} \sigma_{\eta_1}^2\left(\sigma_1^2\left(\sigma_2^2 + \sigma_{\eta_2}^2\right) - \sigma_{12}^2\right) & \sigma_{12}\sigma_{\eta_1}^2\sigma_{\eta_2}^2 \\ \sigma_{12}\sigma_{\eta_2}^2\sigma_{\eta_1}^2 & \sigma_{\eta_2}^2\left(\sigma_2^2\left(\sigma_1^2 + \sigma_{\eta_1}^2\right) - \sigma_{12}^2\right) \end{pmatrix}$$

$$\widehat{\rho} = \frac{\sigma_{12}\sigma_{\eta_2}^2\sigma_{\eta_1}^2}{\sqrt{\sigma_{\eta_1}^2\left(\sigma_1^2\left(\sigma_2^2 + \sigma_{\eta_2}^2\right) - \sigma_{12}^2\right)}\sqrt{\sigma_{\eta_2}^2\left(\sigma_2^2\left(\sigma_1^2 + \sigma_{\eta_1}^2\right) - \sigma_{12}^2\right)}}$$

in which $\rho_1 = \frac{E(\varepsilon_1(\varepsilon_1+\eta_1))}{\sigma_1\sqrt{\sigma_1^2+\sigma_{\eta_1}^2}} = \frac{\sigma_1}{\sqrt{\sigma_1^2+\sigma_{\eta_1}^2}}$ and $\rho_2 = \frac{E(\varepsilon_2(\varepsilon_2+\eta_2))}{\sigma_2\sqrt{\sigma_2^2+\sigma_{\eta_2}^2}} = \frac{\sigma_2}{\sqrt{\sigma_2^2+\sigma_{\eta_2}^2}}$ and $v_1 = (\varepsilon_1 + \eta_1)$ and $v_2 = (\varepsilon_2 + \eta_2)$.

In addition, using these distributions and the properties of the normal distribution we can build the likelihood function for each individual, including nine possible combinations, as we do not have sample separability with a two error term (Hewitt, 1993). For example, consider the case that both levels $(\bar{x}_1^1, \bar{x}_2^1)$ of consumption are below the kink points (x_{11}, x_{22}), then we have that

- $\bar{x}_1^1 < x_{11}$ and $\bar{x}_2^1 < x_{22}$ which is characterized by $\varepsilon_1 + \eta_1 = v_1 = x_{11} - \bar{x}_1^1$, $\varepsilon_2 + \eta_2 = v_2 = x_{22} - \bar{x}_2^1$, $\varepsilon_1 \leq x_{11} - \bar{x}_1^1$ and $\varepsilon_2 \leq x_{22} - \bar{x}_2^1$. Then contribution to the likelihood function of this component is given by

$$\int_{-\infty}^{x_{11}-\bar{x}_1^1} \int_{-\infty}^{x_{22}-\bar{x}_2^1} f_{\varepsilon_1,\varepsilon_2,\varepsilon_1+\eta_1,\varepsilon_2+\eta_2}(.)d\varepsilon_2 d\varepsilon_1 = \int_{-\infty}^{x_{11}-\bar{x}_1^1} \int_{-\infty}^{x_{22}-\bar{x}_2^1} f_{\varepsilon_1+\eta_1,\varepsilon_2+\eta_2}{}^* f_{\varepsilon_1,\varepsilon_2/(\varepsilon_1+\eta_1),(\varepsilon_2+\eta_2)}(.)d\varepsilon_2 d\varepsilon_1$$

$$f_{\varepsilon_1+\eta_1,\varepsilon_2+\eta_2}{}^* \int_{-\infty}^{x_{11}-\bar{x}_1^1} \int_{-\infty}^{x_{22}-\bar{x}_2^1} f_{\varepsilon_1,\varepsilon_2/(\varepsilon_1+\eta_1),(\varepsilon_2+\eta_2)}d\varepsilon_2 d\varepsilon_1$$

Because we know all the distributions, we can evaluate the value of this expression. Similar expressions, although more complicated in some cases, can be derived for each consumption pattern and included in the likelihood function.

Appendix C. Estimation of parameters for analyzing consumption in discrete continuous choice demand models

Estimating the parameters related to linked consumption of resources in a discrete continuous choice demand model requires constructing a likelihood function that represents the probability that each resource experiences the consumption pattern observed in the overall sample. The characteristics of the likelihood function depend both on the assumptions about the distribution of the error terms of the models and on the price structure consumers face. For example, let us call w^* the optimal level of consumption, and assume that the two error terms are independent and normally distributed with mean zero and variance as σ_η and σ_ε. Here the contribution to the likelihood function of an individual in a uniform price system would be:

$$l_i = \ln\left(\frac{1}{\sqrt{2\pi}(\sigma_n + \sigma_\varepsilon)} \exp\left(-\frac{1}{2}\left(\frac{\ln w - \ln w^*}{\sigma_\eta + \sigma_\varepsilon}\right)^2\right)\right) \qquad (C-1)$$

The likelihood function for increasing block prices is more complicated. If there are K blocks, then there exist $K - 1$ kinks in the budget set; let's call them W_k. For each block there exists a virtual income $_k=y+d_k$ that enables individuals to consume inside this block. In order to define d_k and therefore $_k=y+d_k$ the virtual income, we must remember that people pay different marginal prices for different units of consumption. Referring again to Fig. 1 in the main paper, people pay only p1 for units below W^*. Therefore there is an implicit benefit of p2 − p1 for each unit consumed in that range. Because higher levels of consumption are associated with higher marginal prices, a subsidy is created for the initial quantities, which can be consumed at a lower marginal price. Then:

$$d_k = \begin{cases} 0 & if \ k = 1 \\ \sum_{j-1}^{k-1} \left(p_{j+1} - p_j\right)w_k & if \ k > 1 \end{cases} \qquad (C-2)$$

To build the likelihood function, we must consider the probability that consumption lies in each segment defined by all the K blocks and the $K-1$ kinks. Following Olmstead et al. (2007), the likelihood function is:

$$\ln\left[\sum_{k=1}^{K}\left(\frac{1}{\sqrt{2\pi}}\frac{\exp\left(-s_k^2/2\right)}{\sigma_v}\right)\left(\Phi(r_k) - \Phi(n_k)\right) + \sum_{k=1}^{K-1}\left(\frac{1}{\sqrt{2\pi}}\frac{\exp\left(-u_k^2/2\right)}{\sigma_\varepsilon}\right)\left(\Phi(m_k) - \Phi(t_k)\right)\right]$$

$$(C-3)$$

with:

$$v = \eta + \varepsilon \qquad t_k = \left(\ln w_k - \ln w_k^*\right)/\sigma_\eta$$

$$\rho = corr(v,\rho), \qquad r_k = (t_k - \rho s_k)/\sqrt{1-\rho^2}$$

$$s_k = \left(\ln w_i - \ln w_k^*\right)/\sigma_v \qquad m_k = \left(\ln w_k - \ln w_{k+1}^*\right)/\sigma_\eta$$

$$u_k = (\ln w_i - \ln w_k)/\sigma_\varepsilon \qquad n_k = (m_{k-1} - \rho s_k)/\sqrt{1-\rho^2}$$

where $\overset{*}{k}$ is the optimal consumption on block k, w_k is consumption at kink point k.

Appendix D: Data sets for estimating residential energy demand

Discrete continuous choice (DCC) modeling has been relatively rare in the literature because it requires consumer-level information. To protect customer privacy, U.S. electric and natural gas utilities historically have restricted the availability and use of customer-level information regarding utility consumption. Utilities strictly limit access to information about a household from which researchers could derive basic factors that influence consumption (e.g., house size, number of household members, number and types of appliances).

For this study we collected large sets of data on household natural gas and electricity consumption.

D.1 Household electricity and natural gas data

One public source of residential energy use data is the Residential Energy Consumption Survey (RECS), performed every 4 years by the Energy Information Administration (EIA) of the U.S. Department of Energy. RECS is a sample survey that provides household-level information on home energy uses and costs, based on a combination of actual utility bills and household survey information. A subset of these data was identified for California. The latest public use files, from survey year 2001, contain data for 541 California households. The EIA releases a public use file that includes annual consumption and expenditures for electricity, natural gas, and other fuels both in total and as estimated by end use, as well as descriptions of household demographics, equipment and appliances, and various energy use practices. These data can be used to estimate electricity and natural gas elasticities for California, via the construction of monthly estimated bills, as Reiss and White (2005) did for electricity using 1993 and 1997 RECS data. Supplemental data on natural gas and electricity use in California households are available through the Residential Appliance Saturation Survey, covering five participating utilities in 2003, including California's three largest investor-owned utilities.[4]

D.2 Processing of natural gas and electricity data

In order to recover the gas and electricity prices faced by households sampled for the 1993 and 1997 RECS, we used a database constructed by Reiss and White (2005), wherein each individual in the survey was matched to a NOAA weather station. Using GIS techniques, each weather station was assigned a ZIP code and city, then matched to its corresponding natural gas provider. Data on the areas supplied by natural gas providers, as well as prices for 1993 and 1997, were obtained from the websites of the providers and by contacting provider personnel. Data on electricity prices were obtained from the database provided by Reiss and White (2005). Because natural gas prices were obtained only for single-family households, we discarded RECS data on consumers who reported living in condominiums or multi-unit houses, as well as those who did not pay their utility bills. We also discarded records of individuals who reported having no natural gas supplied to their households. Because some socioeconomic variables have different ranges or do not match between the 1993 and 1997 survey years, we had to transform some of them to make the answers comparable and discard others. Given that RECS reports annual energy consumption, we calculated average natural gas prices for each household.

References

Bernstein MA, Griffin J (2006) Regional differences in the price-elasticity of demand for energy. RAND Corporation, Santa Monica, California. NREL/SR-620-39512. February
Bohi DR, Zimmerman MB (1984) An up-date on econometric studies of energy demand behavior. Annu Rev Energ 9:105–154
Borenstein S (2010) To what electricity price do consumer respond? residential demand elasticity under increasing block pricing. NBER working paper. www.econ.yale.edu/seminars/apmicro/.../borenstein-090514.pdf
Cohen R, Nelson B, Wolff G (2004) Energy down the drain: the hidden costs of California water supply. Natural Resources Defense Council. August
[Energy Information Administration (EIA)] (2004) Residential energy consumption survey, 2001: consumption and expenditures data tables. U.S. Department of Energy, Washington, DC
EIA (2008) Residential energy consumption survey, 2005: household characteristics tables. U.S. Department of Energy, Washington, DC
Espey J, Espey M (2004) Turning on the lights: a meta-analysis of residential electricity demand elasticities. J Agr Appl Econ 36(1):65–81

---

[4] California Energy Commission. Residential Appliance Saturation Survey (RASS). www.energy.ca.gov/appliances/rass/index.html. Accessed 20 January 2010.

Garcia-Cerrutti M (2000) Estimating elasticities of residential energy demand from Panel County data using dynamic random variables models with heteroskedastic and correlated error terms. Resource Energ Econ 22:335–366

Hewitt JA, Hanemann WM (1995) A discrete/continuous choice model of demand under block pricing. Land Econ 71(2):173–192

Houthakker HS, Taylor LD (1970) Consumer demand in the United States: analyses and projections, 2nd edn. Harvard University Press, Cambridge

Ito K (2010) How do consumers respond to nonlinear pricing? Evidence from household electricity demand, Job market paper, Department of Agricultural and resource Economics UC Berkeley

Joutz F, Trost RP (2007) An economic analysis of consumer response to higher natural gas prices. June 20, 2007. Presentation ("AGAslides.pdf".) www.aga.org/Events/presentations/invrel/2007/AGAPriceElasticityWebcast.htm

Lee L, Pitt M (1987) Microeconometric models of rationing, imperfect markets, and non-negativity constraints. J Econometrics 36:89–110

Maddala GS, Trost RP, Li H, Joutz F (1997) Estimation of short-run and long-run parameters from panel data using shrinkage estimators. J Bus Econ Stat 15:90–100

Nieswiadomy ML, Molina DJ (1991) Note on price perception in water demand models. Land Econ 67(3)

Olmstead S, Hanemann W, Stavins R (2007) Water demand under alternative price structure. J Environ Econ Manag 54:181–198

Reiss PC, White MW (2005) Household electricity demand, revisited. Rev Econ Stud 72(3):853–883

Simulating the impacts of climate change, prices and population on California's residential electricity consumption

Maximilian Auffhammer · Anin Aroonruengsawat

Received: 9 March 2010 / Accepted: 28 September 2011 / Published online: 24 November 2011
© Springer Science+Business Media B.V. 2011

Abstract This study simulates the impacts of higher temperatures resulting from anthropogenic climate change on residential electricity consumption for California. Flexible temperature response functions are estimated by climate zone, which allow for differential effects of days in different temperature bins on households' electricity consumption. The estimation uses a comprehensive household level dataset of electricity bills for California's three investor-owned utilities (Pacific Gas and Electric, San Diego Gas and Electric, and Southern California Edison). The results suggest that the temperature response varies greatly across climate zones. Simulation results using a downscaled version of the National Center for Atmospheric Research global circulation model suggest that holding population constant, total consumption for the households considered may increase by up to 55% by the end of the century. The study further simulates the impacts of higher electricity prices and different scenarios of population growth. Finally, simulations were conducted consistent with higher adoption of cooling equipment in areas which are not yet saturated, as well as gains in efficiency due to aggressive energy efficiency policies.

Keywords Climate change · Adaptation · Impacts estimation · Electricity consumption

1 Introduction

Forecasts of electricity demand are of central importance to policy makers and utilities for purposes of adequately planning future investments in new generating

M. Auffhammer (✉)
UC Berkeley ARE/IAS, 207 Giannini Hall, Berkeley, CA 94720-3310, USA
e-mail: auffhammer@berkeley.edu

A. Aroonruengsawat
Faculty of Economics, Thammasat University, Bangkok, Thailand
e-mail: anin@econ.tu.ac.th

capacity. Total electricity consumption in California has more than quadrupled since 1960, and the share of residential consumption has grown from 26% to 34% (EIA SEDS 2008). Today, California's residential sector alone consumes as much electricity as Argentina, Finland, or roughly half of Mexico. The majority of electricity in California is delivered by three investor-owned utilities and more than one a hundred municipal utilities.

California's energy system faces several challenges in attempting to meet future demand (CEC 2005). In addition to rapid population growth, economic growth and an uncertain regulatory environment, the threat of significant global climate change has recently emerged as a factor influencing the long term planning of electricity supply. The electric power sector will be affected by climate change through higher cooling demand, lower heating demand, and potentially stringent regulations designed to curb emissions from the sector.

This paper simulates how the residential sector's electricity consumption will be affected by different scenarios of climate change. We make four specific contributions to the literature on simulating the impacts of climate change on residential electricity consumption. First, through an unprecedented opportunity to access the complete billing data of California's three major investor-owned utilities, we are able to provide empirical estimates of the temperature responsiveness of electricity consumption based on micro-data. Second, we allow for a geographically specific response of electricity consumption to changes in weather.[1] Third, we provide simulations of future electricity consumption under constant and changing climate, electricity price, and population scenarios. Finally, we provide worst and best case simulation results, assuming uniform "best" and "worst" climate sensitivities for the entire state based on our estimation results. These simulations provide us with upper and lower bound estimates from different adaptation scenarios.

2 Literature review

The historical focus of the literature forecasting electricity demand has been on the role of changing technology, prices, income, and population growth (e.g., Fisher and Kaysen 1962). Early studies in demand estimation have controlled for weather, which leads to estimation efficiency gains (e.g., Houthakker and Taylor 1970). Simulations based on econometrically estimated demand functions had therefore focused on different price, income, and population scenarios, while assuming a stationary climate system. The onset of anthropogenic climate change has added a new and important dimension of uncertainty over future demand, which has spawned a small academic literature on climate change impacts estimation, which can be divided into two approaches.

In the engineering literature, large-scale bottom-up simulation models are utilized to simulate future electricity demand under varying climate scenarios. The advantage

---

[1]Two closely related and in parts overlapping companion papers contain more detail on the weather data extrapolation. Aroonruengsawat and Auffhammer (2009) provides simulation results for the households in the low income CARE program. Aroonruengsawat and Auffhammer (2011) further explore the sources of heterogeneity in the response function due to differences in the characteristics of the population and housing stock.

of the simulation model approach is that it allows one to simulate the effects of climate change given a wide variety of technological and policy responses. The drawback to these models is that they contain a large number of response coefficients and make a number of specific and often untestable assumptions about the evolution of the capital stock and its usage. Cline (1992) finds that increases in annual temperatures ranging from 1.0–1.4°C (1.8–2.5°F) in 2010 would result in demand of 9% to 19% above estimated new capacity requirements (peak load and base load) in the absence of climate change. The estimated impacts rise to 14% and 23% for the year 2055 and an assumed 3.7°C (6.7°F) temperature increase.

Baxter and Calandri (1992) project electricity demand to the year 2010 under two global warming scenarios: a rise in average annual temperature of 0.6°C (1.1°F) (Low scenario) and of 1.9°C (3.4°F) (High scenario). They find that electricity use increases from the constant climate scenario by 0.6% to 2.6%, while peak demand increases from the baseline scenario by 1.8% to 3.7%. Rosenthal et al. (1995) estimate that a 1°C (1.8°F) increase in temperature will reduce U.S. energy expenditures in 2010 by $5.5 billion (1991 dollars).

The economics literature has favored the econometric approach to impacts estimation, which is the approach we adopt in the current study. While there is a large literature on econometric estimation of electricity demand, the literature on climate change impacts estimation is small and relies on panel estimation of heavily aggregated data or cross-sectional analysis of more micro-level data. The first set of papers attempts to explain variation in a cross section of energy expenditures based on survey data to estimate the impact of climate change on fuel consumption choices. Mansur et al. (2008) endogenize fuel choice, which is usually assumed to be exogenous. They find that warming will result in fuel switching towards electricity. The drawback of the cross sectional approach is that one cannot econometrically control for unobservable differences across firms and households, which may be correlated with weather/climate potentially leading to biased coefficients.

Franco and Sanstad (2008) explain pure time series variation in hourly electricity load at the grid level over the course of a year. They use data reported by the California Independent System Operator (CalISO) for 2004 and regress it on a population weighted average of daily temperature. Their estimates show a nonlinear impact of average temperature on electricity load, and a linear impact of maximum temperature on peak demand. Relative to the 1961–1990 base period, the range of increases in electricity and peak load demands are 0.9–20.3% and 1.0–19.3%, respectively. Crowley and Joutz (2003) use a similar approach where they estimate the impact of temperature on electricity load using hourly data in the Pennsylvania, New Jersey, and Maryland Interconnection. They find that a 2°C (3.6°F) increase in temperature results in an increase in energy consumption of 3.8% of actual consumption.

Deschênes and Greenstone (2011) provide the first panel data-based approach to estimating the impacts of climate change on residential electricity consumption. They explain variation in U.S. state-level annual panel data of residential electricity consumption using flexible functional forms of daily mean temperatures. The identification strategy behind their paper, which is one we will adopt here as well, relies on random fluctuations in weather to identify climate effects on electricity consumption. The model includes state fixed effects, census division by year fixed effects, and controls for precipitation, population, and income. The temperature data enter the model as the number of days in 20 predetermined temperature intervals.

The authors find a U-shaped response function where electricity consumption is higher on very cold and hot days. The impact of climate change on annual electricity consumption by 2099 is in the range of 15–30% of the baseline estimation or 15 to 35 billion (2006 US$). The panel data approach allows one to control for differences in unobservables across the units of observation, resulting in consistent estimates of the coefficients on temperature.

The current paper is the first paper using a panel of household level electricity billing data to examine the impact of climate change on residential electricity consumption. Through a unique agreement with California's three largest investor-owned utilities, we gained access to their complete billing data for the years 2003–2006. We identify the effect of temperature on electricity consumption using within household variation in temperature, which is made possible through variation in the start dates and lengths of billing periods across households. Since our dataset is a panel, we can control for household fixed effects, month fixed effects, and year fixed effects. The drawback of this dataset is that the only other reliable information we have about each individual household is price and the five-digit ZIP code location.

3 Data

3.1 Residential billing data

The University of California Energy Institute jointly with California's investor-owned utilities established a confidential data center, which contains the complete billing history for all households serviced by Pacific Gas and Electric, Southern California Edison, and San Diego Gas and Electric for the years 2003–2006. These three utilities provide electricity to roughly 80% of California households.

The data set contains the complete information for each residential customer's bills over the four year period. Specifically, we observe an ID for the physical location, a service account number, bill start-date, bill end-date, total electricity consumption (in kilowatt-hours, kWh) and the total amount of the bill (in $) for each billing cycle as well as the five-digit ZIP code of the premises. Only customers who were individually metered are included in the data set. For the purpose of this paper, we define a customer as a unique combination of premise and service account number. It is important to note that each billing cycle does not follow the calendar month and the length of the billing cycle varies across households with the vast majority of households being billed on a 25–35 day cycle. While we have data covering additional years for two of the utilities, we limit the study to the years 2003 to 2006, to obtain equal coverage. Figure 1 displays the ZIP codes we have these data for, which is the majority of the state.

Due to the difference in climate conditions across the state, California is divided into 16 building climate zones, each of which require different minimum efficiency building standards specified in an energy code. We expect this difference in building standards to lead to a different impact of temperature change on electricity consumption across climate zones. We will therefore estimate the impact of mean daily temperature on electricity consumption separately for each climate zone. There is no guarantee that the exogenously specified climate zones are the best division to

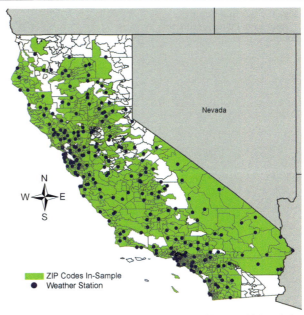

Fig. 1 Observed residential electricity consumption 2003–2006 and NOAA cooperative weather stations

Note: The map displays five-digit zip codes with available geographic boundaries.

use for this analysis, yet a redesign of the climate zones is beyond the scope of this project. The CEC climate zones are depicted in Fig. 2.

The billing data set contains 300 million observations, which exceeds our ability to conduct estimation using standard statistical software. We therefore resort to sampling from the population of residential households to conduct econometric estimation. We designed the following sampling strategy. First we only sample from households with regular billing cycles, namely 25–35 days in each billing cycle and which have at least 35 bills over the period of 2003–2006. We removed households on California's low income CARE program, which we examine in Aroonruengsawat and Auffhammer (2009). We also removed bills with an average daily consumption less than 2 kWh or more than 80 kWh. The reason for this is our concern that these outliers are not residential homes, but rather vacation homes and small scale "home based manufacturing and agricultural facilities". Combined with the fact that our data does not contain single-metered multi-family homes, our sampling strategy is likely to result in a slight under representation of multifamily and smaller single family homes. These are more likely to be rental properties than larger single family units. Our results should be interpreted keeping this in mind.[2]

From the population subject to the restrictions above, we take a random sample from each ZIP code, making sure that the relative sample sizes reflect the relative

---

[2]After removing outlier bills, we compared the population average daily consumption of bills with billing cycles ranging from 25–35 days to the average daily consumption of bills for any length. The average daily consumption by climate zone in the subset of bills we sample from is roughly $\frac{1}{10}$th of a standard deviation higher than the mean daily consumption of the complete population including bills of any length.

Fig. 2 California Energy Commission building climate zones

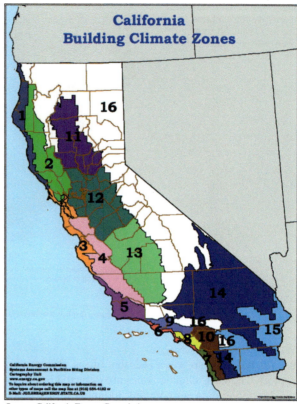

Source: California Energy Commission.

sizes of the population by ZIP code. We draw the largest possible representative sample from this population given our computational constraints. For each climate zone we test whether the mean daily consumption across bills for our sample is different from the population mean and fail to reject the null of equality, suggesting that our sampling is indeed random, subject to the sample restrictions discussed above. We proceed with estimation of our models by climate zone, which makes concerns about sampling weights moot. No single ZIP code is responsible for more than 0.5% of total consumption. Table 1 lists the summary statistics for our sample. Table 1 displays the summary statistics of our consumption sample by climate zone. There is great variability in average usage across climate zones, with the central coast's (zone 3) average consumption per bill at roughly 60% that of the interior southern zone 15. The average electricity price is almost identical across zones, at 13 cents per kWh.

3.2 Weather data

To generate daily weather observations to be matched with the household electricity consumption data, we use the Cooperative Station Dataset published by National Oceanic and Atmospheric Administration's (NOAA) National Climate Data Center (NCDC). The dataset contains daily observations from more than 20,000 cooperative

Table 1 Summary statistics for non-CARE households

	No. of obs.	No. of HH	Usage per bill per billing cycle (Kwh)		Average price per billing cycle ($/Kwh)		Percentiles daily mean temperature distribution in sample (°F)				
			Mean	S.D.	Mean	S.D.	1	5	50	95	99
Zone 1	1,459,578	31,879	550	354	0.13	0.03	34.5	37.5	54.7	77.0	80.0
Zone 2	2,999,408	65,539	612	385	0.13	0.03	36.0	39.0	55.5	77.5	80.5
Zone 3	3,200,851	69,875	469	307	0.13	0.02	42.0	44.3	57.0	75.0	78.0
Zone 4	4,232,465	92,294	605	362	0.13	0.03	40.5	42.8	57.8	81.4	85.5
Zone 5	2,621,344	57,123	504	317	0.13	0.03	42.0	44.3	58.8	76.0	78.5
Zone 6	2,970,138	64,145	529	334	0.13	0.03	48.5	50.4	62.0	78.0	81.0
Zone 7	3,886,347	85,169	501	327	0.15	0.04	47.0	48.9	61.5	77.5	80.0
Zone 8	2,324,653	50,373	583	364	0.14	0.03	49.5	51.5	63.3	80.6	83.3
Zone 9	3,067,787	66,231	632	389	0.13	0.03	48.0	50.3	63.0	81.0	83.5
Zone 10	3,202,615	70,088	700	416	0.14	0.03	35.5	39.0	61.0	81.8	84.5
Zone 11	4,106,432	90,245	795	455	0.13	0.03	28.5	32.8	54.8	84.3	87.0
Zone 12	3,123,404	68,342	721	420	0.13	0.03	38.5	40.8	58.5	84.0	87.0
Zone 13	3,827,483	84,493	780	464	0.13	0.03	36.6	39.3	59.0	87.8	90.0
Zone 14	4,028,225	88,086	714	413	0.13	0.03	32.0	35.0	57.5	91.3	95.0
Zone 15	2,456,562	54,895	746	532	0.13	0.03	34.5	37.8	63.8	97.0	99.5
Zone 16	3,401,519	74,644	589	409	0.13	0.02	22.5	26.5	52.3	83.0	86.5

Notes The table displays summary statistics for residential electricity consumption for the sample used in the estimation

weather stations in the United States, U.S. Caribbean Islands, U.S. Pacific Islands, and Puerto Rico. Data coverage varies by station. Since our electricity data cover the state of California for the years 2003–2006, the dataset contains 370 weather stations reporting daily data. In the dataset we observe daily minimum and maximum temperature as well as total daily precipitation and snowfall. Since the closest meaningful geographic identifier of our households is the five-digit postal ZIP code, we select stations as follows. First, we exclude any stations not reporting data in all years. Further we exclude stations reporting fewer than 300 observations in any single year and stations at elevations more than 7,000 feet above sea level, which leaves us with 269 "qualifying" weather stations. Figure 1 displays the distribution of these weather stations across the state. While there is good geographic coverage of weather stations for our sample, we do not have a unique weather station reporting data for each ZIP code. To assign a daily value for temperature and rainfall, we need to assign a weather station to each ZIP code. We calculate the distance of a ZIP code's centroid to all qualifying weather stations and assign the closest weather station to that ZIP code. As a consequence of this procedure, each weather station on average provides data for approximately ten ZIP codes. An alternate strategy would be to use downscaled weather data. We do not have access to such data at a daily frequency for the years in our sample. Our econometric strategy controls for time invariant differences at the household level via fixed effects to offset some of the issues arising from the imperfect measurement of temperature.

Since we do not observe daily electricity consumption by household, but rather monthly bills for billing periods of differing length, we require a complete set of daily weather observations. The NCDC data have a number of missing values, which we fill in using the algorithm used by Auffhammer and Kellogg (2011). We end up with a complete set of time series for minimum temperature, maximum temperature and precipitation for the 269 weather stations in our sample. For the remainder of our empirical analysis, we use these patched series as our observations of weather.

3.3 Other data

In addition to the quantity consumed and average bill amount, all we know about the households is the five-digit ZIP code in which they are located. We purchased socio demographics at the ZIP code level from a firm aggregating this information from census estimates (zip-codes.com). We only observe these data for a single year (2006). The variables we will make use of are total population and average household income. The final sample used for estimation comprises households in ZIP codes which make up 81% of California's population. We observe households for 1,325 ZIP codes and do not observe households for 239 ZIP codes. The 239 ZIP codes are not served by the three utilities, which provided us with access to their billing data. ZIP codes in our sample are slightly more populated, have larger households, are wealthier, and are at lower elevations. There is no statistically significant difference in population, median age, or land area. Taking these differences into consideration is important when judging the external validity of our estimation and simulation results.

4 Econometric estimation

As discussed in the previous section, we observed each household's monthly electricity bill for the period 2003–2006. Equation 1 below shows our main estimating equation, which is a simple log-linear specification commonly employed in aggregate electricity demand and climate change impacts estimation (e.g., Deschênes and Greenstone 2011).

$$\log(q_{it}) = \sum_{p=1}^{k} \beta_p D_{\text{pit}} + \gamma Z_{it} + \alpha_i + \phi_m + \gamma_y + \varepsilon_{it} \qquad (1)$$

$\log(q_{it})$ is the natural logarithm of household i's electricity consumed in kilowatt-hours during billing period t, D_{pit} are binned weather observations, Z_{it} are time varying confounders at the household level, α_i are household fixed effects, ϕ_m are month-of-year fixed effetcs, γ_y are year fixed effects and ε_{it} is a stochastic disturbance term. In this section we explain these variables in detail.

For estimation purposes our unit of observation is a unique combination of premise and service account number, which is associated with an individual and structure. We thereby avoid the issue of having individuals moving to different structures with more or less efficient capital or residents with different preferences over electricity consumption moving in and out of a given structure. California's housing stock varies greatly across climate zones in its energy efficiency and installed energy consuming capital. We estimate Eq. 1 separately for each of the sixteen climate zones discussed in the data section, which are also displayed in Fig. 2. The motivation for doing so is that we would expect the relationship between consumption and temperature to vary across these zones, as there is a stronger tendency to heat in the more northern and higher altitude zones and a stronger tendency to cool, but little heating taking place in the hotter interior zones of California.

The main variables of interest in this paper are those measuring temperature. The last five columns of Table 1 display the median, first, fifth, ninetieth, and ninety-fifth percentile of the mean daily temperature distribution by climate zone. The table shows the tremendous differences in this distribution across climate zones. The south eastern areas of the state for example, are significantly hotter on average, yet also have greater variances.

Following recent trends in the literature we include our temperature variables in a way that imposes a minimal number of functional form restrictions in order to capture potentially important nonlinearities of the outcome of interest in weather (e.g., Schlenker and Roberts 2009). We achieve this by sorting each day's mean temperature experienced by household i into one of k temperature bins. In order to define a set of temperature bins, there are two options found in the literature. The first is to sort each day into a bin defined by specific equidistant (e.g., 5°F) cutoffs. The second approach is to split each of the sixteen zones' temperature distributions into a set of percentiles and use those as the bins for sorting. The latter strategy allows for more precisely estimated coefficients, since there is guaranteed coverage in each bin.

There is no clear guidance in the literature on which approach provides better estimates and we therefore conduct our simulations using both approaches. For the percentile strategy, we split the temperature distribution into deciles, yet break down the upper and bottom decile further to include buckets for the first, fifth, ninety-fifth, and ninety-ninth percentile to account for extreme cold/heat days. We therefore have a set of 14 buckets for each of the sixteen climate zones. The thresholds for each vary by climate zone. For the equidistant bins approach, we split the mean daily temperature for each household into a set of 5° bins. In order to avoid the problem of imprecise estimation at the tails due to insufficient data coverage, we require that each bin have at least 1% of the data values in it for the highest and lowest bin. The highest and lowest bins in each zone therefore contain a few values which exceed the 5° threshold.

For each household, bin definition and billing period we then counted the number of days the mean daily temperature falls into each bin and recorded this as D_{pit}. The main coefficients of interest to the later simulation exercise are the β_p's, which measure the impact of one more day with a mean temperature falling into bin p on the log of household electricity consumption. For small values, β_p's interpretation is approximately the percent change in household electricity consumption due to experiencing one additional day in that temperature bin.

Z_{it} is a vector of observable confounding variables which vary across billing periods and households. The first of two major confounders we observe at the household level are the average electricity price for each household for a given billing period. California utilities price residential electricity on a block rate structure. The average price experienced by each household in a given period is therefore not exogenous, since marginal price depends on consumption (q_{it}). Identifying the price elasticity of demand in this setting is problematic, and a variety of approaches have been proposed (e.g., Hanemann 1984). The maximum likelihood approaches are computationally intensive and given our sample size cannot be feasibly implemented here. More importantly however, we do not observe other important characteristics of households (e.g., income) which would allow us to provide credible estimates of these elasticities. For later simulation we will rely on the income specific price elasticities provided by Reiss and White (2005), who used a smaller sample of more detailed data based on the national level RECS survey. We have run our models by including price directly, instrumenting for it using lagged prices and omitting it from estimation. The estimation results are almost identical for all three approaches, which is reassuring. While one could tell a story that higher temperatures lead to higher consumption and therefore higher marginal prices for some households, this bias seems to be negligible given our estimation results. In the estimation and simulation results presented in this paper, we omit the average price from our main regression. The second major time varying confounder is precipitation in the form of rainfall. We control for rainfall using a second order polynomial in all regressions.

The α_i are household fixed effects, which control for time invariant unobservables for each household. The ϕ_m are month-specific fixed effects, which control for unobservable shocks to electricity consumption common to all households. The γ_y are year fixed effects which control for yearly shocks common to all households. To credibly identify the effects of temperature on the log of electricity consumption, we require that the residuals conditional on all right hand side variables be orthogonal to the temperature variables, which can be expressed as $E[\varepsilon_{it} D_{pit} | D_{-pit}, Z_{it}, \alpha_i, \phi_m, \gamma_y] = 0$. Since we control for household fixed effects, identification comes from within household variation in daily temperature after controlling for shocks common to all households, rainfall, and average prices.

We estimate Eq. 1 for each climate zone using a least squares fitting criterion and a variance covariance matrix clustered at the zip code. Figure 3 plots the estimated temperature response coefficients for each of the climate zones against the midpoints of the bins for the percentile and equidistant bin approaches. The coefficient estimates are almost identical, which is reassuring. We do not display the confidence intervals around the estimated coefficients. The coefficients are so tightly estimated that for visual appearance, displaying the confidence intervals simply makes the lines appear thick. From this figure, several things stand out. First, there is tremendous heterogeneity in the shape of the temperature response of electricity consumption across climate zones. Many zones have almost flat temperature response functions, such as southern coastal zones (5, 6, and 7). Other zones display a very slight negative slope at lower temperatures, especially the northern areas of the state (1, 2, and 11), indicating a decreased consumption for space heating as temperatures increase. On the other end of the spectrum, for most zones in the interior and southern part of the state we note a significant increase in electricity consumption in the highest temperature bins (4, 8, 9, 10, 11, 12, 13, and 15). We further note that the relative magnitude of this approximate percent increase in household electricity consumption in the higher temperature bins varies greatly across zones as indicated by the differential in slopes at the higher temperatures across zones. Zones 15 and 13 have a much higher temperature response than the mountainous zone 14. As far as inflection points are concerned, it is quite reassuring that with the exception of zone 14, they are all at 65°, which is the usual reference temperature to calculate degree days.

5 Simulations

In this section we simulate the impacts of climate change on electricity consumption under two different scenarios of climate change, three different electricity price scenarios, and three different population growth scenarios. We calculate a simulated trajectory of aggregate electricity consumption from the residential sector until the year 2100, which is standard in the climate change literature.

5.1 Temperature simulations

To simulate the effect of a changing climate on residential electricity consumption, we require estimates of the climate sensitivity of residential electricity consumption as well as a counterfactual climate. In the simulation for this section we use the estimated climate response parameters shown in Fig. 3. Using these estimates as the basis of our simulation has several strong implications. First, using the estimated β_p parameters implies that the climate responsiveness of consumption within climate

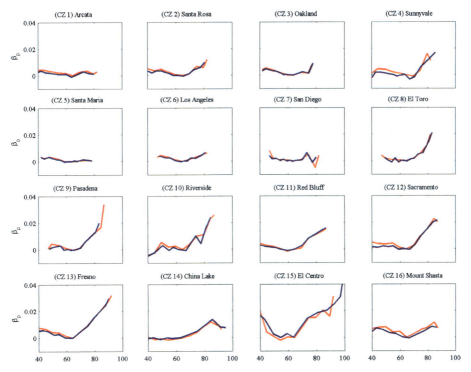

Notes: The panels display the estimated temperature slope coefficients for each of the fourteen percentile bins (blue) and the equidistant bins (red) against the midpoint of each bin. The plots were normalized using the coefficient estimate for the 60–65 temperature bin. The title of each panel displays the name of a representative city for that climate zone.

Fig. 3 Estimated climate response functions

zones remains constant throughout the century. This is a strong assumption, since we would expect that households in zones which currently do not require cooling equipment may potentially invest in such equipment if the climate becomes warmer. This would lead us to believe that the temperature responsiveness in higher temperature bins would increase over time. On the other hand, one could potentially foresee policy actions such as more stringent appliance standards, which improve the energy efficiency of appliances such as air conditioners. This would decrease the electricity per cooling unit required and shift the temperature response curve downwards in the higher buckets. We will deal with this issue explicitly in Section 5.4.

As is standard in this literature, the counterfactual climate is generated by a General Circulation Model (GCM). These numerical simulation models generate predictions of past and future climate under different scenarios of atmospheric greenhouse gas (GHG) concentrations. The quantitative projections of global climate change conducted under the auspices of the IPCC and applied in this study are driven by modeled simulations of two sets of projections of twenty-first century social and economic development around the world, the so-called "A2" and "B1" storylines in the 2000 Special Report on Emissions Scenarios (SRES) (IPCC 2000). The A2 and B1 storylines and their quantitative representations represent two quite different possible trajectories for the world economy, society, and energy system, and imply divergent future anthropogenic emissions, with projected emissions in the A2 being substantially higher.

We simulate consumption for each scenario using the National Center for Atmospheric Research Parallel Climate Model 1 (NCAR). These models were provided to us in their downscaled version for California using the Bias Correction and Spatial Downscaling (BCSD) and the Constructed Analogues (CA) algorithms (Cayan et al. 2009). There is no clear guidance in the literature as to which algorithm is preferable for impacts estimation. We therefore provide simulation results using both methods. To obtain estimates for a percent increase in electricity consumption for the representative household in ZIP code j and period $t + h$, we use the following relation:

$$
\frac{q_{j,t+h}}{q_{j,t}} = \frac{\exp\left(\sum_{p=1}^{k} \hat{\beta}_{pj} D_{pj,t+h}\right)}{\exp\left(\sum_{p=1}^{k} \hat{\beta}_{pj} D_{pj,t}\right)}
\tag{2}
$$

We implicitly assume that the year fixed effect and remaining right hand side variables are the same for period $t + h$ and period t, which is a standard assumption made in the majority of the impacts literature. As Aroonruengsawat and Auffhammer (2009, 2011) show, the areas with the steepest response functions at higher temperature bins happen to be the locations with highest increases in the number of high and extremely high temperature days. While this is not surprising, this correspondence leads to very large increases in electricity consumption in areas of the state experiencing the largest increases in temperature, which also happen to be the most temperature sensitive in consumption—essentially the southeastern parts of the state and the Central Valley.

The first simulation of interest generates counterfactuals for the percent increase in residential electricity consumption by a representative household in each ZIP code. We feed each of the two climate model scenarios through Eq. 2 using the 1980–1999 average number of days in each temperature bin as the baseline. Figure 4 displays the predicted percent increase in per household consumption for the periods 2020–2039, 2040–2059, 2060–2079 and 2080–2099 using the NCAR PCM model forced by the A2 scenario using the percentile bins. Figure 5 displays the simulation results for the SRES forcing scenario B1.

Changes in per household consumption are driven by two factors: the shape of the weather-consumption relationship and the change in projected climate relative to the 1980–1999 period. The maps show that for most of California, electricity consumption at the household level will increase. The increases are largest for the Central Valley and areas in south eastern California, which have a very steep temperature response of consumption and large projected increases in extreme heat days. Simulation results for this model and scenario suggest that some ZIP codes in the Central Valley by the end of the century may see increases in household consumption in excess of 100%. The map also shows that a significant number of ZIP codes are

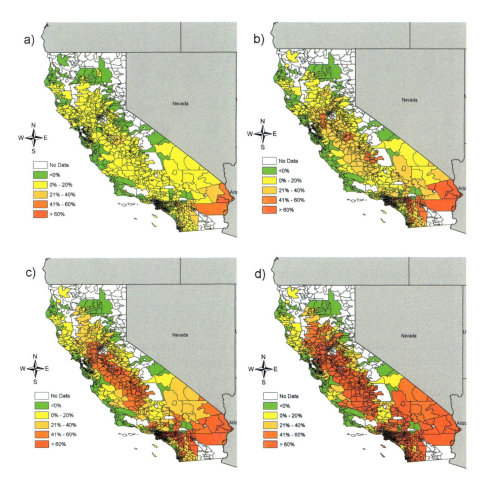

Fig. 4 Simulated increase in household electricity consumption by zip code for the periods 2020–2039 (**a**), 2040–2059 (**b**), 2060–2079 (**c**), and 2080–2099 (**d**) in percent over 1980–1999 simulated consumption. Model NCAR PCM forced by IPCC SRES A2

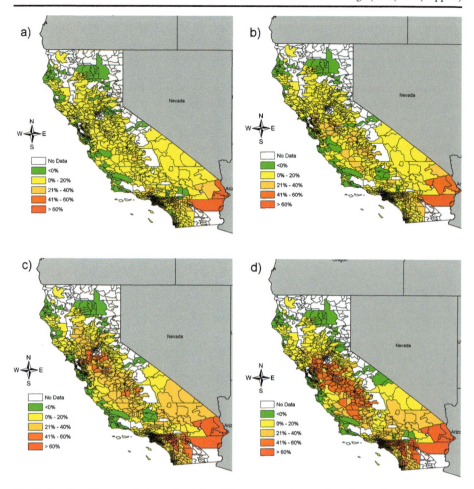

Fig. 5 Simulated increase in household electricity consumption by zip code for the periods 2020–2039 (**a**), 2040–2059 (**b**), 2060–2079 (**c**), and 2080–2099 (**d**) in percent over 1980–1999 simulated consumption. Model NCAR PCM forced by IPCC SRES B1

expected to see drops in household level electricity consumption-even at the end of the current century. It is important to keep in mind that the current projections assume no change in the temperature electricity response curve. Specifically, the current simulation rules out an increased penetration of air conditioners in areas with currently low penetration rates (e.g., Santa Barbara) or improvements in the efficiency of these devices. The projected drops essentially arise from slightly reduced heating demand. We conduct a simulation below, which addresses this concern.

While changes in per household consumption are interesting, from a capacity planning perspective it is overall consumption that is of central interest from this simulation. We use the projected percent increase in household consumption by ZIP code and calculate the weighted overall average increase, using the number of households by ZIP code as weights, in order to arrive at an aggregate percent increase in consumption. The top panel of Table 2 displays these simulation results for aggregate consumption. Predicted aggregate consumption across all ZIP codes in our dataset ranges from an 18% increase in total consumption to 55% increase in total consumption by the end of the century. To put this into perspective, this represents an annual growth rate of aggregate electricity consumption between 0.17% and 0.44%, if all other factors are equal. These growth rates accelerate from period to period, as the number of extreme heat days predicted from the GCMs increases in a slightly non-linear fashion. For the first 20-year period, the simulated annual growth rates range from 0.10% per year to 0.29% per year.

These simulations hold population constant at current (2006) levels. This leads to an interesting comparison in per capita terms. The average annual growth rate in per capita consumption during 1960–1973 was approximately 7% and slowed down to a remarkable 0.29% during 1974–1995. Growth rates during the last decade of available data have increased to a higher rate of 0.63%, and this difference in growth rates is statistically significant. The estimates from our simulation for the 2000–2019 period suggest that 26–60% of this increased growth rate may be due to changing climate.

5.2 Temperature and price simulations

The assumed flat prices from the previous section should be interpreted as a comparison benchmark. It is meaningful and informative to imagine climate change imposed on today's conditions. It is worth pointing out, however, that real residential electricity prices in California have been on average flat since the early-mid 1970s spike. In this section we will relax the assumption of constant prices and provide simulation results for increasing electricity prices under a changing climate.

Table 2 Simulated percent increase in residential electricity consumption relative to 1980–2000 for the constant, low price, and high price scenarios

Bin type downscaling IPCC scenario	Price increase (%)	Equidistant (%)				Percentile (%)			
		BCSD		CA		BCSD		CA	
		A2	B1	A2	B1	A2	B1	A2	B1
2000-19	±0	5	2	5	3	6	3	5	3
2020-39	±0	5	8	7	8	6	9	7	8
2040-59	±0	15	9	17	10	17	11	17	10
2060-79	±0	24	15	28	16	28	17	28	16
2080-99	±0	48	18	50	20	55	21	50	20
2000-19	±0	5	2	5	3	6	3	5	3
2020-39	+30	−6	−3	−5	−4	−5	−3	−4	−3
2040-59	+30	3	−2	3	−2	6	−1	5	−1
2060-79	+30	11	3	11	2	15	5	15	4
2080-99	+30	33	6	29	4	39	9	35	7
2000-19	±0	5	2	5	3	6	3	5	3
2020-39	+30	−6	−3	−5	−4	−5	−3	−4	−3
2040-59	+60	−9	−13	−8	−13	−6	−12	−7	−12
2060-79	+60	−1	−9	−1	−10	2	−7	2	−8
2080-99	+60	18	−6	15	−7	24	−4	20	−5

While we have no guidance on what will happen to retail electricity prices 20 years or further out into the future, we construct two scenarios. The first scenario we consider is a discrete 30% increase in real prices starting in 2020 and remaining at that level for the remainder of the century. This scenario is based upon current estimates of the average statewide electricity rate impact by 2020 of AB 32 compliance combined with natural gas prices to generators within the electric power sector. These estimates are based on analysis commissioned by the California Public Utilities Commission. This scenario represents the minimum to which California is committed in the realm of electricity rates. This scenario could be interpreted as one assuming very optimistic technological developments post 2030, implying that radical CO_2 reduction does not entail any cost increases, or as a California and worldwide failure to pursue dramatic CO_2 reductions such that California's AB 32 effort is not expanded. The second scenario we consider is one where electricity prices increase by 30% in 2020 and again by 30% in 2040 and remain at that level thereafter. We consider the additional increase in mid-century price in essence as an "increasing marginal abatement cost" story. Under this scenario, AB 32 is successfully implemented and a path towards achieving the 2050 targets is put in place. These additional steps are assumed to be proportionally more expensive.

To simulate the effects of price changes on electricity consumption, we require estimates of the price elasticity of demand. In this paper we rely on the estimates of mean price elasticity provided by Reiss and White (2005). Specifically, they provide a set of average price elasticities for different income groups, which we adopt here. Since we do not observe household income, we assign a value of price elasticity to each ZIP code based on the average household income for that ZIP code. Households are separated into four buckets, delineated by $18,000, $37,000, $60,000 with estimated price elasticities of −0.49, −0.34, −0.37, and −0.29 respectively. It is important to note that these price elasticities are short-run price elasticities. These are valid if one assumes a sudden increase in prices, as we do in this paper. To our knowledge, reliable long-term price elasticities based on micro data for California are not available. As Houthakker and Taylor (1970) and Espey and Espey (2004) point out, the long run elasticities in theory are larger than the elasticities used in this paper, suggesting potentially larger price effects.

The second panel in Table 2 presents the simulation results under the two different scenarios of climate change given a sudden persistent increase in electricity prices in the year 2020. Given the range of price elasticity estimates, it is not surprising that the simulated increases in residential electricity consumption for the first period after the price increase are roughly 6–12% lower than the predicted increases given constant prices. For the NCAR model under both considered forcing scenarios and both downscaling algorithms, the path of electricity consumption under these price scenarios returns to levels below its 1980–2000 mean for the 2020–2040 period, given this assumed price trajectory.

The third panel in Table 2 presents the simulation results for both forcing scenarios and downscaling methods given the high price scenario. Given the significant increase in prices after 2020 and again in 2040, the consumption trajectory stays flat for the entire simulation period using the NCAR model for the B1 scenario. The higher forcing scenario A2 shows a relatively flat trajectory, yet still predicts significant increases in consumption for the last decades of the century-even in the face of these higher prices. It is important to note that these effects are conditional on the estimated price elasticities being correct. Smaller elasticities would translate into price based policies, such as taxes or cap and trade systems, being less effective at curbing demand compared to standards.

5.3 Temperature and population

California has experienced an almost seven-fold increase in its population since 1929 (BEA 2008). California's population growth rate over that period (2.45%) was more than twice that of the national average (1.17%). Over the past 50 years California's population has grown by 22 million people to almost 37 million in 2007 (BEA 2008). To predict what the trajectory of California's population will look like until the year 2100, many factors have to be taken into account. The four key components driving future population are net international migration, net domestic migration, mortality rates, and fertility rates. The State of California provides forecasts fifty-five years out of sample, which is problematic since we are interested in simulating end-of-century electricity consumption. The Public Policy Institute of California has generated a set of population projections until 2100 at the county level, which are discussed in Sanstad et al. (2009).

The three sets of projections developed for California and its counties are designed to provide a subjective assessment of the uncertainty of the state's future population. The projections present three very different demographic futures. In the low series, population growth slows as birth rates decline, migration out of the state accelerates, and mortality rates show little improvement. In the high series, population growth accelerates as birth rates increase, migration increases, and mortality declines. The middle series, consistent with (but not identical to) the California Department of Finance projections assumes future growth in California will be similar to patterns observed over the state's recent history, patterns that include a moderation of previous growth rates but still large absolute changes in the state's population. In the middle series, international migration flows to California remain strong to mid-century and then subside, net domestic migration remains negative but of small magnitude, fertility levels (as measured by total fertility rates) decline slightly, and age-specific mortality rates continue to improve. The high projection is equivalent to an overall growth rate of 1.47% per year and results in a quadrupling of population to 148 million by the end of the century. The middle series results in a 0.88% annual growth rate and 2.3-fold increase in total population. The low series is equivalent to a 0.18% growth rate and results in a population 18% higher than today's. Projections are available at the county level and not at the ZIP code level. We therefore assume that each ZIP code in the same county experiences an identical growth rate.

Table 3 displays the simulated aggregate electricity consumption given the three population growth scenarios. This table holds prices constant at the current level and therefore presents a "worst case scenario". It is not surprising to see that population uncertainty has much larger consequences for simulated total electricity consumption compared to uncertainty over climate or uncertainty over prices. The simulations for the low forcing scenario B1 and the low population growth scenario show 65–70% increase in residential electricity consumption. The same figure for the medium growth scenario predicts a 179–189% increase in consumption and the worst case scenario predicts a 350% increase in consumption. If we consider the A2 forcing, the predicted low population average increase in consumption is a 118% increase or a 478% increase for the high population growth scenario.

Table 3 Simulated percent increase in residential electricity consumption relative to 1980–2000 for the low, middle, and high population scenarios

Bin type	Equidistant (%)				Percentile (%)			
downscaling	BCSD		CA		BCSD		CA	
IPCC scenario	A2	B1	A2	B1	A2	B1	A2	B1
Low population growth scenario								
2000-19	17	13	16	14	18	14	16	15
2020-39	31	34	33	34	32	35	34	35
2040-59	48	41	50	41	52	42	53	42
2060-79	66	52	68	51	72	55	73	54
2080-99	113	65	113	65	124	70	123	70
Medium population growth scenario								
2000-19	19	15	18	16	19	16	18	16
2020-39	48	52	51	52	50	54	52	53
2040-59	99	88	101	89	104	91	105	91
2060-79	154	133	157	133	164	139	166	138
2080-99	258	179	257	189	277	189	275	188
High population growth scenario								
2000-19	23	19	22	20	23	20	22	20
2020-39	64	68	66	68	66	70	68	69
2040-59	135	123	137	123	141	126	142	125
2060-79	240	212	243	211	252	219	254	218
2080-99	464	342	462	342	495	357	490	356

5.4 Adaptation simulations

As mentioned at the beginning of this chapter, all previous simulation exercises we have conducted assumed that the temperature response functions are fixed for each climate zone until the end of the century. This implicitly assumes that people do not adapt to a changing climate, which is a crucial and potentially non-credible assumption. If the coastal areas of California will experience higher mean temperatures and more frequent extreme heat events, it is likely that newly constructed homes will have built in central air conditioning. Further, owners of existing homes may install air conditioning equipment ex-post. This type of adaptation would result in a stronger temperature response at higher temperatures, whereby the temperature elasticity in the highest bins would increase over time. On the other hand, forward-looking planners and policy makers may put in place more stringent building codes for new construction combined with more stringent appliance standards, which would decrease the energy intensity of existing capital and homes. California has a long history of these policies and is considered a worldwide leader in these energy efficiency policies. These more stringent policies are designed to offset future increases in consumption. Bottom up engineering models by design can capture the impact of building- and device-specific changes due to regulations on energy consumption. Their drawback, however, is that they have to rely on a large number of assumptions regarding the composition of the housing stock and appliances, as well as making behavioral assumptions about the individuals using them.

In order to conduct meaningful simulations, it is imperative to obtain micro level data on heating technology and air conditioner penetration, which currently are not available on a large scale. While we cannot conduct such a detailed simulation

Table 4 Simulated percent increase in residential electricity consumption relative to 1980–2000 assuming a common low (zone 7) and high (zone 12) temperature response function

Forcing	Zone 7 (%)		Zone 12 (%)	
	A2	B1	A2	B1
2000-19	1	−1	13	7
2020-39	1	1	5	7
2040-59	2	1	29	13
2060-79	2	1	57	28
2080-99	3	0	122	40

incorporating specific technology changes, we conduct the following thought experiment. Our baseline simulation has assumed that each climate zone maintains its specific response function throughout the remainder of the century. To bound how important the heterogeneity in the response function is to the aggregate simulation results, we design an "almost best case" scenario, where we assume that all zones have the response function of coastal San Diego (Zone 7). This zone's response function is relatively flat. The left panel of Table 4 shows the simulation results assuming this optimistic scenario.

Worst-case increases under forcing A2 results in a 3% increase in electricity consumption by the end of the century, which is essentially flat compared to the baseline simulation shown in Table 2. Next we come up with an "almost worst case" scenario, where we let all of California adopt the response function of Zone 12, the Central Valley. The right panel of Table 4 shows the results from this simulation. The overall increases in simulated electricity consumption are more than twice those of the baseline scenario across simulations considered in Table 2. The large impact of the assumed temperature responsiveness function on overall simulated residential electricity consumption underlines the importance of improving energy efficiency of buildings and appliances.

6 Conclusions

This study has provided the first estimates of California's residential electricity consumption under climate change based on a large set of panel micro-data. We use random and therefore exogenous weather shocks to identify the effect of weather on household electricity consumption. We link climate zone specific weather response functions to a state of the art downscaled global circulation model to simulate growth in aggregate electricity consumption. We further incorporate potentially higher prices and population levels to provide estimates of the relative sensitivity of aggregate consumption to changes in these factors. Finally we show estimates of aggregate consumption under an optimistic and pessimistic scenario of temperature response.

There are several novel findings from this paper. First, simulation results suggest much larger effects of climate change on electricity consumption than previous studies. This is largely due to the highly non-linear response of consumption at higher temperatures. Our results are consistent with the findings by Deschênes and Greenstone (2011). They find a slightly smaller effect using national data. It is not surprising that impacts for California, a state with a smaller heating demand (electric or otherwise), would be bigger. Second, temperature response varies greatly across the climate zones in California—from flat to U-shaped to hockey stick shaped. This suggests that aggregating data over the entire state may ignore important nonlinearities, which combined with heterogeneous climate changes across the state may lead to underestimates of future electricity consumption. Third, population uncertainty leads to larger uncertainty over consumption than uncertainty over climate change. Finally, policies aimed at reducing the weather sensitivity of consumption can play a large role in reducing future electricity consumption. Specifically, region specific HVAC standards may play a significant role in offsetting some of the projecting increases in consumption.

References

Aroonruengsawat A, Auffhammer M (2009) Impacts of climate change on residential electricity consumption: evidence from billing data. California Energy Commission PIER Report CEC-500-2009-018-D

Aroonruengsawat A, Auffhammer M (2011) Impacts of climate change on residential electricity consumption: evidence from billing data. In: Libecap G, Steckel RH (eds) The economics of climate change: past and present. University of Chicago Press

Auffhammer M, Kellogg R (2011) Clearing the air? The effects of gasoline content regulation on air quality. Am Econ Rev 101(6):2687–2722

Baxter LW, Calandri K (1992) Global warming and electricity demand: a study of California. Energy Policy 20(3):233–244

Bureau of Economic Analysis (2008) Regional economic accounts. Washington, DC. www.bea.gov/regional/spi/default.cfm?series=summary

California Energy Commission (2005) Integrated energy policy report. Sacramento, California

Cayan D, Tyree M, Dettinger M, Hidalgo H, Das T, Maurer E, Bromirski P, Graham N, Flick R (2009) Climate change scenarios and sea level rise estimates for the California 2009 climate change scenarios assessment. California Energy Commission PIER Report CEC-500-2009-014-F

Cline WR (1992) The economics of global warming. Institute for International Economics, Washington

Crowley C, Joutz F (2003) Hourly electricity loads: temperature elasticities and climate change. In: 23rd US Association of Energy Economics North American conference

Deschênes O, Greenstone M (2011) Climate change, mortality, and adaptation: evidence from annual fluctuations in weather in the US. American Economic Journal: Applied Economics, 3(4): 152-185

Energy Information Administration (EIA) (2008) State energy data system. Washington, DC

Espey JA, Espey M (2004) Turning on the lights: a meta-analysis of residential electricity demand elasticities. J Agric Appl Econ 36(1):65–81

Fisher FM, Kaysen C (1962) A study in econometrics: the demand for electricity in the US. North Holland Publishing Company

Franco G, Sanstad A (2008) Climate change and electricity demand in California. Clim Change 87:139–151

Hanemann W (1984) Discrete/continuous models of consumer demand. Econometrica 52:541–561

Houthakker HS, Taylor LD (1970) Consumer demand in the United States: analyses and projections. Harvard University Press, Cambridge

Intergovernmental Panel on Climate Change (IPCC) (2000) Emissions scenarios. Cambridge University Press, Cambridge

Mansur E, Mendelsohn R, Morrison W (2008) Climate change adaptation: a study of fuel choice and consumption in the US energy sector. J Environ Econ Manage 55(2):175–193

Reiss PC, White WM (2005) Household electricity demand revisited. Rev Econ Stud 72:853–883

Rosenthal D, Gruenspecht H, Moran E (1995) Effects of global warming on energy use for space heating and cooling in the United States. Energy J 16:77–96

Sanstad AH, Johnson H, Goldstein N, Franco G (2009) Long-run socioeconomic and demographic scenarios for California. Prepared for the Public Interest Energy Research Program, California Energy Commission, CEC-500-2009-013-F

Schlenker W, Roberts M (2009) Nonlinear temperature effects indicate severe damages to US crop yields under climate change. Proc Natl Acad Sci US Am 106(37):15594–15598

Erratum to: Simulating the impacts of climate change, prices and population on California's residential electricity consumption

Maximilian Auffhammer · Anin Aroonruengsawat

Published online: 29 June 2012
© Springer Science+Business Media B.V. 2012

Erratum to: Climatic Change (2011) 109 (Suppl 1):S191–S210
DOI 10.1007/s10584-011-0299-y

Abstract We discovered an error in the computer code generating the simulation results in section 5 of Auffhammer and Aroonruengsawat (Clim Chang 109(Supplement 1):191–210, 2011). While four out of five main findings are unaffected, the simulated impacts of climate change on annual residential electricity consumption are an order of magnitude smaller, which is consistent with findings in the previous literature.

1 Corrected simulation results

The econometric model based on equation (1) in Auffhammer and Aroonruengsawat (2011) uses counts of days in 14 discrete weather bins during a billing period, which range from 25 to 35 days in length. The simulation exercise, inexcusably, did not scale the climate model output to the average billing period length of 30 days but used annual counts instead. As the estimated equation (1) is log linear in nature, the simulation results based on equation (2) in the paper are incorrect, as $\frac{e^{ax}}{e^{ay}} \neq \frac{e^{x}}{e^{y}}$. This error does not affect the econometric estimation results in section (4), yet significantly changes the results of the simulations conducted in section (5) of the paper.

We have posted a corrected manuscript of the entire paper at http://are.berkeley.edu/~auffham.

The online version of the original article can be found at http://dx.doi.org/10.1007/s10584-011-0299-y.

M. Auffhammer (✉)
UC Berkeley ARE/IAS, 207 Giannini Hall, Berkeley, CA 94720-3310, USA
e-mail: auffhammer@berkeley.edu

A. Aroonruengsawat
Faculty of Economics, Thammasat University, Bangkok, Thailand
e-mail: anin@econ.tu.ac.th

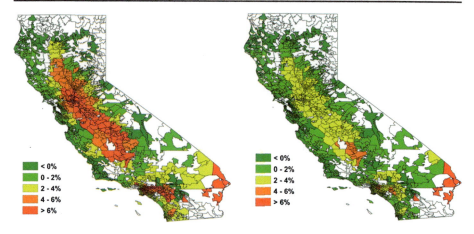

Fig. 1 Simulated increase in household electricity consumption by zip code for the period 2080–99 in percent over 1961–1990 simulated consumption. Model NCAR PCM forced by IPCC SRES A2 (*left*) and IPCC SRES B1 (*right*)

Figure 1 below displays the corrected versions of Figures (3d) and (4d) in the original paper. The spatial pattern of the corrected household level impacts is almost identical to the distribution shown in the paper, yet the scale is different by an order of magnitude. This change has implications for the predicted increases of aggregate residential electricity consumption for the state. Table 2 below shows the corrected table 2 in the paper. The predicted increases in residential electricity consumption by end of century - without

Table 2 Simulated percent increase in residential electricity consumption relative to 1961–1990 for the constant, low price and high price scenarios

Bin type downscaling	Price Increase (%)	Equidistant (%)				Percentile (%)			
		BCSD		CA		BCSD		CA	
IPCC scenario		A2	B1	A2	B1	A2	B1	A2	B1
2000–19	±0	1 %	0 %	1 %	0 %	1 %	0 %	1 %	0 %
2020–39	±0	0 %	1 %	0 %	1 %	1 %	1 %	1 %	1 %
2040–59	±0	1 %	1 %	1 %	1 %	1 %	1 %	1 %	1 %
2060–79	±0	2 %	1 %	2 %	1 %	2 %	1 %	2 %	1 %
2080–99	±0	3 %	1 %	3 %	1 %	3 %	1 %	3 %	1 %
2000–19	±0	1 %	0 %	1 %	0 %	1 %	0 %	1 %	0 %
1020–39	+30	−10 %	−10 %	−10 %	−10 %	−10 %	−10 %	−10 %	−10 %
2040–59	+30	−9 %	−10 %	−9 %	−10 %	−9 %	−9 %	−9 %	−9 %
2060–79	+30	−9 %	−9 %	−9 %	−9 %	−8 %	−9 %	−8 %	−9 %
2080–99	+30	−8 %	−9 %	−8 %	−9 %	−7 %	−9 %	−7 %	−9 %
2000–19	±0	1 %	0 %	1 %	0 %	1 %	0 %	1 %	0 %
1020–39	+30	−10 %	−10 %	−10 %	−10 %	−10 %	−10 %	−10 %	−10 %
2040–59	+60	−19 %	−20 %	−19 %	−20 %	−19 %	−20 %	−19 %	−20 %
2060–79	+60	−19 %	−20 %	−19 %	−20 %	−19 %	−19 %	−19 %	−19 %
2080–99	+60	−18 %	−19 %	−18 %	−19 %	−18 %	−19 %	−18 %	−19 %

accounting for population growth or price increases - range from 1 to 3 % using the NCAR PCM model, which is slightly lower than the 3 to 5 % range for all sectors using aggregate load in CalISO suggested by Franco and Sanstad (2008). The price simulations using the corrected climate simulations suggest that, subject to the caveats in the paper, the aggressive price scenario is consistent with an 18–19 % decrease in electricity consumption over baseline, which is significant. Table 3 corrects the consumption estimates taking into account population growth and climate change. For the medium population growth scenario, aggregate consumption is consistent with 133–139 % increase in consumption. The high population growth scenario suggests increases in consumption by between 272 and 280 %.

2 Implications

There were five main conclusions in the paper. The first four findings are unaffected by our coding error. First, the econometrically estimated response of residential electricity consumption to temperature is spatially heterogeneous. Second, the simulated impacts of climate change on household level electricity consumption are also spatially heterogeneous, with the Central Valley and South Eastern parts of the state predicted to experience the largest increases. Third, two sequential 30 % increases in electricity price are simulated to significantly decrease electricity consumption from this sector. Fourth, increases in

Table 3 Simulated percent increase in residential electricity consumption relative to 1961–1990 for the low, medium and high population growth scenarios

Bin type downscaling IPCC scenario	Price Increase (%)	Equidistant (%) BCSD A2	B1	CA A2	B1	Percentile (%) BCSD A2	B1	CA A2	B1
Low Population Growth Scenario									
2000–19	±0	12 %	11 %	12 %	11 %	12 %	11 %	12 %	11 %
2020–39	±0	25 %	25 %	25 %	25 %	25 %	25 %	25 %	25 %
2040–59	±0	29 %	28 %	29 %	28 %	29 %	28 %	29 %	28 %
2060–79	±0	31 %	30 %	31 %	30 %	32 %	31 %	32 %	31 %
2080–99	±0	39 %	37 %	39 %	37 %	40 %	37 %	40 %	37 %
Medium Population Growth Scenario									
2000–19	±0	13 %	13 %	13 %	13 %	13 %	13 %	13 %	13 %
1020–39	±0	42 %	42 %	42 %	42 %	42 %	42 %	42 %	42 %
2040–59	±0	72 %	72 %	72 %	72 %	73 %	72 %	73 %	72 %
2060–79	±0	103 %	101 %	103 %	101 %	103 %	102 %	103 %	102 %
2080–99	±0	138 %	133 %	138 %	133 %	139 %	134 %	139 %	134 %
High Population Growth Scenario									
2000–19	±0	17 %	17 %	17 %	17 %	17 %	17 %	17 %	17 %
1020–39	±0	57 %	57 %	57 %	57 %	57 %	57 %	57 %	57 %
2040–59	±0	105 %	104 %	105 %	104 %	105 %	104 %	105 %	104 %
2060–79	±0	173 %	171 %	173 %	171 %	173 %	171 %	173 %	171 %
2080–99	±0	278 %	272 %	278 %	272 %	280 %	273 %	280 %	273 %

population will have significantly larger impacts on increases in consumption than climate change, and these increases are likely much larger than the aggressive price scenario can offset.

The finding that has changed significantly is the magnitude of the impacts of climate change at the household and aggregate level. They are an order of magnitude smaller than previously stated, which means that for annual consumption based on our simulation without adaptation, climate change is predicted to have minor effects on annual electricity consumption. This does not rule out significant impacts during peak times.

References

Auffhammer M, Aroonruengsawat A (2011) Simulating California's future residential electricity consumption under different scenarios of climate change. Clim Chang 109(Supplement 1):191–210

Franco G, Sanstad A (2008) Climate change and electricity demand in California. Clim Chang 87:139–151

Effects of climate change and wave direction on longshore sediment transport patterns in Southern California

Peter N. Adams · Douglas L. Inman · Jessica L. Lovering

Received: 21 September 2011 / Accepted: 26 September 2011 / Published online: 24 November 2011
© Springer Science+Business Media B.V. 2011

Abstract Changes in deep-water wave climate drive coastal morphologic change according to unique shoaling transformation patterns of waves over local shelf bathymetry. The Southern California Bight has a particularly complex shelf configuration, of tectonic origin, which poses a challenge to predictions of wave driven, morphologic coastal change. Northward shifts in cyclonic activity in the central Pacific Ocean, which may arise due to global climate change, will significantly alter the heights, periods, and directions of waves approaching the California coasts. In this paper, we present the results of a series of numerical experiments that explore the sensitivity of longshore sediment transport patterns to changes in deep water wave direction, for several wave height and period scenarios. We outline a numerical modeling procedure, which links a spectral wave transformation model (SWAN) with a calculation of gradients in potential longshore sediment transport rate (CGEM), to project magnitudes of potential coastal erosion and accretion, under proscribed deep water wave conditions. The sediment transport model employs two significant assumptions: (1) quantity of sediment movement is calculated for the transport-limited case, as opposed to supply-limited case, and (2) nearshore wave conditions used to evaluate transport are calculated at the 5-meter isobath, as opposed to the wave break point. To illustrate the sensitivity of the sedimentary system to changes in deep-water wave direction, we apply this modeling procedure to two sites that represent two different coastal exposures and bathymetric configurations. The Santa Barbara site, oriented with a roughly west-to-east trending coastline, provides an example where the behavior of the coastal erosional/accretional character is exacerbated by deep-water wave climate intensification. Where sheltered, an increase in wave height enhances accretion, and where exposed, increases in wave height and period enhance erosion. In contrast, all simulations run for the Torrey

P. N. Adams (✉)
Department of Geological Sciences, University of Florida, Gainesville, FL 32611, USA
e-mail: adamsp@ufl.edu

D. L. Inman
Scripps Institution of Oceanography, University of California, San Diego, La Jolla, CA 92093, USA

J. L. Lovering
Department of Geological Sciences, University of Florida, Gainesville, FL 32611, USA

Pines site, oriented with a north-to-south trending coastline, resulted in erosion, the magnitude of which was strongly influenced by wave height and less so by wave period. At both sites, the absolute value of coastal accretion or erosion strongly increases with a shift from northwesterly to westerly waves. These results provide some examples of the potential outcomes, which may result from increases in cyclonic activity, El Niño frequency, or other changes in ocean storminess that may accompany global climate change.

1 Introduction

In California, 80% of the state's residents live within 30 miles of the coast (Griggs et al. 2005). To mitigate the effects of climate change on coastal communities, it is necessary to assess the oceanographic and geomorphic changes expected within the coastal zone. Effective planning for the future of the California coast will need to draw on climate models that predict the forcing scenarios and coastal change models that predict the coast's response.

Coastal landforms exhibit dynamic equilibrium by adjusting their morphology in response to changes in sea level, sediment supply, and ocean wave climate. Global climate change exerts varying degrees of influence on each of these factors. Proxy records indicate that wave climate has influenced coastal sedimentary accretion throughout the Holocene (Masters 2006) and, although the causative links between climate change and severe storms are not reconciled in the scientific literature (Emanuel 2005; Emanuel et al. 2008), it is well-accepted that changes in ocean wave climate (i.e. locations, frequency, and severity of open ocean storms) will bring about changes in the locations and magnitudes of coastal erosion and accretion in the future (Slott et al. 2006). Numerous studies indicate that changes in ocean wave climate are detectable (Gulev and Hasse 1999; Aumann et al. 2008; Komar and Allan 2008; Wang et al. 2009), but translating these open ocean changes to nearshore erosional driving forces is complex, and requires an understanding of the interactions between wave fields and the bathymetry of the continental shelf. Along the southern California coast, from Pt. Conception to the U.S./Mexico border, the situation is further complicated by the intricate shelf bathymetry and the presence of the Channel Islands, which prominently interfere with incoming wave field (Shepard and Emery 1941).

In this paper we investigate potential effects of changes in ocean wave climate (wave height, period, and direction) on the magnitudes of coastal erosion and accretion at two, physiographically different sites within the Southern California Bight (SCB). We consider the physical setting of this location, describe our mathematical modeling approach, and present the results of a series of numerical experiments that explore a range of wave climates. Lastly, we discuss the implications of the modeled coastal behavior in light of some possible scenarios of global climate change.

2 Geomorphic and oceanographic setting

For the purposes of this study, we consider the SCB to extend from a northwestern boundary at Point Arguello (34.58° N, 120.65° W, Fig. 1) to the U.S.-Mexico border (32.54° N, 117.12° W, Fig. 1) south of San Diego. Tectonic processes along this active margin, between the Pacific and North American plates, are responsible for shaping the shallow ocean basins, continental shelf, and large-scale terrestrial landmasses (Christiansen and Yeats 1992). The edges of the plates on either side of the boundary have been folded

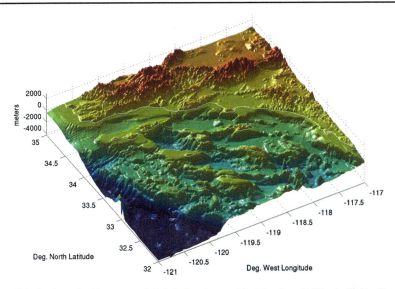

Fig. 1 Hillshade view of a 30-arc second digital elevation model of Southern California Bight with modern shoreline shown as a *white line*. Note complex appearance of high relief terrestrial landscape and irregular submarine basin bathymetry. Digital elevation data obtained from the NOAA National Geophysical Data Center Coastal Relief Model (http://www.ngdc.noaa.gov/mgg/coastal/startcrm.htm). The mean sea level shoreline and 5-m isobath were interpolated from this data set as well

and fractured by transpressional plate motions, creating the high relief terrestrial landscape, pocket beaches backed by resistant bedrock sea cliffs, narrow continental shelf, deeply incised submarine canyons, and irregularly shaped submarine basins that are characteristic of a collisional coasts, as classified by (Inman and Nordstrom 1971). In particular, the coastal mountain ranges and local shelf basins have been constructed by crustal displacement along a network of subparallel strike-slip faults, which characterize the plate interface (Hogarth et al. 2007). In general, these motions have resulted in the highly irregular, complex bathymetry that makes up the California Borderlands (Legg 1991), that feature the Channel Islands, as well as numerous submerged seamounts and troughs (Fig. 1).

The wave climate of Southern California has been extensively studied since the pioneering investigations that applied the theoretical relationships of wave transformation to predict breakers and surf along the beaches of La Jolla, California (Munk and Traylor 1947; Sverdrup and Munk 1947). Buoys maintained by the National Oceanic and Atmospheric Administration (NOAA) have greatly assisted understanding of deep-water wave conditions within the SCB (O'Reilly et al. 1996). Monitoring efforts continued to be improved through the development of the Coastal Data Information Page (CDIP) program, Scripps Institution of Oceanography, which provides modeled forecasts at a number of locations. Within the SCB, the presence of the Channel Islands (Fig. 1) significantly alters the deep-water (open ocean) wave climate to a more complicated nearshore wave field along the Southern California coast. The islands intercept waves approaching from almost any direction and the shallow water bathymetry adjacent to the islands refracts and reorients wave rays to produce a complicated wave energy distribution along the coast of the Southern California mainland. Several studies have targeted the sheltering effect of the Channel Islands within the SCB and the complexity of modeling wave transformation through such a complicated bathymetry (Pawka et al. 1984; O'Reilly and Guza 1993; Rogers et al. 2007). It has also been documented that wave reflection off sheer cliff faces in the Channel Islands can be a very important process in the alteration of wave energy along the mainland coast (O'Reilly et al. 1999). The resulting distribution of wave energy at the coast consists of dramatic longshore variability in wave energy flux and radiation stress. These factors are considered to be fundamental in generating the nearshore currents responsible for longshore sediment transport and the maintenance of sandy beaches. Some studies highlight evidence for changing storminess and wave climate in the northeast Pacific Ocean (Bromirski et al. 2003, 2005). Recently, (Adams et al. 2008) examined a 50-year numerical hindcast of deep-water, winter wave climate in the bight, to understand the correlation of decadal-to-interannual climate variability with offshore wave fields. Their study found that El Niño winters during Pacific Decadal Oscillation (PDO) warm phase have significantly more energetic wave fields than those during PDO-cool phase, suggesting an interesting connection between global climate change and coastal evolution, based on patterns of storminess.

3 Model description

The numerical model employed to evaluate potential coastal change consists of two components: (1) a spectral wave transformation model, known as SWAN (Booij et al. 1999), that calculates shoaling and refraction of a proscribed deep water wave field over a defined bathymetric grid, and reports coastal wave conditions, and (2) an empirically-derived longshore sediment transport formulation, referred to herein as CGEM (Coastal Geomorphic Erosion Model), that utilizes the coastal wave conditions derived by the wave transformation model to compute divergence of volumetric transport rates of nearshore sediment, also known as *divergence of drift* (Inman 1987; Inman and Jenkins 2003). This divergence of drift is the difference between downdrift and updrift volumetric transport rates (sediment outflow minus sediment inflow), and represents the volume of sedimentary erosion or accretion at a coastal compartment over the model time step. The interaction between the two components of the model is shown schematically in Fig. 2.

Two significant assumptions, and one model limitation, are invoked to simplify calculations. First, wave transformation is calculated to the fixed 5-meter isobath, which is usually seaward of the wave break point. Although the sediment transport model calls for wave conditions at the break point, wave breaking proceeds over a *breaker zone*, that can be several tens of meters wide, depending on the slope of the beach. Through several tests of the SWAN wave model, we have determined that, under vigorous deep water wave

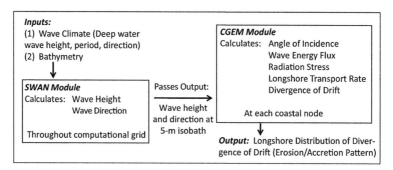

Fig. 2 Schematic diagram of numerical modeling procedure used in this study, showing relationship between SWAN and CGEM components

conditions (i.e. H_s=4–5 m and T=14 s), wave breaking initiates over a depth of 9–10 m with the percentage of wave breaking increasing shoreward. We consider the mean water depth within this breaker zone (4.5–5 m) to be a reasonable representative value to use for breakpoint conditions in our modeling procedure. Second, the longshore sediment transport formulation assumes a transport-limited case, as opposed to a supply-limited case. The transport-limited assumption results in the calculation of *potential* divergence of drift, which would reflect the case of an inexhaustible supply of nearshore sediment. If terrestrial sediment supply to the coast is limited, from decreased riverine inputs for example, this assumption may be challenged. Lastly, the two components of the model are not yet backward coupled, meaning that nearshore bathymetry is not updated after a calculation of divergence of drift is conducted. Hence, the model should not be run over a time series of changing wave conditions to simulate the evolutionary behavior of a coast. In this paper, we present the results of instantaneous scenarios of divergence of drift for individual sets of deep-water wave conditions.

3.1 Wave transformation model

SWAN, short for *Simulating WAves Nearshore*, is a 3rd generation, finite-difference, wave model that operates on the principle of a wave action balance (energy density divided by relative frequency) (Holthuijsen et al. 1993; Booij et al. 1999). In the absence of wind forcing, this model requires only two general inputs to perform a computation of the wave field throughout a region—bathymetry and deep-water wave conditions.

Ocean bathymetry exhibits a strong control on the direction and rate of wave energy translation. When the water depth is shallower than half the wavelength, interaction of wave orbitals with the sea floor causes shoaling transformation and refraction of waves, so the spatial pattern of nearshore wave energy depends strongly on the distribution of seafloor elevation. A review of some studies that investigated how bathymetric changes influence shoreline change by altering shoaling and refraction patterns is provided by (Bender and Dean 2003). In this study, we used two bathymetric grids for the wave transformation modeling, both obtained from the National Geophysical Data Center (NOAA) U.S. coastal relief model grid database (http://www.ngdc.noaa.gov/mgg/coastal/coastal.html). A spatially-coarse grid (30 arc-sec) was used to evaluate wave conditions over the entire bight (32°–35°N, 121°–117°W), providing 480×360 cell matrices of wave heights and directions covering approximately 124,000 km^2, in which each value represents the conditions for a 0.72 km^2 area of sea surface. Two smaller, high resolution grids (3 arc-sec), referred to herein as "nests", were used to evaluate local wave conditions on a finer spatial scale. The nests named *SntBrb* and *TorPns*, for portions of the Santa Barbara and Torrey Pines coasts respectively, each occupy approximately 700 km^2 within the domain of the main (coarse) grid (Fig. 3), resulting in each cell of output representing wave conditions over a 0.0072 km^2 area of sea surface.

At the western and southern margins of the coarse grid, deep-water wave conditions are proscribed as boundary conditions, consisting of significant wave height, peak period with a JONSWAP frequency distribution (Hasselmann et al. 1973), and peak direction with a cosine square spread of 15°. The wave field is computed from the grid boundaries over the input bathymetry to the coast. Example results from a SWAN run for the coarse grid are provided in Fig. 3. From the SWAN output of the coarse grid, the boundary conditions for the individual nests were obtained (examples provided in Figs. 4 and 5). Because the goal of the investigation was to examine "snapshots" of longshore distribution of erosion/ accretion patterns resulting from various deep-water wave scenarios, all SWAN runs conducted in this study were performed in stationary mode. Temporally evolving wave

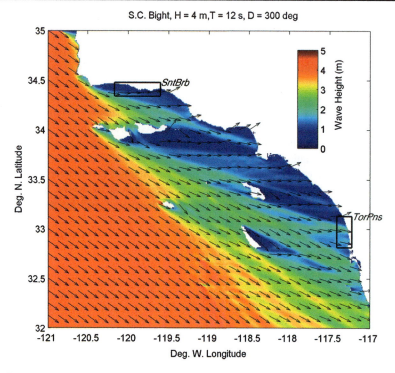

Fig. 3 Example wave height and direction output from SWAN wave transformation model over the entire coarse grid of the Southern California Bight. Locations of the two study sites explored in this study are shown in boxes

fields, such as those produced during a storm were not examined, thereby avoiding the problem of swell arrival timing often associated with stationary runs (Rogers et al. 2007).

3.2 Longshore sediment transport model

The output passed from the wave transformation model to the longshore sediment transport model includes the complete oceanic grids of significant wave height and peak directions at each "wet node" within the particular nest being examined. CGEM queries and interpolates the SWAN output at a known set of locations within the nest that constitute the 5-meter isobath, using an alongshore spacing of 100 m. These longshore sets of wave height and direction form the basis of the sediment transport computation, which originates from a semi-empirical formulation originally proposed by Komar and Inman (1970). This relationship was later modified and named the CERC formula (Rosati et al. 2002), which has been used in several coastal evolution modeling studies recently (Ashton et al. 2001; Ashton and Murray 2006). The CERC formula has been tested and found to provide results in close agreement with processed based models (Haas and Hanes 2004). In its most general form, the volumetric longshore sediment transport rate, Q_l, can be expressed as

$$Q_l = \frac{I_l}{(\rho_s - \rho_w)gN_o} \quad (1)$$

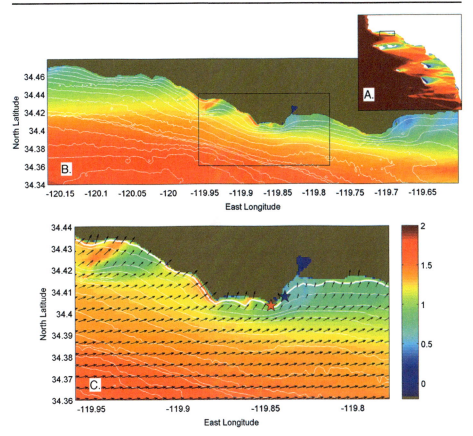

Fig. 4 Example wave height and direction output from SWAN wave transformation model for input conditions of H_{sig}=2 m, T=12 s, and α=270°. **a**. Wave height map showing results over entire Southern California Bight and location of *SntBrb* nest (*box*). **b**. Wave height map showing results over the nested Santa Barbara grid (*SntBrb*), approximately 50 km of west-east trending coast, and location of region of interest (*box*). **c**. Detailed wave height map showing wave direction vectors, bathymetric contours (10 m contour interval shown in thin white lines), location of sites SB-1 (*red star*) and SB-2 (*blue star*). Location of 5-meter isobath shown in *thick white line*. Wave heights on A., B., and C. are plotted with respect to the same colorbar whose units are meters

where I_l is the immersed-weight transport rate (Inman and Bagnold 1963), ρ_s and ρ_w are densities of quartz sediment (2,650 kg/m^3) and seawater (1,024 kg/m^3), respectively, g is the gravitational acceleration constant (9.81 m/s^2), and N_o is the volume concentration of solid grains (1—porosity), set to 0.6, for all numerical experiments in this study. The subscript l is used to represent the *longshore component* of any variable with which it is associated.

3.2.1 Angle of incidence

The angle of incidence is of primary importance in the calculation of longshore sediment transport. If wave rays approach the beach at an angle perfectly orthogonal to the trend of the coast, the longshore component of wave energy flux is zero, and there is no net longshore current to drive longshore sediment transport. If wave rays approach the beach at an oblique angle (somewhere between orthogonal and parallel),

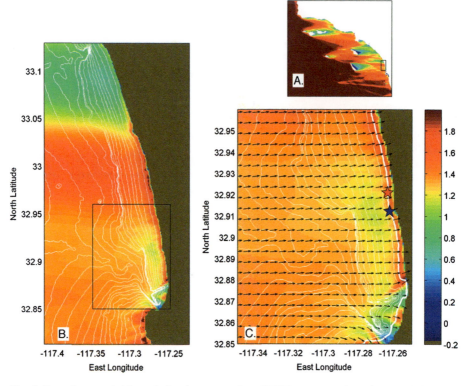

Fig. 5 Example wave height and direction output from SWAN wave transformation model for input conditions of H_{sig}=2 m, T=12 s, and α=270°. **a**. Wave height map showing results over entire Southern California Bight and location of *TorPns* nest (*box*). **b**. Wave height map showing results over the nested Torrey Pines grid (*TorPns*), approximately 40 km of north–south trending coast, and location of region of interest (*box*). **c**. Detailed wave height map showing wave direction vectors, bathymetric contours (10 m contour interval shown in *thin white lines*), location of sites TP-1 (*red star*) and TP-2 (*blue star*). Location of 5-meter isobath shown in *thick white line*. Wave heights on **a**., **b**., and **c**. are plotted with respect to the same colorbar whose units are meters

there is a component of wave energy flux parallel to the shoreline, which drives longshore sediment transport.

Along the 5-meter isobath, the coastal orientation is computed by a downcoast-moving sliding window computation, which fits a trendline to 5 adjacent points and reports an azimuth value for the midpoint of the segment. The coastal orientation is subtracted from the queried wave direction along the isobath to provide an angle of incidence.

3.2.2 Wave energy flux

The longshore component of wave energy flux is considered to provide the fluid thrust required to move sediment under the influence of the breaking wave bore. The governing equation is

$$P_l = ECn = \frac{1}{8}\rho_w g H^2 Cn \sin\alpha \cos\alpha \qquad (2)$$

where E is wave energy density, C is nearshore wave celerity, which is depth controlled, n is ratio of group to individual wave speed (~1 in shallow water and 1/2 in deep water), H is nearshore (breaking) wave height, α is the angle of incidence, the trigonometric components of which result from the tensor transformation of the onshore flux of longshore directed momentum (Longuet-Higgins and Stewart 1964). As noted above, we estimate breaking wave height by evaluating this quantity at the 5-meter isobath. The immersed weight transport rate is simply a scaled version of the longshore component of wave energy flux

$$I_l = KP_l \tag{3}$$

where the scaling parameter, K, is set to 0.8 in this study. It is noted that Komar and Inman (1970) tested this relationship at different locations, where beach sediment sizes differed, but found that the relationship held, irrespective of sedimentary texture.

3.2.3 Divergence of drift

After presenting the relationships for the longshore component of wave energy flux, immersed weight transport rate, and volumetric longshore sediment transport rate, we can show that a simple relationship for the divergence of drift is obtained by applying the vector differential operator to the volumetric longshore sediment transport rate using a dot product,

$$\nabla \cdot Q_l = \frac{\partial Q_l}{\partial x} \tag{4}$$

where x is the position along the coast or, in CGEM, position along the 5-meter isobath.

This quantity is, effectively, the calculated change in sediment volume over the longshore reach dx, during the time interval that these wave conditions are applied. Herein, we adopt the sign convention that divergence is defined as the net difference between sediment inflow and sediment outflow, making positive divergence of drift (where inflow exceeds outflow) result in accretion at a site, whereas negative divergence of drift (where outflow exceeds inflow) results in erosion. The longshore pattern of divergence of drift is therefore out of phase with longshore sediment transport, as expected. At longshore positions where transport is increasing at the greatest rate (positively sloping inflection points), divergence of drift is at a local minimum; where transport is decreasing at the greatest rate (negatively sloping inflection points), divergence of drift is at a local maximum; where transport is at a local minimum or maximum, divergence of drift should be zero.

4 Numerical experiments and results

To provide insight on how climate change-driven alteration of deep water wave conditions might affect the magnitude of erosion and accretion along the Southern California coast, we conducted a series of controlled numerical experiments at two physiographically-distinct, reaches of the Southern California coast, which we refer to as the Santa Barbara (*SntBrb* nest) and Torrey Pines (*TorPns* nest) sites. The goal of these experiments is to test the hypothesis that deep water wave direction exhibits critical control on the longshore sediment transport patterns at coastal sites within the bathymetrically complex SCB. Each of the locations chosen witness unique swell patterns resulting largely from their relative orientations with respect to the deep water wave field of the North Pacific Ocean, each site's local shelf bathymetry, and the blocking patterns that the Channel Islands provide to each site (Fig. 3). The two sites represent end member coastal orientations within the Bight; the *SntBrb* nest exhibits a west-to-east general shoreline orientation, whereas the *TorPns* nest exhibits a north-to-south

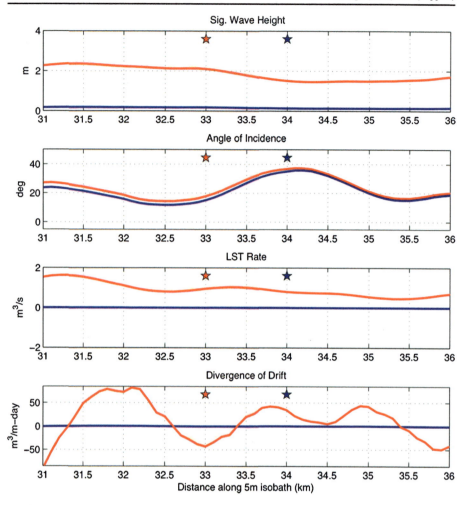

Fig. 6 Example CGEM output along 5-meter isobath within the nested Santa Barbara Grid (*SntBrb*) for two sets of deep water wave conditions, which differ only in wave direction. *Blue lines* show results of deep water conditions H_{sig}=4 m, T=16 s, and α=320°. *Red lines* show results of deep water conditions H_{sig}=4 m, T=16 s, and α=270°. Red and blue stars show locations of SB-1 and SB-2, used for numerical experiments. LST is an abbreviation of longshore sediment transport

general shoreline orientation. For each experiment, we conducted 104 SWAN-CGEM simulations that vary deep water wave direction from 260° to 320° in 5° intervals for four pairs of wave height period scenarios: (1) H=2 m, T=12 s, (2) H=2 m, T=16 s, (3) H=4 m, T=12 s, (4) H=4 m, T=16 s. These ranges span the distributions of deep water wave conditions documented for the SCB by Adams et al. (2008).

4.1 Site 1—Santa Barbara

The western end of Goleta Beach, adjacent the UCSB campus in southeastern Santa Barbara county, California, has been the site of regular nourishment due to chronic sand loss and the community desire to maintain a recreational beach. The site has witnessed profound changes in beach width, morphology, and sediment volume over the past 30 years with anecdotal photo histories documenting wide, well-vegetated, sandy beaches in the 1970's, which were fully inundated during the El Niño winters of 1982–83 and 1997–98 (Sylvester 2010). Waves entering the Santa Barbara channel, as swell, have been modeled by (Guza et al. 2001), and are regularly forecasted by the Coastal Data Information Program, CDIP.

4.1.1 SWAN transformed wave field

SWAN output from a moderate, westerly swell (H=2 m, T=12 s, α=270°) is shown in Fig. 4. The wave height distribution pattern for the entire SCB (Fig. 4a) illustrates the blocking effect of the Channel Islands. Figure 4b shows the decrease (by more than half) in wave height as waves enter shallow water. Figure 4c shows the significant amount of refraction that occurs as waves approach the nearshore and the significant sheltering experienced by the Goleta Beach site, herein referred to as SB-2 (blue star), as compared to the exposed site SB-1 (red star) located immediately west of Goleta Point, 1 km from SB-2.

4.1.2 CGEM results

It is instructional to observe the results of two CGEM simulations plotted along shore in the vicinity of SB-1 and SB-2. Figure 6 shows the strong influence exerted by wave direction at the Santa Barbara site. Output from two SWAN simulations are passed to CGEM to examine longshore patterns of potential sediment transport rate and divergence of drift. For each SWAN simulation, deep water wave height and period are set to 4.0 m and 16 s, but the deep water wave direction is 320° (northwesterly) in case A, representative typical La Niña storm wave conditions, as opposed to 270° (westerly) in case B, representative of El Niño storm wave conditions, during which time the jet stream occupies a more southern position than usual due to the anomalous atmospheric pressure distribution (Storlazzi and Griggs 2000). Comparison of the two deep water input cases is as follows. Along the 5-meter isobath, the significant wave height for Case A (northwesterly) is very small (<0.5 m), in comparison to deep water inputs conditions (H_{sig}=4 m), everywhere along the 5 km reach. For Case B, the wave heights are substantially higher, approximately 2°m everywhere along the reach. The angle of incidence varies for both Case A and Case B, but in the same pattern, developing a sizable longshore component of wave energy flux. The longshore sediment transport rates vary in much the same manner as wave heights for the two cases, with appreciable transport in Case B, and negligible transport in Case A. The potential divergence of drift pattern for Case B shows loss of sediment (erosion) at SB-1 (red star) and gain of sediment (accretion) at SB-2 (blue star). Potential divergence of drift is negligible along the length of the 5 km reach for Case A.

4.1.3 SntBrb experiment

The 104 SWAN-CGEM simulations which were run for the *SntBrb* experiment produced divergence of drift patterns which are reported for SB-1 and SB-2 in Fig. 7. This compendium of experiment results illustrates that the exposed SB-1 site experiences increasing erosion as wave conditions become more westerly. This is in contrast to the sheltered SB-2 site, which becomes more accretionary as wave direction becomes more westerly. Increasing deep-water wave height causes enhancement of erosional or accretional behavior at both SB-1 and SB-2, depending on the site tendency under milder conditions. Increasing wave period causes enhanced erosion at SB-1, but causes decreased accretion at SB-2.

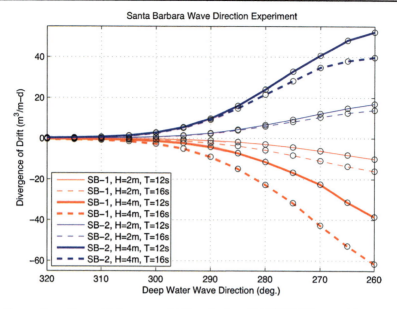

Fig. 7 Compendium of potential divergence of drift results from 104 SWAN-CGEM model simulations for the *SntBrb* nest at sites SB-1 and SB-2. Positive values of divergence of drift represent accretion and negative values represent erosion

4.2 Site 2—Torrey pines

Torrey Pines beach, located approximately 7 km north of Scripps Pier in La Jolla, California, has been the site of many scientific inquiries in the field of coastal processes (Thornton and Guza 1983; Seymour et al. 2005; Yates et al. 2009). The relatively straight, north–south trending reach resides within the Oceanside littoral cell and owes any longshore variation in wave energy flux to the blocking effects of the Channel Islands, rather than to complexities of nearshore bathymetry, save for the areas around the Scripps and La Jolla submarine canyons in the southern portion of the *TorPns* nest.

4.2.1 SWAN transformed wave field

As for the Santa Barbara site discussed above, we show the behavior of the Torrey Pines site to a SWAN simulation for a moderate, westerly swell ($H=2$ m, $T=12$ s, $\alpha=270°$) in Fig. 5. The demonstrable change in wave height visible around 33.05° north latitude in Fig. 5b is a result of waves penetrating through a window between Santa Catalina and San Clemente Islands during periods of westerly swell. The general shore-normal orientation of the wave field (for input conditions shown) promotes nearshore wave height increase as a result of shoaling in the absence of refraction.

4.2.2 CGEM results

Comparative examples of CGEM simulations from the *TorPns* nest for Cases A and B (described above in Section 4.1.2) are given in Fig. 8. Just as for the *SntBrb* nest, the significant wave height for Case A along the 5-meter isobath in the vicinity of Torrey Pines

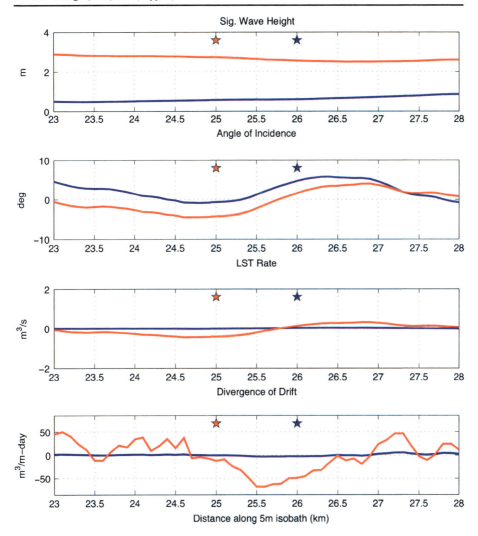

Fig. 8 Example CGEM output along 5-meter isobath within the nested Torrey Pines Grid (*TorPns*) for two sets of deep water wave conditions, which differ only in wave direction. *Blue lines* show results of deep water conditions H_{sig}=4 m, T=16 s, and α=320°. *Red lines* show results of deep water conditions H_{sig}=4 m, T=16 s, and α=270°. *Red and blue stars* show locations of TP-1 and TP-2, used for numerical experiments. LST is an abbreviation of longshore sediment transport

sites TP-1 and TP-2 is very small (<1.0 m). However, in the *TorPns* nest, this may be due to blockage of waves by the Channel Islands rather than to severe refraction as in the *SntBrb* nest. Angles of incidence for both case A and B at *TorPns* are small, approximately less than +/−5°. For case B, between kilometer markers 23 and 25.7, angle of incidence is negative which results in northward-directed longshore sediment transport pattern in this region, as opposed to the southward directed transport elsewhere in the nest. For case A, divergence of drift pattern is negligible everywhere within the 5 km span surrounding TP-1 and TP-2, whereas for case B, a strongly negative potential divergence of drift (erosion) emerges at Torrey Pines Beach, near TP-1 and TP2.

4.2.3 TorPns experiment

As for the *SntBrb* nest discussed in Section 4.1.3, the 104 SWAN-CGEM simulations, which were run for the *TorPns* experiment produced divergence of drift patterns which are reported for TP-1 and TP-2 in Fig. 9. All simulations run for the Torrey Pines site resulted in erosion. At TP-1, the peak in magnitude of potential divergence of drift occurs when waves are just north of westerly (275°–290°). Increases in wave period slightly increased erosion for 2°m waves, and shifted the peak direction for maximum erosion for 4°m waves, at TP-1. At TP-2, wave height has a profound influence; doubling of the deep water wave height causes potential divergence of drift to approximately triple across the directional range. Increasing wave period, however, had negligible effect at site TP-2.

5 Summary and implications

A numerical modeling procedure for assessing the patterns of littoral sediment transport in Southern California has been presented. The procedure combines a spectral wave transformation model with a calculation of gradients (divergence) in longshore sediment transport rates, assuming transport-limited conditions. To illustrate some specific coastal impacts resulting from climate change, we have applied this procedure to two physically-distinct sites within the SCB. We conducted a sensitivity analysis at the two study sites, whereby effects of variability in deep water wave direction were explored for four wave height/wave period combinations.

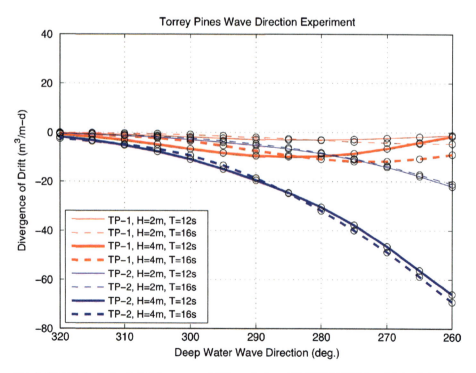

Fig. 9 Compendium of potential divergence of drift results from 104 SWAN-CGEM model simulations for the *TorPns* nest at sites TP-1 and TP-2

This study demonstrates that the longshore sediment transport patterns in the littoral zone, and therefore the locations of erosional hotspots, along the Southern California coast are extremely sensitive to deep water wave direction. We speculate that this sensitivity is due to two principle reasons: (1) the severe refraction required by northwesterly waves in order to be incident upon south facing coasts (e.g. *SntBrb* nest), which could also be considered a sheltering effect of Pt. Arguello, and (2) the sheltering effects of the Channel Islands (blocking of incoming swells and additional refraction of grazing waves) on west facing coasts (*TorPns* nest). Observations of wave interruption by islands were made decades ago by Arthur (1951).

Although the cross-shore transport of sediment in the littoral zone is not specifically addressed in these numerical experiments, we acknowledge its potential importance during large wave events. Long-period swells, often associated with large wave events, will increase refraction, which increases the cross-shore component of wave energy flux. Associated high wave set-up can promotes off-shore transport and the temporary storage of littoral sediment in offshore bars. Despite the fact that cross-shore transport can be strong during events, the associated "erosion" is often temporary, as recovery proceeds relatively rapidly after a large wave event via onshore transport of bar sediment.

It is noteworthy that the Santa Barbara wave direction experiment illustrated how the divergence of drift at a site exposed to the open ocean (SB-1) experiences enhancement of erosion as the wave field intensifies (increased wave heights and periods), whereas the divergence of drift at a site locally sheltered by a headland (SB-2) experiences enhancement of accretion as the deep water wave heights are increased. The fact that the sheltered site experienced slight decreases in accretion for increased wave periods may be due to the higher degree of refraction that the longer period swells undergo before reaching the coast.

The Torrey Pines experiment reveals interesting behavior regarding the direction of longshore transport and erosion. As shown in Fig. 8, for a westerly wave field of strong intensity (H_s=4 m, T=16 s, a=270°), the angle of incidence between position markers 23–25.75 km is negative, but changes to positive between position markers 25.75–28 km. The result of this change in angle of incidence is a change in direction of longshore transport from northward to southward. However, divergence of drift is negative for this set of deep water wave conditions at both TP-1, where transport direction is northward, and TP-2, where transport direction is southward. This persistence of erosional character at this site, irrespective of transport direction, suggests that continental shelf bathymetry may exert a unique control on nearshore wave fields at this site.

It is observed that at both SB-1 (*SntBrb* nest) and TP-2 (*TorPns* nest), negative divergence of drift (erosion) appears to operate under all deep water conditions simulated. This brings up two questions: (1) Why does the coast at SB-1 protrude seaward relative to the coast at SB-2, if the SB-1 shoreline is retreating under all simulated conditions? (2) Why does the Torrey Pines coastline maintain a relatively straight appearance if the shoreline at TP-2 is consistently retreating more rapidly than the shoreline at TP-1? Several explanations are offered to address these discrepancies and we suspect the answer is a combination of these. First, as mentioned above, the erosion/accretion portion of the CGEM model assumes transport-limited conditions, meaning that deficiencies in sediment supply are not considered to play a role in coastal landform evolution. If sediment supply is limited at these sites, then the model may overestimate the magnitude of divergence of drift. Second, the model only considers the movement of sediment alongshore at these coastal sites and does not address the rocky cliffs that back the beaches along much of the Southern California coast. During high wave conditions when sediment supply is limited, it is quite likely that beach sand is temporarily stored offshore in bars and waves directly impact the bedrock cliffs, whose retreat is not governed by Eq. (4) above. To simulate shoreline retreat of an exposed cliffed coast, bare bedrock cutting processes must be adequately modeled. Neither of these caveats, however, changes the fact that the gradients in potential longshore sediment transport patterns are highly sensitive to deep water wave conditions, and particularly, to wave direction.

These results have implications regarding climate change, longshore sediment transport patterns, and the distribution of hotspots of coastal erosion within the SCB. Our numerical simulations illustrate a dramatic increase in absolute value of divergence of drift as wave climate approaches westerly deep water wave directions. It has been documented that during El Niño winters, waves entering the Southern California Bight tend to be more westerly than during non El Niño winters (Adams et al. 2008). Therefore, model results presented herein are consistent with the observations of severe coastal change in California during the El Niño winters of 1982–83, and 1997–98 (Storlazzi et al. 2000). Recent research investigating tropical cyclonic behavior during the early Pliocene (5–3 ma) reveals a feedback that may serve to increase hurricane frequency and intensity in the central Pacific during warmer intervals (Federov et al. 2010). This is relevant to Earth's current climate trend because the early Pliocene is considered a possible analogue to modern greenhouse conditions. Federov et al. (2010) provide results of numerical simulations of tropical cyclone tracks in early Pliocene climate, which illustrate a dramatic poleward shift in sustained hurricane strength within the eastern Pacific Ocean. This pattern results in increased westerly storminess, implying that warmer climates will cause the SCB to witness greater absolute values of divergence of drift. In other words, a more volatile coastline, exhibiting higher magnitude erosion and accretion, might be expected as a result of increased frequency of strong westerly waves.

Acknowledgements This manuscript benefitted from thoughtful comments of three anonymous reviewers as well as conversations with Shaun Kline. This research was funded by the California Energy Commission's (CEC) Public Interest Energy Research Program. Special thanks are due to Guido Franco at the CEC, and the other guest editors of this special issue.

References

Adams P, Inman D, Graham N (2008) Southern California deep-water wave climate: characterization and application to coastal processes. J Coast Res 24(4):1022–1035

Arthur R (1951) The effect of islands on surface waves. Bulletin of the Scripps Institution of Oceanography p 6 no 1

Ashton A, Murray A (2006) High-angle wave instability and emergent shoreline shapes: 1. modeling of sand waves, flying spits, and capes. J Geophys Res Earth Surf 111(F4):F04,011

Ashton A, Murray A, Arnoult O (2001) Formation of coastline features by large-scale instabilities induced by high-angle waves. Nature 414(6861):296–300

Aumann H, Ruzmaikin A, Teixeira J (2008) Frequency of severe storms and global warming. Geophys Res Lett 35(19):1516

Bender C, Dean R (2003) Wave field modification by bathymetric anomalies and resulting shoreline changes: a review with recent results. Coast Eng 49(1–2):125–153

Booij N, Ris R, Holthuijsen L (1999) A third-generation wave model for coastal regions. 1. Model description and validation. J Geophys Res 104(C4):7649–7666

Bromirski P, Flick R, Cayan D (2003) Storminess variability along the California coast: 1858–2000. J Clim 16(6):982–993

Bromirski P, Cayan D, Flick R (2005) Wave spectral energy variability in the northeast pacific. J Geophys Res 110:C03,005

Christiansen RL, Yeats RS (1992) Post-Laramide geology of the U.S. Cordilleran region. In: Burchfiel BC, Lipman PW, and Zoback ML (eds), The Cordilleran orogen: conterminous U.S.: the geology of North America [DNAG] Vol. G-3: geological society of America, p. 261–406

Emanuel K (2005) Increasing destructiveness of tropical cyclones over the past 30 years. Nature 436 (7051):686–688
Emanuel K, Sundararajan R, Williams J (2008) Hurricanes and global warming. Bull Am Meteorol Soc 89 (3):347–367
Federov AV, Brierley CM, Emanuel K (2010) Tropical cyclones and permanent El Niño in the early Pliocene epoch. Nature 463:1066–1070
Griggs GB, Patsch K, Savoy LE (2005) Living with the changing California coast p 540
Gulev S, Hasse L (1999) Changes of wind waves in the north atlantic over the last 30 years. Int J Climatol 19 (10):1091–1117
Guza R, O'Reilly W, Region USMMSPO of California U Institute SBMS (2001) Wave prediction in the Santa Barbara channel: final technical summary, final technical report. US Dept of the Interior, Minerals Management Service, Pacific OCS Region p 16
Haas K, Hanes DM (2004) Process based modeling of total longshore sediment transport. J Coast Res 20 (3):853–861
Hasselmann K, Barnett T, Bouws E, Carlson H, Cartwright D, Enke K, Ewing J, Gienapp H, Hasselmann DE, Meerburg A, Mller P, Olbers D, Richter K, Sell W, Walden H (1973) Measurements of wind-wave growth and swell decay during the joint north sea wave project (JONSWAP). Ergnzungsheft zur Deutschen Hydrographischen Zeitschrift Reihe 8(12):95
Hogarth L, Babcock J, Driscoll N, Dantec N, Haas J, Inman D, Masters P (2007) Long-term tectonic control on Holocene shelf sedimentation offshore La Jolla, California. Geology 35(3):275
Holthuijsen LH, Booij N, Ris RC (1993) A spectral wave model for the coastal zone. Proceedings of the 2nd international symposium on ocean wave measurement and analysis, New Orleans, LA, pp. 630–641
Inman DL (1987) Accretion and erosion waves on beaches. Shore Beach 61(3&4)
Inman DL, Bagnold RA (1963) Littoral processes. The Sea 6:529–553
Inman D, Jenkins SA (2003) Accretion and erosion waves on beaches. Encyclopedia Coast Sci pp 1–4
Inman D, Nordstrom C (1971) On the tectonic and morphologic classification of coasts. J Geol 79(1):1–21
Komar P, Allan J (2008) Increasing hurricane-generated wave heights along the US east coast and their climate controls. J Coast Res 24(2):479–488
Komar P, Inman D (1970) Longshore sand transport on beaches. J Geophys Res Oceans 70(30):5914–5927
Legg MR (1991) Developments in understanding the tectonic evolution of the California continental borderland. From Shoreline to Abyss: contributions in marine geology in honor of Francis Parker Shepard (Society for Sedimentary Geology):291–312
Longuet-Higgins M, Stewart R (1964) Radiation stresses in water waves; a physical discussion, with applications. Deep-Sea Res 11(4):529–562
Masters P (2006) Holocene sand beaches of southern California: ENSO forcing and coastal processes on millennial scales. Palaeogeogr Palaeoclimatol Palaeoecol 232(1):73–95
Munk W, Traylor M (1947) Refraction of ocean waves: a process linking underwater topography to beach erosion. J Geol
O'Reilly WC, Guza RT (1993) Comparison of two spectral wave models in the Southern California Bight. Coast Eng 19(3):263–282
O'Reilly WC, Herbers THC, Seymour RJ, Guza RT (1996) A comparison of directional buoy and fixed platform measurements of Pacific swell. J Atmos Ocean Technol 13(1):231–238
O'Reilly WC, Guza RT, Seymour RJ (1999) Wave prediction in the Santa Barbara Channel, Proc. 5th California Islands symposium, mineral management service, Santa Barbara CA, March 29–31
Pawka SS, Inman DL, Guza RT (1984) Island sheltering of surface gravity waves: model and experiment. Cont Shelf Res 3:35–53
Rogers W, Kaihatu J, Hsu L, Jensen R, Dykes J, Holland K (2007) Forecasting and hindcasting waves with the SWAN model in the southern California Bight. Coast Eng 54(1):1–15
Rosati J, Walton TL, Bodge K (2002) Longshore sediment transport. Coastal engineering manual, Part III, coastal sediment processes chapter 2–3, volume engineer manual 1110-2-1100 U.S. Army corps of engineers
Seymour R, Guza R, O'Reilly W, Elgar S (2005) Rapid erosion of a small Southern California beach fill. Coast Eng 52:151–158
Shepard FP, Emery KO (1941) Submarine topography off the California coast: Canyons and tectonic interpretation. US Geological Survey Special Paper p 171
Slott J, Murray A, Ashton A, Crowley T (2006) Coastline responses to changing storm patterns. Geophys Res Lett 33(18):L18,404
Storlazzi C, Griggs G (2000) Influence of El Niño–Southern Oscillation (ENSO) events on the evolution of central California's shoreline. Geol Soc Am Bull 112(2):236
Storlazzi C, Willis C, Griggs G (2000) Comparative impacts of the 1982–83 and 1997–98 El Niño winters on the central California coast. J Coast Res 16(4):1022–1036

Sverdrup HU, Munk WH (1947) Wind, sea, and swell: theory of relations for forecasting. US Navy Dept, Hydrographic Office, HO Pub (601):1–44

Sylvester AG (2010) UCSB beach: 41 years of waxing and waning. http://wwwgeolucsbedu/faculty/sylvester/UCSBbeacheshtml

Thornton E, Guza R (1983) Transformation of wave height distribution. J Geophys Res 88(10):5925–5938

Wang X, Swail V, Zwiers F, Zhang X, Feng Y (2009) Detection of external influence on trends of atmospheric storminess and northern oceans wave heights. Clim Dyn 32(2):189–203

Yates M, Guza R, O'Reilly W, Seymour R (2009) Seasonal persistence of a small Southern California beach fill. Coast Eng 56:559–564

Potential impacts of increased coastal flooding in California due to sea-level rise

Matthew Heberger · Heather Cooley · Pablo Herrera · Peter H. Gleick · Eli Moore

Received: 12 February 2010 / Accepted: 21 September 2011 / Published online: 24 November 2011
© Springer Science+Business Media B.V. 2011

Abstract California is likely to experience increased coastal flooding and erosion caused by sea-level rise over the next century, affecting the state's population, infrastructure, and environment. As part of a set of studies on climate change impacts to California, this paper analyzes the potential impacts from projected sea-level rise if no actions are taken to protect the coast (a "no-adaptation scenario"), focusing on impacts to the state's population and infrastructure. Heberger et al. (2009) also covered effects on wetlands, costs of coastal defenses, and social and environmental justice related to sea-level rise. We analyzed the effect of a medium-high greenhouse gas emissions scenario (Special Report on Emissions Scenarios A2 in IPCC 2000) and included updated projections of sea-level rise based on work by Rahmstorf (Science 315(5810): 368, 2007). Under this scenario, sea levels rise by 1.4 m by the year 2100, far exceeding historical observed water level increases. By the end of this century, coastal flooding would, under this scenario, threaten regions that currently are home to approximately 480,000 people and $100 billion worth of property. Among those especially vulnerable are large numbers of low-income people and communities of color. A wide range of critical infrastructure, such as roads, hospitals, schools, emergency facilities, wastewater treatment plants, and power plants will also be at risk. Sea-level rise will inevitably change the character of California's coast; practices and policies should be put in place to mitigate the potentially costly and life-threatening impacts of sea-level rise.

M. Heberger (✉) · H. Cooley · P. Herrera · P. H. Gleick · E. Moore
Pacific Institute, 654 13th Street, Oakland, CA 94612, USA
e-mail: mheberger@pacinst.org

H. Cooley
e-mail: hcooley@pacinst.org

P. Herrera
e-mail: pt_herrera@gmail.com

P. H. Gleick
e-mail: pgleick@pacinst.org

E. Moore
e-mail: emoore@pacinst.org

1 Introduction

The residents of California's coast are already familiar with disaster and live with the risk of flooding from coastal storms and tsunamis, and landslides and property damage due to coastal erosion. In spite of these risks, development along California's coast is extensive. It was estimated that, in 2003, 31 million people lived in the state's 20 coastal counties. In fact, six of the ten fastest growing coastal counties in the United States between 1980 and 2003 were in California (NOAA 2004). Major transportation corridors and other critical infrastructure are also concentrated near California's coast, including oil, natural gas, and nuclear energy facilities, as well as major ports, harbors, and wastewater treatment plants.

In 2008, a set of comprehensive climate scenarios were prepared for the California Energy Commission's Public Interest Energy Research (PIER) Climate Change Research Program, including estimates of future sea-level rise (Cayan et al. 2009). Under medium to medium-high emissions scenarios, researchers estimated that mean sea levels could rise by between 1 m and 1.4 m by the year 2100. Rising sea level, combined with the associated storm surge, wave runup, and related factors will have two important effects: first, it exposes areas that were previously considered safe from flooding to new risks; second, in areas that are already at risk, it will increase the frequency and severity of flooding. In areas where the coast erodes easily, sea-level rise is also likely to accelerate shoreline recession due to erosion. Erosion of some barrier dunes may also expose previously protected areas to flooding. Erosion risks to California's costs are described in detail elsewhere (Heberger et al. 2009; Revell et al. 2011).

National studies on the economic cost of sea-level rise suggest that while adapting to climate change will be expensive, so are the costs of doing nothing (Titus et al. 1992 and Yohe et al. 1996). Flooding in the United States is currently responsible for an average of 140 deaths per year and $6 billion in property losses (USGS 2006). Continued development near coastlines and sea-level rise threaten to worsen vulnerability and increase future losses. Because flood damages and cost are highly site-specific, regional analyses are critical for guiding land-use decisions and evaluating adaptive strategies.

A previous study of the San Francisco Bay area (Gleick and Maurer 1990) concluded that a 1-meter sea-level rise would threaten existing commercial, residential, and industrial structures around San Francisco Bay valued at $48 billion (in year 1990 dollars). Building or strengthening levees and seawalls to protect existing high-value development was estimated to require a capital investment of approximately $1 billion (in year 1990 dollars) and an additional $100 million per year for ongoing maintenance. Gleick and Maurer also noted that substantial areas of the San Francisco Bay, especially wetlands and marshes, could not be protected and would likely be damaged or lost. A more recent analysis by Neumann et al. (2003) found that the economic cost of a 1.0 m sea level rise along the entire California coast would range from $148 million to $635 million (in year 2000 dollars), which includes the cost of protecting existing structures using beach nourishment, levees, and seawalls.

In this study, we analyzed the threats to California's 2,000 miles of coast from increased flooding and erosion caused by climate change induced sea-level rise, together with the associated storm and wave effects. While this article summarizes threats to the state's population and infrastructure from a single scenario of sea-level rise, more detail on the timing and degree of vulnerability for different levels of rise are included in the comprehensive report. The full study also covered erosion impacts, effects on wetlands, costs of coastal defenses, and issues of social and environmental justice related to sea-level rise (Heberger et al. 2009). No reliable estimates of how climate changes would alter El Niño or La Niña events are yet available, and we did not include them here. Future assessments could integrate that information when it becomes available.

2 Methods

Numerous studies have attempted to quantify the cost of sea-level rise and have been based primarily on a framework developed in Yohe (1989) and refined in Yohe et al. (1996) and Yohe and Schlesinger (1998). Yohe used a cost-benefit model to evaluate the property at risk and the cost of protecting or abandoning that property. He assumed that property will be protected if its value exceeds the cost to protect it at the time of flooding. Protection costs were based on the construction cost of a protective structure such as a seawall. If the value of the property does not exceed the cost of protection, Yohe assumed that the property would be abandoned, incurring a cost equal to the value of the land and structure at the time of inundation. The total cost to society of sea-level rise using this approach is the sum of the protection cost plus the value of the lost property.

To determine the value of lost property, the Yohe approach considers land and structure values separately. In most locations, coastal land commands a premium price, with the price declining as one moves inland. With inundation, the Yohe method assumes that land values will simply migrate inland, and thus, the economic value of lost land is equal to the economic value of interior land. The value of structures is calculated under two conditions: with and without foresight. With perfect foresight, the economic value of structures is assumed to depreciate over time as the "impending inundation and abandonment become known" (Yohe and Schlesinger 1998), approaching $0 at the time of inundation. Without foresight, the structure value does not depreciate.

There are several important shortcomings to this approach; the just-in-time approach to coastal protection is unlikely, and prioritizing protection based solely on property value fails to reflect a range of other societal concerns for public access, habitat, scenery, and social justice. This analysis used a different approach to estimate a value of assets potentially at risk from sea-level rise. We performed a planning-level estimate of economic vulnerability by summing the replacement value of property that will be vulnerable to damaging floods in the future, assuming no adaptation mechanisms are undertaken. Actions taken to defend the coast, expand wetland buffers, or floodproof structures, if taken on time, are likely to prevent potential damages. We also note that a number of potential costs of sea-level rise are excluded from this analysis, such as relocation expenses, lost wages and business revenue, and value lost by lost or degraded coastal ecosystems.

We based our analysis on a 1%-annual chance coastal flood, or the so-called 100-year flood. The terminology used to describe the recurrence interval can be misleading and is often misinterpreted. A "100-year flood" does not refer to a flood level that occurs every 100 years. Rather, it refers to a flood that has a 1/100, or 1%, chance of occurring in any year. Over the course of a typical 30-year mortgage a 100-year flood has a 26% chance of occurring one or more times. We used scenarios of sea-level rise and mapped areas likely to be inundated by a 100-year flood under current conditions, and conditions in the year 2100. Additional temporal and spatial estimates and details are provided in Heberger et al. (2009). Geographic layers depicting flood extents were overlaid with geospatial data using GIS software to produce quantitative estimates of the population, infrastructure, and replacement value of property at risk from sea-level rise, as well as the impacts on harder-to-quantify coastal ecosystems. Our estimates of populations at risk are based on current population data, not a projection of populations that might be at risk in the future. If no policies are put in place to limit new exposure in areas at risk of rising seas, our estimates will underestimate impacts following years of population growth and development. If, however, policymakers are proactive about reducing coastal risks in coming decades, the levels of risk could be substantially reduced.

The study area spans approximately 1,800 km (1,100 miles) of California's Pacific coast and 1,600 km (1,000 miles) of shoreline along the inside perimeter of the San Francisco Bay. The San Francisco Bay study area extends from the Golden Gate in the west to Pittsburg, California, in the east and San Jose in the south. The eastern boundary of the San Francisco Bay study was set according to where United States Geological Survey (USGS) researchers were able to accurately model flood elevations in the Bay.

2.1 Sea-level rise projections

Sea levels are constantly in flux, subject to the influence of astronomical forces from the sun, moon, and earth, as well as meteorological effects like El Niño. Measurements at tide gages around the world indicate that the global mean sea level is rising. Water level measurements from the San Francisco gage (NOAA 2009), shown in Fig. 1, indicate that mean sea level rose by an average of 2.01±0.21 mm per year from 1897 to 2006, equivalent to a change of 20 cm (8 inches) in the last century. (The solid vertical line coincides with the San Francisco earthquake of 1906. NOAA researchers have fit separate trendlines before and after an apparent datum shift that occurred in 1897 to account for possible vertical movement of the land surface where the gages is located, disrupting consistent measurements.)

Sea levels are expected to continue to rise, and the rate of increase will likely accelerate. The Intergovernmental Panel on Climate Change (IPCC), in its Fourth Assessment Report (Meehl et al. 2007), estimated that sea levels may rise by 0.2 m to 0.6 m by 2100, relative to a baseline of 1980–1999, in response to changes in oceanic temperature and the exchange of water between oceans and land-based reservoirs, such as glaciers and ice sheets (Meehl et al. 2007). More recent research indicates that sea-level rise from 1993 to 2006 has outpaced the IPCC projections (Rahmstorf 2007; Allison et al. 2009). Previous models failed to include ice-melt contributions from the Greenland and Antarctic ice sheets and may underestimate the change in volume of the world's oceans.

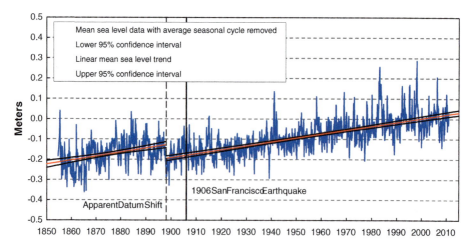

Fig. 1 Trend in monthly mean sea level at the San Francisco tide station from 1897 to 2006 (records begin in 1854; the solid black line represents the major earthquake in 1906). Redrawn from NOAA Sea Levels Online, http://co-ops.nos.noaa.gov/sltrends/sltrends_station.shtml?stnid=9414290

To address these new factors, the California Climate Impacts Study developed sea-level rise forecasts using a methodology developed by Rahmstorf (2007). Cayan et al. (2009) produced global sea-level estimates based on projected surface air temperatures from global climate simulations for both the IPCC A2 and B1 scenarios. The A2 storyline is characterized by "self-reliance and preservation of local identities" (IPCC 2000). Population is expected to continuously increase, but economic growth and technological development are expected to be slow. The B1 storyline has the same population projections as the A1 storyline but "rapid changes in economic structures toward a service and information economy, with reductions in material intensity, and the introduction of clean and resource-efficient technologies." Additionally, Cayan et al. (2009) modified the sea-level rise estimates to account for water trapped in dams and reservoirs that artificially reduced runoff into the oceans during the 20th century (Chao et al. 2008).

Cayan et al. estimate that mean sea level along the California coast will rise by 1.0 m under the B1 scenario by the year 2100, and 1.4 m under the A2 scenario, as shown in Fig. 2. The highest scenario, A1FI assumes continued high use of fossil fuels; it was not used in this analysis, but is shown for comparative purposes.

2.2 Mapping the Pacific coast

Sea-level rise increases the risk of flooding in low-lying coastal areas. For the California coast, we used GIS software (ESRI's ArcGIS Desktop 9.2) to produce maps of the areas at risk of inundation from a 1.4 m sea-level rise. For the Pacific coast, we approximate the potential future flood impact by adding projected sea-level rise estimates to water levels associated with a 100-year flood, i.e., current flood elevations for the 100-year flood are increased by 1.4 m, the projected increase in sea level by 2100 under the A2 scenario.

Existing flood levels were based on estimates of the 100-year flood elevation (also called the *base flood elevation* or BFE) from Flood Insurance Studies published by the Federal Emergency Management Agency (FEMA). Flood elevations are a function of a number of local factors, and vary considerably even over a few miles of coast. In California, coastal base flood elevations range from 10.5 m in Mendocino County in the north to 2.3 m in San Diego Harbor in the south (all elevations for the study are reported relative to the North American Vertical Datum of 1988 or NAVD88). As part of this project, we worked with researchers and coastal engineers to develop a GIS layer of approximate 100-year flood elevations for the entire California coast. To develop this new dataset, Battalio et al. (2008) performed the following tasks:

1. Compiled available coastal flood BFEs published by FEMA for the California coast.
2. Estimated BFEs where FEMA estimates are not available using professional judgment.

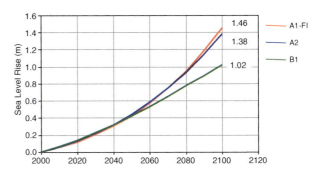

Fig. 2 Scenarios of sea-level rise to 2100. Source: Dan Cayan, Scripps Institution of Oceanography. Based on simulations using the NCAR CCSM3 general circulation model. Sea level changes forecast using the method described in Rahmstorf, 2007

3. Converted elevations to the North American Vertical Datum of 1988 (NAVD88).
4. Adjusted elevations to nearest half foot based on observed sea-level rise to present day.

Our approach assumes that all tide datums, e.g., mean high tide and flood elevations, will increase by the same amount as mean sea level. There is some evidence that this assumption may not always hold true. For example, Flick et al. (1999) found that, in San Francisco, one measure of high tide, mean higher high water (MHHW), was increasing faster than mean sea level.

We used automated mapping methods in GIS to delineate areas inundated by the current and future flood elevations. The key inputs to this analysis are digital elevation models (DEMs), gridded datasets that contain values representing elevations of the earth's surface. We used the most accurate, high-resolution, up-to-date terrain data available.

The elevation datasets used for this project are summarized in Table 1. For much of the Central and Northern California coast, high-accuracy Light Detection and Ranging (LIDAR) data were available from Airborne LIDAR Assessment of Coastal Erosion (ALACE) project, a partnership between NOAA, the National Aeronautics and Space Administration (NASA), and USGS. The ALACE project emphasized shoreline change, and so the data were available for a relatively narrow swath of the coast. The coverage did not always extend inland far enough to fully map the coastal floodplain. In addition, there were several gaps in coverage along the entire coast.

We supplemented the LIDAR data, filling in gaps in coverage with topographic information from the USGS National Elevation dataset. Although these data are at a lower resolution and accuracy, they allowed us to map the entire coast. For portions of the Southern California coast, Interferometric Synthetic Aperture Radar (IfSAR) data were available from NOAA. These data are of coarser resolution than the LIDAR data described above, i.e., 3-meter pixel resolution compared to 2-meter, and have less vertical accuracy, i.e., ±2.2 m for the IFSAR data compared to ±0.07 m for the LIDAR data. Additional details on the GIS data sources and processing steps are summarized in Heberger et al. (2009).

GIS raster math tools were used to compare the elevation of land surfaces with the adjacent flood elevation to determine the extent of flooding. The resulting inundation grids were boundary-smoothed and small isolated ponds and islands were removed. The raster datasets were then converted to vector polygons and merged so they could be used in the social and economic analyses.

Table 1 Elevation datasets used for mapping coastal flood risks

Dataset	National Elevation Dataset	ALACE 1998	ALACE 2002	So. Cal. IFSAR
Source/Mission	USGS	NASA, NOAA, USGS	NASA, NOAA, USGS	NOAA
Geographic coverage	National	Stinson Beach to Santa Barbara	Northern border of California to Stinson Beach	Santa Barbara to Mexican border
Data collection method	Various	LIDAR	LIDAR	IFSAR
Horizontal Resolution	10 m	3 m	2 m	3 m
Year collected	Various	1996-2000	2002	2002-2003
Stated vertical accuracy	±7.5 m	±0.15 m	±0.20 m	±2.2 m

2.3 Mapping San Francisco Bay

Sea-level rise inundation maps were generated from the climate scenarios by the USGS (Knowles 2009) using a suite of computer models under the CASCADE project that simulate the hydrodynamics of San Francisco Bay under future climate scenarios. The CASCADE project allowed us to conduct a more detailed analysis of impacts along the margins of the San Francisco Bay. Similar models have not yet been developed for other sections of the coast.

To estimate inundated areas in the Bay, "the highest resolution elevation data available were assembled from various sources and mosaicked to cover the land surfaces of the San Francisco Bay region. Next, to quantify high water levels throughout the Bay, a hydrodynamic model of the San Francisco Estuary was driven by a projection of hourly water levels at the Presidio. This projection was based on a combination of climate model outputs and empirical models and incorporates astronomical, storm surge, El Niño, and long-term sea level rise influences" (Knowles 2009). The Bay computer model simulates the water surface elevation for each hour from 2000 to 2009. Inputs to the model include both upstream inflows and downstream water surface elevations. The Bay model simulates the water surface elevation for each hour from 2000 to 2099 and is driven by both upstream and downstream boundary conditions. Based a statistical analysis of this output, flood layers of a 0.5 m, 1.0 m, and 1.4 m sea-level rise were produced for five flood recurrence intervals for each of four years between 2000 and 2099. The analysis presented in Heberger et al. (2009) reports results for each sea-level rise scenario. In this paper, we report current risk based on year 2000 conditions, and those in the year 2100, based on a 1.4-meter sea-level rise.

2.4 Estimating population impacts

To determine populations at risk if no adaptation actions are taken, we intersected the inundation layers with the census block boundaries (United States Census Bureau 2000) in GIS. We used year-2000 population data aggregated by census block, the smallest geographic unit for which the Census Bureau reports data collected from all households. We make the assumption common in GIS analyses that the population is distributed evenly within a block's boundaries. So if our mapping shows that 50% of a 500-person census block is inundated by a flood, we estimate that 250 people are at risk. This method may result in an underestimate (where the houses are clustered on the coast) or an overestimate (when the houses are set back from the coast). This is a limitation to many demographic analyses done in a GIS, and despite its lack of accuracy, is suitable for a state-wide planning-level estimate.

We investigated the race, income, and other characteristics of the population vulnerable to current and future coastal flooding through data from the 2000 US Census, also aggregated at the census block level. In the environmental justice literature, "of color" refers to those who reported their race as other than white in the 2000 US Census; this included American Indian or Alaska Native; Asian; Black or African American; Native Hawaiian or Other Pacific Islander.

2.5 Determining infrastructure impacts

Data for the replacement value of buildings and contents was taken from datasets supplied with the HAZUS model. HAZUS is a computer model for conducting standardized, nationally applicable natural hazards loss estimation developed for FEMA's Mitigation Division by the National Institute of Building Sciences (FEMA 2006). HAZUS uses a database called the "General Building Stock Inventory" that contains the replacement value of buildings and contents in each Census block. Replacement values are based on data from a number of sources including the U.S. Census Bureau, Dun & Bradstreet (a business listing service), and the U.S. Department of Energy. The HAZUS model estimates direct economic losses based on the repair and replacement of damaged or destroyed buildings and their contents, and includes: a) cost of repair and replacement of damaged and destroyed buildings, b) cost of damage to building contents, and c) losses of building inventory (contents related to business activities).

To determine the replacement value for structures in the areas at risk, we intersected the inundation layers with year 2000 census block data. As with the demographic analysis, we assumed that the building value is distributed uniformly over a census block's area. It should be noted that replacement value is almost always lower than the actual market value of a building. We compared replacement costs and the market value of one class of building, single-family homes, at a few locations along the California coast and found that the replacement costs in HAZUS may substantially underestimate actual market values for residential properties. According to the HAZUS database, the median home replacement values range from $63,000 in Del Norte County to $135,000 in San Mateo County. In comparison, the median home price in California was $286,000 in November 2008. In Northern California, the median price was $307,000, and in the San Francisco Bay Area, the median price was $474,000. This underscores the fact that a building's market value is usually greater than its replacement cost.

Important transportation infrastructure is also at risk of flooding and erosion from projected increases in sea-level rise. We estimated the miles of roadways and railroads at risk by overlaying the GIS inundation and erosion hazard layers with transportation data published by TeleAtlas (2008). The polylines in the TeleAtlas roads GIS database are two-dimensional; since we do not have any additional information on roadways' elevations, we assume they are clamped to the ground as it is represented by our terrain data. This assumption may be violated for some elevated roadways, as well as bridges and tunnels. Additionally, the railroad file does not provide information on the number of tracks (e.g., single or double), so the miles of railroad affected may underestimate actual miles of track.

We did not attempt to quantify the cost of flooding on roads and railways. In some cases, damages may be minor, resulting in temporary closures and modest repairs. As the frequency and intensity of flooding increases, however, closures may become longer and the cost of repair may rise. Eventually, roads and railways may need to be raised or rerouted. The cost of repairing, moving, or raising roads and railways is highly site-specific and dependent on the level of damage that is sustained. Furthermore, flooding and closure of roads and railways can have significant impacts on the local, state, and national economy. Railways are particularly important for moving goods in and out of California ports. In addition, road closures can prevent people from getting to work, causing major economic disruptions. Thus, the information on roads and railways is presented as miles of structures at risk rather than value.

A number of other facilities along the coast are also at risk of flooding and erosion. We evaluated the sites and facilities at risk by overlaying the GIS inundation layer with the future hazard zones. Data on the locations of schools and emergency facilities come from the HAZUS geographic database (FEMA 2006). Data on licensed healthcare facilities were obtained from the California Office of Statewide Health Planning and Development (2006). Data on coastal power plants were provided by the California Energy Commission.

We obtained data on U.S. EPA-monitored hazardous materials sites from the U.S. EPA Geospatial Data Access Project (US EPA 2008), including Superfund sites, hazardous waste generators, facilities required to report emissions for the Toxics Release Inventory, facilities regulated under the National Pollutant Discharge Elimination System (NPDES), major dischargers of air pollutants with Title V permits, and brownfields (abandoned industrial sites, many of which have polluted soil and groundwater).

We developed a custom GIS layer of wastewater treatment plants based on data in the U. S. EPA's Permit Compliance System (PCS) database. The coordinates were inaccurate, so we adjusted the location of plants based on aerial photos. Also, we noticed that a few facilities were missing, so we added facilities based on telephone and Internet research.

3 Results

In this analysis, we used the extent of the 100-year coastal floodplain to evaluate vulnerability to inundation. Our analysis shows that a significant amount of land, infrastructure, and property are already located in the 100-year floodplain. Under year-2000 conditions, we estimate that 1,200 km^2 (460 square miles) are in the coastal floodplain; with a 1.4-m sea level rise, the floodplain could expand to 1,500 km^2 (570 square miles). The following sections summarize the vulnerable population and infrastructure located in the floodplain areas.

3.1 Population at risk

We estimated 260,000 people, or about 1% of the population of California's coastal counties, live in areas that are currently vulnerable to a 100-year flood event. As sea levels rise, the area and the number of people vulnerable to flooding also rise. We estimate that a 1.4 m sea-level rise will put around 480,000 people at risk from a 100-year flood event. Table 2 reports the population vulnerable to a 100-year flood event along California's coast by county. Populations in San Mateo and Orange Counties are especially vulnerable, accounting for about half of those at risk with a 1.4 m sea-level rise. Large numbers of residents in Alameda, Marin, and Santa Clara counties are also at risk.

An analysis of the vulnerable population's racial makeup revealed that sea-level rise induced flooding disproportionately affects whites in 10 of 20 counties along the coast. In Los Angeles County, for example, 72% of those affected are white, while only 31% of the population in the county is white. Conversely, in 10 of the 20 counties studied, communities of color are disproportionately impacted, including every county around San Francisco Bay. The greater proportion of people of color in areas affected by sea-level rise highlights the need for these counties to take concerted efforts to understand and mitigate potential environmental injustice. A more detailed demographic assessment and discussion of environmental justice is available in Heberger et al. 2009.

3.2 Emergency and healthcare facilities

Table 3 shows the schools and emergency and healthcare facilities that are currently at risk from a 100-year flood event and that will be at risk following a 1.4 m sea-level rise. Numerous schools are vulnerable as well. In 2000, 65 schools were vulnerable to a 100-year flood event. With a 1.4 m sea-level rise, however, the number of schools at risk doubles, rising to 137 schools. Significant numbers of healthcare facilities are also at risk.

Table 2 Population vulnerable to a coastal 100-year flood in California, by county (* denotes counties on San Francisco Bay)

County	Currently at risk	At risk with 1.4 m SLR	Percent increase
Alameda*	12,000	66,000	450
Contra Costa*	840	5,800	590
Del Norte	1,700	2,500	47
Humboldt	3,600	7,500	110
Los Angeles	3,600	13,000	260
Marin*	26,000	40,000	54
Mendocino	520	630	21
Monterey	10,000	14,000	40
Napa*	760	1,500	97
Orange	70,000	110,000	57
San Diego	570	690	21
San Francisco*	3,600	10,000	180
San Luis Obispo	4,600	6,300	37
San Mateo*	91,000	130,000	43
Santa Barbara	660	1,300	97
Santa Clara*	13,000	31,000	140
Santa Cruz	4,500	5,600	24
Solano*	3,700	12,000	220
Sonoma*	3,200	9,600	200
Ventura	7,000	16,000	130
State Total	260,000	480,000	85

In 2000, there were 20 healthcare facilities at risk of a 100-year flood. With a 1.4 m sea-level rise, however, the number of healthcare facilities at risk rises to 55.

3.3 Hazardous materials sites

The presence of land or facilities containing hazardous materials in areas at risk of inundation increases the risk of release of these materials into the environment and exposure to toxic chemicals for nearby residents and ecosystems. For example, sediment samples in New Orleans taken 1 month after Hurricane Katrina found excess levels of arsenic, lead, and the gasoline constituent benzene, all considered toxic pollutants by the U.S. EPA (Adams et al. 2007). Those living or working near these facilities may be affected by the potential release and spreading of contamination through floodwaters or through flood-related facility malfunctions.

Table 3 Schools, emergency and healthcare facilities at risk from a 100-year coastal flood following a 1.4 m sea-level rise

Facility	Current risk	Risk with 1.4 m sea-level rise
Schools	65	137
Healthcare facilities	20	55
Fire stations and training facilities	8	17
Police stations	9	17

We evaluated sites containing hazardous materials at risk of flooding along the Pacific coast and San Francisco Bay. Here, we report on a range of sites monitored by the U.S. EPA, including Superfund sites; hazardous waste generators; facilities required to report emissions for the Toxics Release Inventory; facilities regulated under the National Pollutant Discharge Elimination System (NPDES); major dischargers of air pollutants with Title V permits; and brownfield properties. An estimated 130 U.S. EPA-regulated sites are currently vulnerable to a 100-year flood event, reported in Table 4. Nearly 60% of these facilities are located in San Mateo and Santa Clara counties, in the area known as Silicon Valley.

The number of facilities at risk increases by 250% with a 1.4 sea-level rise, with more than 330 facilities at risk of a 100-year flood event. San Mateo, Alameda, and Santa Clara counties have the highest numbers of U.S. EPA-regulated sites within future flood areas.

3.4 Roads and railways

There are many roads and railways that are vulnerable today and with sea-level rise. Under year 2000 conditions, 1,600 miles of roads and highways are at risk from a 100-year flood (of these, 220 miles are highways). With a 1.4 m sea-level rise, the mileage doubles to 3,500 miles, of which 430 miles are highway. The mileage of at-risk railroads increases from 140 under current conditions to 280 miles by 2100. About 50% of the roadways and 60% of the railways at risk are concentrated around the San Francisco Bay. Much of this infrastructure is protected by levees, seawalls, and other structures, which are not likely to provide adequate protection against higher seas unless they are raised and strengthened. Note that we do not provide estimates of the value of the transportation infrastructure at

Table 4 US EPA-regulated sites within areas vulnerable to a 100-year coastal flood event following a 1.4 m sea-level rise

County	Sites currently at risk	At risk with 1.4 m sea-level rise
Alameda	6	63
Contra Costa	4	22
Del Norte	1	3
Humboldt	10	13
Los Angeles	13	26
Marin	1	6
Monterey	1	1
Napa	1	2
Orange	4	16
San Diego	–	13
San Francisco	–	4
San Luis Obispo	–	1
San Mateo	39	78
Santa Barbara	1	5
Santa Clara	41	53
Santa Cruz	5	6
Solano	2	5
Sonoma	–	2
Ventura	5	13
Total	134	332

Data Source: EPA Geospatial Data Access Project 2008

risk. The economic value of roads and railroads is a complicated subject, and reliable information was not readily available for our study.

3.5 Power plants

Many of California's thermoelectric power plants are located on the coast, as they make use of seawater for cooling, and a number of them may be vulnerable to sea-level rise. In some cases, actual power generating infrastructure is at risk; in others, intake or other peripheral structures are vulnerable. Specific site assessments are needed at each coastal plant to determine the actual risk. We identified 30 coastal power plants that are potentially at risk, with a combined capacity of more than 10,000 megawatts (MW), from a 100-year flood with a 1.4 m sea-level rise. The capacities of the vulnerable power plants range from a relatively small 0.2 MW plant to one that is more than 2,000 MW. The majority of vulnerable plants are located in Southern California and along the San Francisco Bay. Figure 3 shows the locations of vulnerable power plants in Southern California.

3.6 Wastewater treatment plants

We identified a total of 28 vulnerable wastewater treatment plants: 21 on the San Francisco Bay and 7 on the Pacific coast. The combined capacity of these plants is 530 million gallons per day (MGD). Figure 4 shows the locations of the plants at risk in the Bay Area. Inundation from floods could damage pumps and other equipment, and lead to untreated sewage discharges. Besides the flood risk to plants, higher water levels could interfere with discharge from outfalls sited on the coast. This could require retrofitting discharge systems with pumps, redesigning outfalls, or other adaptation responses. Cities and sanitation districts should begin to assess how higher water levels will affect plant operations and plan for future conditions.

3.7 Seaports

Goods movement in California, and especially the San Francisco Bay Area, is critically important to the state's economy. A recent report by the Metropolitan Transportation Commission stated that "over 37% of Bay Area economic output is in manufacturing, freight transportation, and warehouse and distribution businesses. Collectively, these goods-movement-dependent businesses spend approximately $6.6 billion [annually] on transportation services. The businesses providing these services also play a critical role as generators of jobs and economic activity in their own right" (Metropolitan Transportation Commission 2004).

Our assessment of future flood risk with sea-level rise show significant flooding is possible at California's major ports in Oakland, Los Angeles, and Long Beach. These ports are important not only to the economy of California, but also to the nation. The Port of Los Angeles-Long Beach, for example, handles 45–50% of the containers shipped into the United States. Of these containers, 77% leave the state; half by train and half by truck (Christensen 2008). Many port managers have already experienced how disasters can affect their operations. Following the Loma Prieta earthquake in 1989, for example, the Port of Oakland sustained damages that interrupted business for 18 months. These disruptions have global economic implications, as evident by a 2002 contract dispute that resulted in a work slowdown at west coast ports and cost the U.S. economy an estimated $1 billion to $2 billion per day. Others speculated that Japan and China would lose several percentage points off their gross domestic product if California's ports closed for longer than a week (Farris 2008).

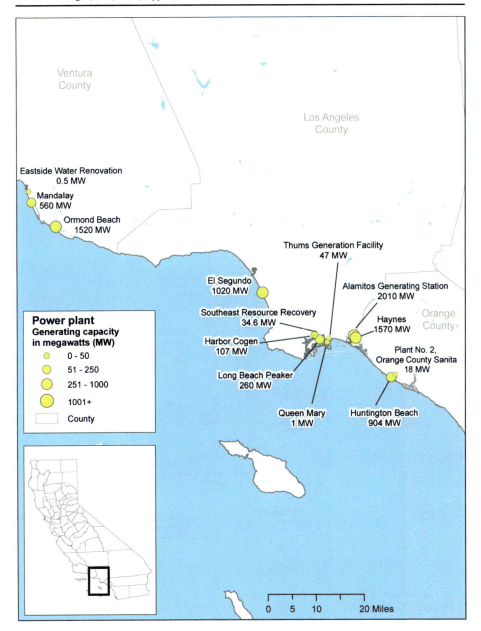

Fig. 3 Southern California power plants vulnerable to a 100-year coastal flood with a 1.4 m sea-level rise

In addition to directly affecting port operations, sea-level rise may cause other interruptions to goods movement at ports. Sea-level rise can reduce bridge clearance, thereby reducing the size of ships able to pass or restricting their movements to times of low tide. Higher seas may cause ships to sit higher in the water, possibly resulting in less efficient port operations (National Research Council 1987). These impacts are highly site specific, and somewhat speculative, requiring detailed local study to verify.

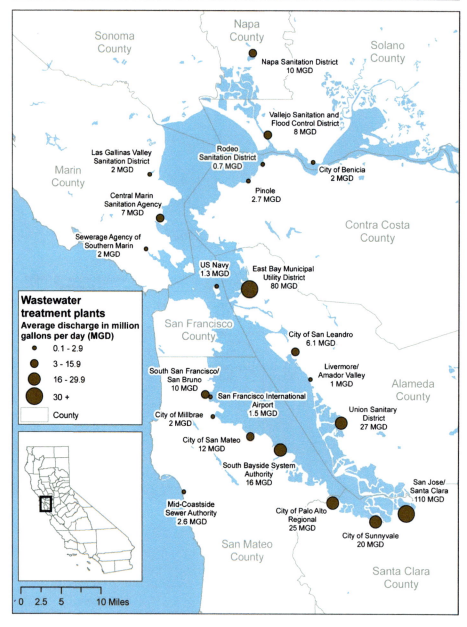

Fig. 4 Wastewater treatment plants on the San Francisco Bay vulnerable to a 100 year flood with a 1.4 m sea-level rise

3.8 Airports

The San Francisco and Oakland airports are vulnerable to flooding with a 1.4-meter sea-level rise. Other major airports near the coast, such as the San Diego, San Jose, and Los Angeles airports, were not identified as vulnerable in our analysis.

The economic impact of a disruption in airport traffic in San Francisco and Oakland is potentially large, and it would have significant effects on the state and regional economy. In 2007, the Oakland International airport transported 15 million passengers and 647,000 metric tons of freight. Activity at the San Francisco International airport is even greater than in Oakland. The San Francisco International Airport is the nation's thirteenth busiest airport, transporting 36 million people in 2007 (Airports Council International 2008). It also plays a significant role in the movement of goods regionally and internationally. In 2007, the San Francisco airport handled 560,000 metric tons of freight. San Francisco Airport ranked twelfth among foreign trade freight gateways by value of shipments in 2005, handling $25 billion in exports and $32 billion in imports (US Department of Transportation 2006), more than double that of the $23.7 billion handled by vessels at the Port of Oakland.

3.9 Property (buildings and contents)

Significant property is at risk of flooding from 100-year flood events as a result of a 1.4 m sea-level rise. Costs cited in the following paragraphs are replacement costs of buildings and contents, and not current market value of the land or buildings. For much of coastal California, market value is significantly greater than construction costs alone. The property at risk from a 100-year flood increases from $51 billion under baseline conditions to $99 billion with a 1.4 m sea-level rise. Two-thirds of the property at risk is concentrated on San Francisco Bay (Table 5) indicating that this region is particularly vulnerable to impacts associated with sea-level rise due to extensive development on the margins of the Bay.

Table 5 Replacement value of buildings and contents (millions of year-2000 dollars) at risk from a 100-year coastal flood, by county

County	Current risk	Risk with 1.4 m SLR
Alameda	3.3	15
Contra Costa	0.19	0.98
Del Norte	0.24	0.35
Humboldt	0.68	1.4
Los Angeles	1.4	3.8
Marin	5	8.7
Mendocino	0.12	0.15
Monterey	1.7	2.2
Napa	0.22	0.41
Orange	11	17
San Diego	0.69	2
San Francisco	0.78	4.9
San Luis Obispo	0.22	0.36
San Mateo	17	24
Santa Barbara	0.46	1.1
Santa Clara	3.7	7.8
Santa Cruz	2.4	3.3
Solano	0.62	1.9
Sonoma	0.32	0.48
Ventura	0.98	2.2
Total	51	99

Figure 5 shows the value of property in each county vulnerable to sea-level rise, with the size of the circle proportional to the value.

Within each region, vulnerability to sea-level rise is highly variable. The risk is greatest in San Mateo County, where $24 billion in property is vulnerable. About $17 billion of property, or about 50% of the total property at risk, is in Orange County. In the San

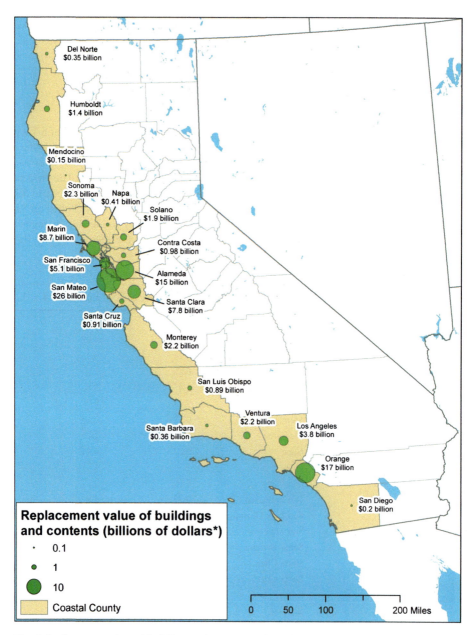

Fig. 5 Replacement value of buildings and contents vulnerable to a 100-year coastal flood with a 1.4 m sea-level rise

Francisco Bay area, Alameda, Marin, Santa Clara, and San Francisco counties are also exposed to a high degree of risk, with potential damages in the billions.

All economic sectors are vulnerable to impacts associated with sea-level rise. Table 6 reports the value of buildings and contents at risk of flooding by major economic sector. About 59% of the assets at risk are residential. The commercial sector, accounting for 27% of the value at risk, will also encounter significant costs. Agriculture, education, religion, and government each account for about 1% of the assets at risk, thus, their exposure to risk is relatively small.

4 Limitations of the analysis

The study covered a large area using approximate methods, and it is important to keep in mind its limitations when interpreting the results. First, there are uncertainties associated with the level of sea-level rise that will be experienced on the California coast in coming decades. The estimate used in this article comes from one method of predicting sea level rise, given a certain emissions scenario, as described by Cayan et al. (2009). Second, there are a number of uncertainties associated with estimating the current and future flood elevations. We used 100-year coastal flood elevations published by FEMA in flood insurance studies for a number of coastal communities. These estimates were developed by different parties over the past 20 years using a variety of methods that vary in detail and accuracy. Third, our method for using flood elevations to delineate floodplains is subject to further uncertainties. This is due to inaccuracies in the digital elevation data used in the analysis, and the mapping methods. We did not use the detailed methods that would be expected in a site-specific floodplain delineation, for example using a detailed hydrody-namic model to estimate how floodwaters will spread overland. Fourth, we made the simplifying assumption that the shoreline profile is constant, and will not change as a result of sea-level rise, erosion, or flooding. In reality, natural shorelines (and even some protected man-made shorelines) are constantly in flux. However, it remains extremely difficult to accurately predict how shorelines will migrate or change in response to rising sea levels (Pilkey and Cooper 2004).

Further uncertainty is introduced by the approximate methods for translating floodplain extents into physical impacts. A well-established method to estimate flood damages to buildings is the use of "depth-damage curves" that relate the depth of flooding to the percentage of a structure's value that is lost in a flood (FEMA 2006). The depth-damage curve method was inappropriate for our analysis for two reasons. First, we are not aware of

Sector	Current risk	Risk + 1.4 m SLR	% of total value at risk in future
Residential	43	58	59%
Commercial	15	27	27%
Industrial	5.8	11.0	11%
Religion	0.61	1.00	1.0%
Government	0.44	0.86	0.9%
Education	0.57	0.85	0.9%
Agriculture	0.34	0.42	0.4%
Total	66	99	100%

Table 6 Value of buildings and contents at risk from a 100-year flood after a 1.4 m sea-level rise, by economic sector (millions of year-2000 dollars)

existing datasets giving the first-floor elevation for buildings near California's coast. While we could have assumed that structures were built "at grade" (e.g. clamped to the earth's surface) or at a constant elevation, there was insufficient data to support this or a similar assumption. Second, the magnitude of damages caused by flooding are affected by factors other than the depth of flooding, such as the velocity of floodwaters, duration of flooding, and whether there is floating debris that can act as a "battering ram." As our study was meant to broadly identify risk over a large area, we did not perform the detailed modeling to quantify these parameters.

For the flood analysis, we estimated the economic cost of varying levels of sea-level rise (and associated storm surge and flooding) based on estimates of the replacement value of buildings and their contents. Only a single scenario is summarized here, though more detailed results can be found in Heberger et al. (2009). We did not include the value of lost land, which should be included if inundation is permanent or leads the abandonment of property. Flooding can also cause serious economic and social disruptions that are not captured in estimates of the buildings and infrastructure. For example, flooding events can cause deaths and injuries. When roadways are flooded or eroded, it can cause a cascade of impacts, such as preventing people from driving to work, blocking evacuation routes, or interfering with the movement of emergency vehicles. It is difficult to put a price tag on any one of these impacts. A more detailed economic analysis could include transportation risks, lost work days, health issues, or impacts on migratory bird habitat.

We also did not factor in any expected changes in population density or the level of development in the regions at risk over the next century; these are largely unknown and will be determined by future policies. If policies are put in place to reduce development in flood-prone regions, society could reduce future risks, and future costs. While limiting coastal development (an institutional adaptation) is likely the most effective way to reduce risk, this approach can also incur costs. If current population trends continue, many more people and places will be affected. We make no estimates of these changes, but future research could look at different scenarios for growth and coastal development and integrate them into the assessment framework similar to the one developed here.

Lastly, we note that we analyzed only a "do nothing" scenario which does not take into account potential adaptation mechanisms. In this regard, the study presents the maximum vulnerability under the climate change scenarios we considered. Future damages from coastal flooding can be lessened by taking appropriate actions; these may include coastal defenses, wetland buffers, and floodproofing buildings. Future research should better characterize and if possible quantify how various programs and policies may change the risk California coastal communities may face.

5 Conclusions

Rising sea levels and associated coastal flooding and damage will be among the most significant impacts of climate change to California. Sea level will rise as a result of thermal expansion of the oceans and an increase in ocean volume as land ice melts and runs off. Over the past century, sea level has risen nearly 20 cm (8 inches) along the California coast and climate models suggest substantial increases in sea level due to climate change over the coming century and beyond. This study evaluates the current population, infrastructure, and property threatened by projected sea-level rise if no actions are taken to protect the coast. The sea-level rise scenario was developed by the State of California from medium to medium-high greenhouse gas emissions scenarios from the Intergovernmental Panel on Climate Change (IPCC) but does not reflect the worst case sea-level rise that could occur.

We estimate that a 1.4 m sea-level rise will put 480,000 people at risk of a 100-year flood event. Among those affected are large numbers of low-income people and communities of color. Populations in San Mateo and Orange Counties are especially vulnerable, with an estimated 130,000 and 110,000 people are at risk in each, respectively. Large numbers of residents (66,000) in Alameda County are also at risk.

A wide range of critical infrastructure is vulnerable to sea-level rise. This includes: nearly 140 schools; 34 police and fire stations; more than 330 U.S. Environmental Protection Agency (EPA)-regulated hazardous waste facilities or sites; an estimated 3,500 miles of roads and highways and 280 miles of railways; 30 coastal power plants, with a combined capacity of more than 10,000 MW; 28 wastewater treatment plants, 21 on the San Francisco Bay and 7 on the Pacific coast, with a combined capacity of 530 million gallons per day; and the San Francisco and Oakland airports. In addition, $100 billion (in year 2000 dollars) worth of property is threatened by increased probability of coastal flooding.

Climate changes are inevitable, and adaptation to unavoidable impacts must be evaluated, tested, and implemented. Sea levels have risen observably in the past century, and scientists forecast that sea-level rise will continue for centuries, even if we stop emitting greenhouse gases immediately. As a result, coastal areas will be subject to increasing risk of inundation and erosion. A number of structural and non-structural policies and actions could be implemented to reduce these risks. While this paper does not include a discussion of explicit adaptation options and costs, Heberger et al. (2009) do include some initial estimates of physical adaptation options, including building or strengthening seawalls and levees. For example, we estimate that protecting vulnerable areas from flooding by building seawalls and levees will cost $14 billion (in year 2000 dollars), along with an additional $1.4 billion per year (in year 2000 dollars) in maintenance costs (Heberger et al. 2009), and may also incur a range of environmental costs such as loss of beaches, wetlands, and wildlife habitat. Continued development in vulnerable areas will put additional assets at risk and raise protection costs. Determining what to protect, how to pay for it, and how those choices are made raises concerns over equity and environmental justice.

The study, conducted for the 2009 California Climate Impacts Assessment, resulted in more information than can be presented here. There are additional figures, GIS data downloads, and color maps in PDF format at the Pacific Institute website at www.pacinst. org. In addition, an interactive map lets the user zoom to anywhere on the coast and get more information on the geographic data layers discussed in this paper.

Acknowledgments Major funds were provided by the California Energy Commission's Public Interest Energy Research (PIER) Program. Additional support came from the Metropolitan Transportation Commission and the California Ocean Protection Council.

The authors wish to thank the scientists and engineers at Philip Williams and Associates for their analysis on coastal flood and erosion hazards. Thanks to Dr. David L. Revell, Robert Battalio, Jeremy Lowe, Justin Vandever, Brian Spear, and Seungjin Baek.

Thanks also go to Noah Knowles, Dan Cayan, Mary Tyree, and Peter Bromirski, and Reinhard Flick of Scripps Institution of Oceanography for much of the oceanographic data.

We owe thanks to a number of agencies and organizations for sharing data and expertise: Philip Pang at the Army Corps of Engineers, Eric Simmons and Ray Lenaburg at FEMA, Reza Navai, Vahid Nowshiravan, and Barry Padilla at the California Department of Transportation, Jennifer Dare at NOAA, and a number of others.

Finally, we are especially grateful for our reviewers: Michael Hanemann, Arlene Wong, June Gin, and several anonymous reviewers who helped to improve the report on which this paper is based.

References

Adams C, Witt E, Wang J, Shaver D, Summers D, Filali-Meknassi Y, Shi H, Luna R, Anderson N (2007) Chemical quality of depositional sediments and associated soils in New Orleans and the Louisiana Peninsula following Hurricane Katrina. Environ Sci Technol 41(10):3437–3443

Airports Council International (2008) Airport traffic reports. Microsoft Excel spreadsheet document at http://www.aci-na.org/stats/stats_traffic.

Allison I et al (2009) The Copenhagen diagnosis: 2009, updating the world on the latest climate science. The University of New South Wales Climate Change Research Centre (CCRC), Sydney, 60 pp

Battalio R, Baek S, Revell D (2008) California coastal response to sea level rise: coastal base flood elevation estimates. Technical memorandum to the Pacific Institute. Phil Williams and Associates, San Francisco

California Office of Statewide Health Planning and Development (2006) California licensed health care facilities. Digital data files from California Spatial Information Library. http://casil.ucdavis.edu/casil/.

Cayan D, Tyree M, Dettinger M, Hidalgo H, Das T, Maurer E, Bromirski P, Graham N, Flick R (2009) Climate change scenarios and sea level rise estimates for California. 2008 Climate change scenarios assessment. California Climate Change Center paper CEC-500-2009-014-F.

Chao BF, Wu YH, Li YS (2008) Impact of artificial reservoir water impoundment on global sea level. Science 320(5873):212

Christensen M (2008) "Protecting America's busiest port from seismic impacts." Presentation at University of Southern California's Megacities Workshop. Los Angeles, California. November 10, 2008. http://mededonline.hsc.usc.edu/research/workshop-2008/session-5-christensen.htm

Farris MT II (2008) Are you prepared for a devastating port strike in 2008? Transportation Journal, Winter 2008

Federal Emergency Management Agency (FEMA) (2006) Hazards U.S. Multi-Hazard (HAZUS-MH). Computer application and digital data files on 2 CD-ROMs. Jessup, Maryland. www.fema.gov/plan/prevent/hazus/

Flick RE, Murray JF, Ewing L (1999) Trends in U.S. tidal datum statistics and tide range: a data report atlas. Scripps Institution of Oceanography, San Diego

Gleick PH, Maurer EP (1990) Assessing the costs of adapting to sea-level rise: a case study of San Francisco Bay. Pacific Institute, Oakland

Heberger M, Cooley H, Herrera P, Gleick P, Moore E (2009) The impacts of sea level rise on the California coast. California Climate Change Center, Sacramento, California. Paper CEC-500-2009-024-F

Intergovernmental Panel on Climate Change (2000) Special report on emissions scenarios. Cambridge University Press, UK. 570 pp

Knowles N (2009) Potential inundation due to rising sea levels in the San Francisco Bay region. California Climate Change Center paper CEC-500-2009-023-F. Sacramento, California

Meehl GA, Stocker TF, Collins WD, Friedlingstein P, Gaye AT, Gregory JM, Kitoh A, Knutti R, Murphy JM, Noda A, Raper SCB, Watterson IG, Weaver AJ, Zhao Z-C (2007) Global climate projections. In: Solomon S, Qin D, Manning M, Chen Z, Marquis M, Averyt KB, Tignor M, Miller HL (eds) Climate change 2007: the physical science basis. Contribution of Working Group I to the Fourth Assessment Report of the Intergovernmental Panel on Climate Change. Cambridge University Press, Cambridge

Metropolitan Transportation Commission (2004) Regional goods movement study for the San Francisco Bay area. Oakland, California. 26 pages. http://www.mtc.ca.gov/pdf/rgm.pdf

National Oceanic and Atmospheric Administration (NOAA) (2004) Population trends along the Coastal United States: 1980–2008. Washington, DC

National Oceanic and Atmospheric Administration (NOAA) (2009) NOAA Sea Levels Online, http://tidesandcurrents.noaa.gov/sltrends/sltrends_station.shtml?stnid=9414290

National Research Council (1987) Responding to changes in sea level: engineering implications. National Academies Press, Washington, DC, www.nap.edu/openbook/0309037816/html/

Neumann JE, Hudgens D, Herr JL, Kassakian J (2003) Market impacts of sea level rise on California coasts. Appendix VIII in: Global climate change and California: potential implications for ecosystems, health, and the economy. California Energy Commission, PIER program. Publication Number 500-03-058CF

Pilkey OH, Cooper JAG (2004) Climate: society and sea level rise. Science 303(5665):1781–1782

Rahmstorf S (2007) A semi-empirical approach to projecting future sea-level rise. Science 315(5810):368

Revell DL, Battalio R, Spear B, Ruggiero P, Vandever J (2011) A Methodology for Predicting Future Coastal Hazards due to Sea-level Rise on the California Coast. Climatic Change, doi:10.1007/s10584-011-0315-2

TeleAtlas (2008) TeleAtlas Dynamap, Geographic data files. Lebanon, New Hampshire

Titus J, Park R, Leatherman S, Weggle J, Greene M, Brown S, Gaunt C, Trehan M, Yohe G (1992) Greenhouse effect and sea level rise: the cost of holding back the sea. Coast Manag 19:219–233

United States Census Bureau (2000) Census 2000 Summary File 3 (SF3). http://factfinder.census.gov

United States Department of Transportation (2006) Top 50 U.S. Foreign trade freight gateways by value of shipments: 2005. Bureau of Transportation Statistics. Washington, D.C. www.bts.gov/programs/international/

United States Environmental Protection Agency (2008) EPA geospatial data access project. Digital data files, accessed November 2008. http://www.epa.gov/enviro/geo_data.html

United States Geological Survey (2006) Flood hazards—a national threat. 2-page factsheet at http://pubs.usgs.gov/fs/2006/3026/2006-3026.pdf

Yohe G (1989) The cost of not holding back the sea: Phase 1 economic vulnerability. In: The potential effects of global climate change on the United States. Report to Congress. Appendix B: Sea Level Rise. Washington, D.C.: U.S. Environmental Protection Agency. EPA 230-05-89-052

Yohe GW, Schlesinger ME (1998) Sea-level change: the expected economic cost of protection or abandonment in the United States. Clim Chang 38:447–472

Yohe G, Neumann J, Marshall P, Ameden H (1996) The economic cost of greenhouse-induced sea-level rise for developed property in the United States. Clim Chang 32(4):387–410

ERRATUM

Erratum to: Potential impacts of increased coastal flooding in California due to sea-level rise

Matthew Heberger · Heather Cooley · Pablo Herrera · Peter H. Gleick · Eli Moore

Published online: 19 January 2012
© Springer Science+Business Media B.V. 2012

Erratum to: Climatic Change (2011) 109 (Suppl 1):S229–S249
DOI 10.1007/s10584-011-0308-1

In Tables 5 and 6, figures are in billions of dollars, not millions as stated in the caption. In Figure 5, several of the numeric labels are incorrect. The reader should refer to Table 5 for the correct values. Furthermore, some of the figures in Table 6 were incorrect. Please refer to the corrected Table 6 below.

Table 6 Value of buildings and contents at risk from a 100-year flood after a 1.4 m sea-level rise, by economic sector (billions of year-2000 dollars)

Sector	Current Risk	Risk + 1.4 m SLR	% of Total Value at Risk in Future
Residential	33	58	59%
Commercial	11	27	27%
Industrial	4.9	11.0	11%
Religion	0.47	1.00	1.0%
Government	0.32	0.86	0.9%
Education	0.42	0.85	0.9%
Agriculture	0.25	0.42	0.4%
Total	51	99	100%

The online version of the original article can be found at http://dx.doi.org/10.1007/s10584-011-0308-1.

M. Heberger (✉) · H. Cooley · P. Herrera · P. H. Gleick · E. Moore
Pacific Institute, 654 13th Street, Oakland, CA 94612, USA
e-mail: mheberger@pacinst.org

H. Cooley
e-mail: hcooley@pacinst.org

P. Herrera
e-mail: pt_herrera@gmail.com

P. H. Gleick
e-mail: pgleick@pacinst.org

E. Moore
e-mail: emoore@pacinst.org

Climatic Change (2011) 109 (Suppl 1)
DOI 10.1007/s10584-011-0315-2

A methodology for predicting future coastal hazards due to sea-level rise on the California Coast

**David L. Revell · Robert Battalio · Brian Spear ·
Peter Ruggiero · Justin Vandever**

Received: 12 February 2010 / Accepted: 21 September 2011 / Published online: 10 December 2011
© Springer Science+Business Media B.V. 2011

Abstract Sea-level rise will increase the risks associated with coastal hazards of flooding and erosion. Along the active tectonic margin of California, the diversity in coastal morphology complicates the evaluation of future coastal hazards. In this study, we estimate future coastal hazards based on two scenarios generated from a downscaled regional global climate model. We apply new methodologies using statewide data sets to evaluate potential erosion hazards. The erosion method relates shoreline change rates to coastal geology then applies changes in total water levels in exceedance of the toe elevation to predict future erosion hazards. Results predict 214 km^2 of land eroded by 2100 under a 1.4 m sea level rise scenario. Average erosion distances range from 170 m along dune backed shorelines, to a maximum of 600 m. For cliff backed shorelines, potential erosion is projected to average 33 m, with a maximum potential erosion distance of up to 400 m. Erosion along the seacliff backed shorelines was highest in the geologic units of Cretaceous marine (K) and Franciscan complex (KJf). 100-year future flood elevations were estimated using two different methods, a base flood elevation approach extrapolated from existing FEMA flood maps, and a total water level approach based on calculations of astronomical tides and wave run-up. Comparison between the flooding methods shows an average difference of about 1.2 m with the total water level method being routinely lower with wider variability alongshore. While the level of risk (actual amount of future hazards) may vary from projected, this methodology provides coastal managers with a planning tool and actionable information to guide adaptation strategies.

1 Introduction

Climate change will affect many aspects of society including water supply, flooding, recreation, and ecosystem services. Accelerated sea level rise (SLR) associated with global

---

D. L. Revell (✉) · R. Battalio · B. Spear · J. Vandever
Philip Williams and Associates, Ltd., 550 Kearny Street, Suite 900, San Francisco, CA 94108, USA
e-mail: drevell@esassoc.com

P. Ruggiero
Department of Geosciences, Oregon State University, Corvallis, OR 97331, USA

🕿 Springer

warming has been identified as a major concern (IPCC 2007). For increasingly populated coastal communities, accelerated SLR will not only inundate coastal areas, but may also significantly impact patterns of coastal change. Coastal erosion is a complex response to many forcing parameters such as marine processes (water levels, waves, sediment supply and transport, etc.), terrestrial processes (rainfall, runoff, wind, etc.), and other instabilities (seismic, biologic, etc.), as well as geology and antecedent topography (Trenhail 2002; Collins and Sitar 2008). The resulting shoreline recession has the potential to significantly endanger public and private resources. Increased storminess and associated wave heights, documented along the U.S. West Coast north of Central California by Allan and Komar (2006), Graham and Diaz (2001), Ruggiero et al. (2010), and Wingfield and Storlazzi (2007) could also exacerbate the effects of higher sea level.

Prior studies that have examined the effect of SLR on the California coast have been confined largely to the San Francisco Bay (Gleick and Maurer 1990; BCDC 2008; Knowles 2009) with one open coast study in San Diego (San Diego Foundation 2008). These studies have only considered inundation over the static landscape. As increased water levels change the location of wave action, erosion and beach profile adjustment (Bruun 1962) can be expected in areas consisting of unconsolidated or loosely consolidated sediments. Quantifying the impact of SLR beyond simple 'bathtub' inundation models is especially important in California as much of the coast is steep, and more susceptible to erosion than direct inundation. The steep geometry is partly the result of tectonics which have uplifted the coast over geologic time scales, which increases the propensity for wave-induced erosion, resulting in bluff and cliff morphologies (Griggs et al. 2005). In most places, the rate of uplift is much slower than sea level rise but can locally affect the relative rate of SLR (Ryan et al. 2008). As a result, a majority of the coastline is backed by seacliffs or sand dunes (Griggs et al. 2005,), and much of this coastline is actively eroding and experiencing loss of land (Hapke et al. 2006). There are extensive private and public developments close to the edge of cliffs or buried in dunes along much of the California Coast. These properties are susceptible to erosion damages but are above the present-day 100-year flood elevation (e.g., FEMA flood maps). Inundation-only modeling approaches show little increase in coastal hazards along higher elevation cliff and dune backed shorelines. For this reason, the California Coastal Commission requires new development to be setback from the shore for a distance based on historic erosion rates and development life expectancy. Historic change rates can be expected to accelerate in the future with SLR, estimates of increased erosion are critical to coastal zone management. This work was funded by the State of California to develop a first-order estimate of the potential erosion and flood hazards resulting from SLR. This work was done in cooperation with the Pacific Institute, Scripps Institution of Oceanography, and the state-federal Coastal Data Information Program (CDIP), with overall project management by the California Ocean Protection Council.

1.1 Previous studies

Previous approaches for evaluating the effects of SLR on coastal change have focused primarily on dune and coastal barrier erosion (Bruun 1962; Dean and Maurmeyer 1983; Leatherman 1984; Kriebel and Dean 1985). Two approaches for assessing dune erosion include an equilibrium profile approach (Bruun 1962; Komar et al. 1999) and a wave impact approach (Larson et al. 2004).

The equilibrium profile approach assumes that over long time scales, an equilibrium profile shape migrates geometrically upward and inland in response to enhanced water elevations whether from SLR or a storm event (Bruun 1962; Komar et al. 1999). The Bruun approach, which has been widely applied to SLR assessments as well as criticized (Pilkey et al. 1993), is less applicable to California due to its diversity of coastal morphologies (cliffs, landslides) and tectonic controls. The wave impact approach accounts for the erosion contribution from individual waves (Fisher and Overton 1984; Nishi and Kraus 1996; Larson et al. 2004); however, this method has not been applied to the assessment of SLR impacts.

In general, the variables controlling cliff erosion can be divided into marine and terrestrial processes (or intrinsic and extrinsic) (Emery and Kuhn 1982; Sunamura 1992; Benumof and Griggs 1999; Benumof et al. 2000; Hampton and Griggs 2004; Young et al. 2009). Given the lack of information at the statewide study scale on the terrestrial processes we assume that historic erosion rates integrate both terrestrial and marine processes, and we drive cliff erosion based on an escalation of marine processes (Robinson 1977; Carter and Guy 1988). The primary marine processes include water levels, wave height, period, and direction, with additional factors including sediment supply and beach sand levels. Several previous studies have found an increase in coastal cliff erosion related to an increase in wave attack and lower sand elevations (Carter and Guy 1988; Komar and Shih 1993; Benumof et al. 2000; Ruggiero et al. 2001; Sallenger et al. 2002; Collins and Sitar 2008). However, it has been noted throughout the literature that cliff geology and rock hardness play an important role in determining erosion rates (Edil and Vallejo 1980; Benumof and Griggs 1999; Budetta et al. 2000). Cliffs composed of hard rock erode slowly and the erosion rate is controlled by material strength and rock mechanics rather than by marine processes. An assessment by the National Research Council (1987) suggested that rising sea level would have a negligible effect on the retreat rate of California seacliffs. Given the diversity of geologic units, wave exposures, shoreline orientations, and sandy beaches fronting the seacliffs along the California coast, this generalization is likely incorrect.

Two USGS National Assessments of Shoreline Change Projects systematically evaluated historic shoreline positions and cliff edges along California to determine shoreline change trends of sandy shorelines and cliff erosion (Hapke et al. 2006; Hapke and Reid 2007). These statewide studies documented trends in coastal change, as well as relationships between shoreline change and cliff recession for certain cliff types. The highest correlations were found along segments of coast predominately affected by marine processes (Hapke et al. 2009).

1.2 Current study

This study aims to assess potential coastal flood and erosion hazards along much of the California coastline for two different SLR scenarios: 1.4 and 1.0 m of rise by 2100 using the methodology of Rahmstorf (2007) as produced in Cayan et al. (2009). To make these results most useful to the planning community and the State government, we predicted future coastal hazard zones at three planning horizons: future years 2025, 2050, and 2100. This study is the first attempt at systematic statewide predictions of future coastal hazards along California and was part of a larger study evaluating the state's vulnerability to sea level rise (Heberger et al. 2009). This larger study was one in a series of studies funded by the California Energy Commission (CEC) that relied on the same global climate model scenario data downscaled for California (Cayan et al. 2009).

1.3 Study area

The California coastline contains a diverse range of geomorphology that is primarily shaped by the local geology and vertical tectonic movements. Across the state, a wide range of shoreline orientations and wave exposures combine with temperature and precipitation gradients to affect the rates of weathering and erosion. For the flood component of the study, the 1% annual recurrence coastal flood elevations, called the Base Flood Elevations (BFEs), were estimated for the entire state. The erosion study area stretched approximately 1,450 km (900 miles) from the Oregon/California border in the north to the Santa Barbara Harbor in the south. Along this stretch some notable segments of coast were not analyzed, primarily along the Big Sur (Monterey County), Devils Slide (San Mateo County), and the Lost Coast (Humboldt County), where the coast is largely dominated by terrestrial mass wasting and landslide processes and data availability is limited. These sites were noted by Hapke et al. (2009) to have a negative correlation between shoreline change and cliff erosion rates indicating that marine processes were not the dominant forcing mechanism in these areas. The southern California coastline was also excluded from the analysis for two primary reasons. First, the most likely coastal management response to SLR, given the large population centers, is likely to be a range of soft and hard engineering solutions (e.g. Flick and Ewing 2009), limiting the applicability of this analysis on natural coastal systems. Secondly, another project funded by the CEC focused on sea level rise impacts to erosion hotspots in Southern California (Adams and Inman 2009).

2 Methods

In this study, we estimate future coastal hazards of erosion and flooding based on two scenarios generated from a regional downscaled global climate model.

The addition of wave runup (also known as uprush) to ocean water level is called the Total Water Level (TWL), and is used to define coastal flood elevations (FEMA 2005). Changes in water levels due to both SLR and the incident wave climate increase future TWL elevations attained along the coast. We hypothesize that increases in the amount of time that waves impact various backshore features, (i.e. the toe of a seacliff or sand dune), increase erosion rates. This hypothesis is supported for short time frames (storms and seasons) by correlation between TWL and dune erosion (Ruggiero et al. 2001) and soft bluff erosion (Collins and Sitar 2008); and in terms of wave height and erosion rate for hard bluff erosion (Sunamura 1982). Antecedent conditions resulting from the timing and extent of geologic uplift and sea level rise greatly complicates attempts to calculate rock erosion due to elevated wave action (Trenhail 2002). These prior studies support the hypothesis that increased water levels and or wave exposure will likely increase coastal erosion rates of soft and hard backshores.

We hypothesize that the increased intensity and extent of wave action on the back shore, in terms of TWLs, due to sea level rise will increase erosion of the backshore, and that the potential erosion response at a particular site can be calculated with consideration of the backshore type, the geology, and the historic shoreline change trends which integrate both the marine and terrestrial erosion processes. The erosion response leads to accelerated erosion, a landward migration of the shoreline and an erosion of upland, with potential inundation of new areas (Fig. 1).

Two data sets are needed to apply this methodology: a coastal backshore characterization and a time series of TWL. Development of these data sets and the estimates of 1% annual coastal flood elevations (BFEs) are discussed below.

2.1 Backshore characterization—geology and backshore type

In order to characterize the morphologic diversity found along the California coast, we combined several statewide datasets representing the backshore type (cliff or dune), coastal

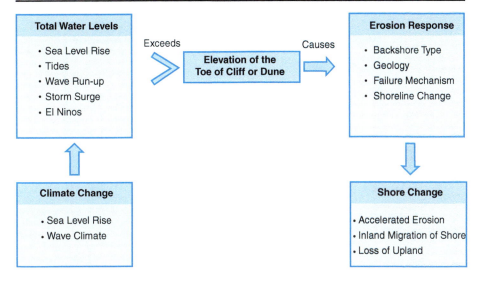

Fig. 1 Conceptual model of coastal response methodology

geology, trends in shoreline change, and various geomorphic features including toe elevations, heights, and beach slopes. These statewide data sets and their source references are described briefly below:

- **Geology**—original data from the California Geological Survey updated by Griggs et al. (2005).
- **Shoreline Inventory—shoreline classification** originally collected by Habel and Armstrong (1978), digitized in 1999, and updated by Griggs et al. (2005).
- **Shoreline Change**—rates from the USGS National Assessment of Shoreline Change in California. These data include: Long term linear regression rates (LRR) for sandy shorelines (1870s to 1998; Hapke et al. 2006), and end point erosion rates (EPR) for cliff backed shorelines (1930s to 1998); along with the 1998 cliff edge delineation used in Hapke and Reid (2007).
- **LIDAR**—shoreline topography collected in April 1998 and October 2002 by NOAA, NASA, and USGS. The primary data set was April 1998, collected to evaluate the storm impacts following the 1997–98 El Niño. The 2002 data set was collected to fill gaps in the 1998 data set and was used where the 1998 data was unavailable.
- **Bathymetry**—10 m contours obtained from California Dept. of Fish and Game. Used to calculate a shoreface beach slope extending from closure depth (estimated at 10 m depth contour) to the toe of the backshore.
- **Foreshore Beach slopes**—Mean high water slopes based on the Stockdon et al. (2002) method. These slopes were used to calculate wave run-up and evaluate erosion from a 100-year storm event.

The data sets above were used to classify the coastline into dune and cliff backed shoreline segments. This classification was applied in GIS to an offshore baseline roughly corresponding with the shoreline inventory of Habel and Armstrong (1978). This backshore baseline was further divided based on geologic units generated during the update of Griggs et al. (2005). Each continuous geologic unit was subdivided into 500 m blocks (Fig. 2). Approximately 2,500 cross shore profiles were extracted from LIDAR topography in

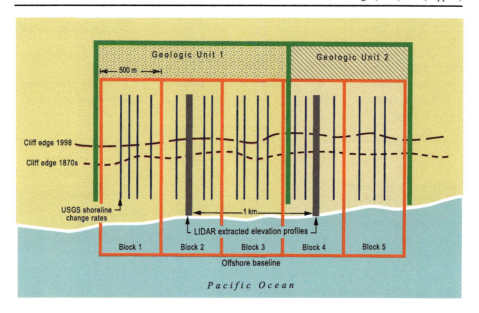

Fig. 2 Method of backshore classification. Backshore classification was based on geologic unit and backshore type. These units were subdivided into 500 m blocks which formed the scale of analysis. For each block, the historic erosion rates from the USGS were averaged and beach slopes and various elevations were attributed to the offshore baseline

ArcGIS©. These profiles were geomorphically interpreted using visual cues to identify significant changes in slope through a custom built interface in MATLAB© to identify toe elevations (Et), foreshore beach slopes, and cliff and dune heights. Attributes necessary to drive the erosion models were averaged over 500 m alongshore segments called blocks and assigned to the backshore baseline. We ground-truthed this data set using a combination of site investigations, oblique air photos from the California Coastal Records Project (www.californiacoastline.org), and topographic information extracted from LIDAR.

Following the initial block averaging only 55% of the blocks had all of the input values necessary to run the erosion models (Table 1). To provide for more complete coverage along the coast, we prioritized the following criteria to fill in missing data gaps for the 500 m spaced blocks:

- Use the block averaged data when available
- Use the geologic unit averaged data continuous with the missing blocks

Table 1 Summary of the backshore characterization comparing USGS published data following application of filtering criteria

	Length (km) (miles)	Total # of blocks	USGS coverage (# of blocks)	USGS% of shore	Current study (# of blocks)	Current% of shore
Cliff	1,142 (710)	3,276	1,607	49.1%	2,897	88.4%
Dune	303 (188)	816	634	77.7%	764	93.6%
Total	1,445 (898)	4,092	2,241	54.8%	3,661*	89.5%

*Blocks with all input parameters necessary to calculate future hazard zones

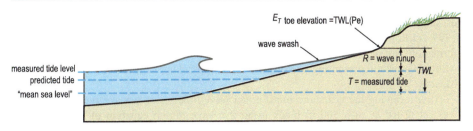

Fig. 3 Total Water Level Definition (after Ruggiero et al. 2001), wave run-up is calculated using the empirical formula of Stockdon et al. 2006

- Use the block averaged LIDAR interpreted profile data
- Use the geologic unit averaged LIDAR interpreted profile data
- Use geologic unit averaged data found in close proximity (+/−15 km)

In lieu of further refinement beyond the scope of the study, the following "filters" were used to identify and eliminate seemingly unrealistic values and values not well handled by the methodology:

- Remove all toe elevations>6 m
- Elevate all toe elevations<1.0 m up to 1.0 m
- Remove all beach slopes>1:4 (height:length)
- Reduce all long term accretion trends >1.5 m/yr to 1.5 m/yr[1]

2.2 Total water levels (TWL)

The TWL is conceptually the combined elevation of wave effects and ocean water levels affected by tides and meteorology. The TWL is determined by the sum of mean sea level, astronomical tides, and wave run-up, and can be affected by other atmospheric forcing such as storm surge and climatic conditions such as El Niño events (Fig. 3; Ruggiero et al. 1996; Ruggiero et al. 2001). The TWL has been identified as a key parameter in quantifying coastal flood and erosion hazards on the U.S. Pacific Coast (Ruggiero et al. 2001; Sallenger et al. 2002; Hampton and Griggs 2004; FEMA 2005; MacArthur et al. 2006), and has been used to assess the implications of climate change along the U.S. Pacific Northwest coast (Ruggiero 2008).

The SLR scenarios used in the California Climate Impact Assessments were generated from a downscaled global climate model (GCM) analyzed at the Scripps Institution of Oceanography (Cayan et al. 2009). The "high scenario" was a 1.4 m rise by 2100 and a "low" scenario of 1.0 m rise by 2100(Cayan et al. 2009). Tides and wave exposures vary with latitude, so deep-water conditions were characterized regionally in the GCM outputs. The GCM predicted a 100-year time series of 3-hour coincidental still water levels (tides, storm effects, and SLR) at two locations—San Francisco and Crescent City; and waves at three locations—Point Conception, San Francisco, and Crescent City (Cayan et al. 2009). The deep water wave time series from the GCM were transformed through a wave refraction model to 140 nearshore locations along the coast to account for the range

---

[1] Hapke et al. (2006) also used the best available LIDAR dataset, a 1998 post El Niño survey. Large sediment deposition associated with high rainfall and discharge was identified as a likely cause of highaccretion rates. Realistic value selected after consultation (Hapke Pers.Comm)

of shoreline orientations, wave exposure, and nearshore bathymetry, (O'Reilly et al. 1993; O'Reilly and Guza 1993; O'Reilly unpublished data). The nearshore transformations are part of the CDIP Monitoring and Prediction (MOP) system (CDIP 2008), with this project representing one of its first widespread applications to evaluating coastal hazards.

After receiving the nearshore transformed wave time series, a wave run-up time series was calculated at each 500 m block using the wave run-up equation of Stockdon et al. (2006). Inputs to the wave run-up equation include deepwater wave height, wave period, and foreshore beach slopes. This wave run-up time series was then added to the still water level (SWL) from the GCM to provide a time series of total water level (TWL=SWL+wave run-up) at each site.

Using the TWL time series, the percent time that the TWL exceeded a specific elevation was calculated. This produced a series of exceedance curves (also called cumulative distributions)for each of the individual backshore segments. Figure 4 compares TWL excedance curves under existing (solid line) and future conditions (dashed line) for the range of wave exposures in the erosion study area.

For each of the ~4,100 blocks, a unique set of TWL exceedance curves was calculated at 10-year intervals (2001–2010, 2011–2020, etc.) from the TWL time series. Changes in TWL exceedance frequency were used to force the erosion response models over the various planning horizons. For each block, the frequency of exceedance of the present day toe elevation (E_t) and the intensity of exceedance (TWL-E_t) were determined. This procedure was repeated for each 10-year interval using two different erosion methodologies for the dune or cliff backshore types.

2.3 Coastal flooding—base flood elevation (BFE)

To evaluate the effects of SLR on coastal flood hazards, we generated two independent estimates of a 100-year flood level. The first estimate focused on updating and expanding the coverage of the 100-year coastal base flood elevations (BFE) originally published by FEMA, while the second estimate was based on calculating a 100-year TWL at each 500 m block.

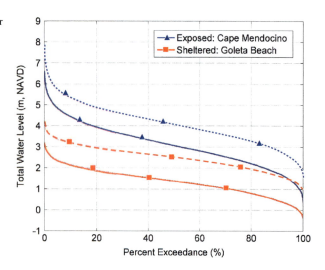

Fig. 4 Example Total Water Level exceedance curves for the most exposed and sheltered coastal segments in the erosion study area. *Solid lines* represent current conditions. *Dashed lines* represent future conditions by 2100

The published FEMA BFEs are an important component in determining flood insurance rates (FIRMs—flood insurance rate maps) and common to land use planning, so this source was chosen to communicate future flood risks. The FEMA BFEs were calculated using methods similar to those used here (based on high ocean water levels coincident with extreme wave runup, see Section 2.2 TWLs), although with less detail consistent with the state of the science in the 1980s when the last studies were done. However, in California, FEMA has not published coastal flood elevations for the majority of the wave-dominated coast, especially along the central and northern portions of the state with low population densities in the 1980s. Sources of published FEMA BFEs included paper maps, FIRMettes, provisional digital FIRMs and effective digital FIRMs. Since the elevations were originally established in the 1980's, elevations were increased by the amount of observed historic sea level rise at the tide gage data for La Jolla, San Francisco and Crescent City gages, representative of southern, central and northern California, respectively. Relative sea level rise was assumed equal to the increase in high tide elevations since the prior tidal epoch (the 19-year period used to calculate tidal datums), which is published by National Ocean Service relative to land datums, and therefore include the effect vertical land motions. We used the most recent tidal epoch 1983–2001, and the prior epoch 1960–1978. A value of 9.1 cm was selected for southern and central California and zero was used for northern California (above Cape Mendocino) since Crescent City showed a relative sea level drop. The elevations were converted to NAVD88 and rounded up to the nearest half foot (15.24 cm) using Corpscon 6.0.1 (ACOE 2005).

The FEMA sources were published at different times and the most recent publication was used for each location. The digital FIRMS are the most recent versions and have been updated from the older National Geodetic Vertical Datum (NGVD) to the contemporary NAVD. Older sources (paper maps) were in NGVD. All published elevations are rounded. The conversion between NGVD and NAVD are not uniform due to different spatial models and data used to develop the datums (hence, the use of Corpscon). Consequently, the combination of different datums, different rounding sequences, and different sea level corrections resulted in site-specific conversions. Each published elevation was assigned to a segment of shoreline within a GIS baseline (different baseline than the coastal backshore characterization).

Where FEMA has not published flood elevations, the flood elevations were estimated using inverse distance weighted interpolation and professional coastal engineering judgment. BFE estimates were informed by the flood elevations published for nearby areas, and consideration of wave exposure. Wave exposure was assessed qualitatively based on the type of land planform (points tend to experience larger waves than embayments due to wave focusing and spreading), shore orientation (the largest wave exposure is from the west to northwest direction), profile steepness (steeper shores have higher run-up elevations), and a general increase with latitude due to the proximity to storms and swells. Small gaps were filled by interpolation unless consideration of wave exposure indicated a different value was appropriate. In some locations, estimates were based on the closest published value with a similar exposure. Results from the BFE methodology were used to evaluate coastal flood impacts (Heberger et al. 2009).

Prior to this study, there was not a single data set of coastal flood elevations in California, and only a few locations had GIS compatible data, and most of the California coast did not have flood maps.

2.4 Coastal backshore response models

2.4.1 Dunes

Dune erosion hazard zones (DHZ) are defined as the coastal band subject to erosion potential over the planning time frame for the sea level rise scenario. DHZ were generated using a location-specific three step methodology (Eq. 1). First, historic erosion rates were multiplied by the planning horizon time step to get the projected baseline erosion. Second, SLR induced shoreline retreat was estimated based on the increased total water level associated with predicted climate change. The third step added a storm-based recession associated with a 100-year storm event occurring at the end of the planning horizon:

$$DHZ = (R_h)^* \Delta t + \Delta TWL/\tan\varphi + (100 - yr\ TWL)/\tan\beta \qquad (1)$$

where

R_h	historic rate of shoreline change
Δt	time step (10 years)
ΔTWL	change in total water level
$\tan\varphi$	shoreface beach slope
$\tan\beta$	foreshore beach slope

Both the shoreline retreat and the 100-year storm event were estimated using the geometric model of dune erosion (Komar et al. 1999; FEMA 2005) but different shore slopes were used. For retreat due to sea level rise, a shoreface slope from MHW to 10 m depth was used. This is similar to the approach by Bruun (1962) and others that the long-term shore recession due to sea level rise relates to the entire zone of wave dissipation (the shore face). In contrast, storm-induced retreat was projected using a foreshore slope (Komar et al. 1999).

Long-term historic shoreline change rates were incorporated to take into account the variety of additional factors such as sediment budget and local geomorphic controls which are not explicitly included in the TWL methodology or resolved in climate modeling. In some cases, the sandy shoreline change rates showed long-term accretion, and in others, localized erosion hotspots. Since the endpoint of the USGS shoreline change study was spring 1998 (following the 1997–98 El Niño), the post El Niño LIDAR data represented an eroded shore. While this eroded condition is consistent with the FEMA guidelines for mapping the 100-year coastal flood (FEMA 2005), the authors consider it less appropriate for the long term changes induced by relative sea level rise. This is because the lower toe elevations indicate a greater TWL exceedance than is normal, as well as other temporary geometries that could affect our baseline from which long term changes are projected. To minimize the influence of the heavy flood and erosion event that occurred just prior to the 1998 LIDAR collection, with its documented impacts of erosion hotspots and large scale beach rotations (Revell et al. 2002; Sallenger et al. 2002; Hapke et al. 2006), we averaged data extracted from the USGS and the LIDAR data sets over each continuous dune stretch and filtered any accretion greater than 1.5 m/yr. We assumed that these localized erosion and accretion signals would be muted as sand dispersed over time along the same stretches of coast that the averaging occurred.

For evaluating the sea level rise impacts, the future toe elevation was established based on the new total water level exceedance curve, assuming the percent exceedance of the TWL above the toe remained constant (Fig. 5a). This was done by moving vertically up from the existing exceedance curve to the future exceedance curve along the present day percent exceedance value (Pe) to intersect at the future toe elevation. This determined the

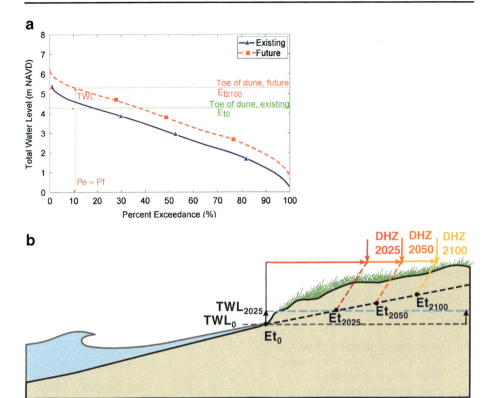

Fig. 5 Dune erosion response model **a** Dune method schematic showing the estimation of future toe elevations and recession from sea level rise **b** Dunes hazard zones in cross section

future elevation of the toe at the specified planning horizon. This future toe elevation was turned into a recession distance by using the average slope of the shoreface calculated from closure depth (estimated as the 10 m contour) to the elevation of the back beach (Bruun 1962; Everts 1985). Finally, the erosion of the toe extended inland through the dune at a

Table 2 Summary of maximum deepwater wave sites and 100-year wave characteristics for each region

Location					TWL Characteristics		
Deepwater Wave Sites	Counties	100-yr TWL m NAVD (ft)	MOP ID	MOP Station	Ho' (m) (ft)	Tp (sec)	SWL m NAVD (ft)
Crescent City	DN, HU, ME	10.7 (35.1')	DN0045	Klamath Spit	12.1 (39.8')	19	2.2 (7.2')
San Francisco	SO, MA, SF, SM, SC, MO	9.8 (32.1')	SM200	Martins Beach	10.5 (34.4')	17	1.6 (5.3')
Point Conception	SL, SB	9.1 (29.9')	B1270	Point Arguello	7.3 (23.9')	19	1.9 (6.3')

Ho' is the deepwater wave height; *Tp* is the peak wave period; SWL is the still water level for each MOP site. These results show the most extreme values for each deepwater wave time series. Mendocino (ME), Humboldt (HU), Del Norte (DN); Monterey (MO), Santa Cruz (SC), San Mateo (SM), San Francisco (SF), Marin (MA) Sonoma Counties (SO); Santa Barbara (SB) and San Luis Obispo (SL) Counties

standard angle of repose of 32°, the angle of stability for dry sand, to the dune height extracted from the LIDAR transects (Fig. 5b).

The maximum 100-yr TWL was selected for the three regions based on the county groupings shown in Table 2. These values are extreme values from the larger data set of TWL estimates calculated for each block.

The progression of the DHZ with sea level rise is shown schematically in Fig. 5b. A example of the mapped DHZ for the three time projections of 2025, 2050 and 2100 for the high SLR scenario is shown in Fig. 6.

2.4.2 Cliffs

Our methodology for predicting the Cliff Erosion Hazard Zones (CHZ) was to increase (i.e. prorate) the historic cliff erosion rates based on the relative increase in time that the TWL exceeded the elevation of the backshore. This method is an evolution of an erosion prediction method developed by Leatherman (1984), which historic rates of erosion and sea level rise are used to calculate future erosion rates based on predicted SLR. Our method substituted the SLR rate with the change in TWL between planning horizons:

$$R_f = R_h + R_h[(P_f - P_e)/P_e] \qquad (2)$$

where

R_f future rate of shoreline change
R_h historic rate of shoreline change
P_f future percent exceedance
P_e existing percent exceedance

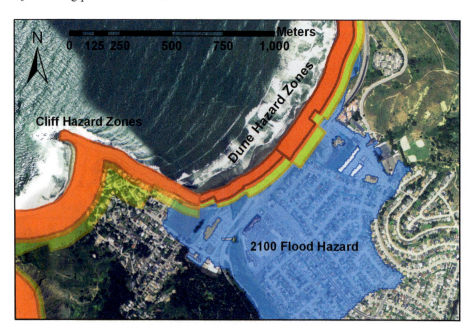

Fig. 6 Example of hazard zones in map view. Cliff hazard zones are on the left, and Dune hazard zones are on the right, with flood hazards extending into the community of Pacifica. Note the steps in the erosion hazard zones demonstrate the differences in alongshore block segment calculations

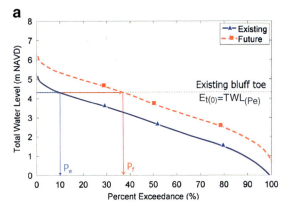

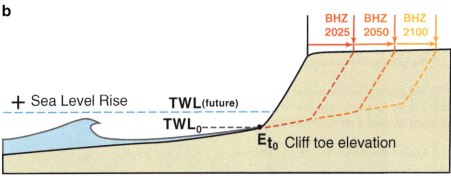

Fig. 7 Cliff erosion response model **a** Cliff method schematic illustrating how changes in percent exceedance are used to prorate the historic erosion rates **b** Cliff hazard zones in cross section

At each 500 m block, we examined Pe and Pf at each planning horizon to determine the change in percent exceedance of the cliff toe (Fig. 7a). This was completed by identifying the present day intersection of the cliff toe elevation (Et) with the exceedance curve. Then moving horizontally, the intersection with the future (next 10 year period) exceedance curve identified the change in percent exceedance assuming a constant toe elevation. This change in percent exceedance was then used to prorate the historic erosion rate at 10 year intervals. For each interval, the prorated erosion rate was turned into a recession distance by multiplying the new prorated erosion rate by the 10 year interval. The overall erosion distance was then the sum of the 10 year recession distances:

$$\text{CHZ} = \Delta t^*[R_{f\_2020} + R_{f\_2030} + R_{f\_2040} + \ldots] + 2^*\Delta t^* s_{Rh} \qquad (3)$$

where

Δt time interval (10 years)
R_{f_YYYY} prorated future erosion rate at year *YYYY*
s_{Rh} standard deviation of the geologic unit historic erosion rate

The third term of Eq. 3 was added to account for alongshore variability in erosion rates within geologic units observed in the historic erosion data. We calculated an alongshore variability factor equal to two standard deviations of the historic erosion rates for each geologic unit. This measure of variability was attributed to each block and incorporated into the projected CHZ erosion distances. While this may over-predict erosion in some areas, the landward extent of the CHZ can be conceptualized as a line that connects the most landward erosion "hotspots" even though the precise locations of those hotspots are not predicted. The progression of the CHZ with sea level rise is shown schematically in Fig. 6. An example of the mapped CHZ for the three time projections of 2025, 2050 and 2100 for the high SLR scenario is shown in Fig. 7b.

3 Results

The results are reasonable approximations of the potential extents of coastal hazards that could result under the modeled climate change scenarios. Increased coastal erosion in California in response to SLR and the potential extent of erosion by 2100 indicates severe vulnerability.

3.1 Backshore characterization

The backshore characterization provided some insights into the variability of geomorphic features for shoreline segments found across California (Table 3). The combination of these datasets enabled a prediction of future coastal hazards.

3.2 Coastal flooding—base flood elevations vs. total water levels

The flood potential along the California coast is mapped in a few areas based primarily on analysis accomplished in the 1980s. We have augmented the available mapping in this study using two data sets and related but different methods. Since both methods use Total Water Level (TWL), we have called one method BFE in

Table 3 Summary statistics on backshore geomorphic and shoreline change parameters

Dunes $n=816$	Min.	Max.	Average	STD.
Toe Elevation, Et (m NAVD)	1.0	5.76	2.17	1.26
Dune Height, Eh (m NAVD)	0	115.74	13.4	15.25
Slope-foreshore ($\tan\beta$)	0	0.229	0.08	0.04
Slope-shore face profile ($\tan\phi$)	0.0075	0.0563	0.0174	0.0073
Shoreline change, R_h (m/yr)	−0.76	0.93	0.12	0.33
Cliffs				
$n=3,276$				
# of geol units=29				
Toe Elevation, Et (m NAVD)	0	6.55	2.32	1.36
Slope-foreshore ($\tan\beta$)	0	0.25	0.08	0.056
Erosion Rate, R_h (m/yr)	−2.78	0	−0.28	0.28
Erosion Rate – Geologic Unit Standard Deviation, s_{Rh} (m/yr)	0	0.77	0.18	0.17

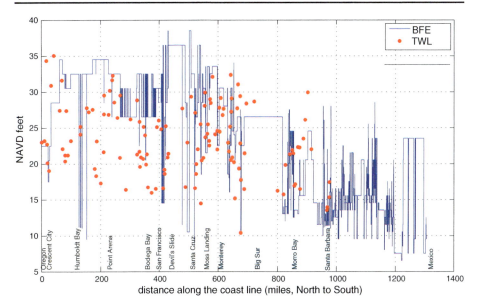

Fig. 8 Comparison of coastal flooding results for the Base Flood Elevation (*lines*) and Total Water level (*dots*) approaches. The TWL approach did not extend into southern California

reference to the FEMA Base Flood Elevation terminology and data source, and use TWL to refer calculations accomplished in this study using GCM water level and wave time series. The calculation of TWL is described in Section 2.2 and the basis for the BFE values is provided in Section 2.3.

The TWL elevations were calculated in this study to force the coastal erosion models and are summarized in Table 2. Figure 8 compares these values to the BFE values. The comparison is an indication of methodology uncertainty because the two sets of estimates entail different data and methods. It should be noted that the BFE is based on FEMA mapping that is accomplished by municipal boundary and calculation points are widely spaced with limited information outside populated areas.

The comparison highlights that flood elevations vary significantly with location. The alongshore variability should be expected since wave heights and wave runup vary substantially along shore due to the variability in bathymetry and topography. Both estimates indicate 100-year flood elevations in northern and central CA mostly between 15 and 35 ft NAVD, which is about 9 to 20 ft above mean higher high water, and lower in southern California.

A direct comparison of the two predictions is undermined by the alongshore variability in values and spacing. However, we estimate that TWL are on average lower than the BFE by about 1.2 m (4 ft). The lower TWL values could indicate that the GCM output is less extreme than historic data in terms of wave conditions and the timing of large wave events with high tides. The lower TWL values may also be more accurate due to the calculation of a 100-year TWL time series rather than less accurate but also less computationally intensive statistical methods used in the 1980's to generate BFEs.

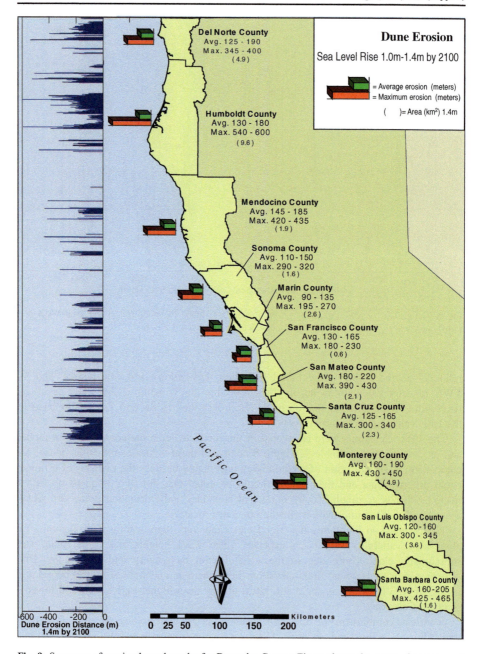

Fig. 9 Summary of erosion hazard results for Dunes by County. Figure shows the range of average and maximum changes for both the 1.0 m and 1.4 m sea level rise scenario. Bar chart on the left of the figure shows the erosion distance calculated at each block for the 1.4 m sea level rise scenario. Predicted erosional area for the 1.4 m scenario is also shown in parentheses

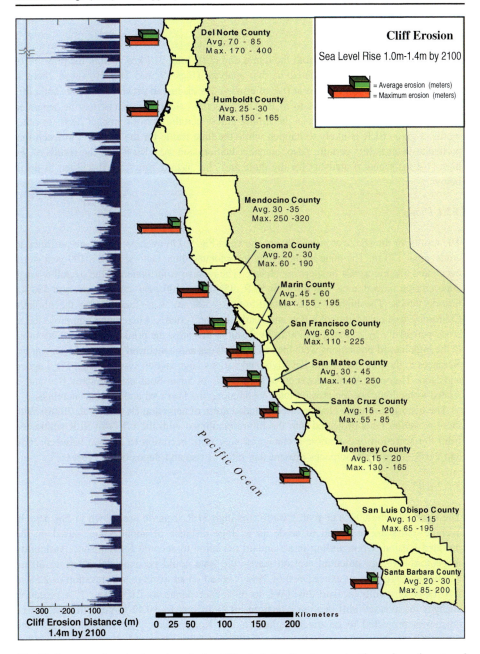

Fig. 10 Summary of erosion hazard results for cliff backed shorelines by county. Figure shows the range of average and maximum changes between the 1.0 m and 1.4 m sea level rise scenario. Bar chart on the left of the figure shows the erosion distance calculated at each block for the 1.4 m sea level rise scenario

3.3 Coastal hazards zones

The erosion hazard zones for both the dunes and cliffs total 214 km^2 within the portions of the 11 coastal counties evaluated in this analysis (Figs. 9 and 10). However, there is significant variation in the areas at risk of erosion. As discussed previously, dunes and cliffs will exhibit different responses to SLR. The potential extent of impacts may be separated from the tabulated time frame, as the timing is less certain than the impact itself. In other words, a given SLR is likely to happen while the time frame for the sea level to reach that particular level is less certain. Also, consider that erosion may lag SLR. The results of the high (1.4 m) scenario analysis for the dune and cliff hazard zones are discussed in detail below.

3.3.1 Dunes

The results for the DHZs at year 2100 are shown in Fig. 9. On average, dunes are projected to erode about 170 m. The dune methodology incorporated three factors, SLR, a 100 year storm event, and historic trends in shoreline change. The relative contribution to the overall average dune erosion hazard zone was 48% for SLR, 45% for a 100 year storm event, and 7% for historic trends. The highest dune hazard zones (>500 m) were found along the northern California coast near Humboldt Bay and nearby lagoon systems, where dunes are projected to erode potentially by as much as 600 m by 2100. In this area, the primary factors contributing to the high hazard predictions were the low lying dunes and ephemeral sandspits located at the entrances to the Mad River and Eel River and various large lagoon systems (e.g., Big Lagoon and Stone Lagoon). Another location of high DHZs occur in Southern Monterey Bay and is related to the high historic erosion rates found along the Fort Ord and Marina shorelines.

The historic shoreline change rates included areas of accretion (Hapke et al. 2006). With these accretion rates projected into the future combined with the effect of SLR subtracted from the accretion, we observe a change in sign from accretion to erosion between 2050 and 2100, when SLR outpaces present day rates of rise and depositional processes.

3.3.2 Cliffs

The results for the CHZs at year 2100 for the high SLR scenario are shown in Fig. 10. On average, cliffs are projected to erode an average distance of about 33 m, based on the prorated erosion. Upon adding in a standard deviation to account for longshore variability in geology average potential erosion across the state could be expected closer to 60 m. Current average percent exceedance of 59% is based on an average toe elevation of 2.59 m NAVD. SLR would escalate this TWL to 69% exceedance by 2100. The 10% proration changes the average cliff erosion rate from −0.31 m/year to −0.42 m/year. In some areas, erosion is projected to be much higher. In Del Norte County, for example, cliffs may potentially erode a maximum distance of 400 m. Cliff erosion is less severe in the other counties along the coast, although remains significant (> 55 m). It should be noted that the historic USGS erosion rates had substantial variation alongshore. A conservative approach incorporated two standard deviations to account for this high variability which greatly escalates potential cliff hazard zones (up to 600 m); however, these conservative results were not reported in the summary result figures (e.g. Fig. 10).

Geologic setting plays an important role in the erosion response of cliff backed shorelines with 29 geologic units identified along the coastline. The geologic units K and Kjf were found to be associated with the highest average cliff erosion rates in California

(−2.8 m/year and −1.8 m/year, respectively). The K unit is a Cretaceous marine unit found predominantly along the northern California coast between Point Arena and Cape Mendocino. This unit is typically a weakly consolidated unit consisting of sandstone, shales and larger conglomerates. The Kjf unit, or Franciscan complex, is found along central and northern California, most notably along some of the landslide backed segments. The Franciscan complex is an accreted terrane (accretionary or tectonic wedge) of heterogeneous rocks. This terrane was scraped off of the Pacific Plate as it subducted under the North American Plate and consists of mafic volcanic rocks including cherts, greywacke, limestones, serpentines and shales all mixed in a relatively chaotic manner. The highest cliff hazards under the high SLR scenario (~400 m by 2100) were located ~3 km south of the Klamath River and are primarily related to the high erosion rates associated with the Franciscan complex.

4 Discussion

The methodology evolved from research along the U.S. West Coast and applied using the best currently available statewide data sets. The results provide an initial assessment of potential future coastal hazards under two possible SLR scenarios. However, uncertainty stems from the limitations of each input data set and the resulting outputs from this study which highlight the need for further research, new data collection, and long term monitoring of geomorphic and physical response variables. Our intent in this discussion is to discuss some of these key findings, identify future work necessary to better predict future erosion, and provide some suggestions on how California and other countries can adapt to climate change.

4.1 Backshore characterization

The backshore was characterized as sandy (dunes) or erosion-resistant cliffs. This classification is not entirely adequate for the complex coastal geology and geomorphology. In some locations, the coastal geology exposed to marine forces changes after some erosion (or accretion) occurs. This is particularly apparent at river mouths with sand spits or dunes fronting cliffs, or in cliffs with multiple geologic units. Some cliffs consist of ancient sand dunes and weakly consolidated sedimentary deposits that can erode as rapidly as sand dunes (e.g. Pacifica; Collins and Sitar 2008). More detailed characterization of the geology and geomorphology (e.g. ranges of toe elevations, rock hardness) would refine the resolution of results. It should also be noted that the geologic units associated with the highest erosion rates (Franciscan complex and Cretaceous marine), are a mix of rock units that are poorly consolidated and often associated with high ground water. Thus, while the erosion models are forced by marine processes, the relation of terrestrial processes with these high erosion rates is important to note. Any new development occurring in these geologic units should require specific study and documentation during permitting to avoid exacerbating groundwater or altering slope stability.

Shoreline armoring was not considered in the erosion analysis. Currently 11% of the California coast is estimated to have shoreline armoring (Griggs et al. 2005) with the majority occurring in southern California. Lack of information on structure heights, condition, and footprint make it nearly impossible to assess the influences of these structures on future conditions at a statewide level. Managers should expect that future use of these structures will come under increasing wave forcing and require higher maintenance costs. Effort to hold the shoreline in place will result in long term losses of beaches, ecology, recreation, and economics. Development of a methodology to specifically address coastal armoring and to allow planning-level evaluation of armored shores should be pursued.

The most comprehensive topographic data available for most of the U.S. West Coast was captured at the end of a major El Niño event in April of 1998. This data set, while spatially comprehensive, represents a single point in time at which the coast was significantly eroded. Data shows areas of eroded conditions characterized by low toe elevations and narrow beaches, as well as other areas which accreted, likely due to sediment deposition following the flood events and erosion of dunes and cliffs. This same LIDAR data was used in the USGS shoreline change studies as the most recent shoreline and likely increased long term erosion rates. The use of this data set for the topographic and geomorphic analysis in this study also likely elevated the prorated erosion rates particularly in the cliff backed shoreline segments due to lower toe elevations. The DHZ calculations may also under-predict the potential erosion hazard. Overall the use of this outdated LIDAR data has introduced unknown levels of uncertainty into this project, but remains the best available statewide data set. This data set is unlikely to represent current conditions at the time of this study. Consequently, recent and routine LIDAR data collection is needed for future analysis. Additional historic shorelines should be incorporated to refine erosion rates.

4.2 Coastal flood elevations

To the authors' knowledge, this is the first time coastal flood estimates have been completed for the entire California coast, and hence it is very difficult to assess their accuracy. The FEMA Pacific Coast Guidelines for Coastal Flood Hazard Analysis and Mapping (FEMA 2005) suggests that one measure of the relative accuracy is to apply multiple methods of prediction that are independent of one another. To that end, the comparison between the TWL and BFE methodologies is insightful, with the TWL method being about 1.2 m (4 ft) lower (Fig. 8). Differences in the TWL method may be due to the use of GCM simulations rather than observed water levels. Additionally, the published BFE values are generally based on studies completed around 20 years ago with sparse resolution and less scientific understanding on wave transformation and run-up modeling. The high variability seen in Fig. 8 compares the two methods and highlights the differing spatial accuracies between the TWL (<500 m) and the BFE (10s of km) estimates.

4.3 Total water levels

The erosion models respond to water level and wave time series output from a downscaled GCM (Cayan et al. 2009). Only GCM output data were used to be consistent with CEC guidance and in order to preserve the implicit coincidence of high water level and waves, and in order to develop an estimate of relative changes. An ensemble of GCM runs could be used to develop a wider range of scenarios with more defined probabilities. More scenarios with different parameters such as changes in wave climate or intensities could also be used to develop a range of erosion and flood responses. For this study only a single wave time series from Crescent City, San Francisco, and Point Conception was used for all scenarios since the GCM did not show any changes to the wave statistics with climate change. However, researchers have shown latitudinal differences in storm wave heights and periods (Graham and Diaz 2001; Allan and Komar 2006; Ruggiero et al. 2010), which was not evident in the GCM output. This would tend to under predict TWLs and hazard zones as one moves north in California. This TWL methodology can account for changes in waves, and their joint probability of occurrence with high water levels, and hence used to explore a wider range of possible climate and management scenarios beyond this present initial application.

While we evaluated over 140 locations along the coast, wave refraction transformations were accomplished only to the 10 m contour. Wave transformations were detailed (using directional spectra), but were condensed to single spectral parameters (wave height and period) for run-up analysis. More detailed wave analysis modules can be used, and may be appropriate for more detailed regional studies.

4.4 Coastal erosion response models

The study purpose was to provide an estimate of *potential* future hazards NOT to predict actual erosion or flooding. Actual erosion from SLR may lag potential erosion, especially for very hard shores and/or rapid supply of littoral material. Future erosion was estimated using conceptual models which included simplified geometric response models for sandy dune backed shorelines based on contributions from SLR, a 100 year storm event, and historic trends in shoreline change. For cliff erosion, a prorated linear increase in historic erosion was related to increased TWL exceeding the toe elevation of the cliff.

For dune backed shorelines, we found that over the next 100 years, SLR had the highest relative contribution to the average hazard zone, followed by the 100 year storm event, and finally the historic trends of shoreline change (48%, 45%, and 7% respectively). These relative contributions indicate that the use of historic trends in calculating development setbacks has limited use for future coastal management.

Improvements for the dune backed shorelines could include linkage with a shoreline evolution model and or sediment budget analysis that could allow extension of the profile-based methodology to include planform changes and feedback on wave transformations.

For cliff backed shorelines, the hazard zones are largely affected by the toe elevation and historic erosion rate. At this statewide scale, we used historic erosion as an integrator for the marine and terrestrial erosion processes relying on the spatial relation between geology and erosion rates, as well as the medium time scales to average out the episodic nature of cliff failures for this initial CHZ assessment. Our investigations identified that one of the highest causes for the variability in potential cliff erosion zones was the variability in toe elevations. When low initial TWL exceedance of the toe was a very small percent, erosion rates would rapidly increase to the point of no longer being valid. This could occur if the TWL exceedance increased from just 1 to 10%. These highlight the sensitivity of the model to the input parameters and the need for more detailed site assessments.

In the results we report only the CHZs derived from the proration of historic erosion rates; however given the alongshore variability in geology and failure mechanisms, we recommend adding a factor of safety to these estimates. By adding a standard deviation in geologic unit erosion rates, to attempt to account for alongshore variability, the average potential erosion distance jumped from 33 m to 65 m. In the most extreme cases, this conservative approach identified potential CHZ extending 600 m inland. The expanded ranges from the standard deviations provide a representation of the uncertainties in these projections and should be used if these results are used to make planning and permitting decisions.

Improvements for cliff backed shorelines would include consideration of the duration of wave impact hours as this is likely an important factor not explicitly evaluated in this analysis. Specifically, we did not evaluate failure mechanisms, angles of repose, rock hardness, or terrestrial erosion processes such as rainfall although it has been recently documented that various feedback mechanisms exist between groundwater seepage, elevated TWLs, cliff failures, and toe protection (Young et al. 2009).

The GCM projections of changes to precipitation could have an influence on sediment yield from the coastal watersheds, however the complexities of the water storage infrastructure, varying projections about precipitation patterns, and limited sediment rating curves was beyond the scope of this initial study, but worth future research efforts.

A key improvement of this overall methodology would be to integrate the erosion and flood hazards, time stepping through the erosion calculations and determining any new flow pathways for flooding. This would enhance the estimate of flooding in low lying areas behind barriers, where the volume of overtopping water is often the primary factor in flooding.

During development of this methodology, we identified several alternative erosion response methods based on a range of prior efforts (e.g. Collins and Sitar 2008) and hybrid variations derived from those published (e.g. Leatherman 1984). However, some of the more processed based approaches are not likely to be practical at such a broad spatial scale. A long-term field data collection effort is needed to develop data sets to allow testing of methods, including the ones used in this study. We expect that there is significant room for improvement, and utility in pursuing model validation to improve predictions. While, these results are approximate order of magnitude estimates; we believe that these results provide the magnitude needed in an initial assessment that can be used to guide adaptation strategies to coastal hazards.

4.5 Management implications

Current and future erosion and inundation hazards have lacked systematic quantitative evaluation and remain largely undefined in California, which undermines key coastal zone management tools such as land use plans, local coastal programs and environmental review documents for specific developments. The methodology described herein can be applied at a regional resolution more appropriate for land use planning and coastal management using more site specific information. The methodology allows substitution of methods and values specific to a site, and can therefore be used as a tool to explore scenarios based on a range of assumptions on sediment budgets, climate changes, management regulations, and land uses. However, without higher resolution statewide data sets, further statewide analysis is probably unnecessary. It is recommended that more focused local pilot studies be funded to refine the predictions and methodology.

The potential erosion and flood summary data indicate that a portion of the coast is likely to be lost to erosion over the next 100 years and provide an initial assessment of the spatial impacts to coastal California (e.g. Fig. 6). These impacts will result in the loss of private and public property and threaten or destroy existing public infrastructure such as roads and utilities. Ecological and recreational losses associated with the loss of beaches should also be expected, although the management responses to SLR and accelerated coastal erosion could greatly affect the extent of ecological and recreational losses.

While the exact timing and extents of these coastal hazards are uncertain, the potential erosion mapped during this project is likely to occur at some point in the future. Even if greenhouse gas emissions were to be stopped today, sea levels are expected to rise for thousands of years given the time scales associated with climate processes and feedbacks (IPCC 2007). Adaptation will mean figuring out how to implement land use policies, value hazardous lands, and develop financing mechanisms to relocate infrastructure and development out of these hazardous areas. Management responses to increasing coastal hazards will determine the future economic and ecological value of the coast of California and the world. Attempting to hold the shoreline in place through shoreline armoring will destroy recreation and ecological values of the coast and should be considered carefully as to their long term effectiveness. The spatial representation of coastal hazards is an important step in evaluating levels of risk, developing adaptation strategies, and raising awareness to communities.

5 Conclusions

Sea level rise as a result of climate changes will increase the risks associated with coastal hazards of flooding and erosion. Previous studies of SLR impacts have primarily focused on the inundation of coastal lands; however, along the active tectonic margin of the U.S. West Coast, the variety of coastal morphologies and geologic formations including seacliffs, landslides, and dunes requires a different approach to evaluating future coastal erosion hazards. In this study, we map future potential coastal hazards of erosion and flooding based on two scenarios generated from a downscaled regional global climate model developed to support California impact assessments to climate change. The erosion method relates backshore types (dune and seacliff), with shoreline change rates and coastal geology, then applies future changes in total water levels in exceedance of the toe elevation to predict future coastal erosion hazards for three planning horizons. The results of the erosion study predict 214 km^2 of land loss by 2100 under a 1.4 m SLR scenario. Average potential erosion distances range from 170 m along dune backed shorelines, to 33 m of average erosion for cliff backed shorelines. Maximum potential erosion distances of up to 400 m are predicted along cliff backed and up to 600 m along dune backed shorelines. Erosion along the seacliff backed shorelines was found to be highest in the geologic units of Cretaceous marine (K) and Franciscan complex (KJf). Areas of historic sandy shoreline accretion found along northern California are observed to reverse sign and become erosion between the 2050 and 2100 planning horizon.

In addition, 100-year future flood elevations were estimated for most of the California coast using two different methods. Each method predicts strong alongshore variability. This variability combined with the different location and spacing between calculation points makes a comparison of the two estimates difficult. The more detailed computations using the GCM average about 1.2 m less than the values based on FEMA maps for existing conditions (without SLR).

While the actual amount or time of occurrence of these future coastal hazards may vary from the predicted, the results highlight the need to develop adaptation strategies. This methodology provides coastal managers with a tool that can be applied regionally using higher resolution data sets. By assessing other climate change and management scenarios, coastal managers can develop actionable information to guide adaptation strategies and effectively communicate potential futures to a wide audience.

Acknowledgements Funding for this study was provided by the California Ocean Protection Council, with support from Dr. Christine Blackburn. This work was part of a larger vulnerability study completed by the Pacific Institute and funded by the California Energy Commission. In particular, we would like to thank Dr. Peter Gleick, Heather Cooley, and Matt Heberger at the Pacific Institute. We also thank Dr. Dan Cayan of USGS and his climate researchers at Scripps—Dr. Reinhard Flick, Dr. Nick Graham, Mary Tyree and Dr. Peter Bromirski. We thank Guido Franco at the California Energy Commission for encouraging this submittal. We would especially like to thank our technical review team who provided timely input— specifically Dr. Cheryl Hapke, Dr. Adam Young, and Dr. Gary Griggs. Dr. Bill O'Reilly of the Coastal Data

Information Program voluntarily provided the nearshore wave transformations necessary to improve the spatial resolution of these hazard predictions. Other members of the scientific community who contributed in different ways included Dr. Patrick Barnard, Dr. Kiki Patsch, Dr. Paul Komar, Brian Fulfrost, Lesley Ewing, Clif Davenport, Kim Sterratt, and Nicole Kinsman. Other members of PWA who supported this work include: Jeremy Lowe, Seungjin Baek, Damien Kunz, and Brad Evans. We would also like to thank 3 anonymous reviewers who provided good comments to improve this contribution. Finally, the authors would like to thank the ocean for inspiring and humbling all of our efforts.

References

Army Corps of Engineers (2005) Corps of Engineers Coordinate Conversion Corpscon Version 6.0. April 26, 2005. http://crunch.tec.army.mil/software/corpscon/corpscon.html
Adams PN, Inman DL, (2009) Climate change and potential hotspots of coastal erosion along the southern California coast. California Climate Change Center. CEC-500-2009-022-F
Allan JC, Komar PD (2006) Climate controls on US West Coast erosion processes. J Coast Res 22(3):511–529
BCDC (2008) San Francisco Bay Conservation and Development Commission (BCDC), 1-meter sea level rise scenarios in San Francisco Bay inundation maps. http://www.bcdc.ca/gov
Benumof BT, Griggs GB (1999) The dependence of seacliff erosion rates, cliff material properties, and physical processes: Sand Diego County, California. Shore and Beach 67(4):29–41
Benumof B, Storlazzi C, Seymour R, Griggs G (2000) The relationship between incident wave energy and seacliff erosion rates: San Diego County, California. J Coast Res 16(4):1167–1178
Bruun P (1962) Sea level rise as a cause of shore erosion. J Waterways Harbors Div 88:117–130
Budetta P, Galietta G, Santo A (2000) A methodology for the study of the relation between coastal cliff erosion and the mechanical strength of soils and rock masses. Eng Geol 56:243–256
Carter CH, Guy DE (1988) Coastal erosion: processes, timing and magnitude at the bluff toe. Mar Geol 84:1–17
Cayan D, Tyree M, Dettinger M, Hidalgo H, Das T, Maurer E, Bromirski P, Graham N, Flick R (2009) Climate change scenarios and sea level rise estimates for California 2008 climate change scenarios assessment. CEC-500-2009-014-F
CDIP (2008) Coastal Data Information Program (CDIP), Scripps Monitoring and Prediction (MOP) System. http://cdip.ucsd.edu/documents/index/product_docs/mops/mop_intro.html
Collins BD, Sitar N (2008) Processes of coastal bluff erosion in weakly lithified sands, Pacifica, California, CA. Geomorphology 97:483–501
Dean RG, Maurmeyer EM (1983) Models for beach profile response. In: Komar P (ed) Handbook of coastal processes and erosion. CRC, Boca Raton, pp 151–165
Edil TB, Vallejo LE (1980) Mechanics of coastal landslides and the influence of slope parameters. Eng Geol 16:83–96
Emery KO, Kuhn GG (1982) Sea cliffs: their processes, profiles, and classification. Geol Soc Am Bull 93:644–654
Everts CH (1985) Sea level rise effects on shoreline position. J Waterw, Port, C-ASCE 111(6):985–999
FEMA (2005) Final draft guidelines, coastal flood hazard analysis and mapping for the Pacific Coast of the United States. Prepared for the U.S. Department of Homeland Security
Fisher JS, Overton MF (1984) Numerical model for dune erosion due to wave uprush. Proceedings of the 19th Coastal Engineering Conference. ASCE pp 1553–1558
Flick RE, Ewing LC (2009) Sand volume needs of southern California beaches as a function of future sea level rise rates. Shore and Beach 77(4):36–45
Gleick PH, Maurer EP (1990) Assessing the costs of adapting to sea-level rise: a case study of San Francisco Bay. Pacific Institute for Studies in Development, Environment, and Security. Berkeley, CA and the Stockholm Environment Institute, Stockholm, Sweden. 57 pp with 2 maps
Graham N, Diaz H (2001) Evidence for intensification of North Pacific winter cyclones since 1948. Bull Am Meteorol Soc 82(9):1869–1893
Griggs G, Patsch K, Savoy L (2005) Living with the changing California Coast. University of California Press, Berkeley and Los Angeles
Habel JS, Armstrong GA (1978) Assessment and atlas of shoreline erosion along the California coast. California Department of Navigation and Ocean Development. 277 p

Hampton MA, Griggs G (2004) Formation, evolution, and stability of coastal cliffs—status and trends. U.S. Geological Survey (USGS) Professional Paper 1693. http://pubs.usgs.gov/pp/pp1693

Hapke CJ, Reid D, Richmond BM, Ruggiero P, List J (2006) National assessment of shoreline change Part 3: Historical shoreline change and associated coastal land loss along sandy shorelines of the California Coast. U.S. Geological Survey Open File Report 2006–1219

Hapke CJ, Reid D (2007) National assessment of shoreline change part 4: Historical Coastal Cliff Retreat along the California Coast. U.S. Geological Survey Open File Report 2007–1133

Hapke CJ, Reid D, Richmond BM (2009) Rates & trends of coastal change in California & the regional behavior of the beach and cliff system. J Coast Res 25(3):603–615

Heberger M, Cooley H, Herrera P, Gleick PH (2009) The impacts of sea level rise on the California Coast. California Climate Change Center . CEC-500-2009-024-F

Intergovernmental Panel on Climate Change (IPCC) (2007) Climate Change 2007: The Physical Science Basis: Summary for Policymakers. Contribution of Working Group I to the Fourth Assessment Report of the Intergovernmental Panel on Climate Change

Knowles N (2009) Potential inundation due to rising sea levels in the San Francisco Bay region. California Climate Change Center. CEC-500-2009-023-D

Komar PD, Shih SM (1993) Cliff erosion along the Oregon coast; a tectonic sea level imprint plus local controls by beach processes. J Coast Res 9:747–765

Komar PD, McDougal WG, Marra JJ, Ruggiero P (1999) The rational analysis of setback distances: applications to the Oregon Coast. Shore and Beach 67(1):41–49

Kriebel DL, Dean RG (1985) Numerical simulation of time dependent beach and dune erosion. Coast Eng 9:221–245

Larson M, Erikson L, Hanson H (2004) An analytical model to predict dune erosion due to wave impact. Coast Eng 51:675–696

Leatherman SP (1984) Coastal geomorphic responses to sea-level rise, Galveston Bay, Texas. In: Barth MC, Titus JG (eds) Greenhouse effect and sea-level rise: a challenge for this generation. Van Nostrand Reinhold, New York

MacArthur RC, Dean R, Battalio RT (2006) Wave processes in nearshore environment for hazard identification. In: Smith JM (ed) Coastal Engineering 2006. Proceedings of the 30th International Conference. San Diego, CA. pp 1775–1787

National Research Council (1987) Responding to changes in sea level: engineering implications. National Academy Press, Washington D.C

Nishi R, Kraus NC (1996) Mechanism and calculation of sand dune erosion by storms. Proceedings of the 25th Coastal Engineering Conference. ASCE, pp 3034–3047

O'Reilly WC, Guza RT (1993) Comparison of two spectral wave models in the Southern California Bight. Coast Eng 19:263–282

O'Reilly WC, Seymour RJ, Guza RT, Castel D (1993) Ocean Wave Measurement and Analysis, Proc 2nd Int Symp July 25–28, 1993. pp 448–457

Pilkey OH, Young RS, Riggs SR, Smith AWS, Wu H, Pilkey WD (1993) The concept of shoreface profile equilibrium: a critical review. J Coast Res 9(1):255–278

Rahmstorf S (2007) A semi-empirical approach to projecting future sea-level rise. Science 315:368–370

Revell DL, Komar PD, Sallenger AH (2002) An application of LIDAR to analyses of El Niño erosion in the Netarts Littoral Cell, Oregon. J Coast Res 18(4):702–801

Robinson LA (1977) Marine erosive processes at the cliff foot. Mar Geol 23:257–271

Ruggiero P, Komar PD, McDougal WG, Beach RA (1996) Extreme water levels, wave run-up and coastal erosion. Proceedings of the 25th International Conference on Coastal Engineering. ASCE 2793–2805 p

Ruggiero P, Komar PD, McDougal WG, Marra JJ, Beach RA (2001) Wave run-up, extreme water levels, and the erosion of properties backing beaches. J Coast Res 17(2):401–419

Ruggiero P (2008) Impacts of climate change on coastal erosion and flood probability in the US Pacific Northwest. In: Solutions to Coastal Disasters 2008. Oahu Hawaii, 158:169

Ruggiero P, Komar PD, Allan JC (2010) Increasing wave heights and extreme value projections: The wave climate of the U.S. Pacific Northwest, Coastal Engineering, doi:10.1016/j.coastaleng.2009.12.005

Ryan HF, Parsons T, Sliter RW (2008) Vertical tectonic deformation associated with the San Andreas fault zone offshore of San Francisco, California, Tectonophysics (2008), doi:10.1016/j.tecto.2008.06.011

Sallenger AH, Krabill W, Brock J, Swift R, Manizade S, Stockdon H (2002) Sea cliff erosion as a function of beach changes and extreme wave run-up during the 1997–1998 El Niño. Mar Geol 187:279–297

San Diego Foundation (2008) Climate change related impacts in the San Diego Region by 2050. Summary prepared for the 2008 Climate Change Impacts Assessment, Second Biennial Science Report to the California Climate Action Team. Prepared by: SAIC, Scripps Institution of Oceanography, and The San Diego Foundation

Stockdon H, Sallenger A, Holman R, List J (2002) Estimation of shoreline position and change using airborne topographic LIDAR data. J Coast Res 18(3):502–513

Stockdon H, Holman R, Howd P, Sallenger AH (2006) Empirical parameterization of setup, swash, and run-up. Coast Eng 53:573–588

Sunamura T (1982) A predictive model for wave-induced erosion, with application to Pacific coasts of Japan. J Geol 90:167–178

Sunamura T (1992) Geomorphology of rocky coasts. Wiley, New York, 302p

Trenhail AS (2002) Modeling the development of marine terraces on tectonically mobile rock coasts. Mar Geol 185:341–363

Wingfield DK, Storlazzi CD (2007) Variability in oceanographic and meteorologic forcing along Central California and its implications on nearshore processes. J Mar Syst 68:457–472

Young AP, Guza RT, Flick RE, O'Reilly WC, Gutierrez R (2009) Rain, waves, and short term evolution of composite seacliffs in southern California. Mar Geol 267:1–7

Climatic Change (2011) 109 (Suppl 1)
DOI 10.1007/s10584-011-0309-0

Estimating the potential economic impacts of climate change on Southern California beaches

Linwood Pendleton · Philip King · Craig Mohn ·
D. G. Webster · Ryan Vaughn · Peter N. Adams

Received: 12 February 2010 / Accepted: 21 September 2011 / Published online: 24 November 2011
© Springer Science+Business Media B.V. 2011

Abstract Climate change could substantially alter the width of beaches in Southern California. Climate-driven sea level rise will have at least two important impacts on beaches: (1) higher sea level will cause all beaches to become more narrow, all things being held constant, and (2) sea level rise may affect patterns of beach erosion and accretion when severe storms combine with higher high tides. To understand the potential economic impacts of these two outcomes, this study examined the physical and economic effects of permanent beach loss caused by inundation due to sea level rise of one meter and of erosion

---

This research was funded by a grant from the California Energy Commission's PIER Program established the California Climate Change Center and the California Department of Boating and Waterways. A version of this paper previously appeared as California Energy Commission Publication CEC-500-2009-033-D.

L. Pendleton (✉)
The Nicholas Institute at Duke University, Duke University, P.O. Box 90335, Durham, NC 27708, USA
e-mail: linwood.pendleton@duke.edu

P. King
Department of Economics, San Francisco State University,
1600 Holloway Avenue, San Francisco, CA 94132, USA
e-mail: pgking@sfsu.edu

C. Mohn
Cascade Econometrics, 4509 230th Way SE, Sammamish, WA 98075, USA
e-mail: craigmohn@earthlink.net

D. G. Webster
Environmental Studies Program, Dartmouth College, 6182 Steele Hall, Hanover, NH 03755, USA
e-mail: d.g.webster@dartmouth.edu

R. Vaughn
Ziman Center, University of California, Los Angeles,
110 Westwood Pl, Los Angeles, CA 90095-0001, USA
e-mail: rkvaughn@ucla.edu

P. N. Adams
Department of Geological Sciences, University of Florida, Gainesville, FL 32611, USA
e-mail: adamsp@ufl.edu

and accretion caused by a single, extremely stormy year (using a model of beach change based on the wave climate conditions of the El Niño year of 1982/1983.) We use a random utility model of beach attendance in Southern California that estimates the impacts of changes on beach width for different types of beach user visiting public beaches in Los Angeles and Orange Counties. The model allows beachgoers to have different preferences for beach width change depending on beach size. We find that the effect of climate-driven beach change differs for users that participate in bike path activities, sand-based activities, and water-based activities. We simulate the effects of climate-related beach loss on attendance patterns at 51 public beaches, beach-related expenditures at those beaches, and the non-market (consumer surplus) value of beach going to those beaches. We estimate that increasing sea level will cause an overall reduction of economic value in beach going, with some beaches experiencing increasing attendance and beach-related earnings while attendance and earnings at other beaches would be lower. We also estimate that the potential annual economic impacts from a single stormy year may be as large as those caused by permanent inundation that would result from a rise in sea level of one meter. The economic impacts of both permanent inundation and storm-related erosion are distributed unevenly across the region. To put the economic impacts of these changes in beach width in perspective, the paper provides simple estimates of the cost of mitigating beach loss by nourishing beaches with sand.

1 Introduction

Always dynamic, California's coasts will certainly be altered by natural forces over the next 100 years. Increases in sea level will likely affect beaches through permanent inundation (the loss of beach due simply to flooding when beaches cannot migrate shoreward) and increasingly intense erosion and accretion when higher high tides interact with severe storms (Cayan et al. 2008). These beaches also are important to home owners who depend on beaches to protect their homes from storm surge, for public infrastructure (especially roads), and to the millions of Californians who use beaches as an important destination for outdoor recreation (Neumann and Hudgens 2006).

Of the many potential economic impacts that may result from the impacts of climate change on beaches, we focus on the effects of sea level rise on the beach-going economy of Southern California. We use a recently developed model of beach choice by day users in Southern California to demonstrate how predictions about future impacts of sea level rise on beaches can be linked to detailed economic models of beach recreation behavior. Our analysis is not intended to provide precise estimates of the impact of climate change on Southern California beach attendance. Our goal, rather, is to develop a framework that demonstrates how to link estimates of beach change (changes in width and volume of sand) caused by sea level rise to economic models of beach attendance, expenditures, and consumer surplus. This is a first step toward evaluating the effects of climate-related beach change on the net economic value of beaches in Southern California.

To illustrate our framework, and the potential magnitude of the economic impacts of sea level rise, we use an economic analysis based on projections of beach width change from permanent inundation due to 1 m (m) of sea level rise and beach width and area change due to an extremely stormy year. We recognize that sea level, and thus beach width, change constantly over the course of a day, a lunar cycle, a year, and over decades. We explore "permanent" inundation as a means of thinking about average loss in beach width that could occur due to a rise in average sea level. Projections of beach width change due to permanent inundation is estimated based on beach slope data (Hapke et al. 2006) and beach width and volume change due to large storms are based on a new model of beach sediment budgets being developed by Adams and Inman (2009), applied to sites in Southern California (Adams et al. 2011). This geomorphic model, the Coastal Erosional Hotspot Potential Model (referred to subsequently as the CEHP) is in its early stages of development. Indeed, we view their results as indicators of the order of magnitude of the potential impacts on beaches that could be associated with extremely stormy years. Our methods can easily be applied to a variety of models that project future beach width.

Beach width has been shown to be an important determinant of where day use visitors go to the beach in Southern California (Pendleton et al. 2008) and is one of the primary explanatory variables in a model of beach choice for Southern California public beaches (originally developed by Hanemann et al. 2005 and recently updated by Pendleton et al. 2011). While tourism to beaches may also be affected by beach width, the models of Hanemann et al. and Pendleton et al. examine beach going only for Southern California residents. Hanemann et al. showed that more than 50% of all households in Los Angeles, Orange, Riverside, and San Bernardino counties had at least one member who went to the beach over the course of a year. This large population of users may account for more than 100 million visits to local beaches annually (Pendleton and Kildow 2006.)

Sea level rise in California, as measured by tide gauges has risen at a rate of 17–20 cm per century (Cayan et. al. 2009), nearly as much as global SLR. Rahmstorf (2007) and others have linked sea level rise to increases in mean surface air temperature, providing a technique for forecasting future increases in sea level rise.

We use Adams and Inman's estimates of potential changes in beach width, and the beach choice model of Pendleton et al. (2011) to model the effects of a 1 m rise in sea level on beach attendance, beach expenditures, and the non-market value of beach going—the economic value of beaches to local beachgoers, beyond what they have to pay to use the beach. To provide perspective for our estimates of the impacts of steady rise in sea level, 1 m over 100 years, we also model the potential impacts on beach width due to a year of unusually intense storm events. In our case, we use the storm events from the El Niño years of 1982 and 1983. In addition to sea level rise, it is thought that deep-water wave fields are changing in accordance with global climate phenomena, such as the El Nino-Southern Oscillation (ENSO) and the Pacific Decadal Oscillation (PDO) (Adams et al. 2008). Because coastal managers may choose to counter permanent inundation and extreme erosion events, we also provide simple, but illustrative, estimates of the costs of physically renourishing beaches, by placing new sand on beaches, following such events.

1.1 Economic value of Southern California beaches

Beaches are an important recreational resource enjoyed by residents of California and many visitors to the state. According to The National Survey of Recreation and the Environment (NSRE) in 2000, nearly 15 million people participated in beach activities in California. This dominates all other forms of marine recreation in the state but is still an underestimate because foreign tourists were not included in the survey.

California residents spend $4 billion annually on beach recreation (Pendleton and Kildow 2006). Many local visitors are able to enjoy the beach at little or no cost, but they enjoy considerable economic benefit from their presence. This benefit is called the *consumer surplus* or *non-market value* of beaches and represents the willingness to pay to visit beaches, beyond what people actually do pay. These non-market values are real and generally come to public attention when beaches are damaged (either through beach loss or deterioration of water quality) or removed from use (e.g., due to an oil spill). The non-market value of beaches has been evaluated numerous times in the literature and has been estimated to contribute more than $2 billion to the economic well-being of Californians (Pendleton and Kildow 2006).

The billions of dollars spent by beachgoers contribute to a number of local economic activities. Day visitors to beaches spend money locally on food, beverages, parking, and beach-related activities and rentals (e.g., body boards, umbrellas). Such purchases partially represent a transfer of expenditures that may have been made elsewhere in the state (e.g., gas and auto), but are largely expenditures that would not have been made in the absence of the beach trip. King (1999) estimated the fiscal impact of beaches in California and reported that in 1998, California's beaches generated $14 billion dollars in direct revenue (King 1999).[1] In two other studies, the average expenditures per person per day trip ($/trip/person) were estimated for visits to California beaches at between $23 and $29 per day (see Pendleton and Kildow 2006 for a review.) Such numbers may appear small when compared to alternative activities, such as amusement parks, but with annual daily visits in the millions, it all adds up to a multi-billion dollar, renewable resource.

1.2 Impact of climate change on beach economies in Southern California

The market and non-market (consumer surplus) values that are generated by beach recreation can be affected by the quality of the coastal environment. Obvious problems such as trash on the beach clearly deter visitors, but beach width is an important factor as well (Lew 2005; Lew and Larson 2006; Bin et al. 2007). Pendleton et al. (2011) show that different users prefer different beach widths, depending on the type of recreation they plan to undertake (e.g., sand-based versus water-based versus pavement-based activities). Changes in the width of beaches due to permanent inundation or storms could change beach attendance substantially. As beach attendance changes, so do local expenditures and non-market value.

For Southern California, climate change may physically affect beaches through at least two mechanisms: (1) permanent beach loss due to inundation caused by sea-level rise, and (2) increased intensity of storms caused by higher high tides (California Coastal Commission 2001; Cayan et al. 2005, 2006). Inherently dynamic, beaches can be eroded very quickly if the rate of sand removal through erosion surpasses the rate of replenishment through accretion. In fact, several studies have indicated a net global loss in beach area over the last 100 years (Bird 1985; NRC 1990; Leatherman 2001; Eurosion 2004) and beaches are expected to shrink more rapidly because of sea-level rise (Brown and McLachlan 2002).

In Southern California, storm events and wave action also contribute substantially to coastal erosion (Flick 1998; Seymour et al. 2005). Storm surges, or waves of extraordinary height that occur during storms (especially storms that coincide with high tides), can be amplified by sea level rise, increasing their destructive power (Cayan et al. 2008). It also is possible that changes in wave climate (e.g., wave direction, height, and period) could have erosional effects on beaches, but evidence suggests that this factor will be much more important at higher latitudes (Allan and Komar 2006; Flick and Bromirski 2008), except when exacerbated by the El Niño Southern Oscillation (ENSO) cycle (e.g., Seymour et al. 1984; Inman and Jenkins 1998).

---

[1] *Direct revenue* is the direct expenditure from people making beach trips for items such as gas and parking, food and drinks from stores, restaurants, equipment rentals, beach sporting goods, beach-related lodging, and incidentals.

While efforts have been made to estimate the overall coastal impacts of climate change in California (see for instance Neumann and Hudgens 2006 and more recently by Heberger et al. 2009), no attempt has been made to examine carefully the impacts of sea level rise due to climate change on the beach going economy of the state. In a report to the California Department of Boating and Waterways, King and Symes (2003) determined that failure to protect Southern California beaches would reduce the California gross state product by over $5.5 billion annually. However, their data reflects changes in use based on the complete absence of beaches in the area, rather than losses specifically due to sea level rise. Cost estimates for previous extremely stormy years, such as the 1997–1998 El Niño, have been estimated at about $1.1 billion for California as a whole (Andrews cited as personal communication in Changnon 2000). In addition to changes in the amount beachgoers spend, climate change-induced alterations of beaches in Southern California could also reduce the consumer surplus that local beachgoers enjoy from having easy access to hundreds of miles of beaches. As noted above, the non-market value of beach going can be quite large.

Section 2 describes our methods. We provide analysis for our results in Section 3. Implications for coastal management in Southern California are presented in Section 4.

2 Methods

Our work links three different types of analysis: a beach attendance model that estimates how beachgoers in Southern California choose among 51 public beaches, the CEHP, which models erosion and accretion patterns for beaches, and an analysis of beach nourishment costs (Fig. 1). The beach model predicts beach attendance patterns based on certain demographic features of potential beachgoers, the cost of travel, and the attributes of beaches, including beach width. Projections of changes in beach width due to permanent inundation are calculated by averaging beach slopes to find the average slope for each of our 51 beaches. We then combine slopes and sea level rise (1 m) to estimate lost beach width. Since sea level rise could increase the erosion and accretion potential of winter

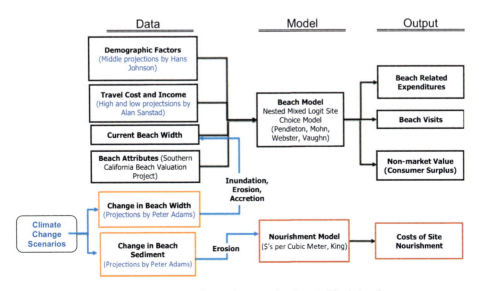

Fig. 1 Dataflow for economic impacts of climate change on Southern California beaches

storms when storms coincide with elevated high tides, we also estimate the impacts of a highly stormy year. To estimate the changes in beach width due to erosion and accretion caused by an extremely stormy year, we use preliminary results from the CEHP model to estimate the effects of the wave climate from El Niño (1982–1983). Finally, we estimate the costs of replacing sand volume lost to permanent inundation or storm-related erosion. In both cases, beach width data and sand volume loss or gain data are provided by estimates from the beach sediment model. We briefly describe each of the three analyses below.

2.1 Beach choice and attendance model

Many factors influence where and when residents of Southern California decide to go to the beach. These include personal factors (e.g., income, age, gender, etc.), the cost of travel, and the different attributes of beaches (e.g., water quality, parking, etc.). Beach width is one of many attributes. Our model predicts the number of visitors, from each census block in four counties in Southern California (Los Angeles, Orange, Riverside, and San Bernardino). Changes in attributes can also affect the visitor's economic well-being of beachgoers by making beaches more or less enjoyable. In simpler terms, non-market benefits measure a visitor's willingness to pay for the experience of the beach. Our paper measures how this non-market value (also referred to as consumer surplus) changes with beach width. Our model also measures the economic impacts of these changes, i.e., how consumer's spending changes when they shift their beach visitation patterns (e.g., buying lunch or gas at Huntington City beach as opposed to another beach).

2.1.1 Data used in model estimation

To simulate the impact of changes in beach width on beach-going activity, Pendleton et al. (2011) modified the original Southern California Beach valuation model of Hanemann et al. 2005. This study contained detailed survey data on beach visitation patterns and demographics. Travel costs were estimated based on respondent location and income was valued at one-half the respondent's hourly income. Data on 46 beach attributes were obtained by site visits (for more detailed discussions of the data see Hanemann et al. 2005 or Pendleton et al. 2011).The authors use beach width measurement data to re-estimate a Random Utility Model (Hanemann et al. 2004) of beach choice (Fig. 2).

Finally, Pendleton et al. (2011) collected data to estimate the width of each beach site from the wet sand to the back of the beach (i.e., the "dry beach"). Using aerial photographs and digital orthophotography quadrangle images from the United States Geological Survey (USGS), the researchers estimated measurements of width (in meters) at 20 m transects along the entire length of each site identified in our study.

2.1.2 Formulation of the beach choice model

As noted above, a user's response to changes in beach width depends on the choice of beach activities. Pendleton et al. (2011) divide the trip data into three activity categories: (1) getting in the water (e.g., swimming), (2) actively using the sand (e.g., volleyball), or (3) activities using paved trails, sidewalks, or beachfront restaurants. A panelist may engage in different activities on different trips.

Pendleton et al. (2011) sequentially model three (nests of) choices for the beachgoers' decision: (1) whether or not to make a trip to the beach, (2) the activity to undertake at the beach, and (3) the beach to visit based on the option which offers the highest utility (see

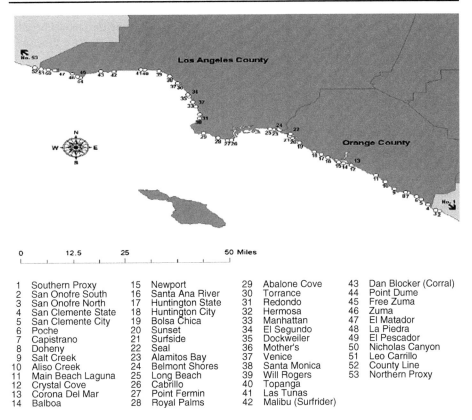

1	Southern Proxy	15	Newport	29	Abalone Cove	43	Dan Blocker (Corral)
2	San Onofre South	16	Santa Ana River	30	Torrance	44	Point Dume
3	San Onofre North	17	Huntington State	31	Redondo	45	Free Zuma
4	San Clemente State	18	Huntington City	32	Hermosa	46	Zuma
5	San Clemente City	19	Bolsa Chica	33	Manhattan	47	El Matador
6	Poche	20	Sunset	34	El Segundo	48	La Piedra
7	Capistrano	21	Surfside	35	Dockweiler	49	El Pescador
8	Doheny	22	Seal	36	Mother's	50	Nicholas Canyon
9	Salt Creek	23	Alamitos Bay	37	Venice	51	Leo Carrillo
10	Aliso Creek	24	Belmont Shores	38	Santa Monica	52	County Line
11	Main Beach Laguna	25	Long Beach	39	Will Rogers	53	Northern Proxy
12	Crystal Cove	26	Cabrillo	40	Topanga		
13	Corona Del Mar	27	Point Fermin	41	Las Tunas		
14	Balboa	28	Royal Palms	42	Malibu (Surfrider)		

Fig. 2 Location of Southern California beaches covered in this study

Fig. 3). The model assumes beachgoers choose the beach that maximizes their utility. The authors use a nested multinomial logit model to analyze the tradeoffs that drive the consumption decision.

Following Train (1998), we use a logarithmically transformed size factor in the application of random utility models to recreational site choice. Pendleton et al. (2011) use

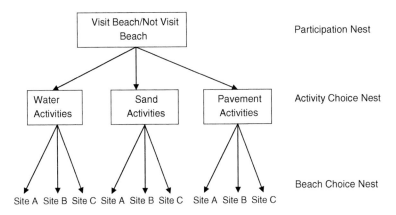

Fig. 3 Beach choice model structure

a simple nested logit structure rather than a mixed-logit (random parameter) model because it provides more control over the choice structure of the model and allows for the use of data for which trip detail may be incomplete (e.g. the respondent may have reported where they visited, but not what they did).

2.2 Width projections

Climate change affects beach width in at least two ways: (1) through permanent beach width loss caused by inundation, and (2) through a change in sediment budgets caused by a changes in deep water wave heights, periods, and directions (a.k.a. wave climate), which alters beach sediment volume. These projections were provided by the CEHP model (Adams and Inman 2009; Adams et al. 2011) and cover scenarios of (1) a 1 m rise in sea level, and (2) the erosion and accretion patterns associated with an extremely stormy year (the El Niño year of 1982/1983).

We used beach slope values, estimated from a Light Detection and Ranging (LIDAR) data set compiled by Hapke et al. (2006) to estimate the potential loss of beach width. We assume permanent inundation occurs gradually across the study period. Adams and Inman (2009) also use computer models to estimate sediment budgets (measured as volume of sand deposited or removed) in cells that are 100 m wide for the length of coast extending from County Line Beach in Ventura County to San Onofre State Park in San Diego County. By 2100, a minimum of 1 m of sea level rise is expected to reduce the average widths of all beaches in Southern California, though some will be affected more than others. Average loss of width for all beaches is estimated to be approximately 9 m, which represents a range from −6 m (County Line) to −16 m (Santa Ana River). Sea level rise also will make the erosion and accretion effects of winter storms more severe

2.3 Demographic and economic projections

Changes in population size, demographics, and the cost of travel could seriously affect beach attendance in the coming century. Our analysis also explores how several demographic and economic factors may interact with sea level rise to alter beach going over time. The most important of these are average household income, gender, race, employment, and projections of estimated travel costs per mile. Hans Johnson (estimates provided by the California Energy Commission) and Alan Sanstad (memo, July 2, 2008) developed scenarios for use in the beach choice model. Table 1 summarizes these contributions and the six scenarios generated for beach attendance and expenditure.

Table 1 Beach choice model width and socioeconomic scenarios

Sea level rise (Adams and Inman)	Demographic changes (Hans Johnson)	Income and travel costs (Alan Sanstad)	
Current sea level (baseline)	Current	Current	Scenario 1
	Midrange predictions	Maximum expected	Scenario 2
		Minimum expected	Scenario 3
Plus 1 m sea level rise (expected with climate change)	Current	Current	Scenario 4
	Midrange predictions	Maximum expected	Scenario 5
		Minimum expected	Scenario 6

Alan Sanstad, from the Lawrence Berkeley National Laboratory, provided projections of two important economic indicators: household income and travel costs. He used the Intergovernmental Panel on Climate Change's *Special Report on Emissions Scenarios* (SRES, Nakicenovic and Stewart 2000) A2 and B1 global scenarios to derive "lower" and "higher" expected values for household income and the cost of driving over 10-year intervals from 2010 to 2100. Both economic indicators affect beach choice.

The population for Los Angeles, Riverside, San Bernardino, and Orange counties is predicted to change dramatically over the course of the next century. Even moderate assumptions regarding fertility and immigration result in a doubling of the regional population from approximately 17 million in 2005 to 32 million in 2100.

3 Results

3.1 Estimated annual economic impact caused by permanent beach loss from inundation due to sea level rise

The economic impacts of permanent beach width loss and even a single stormy year are large and unevenly distributed across the region. Inundation due to sea level rise has only a modest impact on total beach attendance to the region. Holding population and demographic conditions fixed at year 2000 levels, a 1 m rise in sea level increases the total annual attendance at beaches in Los Angeles and Orange counties by 589,000 visits, a relatively small percentage. However, although the relative change in annual visits may be small, the overall change in consumer surplus is substantial (Fig. 4) since the erosion of beach width substantially reduces the recreational value of a beach day and induces some visitors to drive farther to reach a beach with the desired width

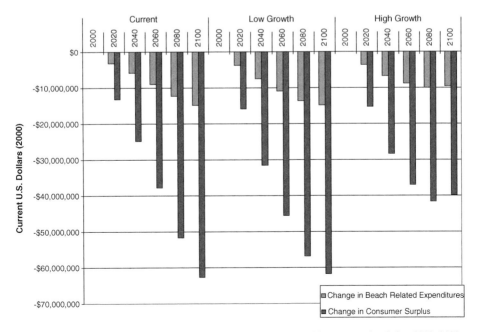

Fig. 4 Change in annual expenditures and annual consumer surplus with +1 m sea level rise: 2000–2100

Changes in the number of beach visits made annually will also affect the amount of money beachgoers spend on beach-related activities and the amount of consumer surplus that they enjoy. Even relatively small overall expected differences in attendance can have a large economic impact. As shown in Fig. 4, compared to scenarios of no sea level rise, estimated direct expenditures on beach-related activities are almost $10 million lower annually under the high growth scenario, and almost $15 million lower annually under the low growth and no change scenarios. Consumer surplus—the reduction in visitors willingness to pay to go to the beach—will be even more affected. Permanent beach loss, caused by sea level rise, may cause the loss in consumer surplus to be as much as $40 million annually under the high growth scenario and more than $60 million under the low growth and no growth scenarios.

3.1.1 The uneven impact of permanent beach loss, due to inundation, on beaches

Beach width and attendance at Southern California beaches has never been uniform; beaches that are more accessible, wider, and provide more amenities tend to draw larger numbers of visitors. Figure 5 depicts the expected beach visits over time with a gradual +1 m sea level rise under the "Current" economic scenario. In this case, all variation results from changes in width due to sea level rise. Even when the absolute loss of beach width is substantial (e.g., >10 m), very large beaches tend to remain large, even with permanent inundation due to sea level rise. As a result, visitors tend to substitute away from already small beaches that have eroded further to beaches that remain large. For example, Newport, Huntington City, and Manhattan beaches will have even higher levels of attendance with sea level rise, while visits to other popular beaches like Huntington State, Venice, and Santa Monica beaches are not expected to differ substantially. Other, relatively small beaches show lower levels of attendance with sea level rise, including Laguna, Bolsa Chica, Torrance, and Redondo beaches.

The differences in beach attendance due to permanent beach loss alone are much more pronounced when we examine the effects at individual beaches. "Winners," or beaches that receive increasing numbers of visitors as sea levels rise, can also expect higher local beach-related expenditures, since people spend about US$25.18 per trip to the beach (year 2000$, Pendleton and Kildow 2006). On the other hand, "Losers," or those beaches where visits are predicted to be fewer, can expect lower beach-related expenditures and thus earnings. The magnitude of such differences is indicated in Table 2, which lists the top five winners and losers when sea level rise is the only factor that is allowed to vary in the model. As a result of complex interactions between beach attributes and sea level rise, beaches like Huntington City and Will Rogers can expect big gains but others such as Laguna and Bolsa Chica can expect big losses with sea level rise.

Losses in the welfare of beachgoers will be also felt differentially across the region. Table 3 lists changes in consumer surplus for residents by their county of origin.

3.1.2 High growth vs. low growth with inundation

As noted above, we do not expect that all other factors will remain constant over the next 100 years. Therefore we also analyzed the impact of sea level rise under two additional socioeconomic scenarios: one in which there is "Low Growth" in income and travel costs and middle projections of growth in population and demographics and another in which there is "Higher Growth." The relative change in attendance, expenditures, and consumer

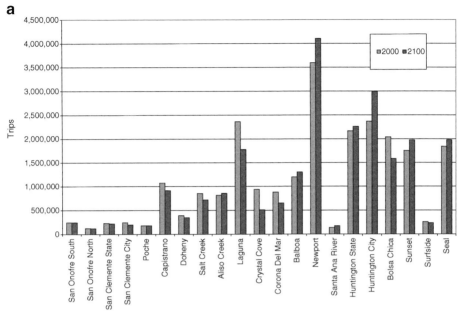

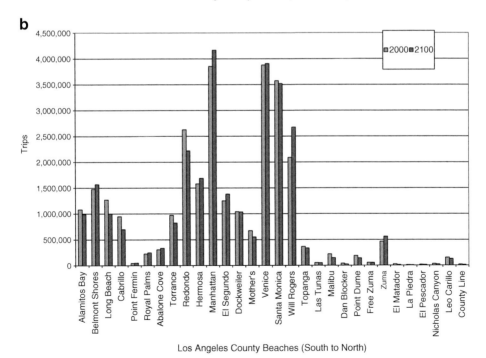

Fig. 5 Attendance by beach width 2100 (projected with 1 m sea level rise) vs. 2000 (measured)

surplus at beaches caused by permanent beach loss follows a similar pattern for all scenarios of economic change, only the magnitude of impacts differs. Gains or losses in

Table 2 Top 5 winners and losers with 1 m sea level rise (current demographic, population, and costs)

Winners	Difference in annual expenditures US$(2000), rounded to nearest million	Losers	Difference in annual expenditures US$(2000), rounded to nearest million
Huntington City	16 million	Main Beach Laguna	−14 million
Will Rogers	15 million	Bolsa Chica	−12 million
Newport	13 million	Crystal Cove	−11 million
Manhattan	8 million	Redondo	−10 million
Sunset	6 million	Long Beach	−7 million

2100 tend to be smaller in the Low Growth scenario than they are in the Current scenario. High economic growth has a greater impact, resulting in even lower final annual attendance estimates.

As before, even under these different scenarios of population and economic change, some beaches have fewer visitors with 1 m of sea level rise while others have more visitors when beaches are permanently inundated due to sea level rise (see Table 4). Losses due to sea level rise are less severe when economic growth is factored into the model. Furthermore, this effect is much more pronounced under the high growth scenario. Compared to either no growth or low growth, change in expenditures by beach due to sea level rise is reduced by half when high growth is assumed.

Changes in expenditures follow similar patterns under the low and high scenarios. Figure 6 compares change in beach-related expenditures due to permanent beach loss caused by sea level rise in the three socioeconomic scenarios. The differences in economic impacts, by county in which the beach is located, are largest with current population, income, and travel costs. Losses are twice as great for Orange County compared to Los Angeles County beaches. However, when economic and demographic projections are factored into the model, the difference between expected expenditures with climate change compared to the baseline is smaller for Orange County. In fact, under either high or low growth, the difference between projected expenditures and the baseline are smaller in the future for Orange County beaches. This is likely due to the difference in population and growth projections for these and neighboring counties.

Table 3 Difference in annual consumer surplus caused by permanent beach loss, due to inundation, from +1 m sea level rise (US$[2000], rounded to nearest million)

	Los Angeles	Orange	Riverside	San Bernardino
Consumer surplus (estimated for county of beachgoer residence)				
2020	−7 million	−4 million	−1 million	−1 million
2040	−12 million	−7 million	−3 million	−2 million
2060	−19 million	−11 million	−4 million	−4 million
2080	−26 million	−15 million	−6 million	−5 million
2100	−31 million	−19 million	−7 million	−6 million

Table 4 Top five winners and losers for permanent beach loss due to 1 m inundation, two socioeconomic scenarios

Winners	Difference in annual expenditures US$[2000], rounded to nearest million	Losers	Difference in annual expenditures US$[2000], rounded to nearest million
Low growth			
Huntington City	14 million	Bolsa Chica	−13 million
Newport	13 million	Laguna	−13 million
Will Rogers	12 million	Redondo	−10 million
Manhattan	8 million	Crystal Cove	−8 million
Sunset	5 million	Long Beach	−7 million
High growth			
Huntington City	7 million	Bolsa Chica	−8 million
Newport	7 million	Laguna	−6 million
Will Rogers	6 million	Redondo	−6 million
Manhattan	5 million	Mother's	−5 million
Sunset	3 million	Long Beach	−5 million

Consumer surplus loss due to permanent beach loss caused by sea level also differs by county (Fig. 7). Residents from Los Angeles County bear the greatest burden in lost consumer surplus, which is over $30 million lower with a 1 m rise in sea level under the current and low growth scenarios and by almost $25 million in the high growth scenario.

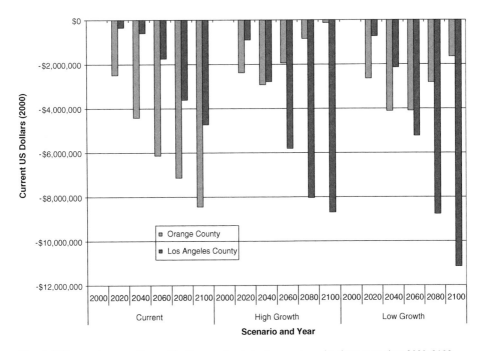

Fig. 6 Difference in expenditure with +1 m sea level rise by county under three scenarios: 2000–2100

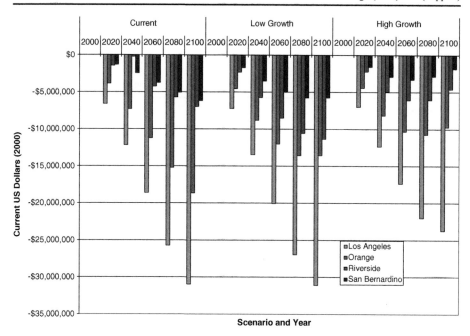

Fig. 7 Difference in consumer surplus due to +1 m rise in sea level under three scenarios: 2000–2100

Orange County experiences smaller differences in consumer surplus in the two growth scenarios, but Riverside County would experience its highest losses under the low growth scenario. San Bernadino suffers least in all three scenarios.

3.2 The economic impact of extremely stormy years

Of course, the effects of sea level rise on beach width are unlikely to occur slowly and evenly across the region. While wave characteristics may not change substantially due to climate change, sea level rise will increase water levels associated with high tides and these elevated water levels are likely to exacerbate the effects of wintertime storms on beach erosion and accretion (Cayan et al. 2008). To explore the potential economic impacts of erosion and accretion caused by extremely stormy years, we investigate the economic impacts of beach change simulated for a year similar to the 1982–1983 El Niño. These estimates are intended only to show how these extreme years compare to the assumption of simple inundation. We do not have good estimates, at this time, of how sea level rise and winter storms will affect beach change over the long run.

Unlike permanent beach loss caused by inundation due to sea level rise, which may occur over 100 years, an extremely stormy period has a large impact in a single year. We assume the effects of a major storm season linger for 1 year. The lasting effects of a storm depend on a number of factors including sediment availability and natural recovery. Back-to-back stormy years could also affect beach sand recovery. Thus, our estimates are intended to give an order of magnitude context for the severity of impacts that could result from increased storm intensity. We find that the effects on beach width and beach use of a single extremely stormy year are on the same order as the effects of 100 years of sea level rise.

The effects of waves, especially from storms, on sediment budgets include both increasing volumes of sand and width at some beaches and the loss of sand and width at other beaches. The most extreme changes in beach width exceed those of permanent inundation due to a full meter of sea level rise. Like the effects of permanent beach loss caused by sea level rise, the net economic effect of beach change due to an extremely stormy year is detrimental with a predicted initial, temporary change of annual visits due to an extremely stormy year like that of El Niño equal to −343,446 visitor days and an associated change in total expenditures of −$8.6 million and a change in consumer surplus equal to −$36.7 million (current US$[2000]). Also, like the effects of permanent beach inundation, the effects of erosion events are uneven across beaches.

Also, as in the case of permanent beach loss, there are winners and losers in terms of expenditures. Table 5 demonstrates the range of change in expenditures that could be experienced by the winners and losers in a year with extreme erosion/accretion events. Interestingly, Laguna and Seal beaches, which lose considerable attendance and expenditure with inundation, are expected to be winners due to beach change caused by extremely stormy years. This is because the CEHP model projects a net loss in width for these beaches due to inundation but a net gain due to an extremely stormy year. On the other hand, beaches like Torrance and Redondo, which were expected to lose some visits and expenditure under inundation are worse off after an extremely stormy year because they lose even more width under conditions of extremely stormy wave conditions.

Finally, a breakdown of the change in consumer surplus due to a year-long change in beach width caused by extreme erosion and accretion events reveals variation in impact among the counties in Southern California. Three of the four counties are net losers under beach change scenarios that result from extremely stormy years (even under current sea level). Residents of Los Angeles County experience a large negative impact after an extreme event, losing over $30 million in consumer surplus annually. Orange County is a distant second, with almost $3 million in losses, followed by San Bernardino at about $2 million. However, Riverside County is a net winner, though only by about $350,000. Los Angeles suffers from having many beachgoers that live near beaches that are likely to be badly damaged during an extremely stormy year.

3.3 The costs of mitigating beach loss through nourishment

3.3.1 Estimates of the cost of nourishment

One of the most common ways to combat loss of beach width due to sea level rise is to nourish eroded beaches. There are two major methods: trucking and dredging. Bringing in sand on trucks is cost-effective for small nourishment projects. Dredging, which requires

Table 5 Top five winners and losers with 1 m sea level rise, an extreme storm year

Winners	Difference in annual expenditures (US$[2000], rounded to nearest million)	Losers	Change in expenditure (current US$[2000])
Main Beach Laguna	20 million	Redondo	−25 million
Seal	16 million	Torrance	−21 million
Aliso Creek	1 million	Salt Creek	−19 million
San Onofre South	9 million	Santa Monica	−9 million
Venice	8 million	Doheny	−9 million

Table 6 Estimated costs for hopper dredge nourishment

Mobilization or demobilization	$585,000
Mobilization or demobilization for additional sites	$60,000
Additional cost per cubic meter	$26

expensive equipment has high fixed costs but significant economies of scale. Consequently, our estimates of nourishment costs depend on the amount of erosion. Nourished beaches require periodic maintenance, since waves and currents constantly move sand in the alongshore and cross-shore directions. Sand may also be removed due to persistent background erosion or a storm. Typical re-nourishment intervals under past sea-level conditions range from 2 to 5 years. Between increasing average erosion rates due to sea level rise and the potential that winter times storms will have greater impacts due to increased high tides (Cayan et al. 2008) and longer storm seasons (Peter Bromirski, personal communication), nourishment may need to be undertaken more often in the future to maintain the current quality of Southern California beaches.

Our model captures this effect by providing simple cost estimates for nourishment to counter simple flooding (permanent inundation) due to one meter rise in sea level and the erosional losses that could occur during an extremely stormy year. Recent cost estimates for beach nourishment were estimated for the Los Angeles County Department of Beaches and Harbors (2007).[2] The study developed a general cost structure for nourishment projects in Los Angeles County.

Table 6 indicates that the variable costs of beach nourishment are $26 per cubic meter, including placing sand on the beach and bulldozing. (Fixed) mobilization/demobilization costs were estimated at $585,000 for one project. For additional projects, the mobilization/demobilization costs are much lower, approximately $60,000 for each additional project. Thus, if one is able to schedule a number of projects together, the fixed costs of mobilization and demobilization, as a percentage of the total costs, may be quite low.

Using these cost parameters, the total cost of a nourishment project per beach was estimated for two different scenarios: (1) replacing beach lost due to inundation, and (2) replacing sand lost due to change in beach volume caused by an extremely stormy year at current sea level. The numerical results of our estimates are summarized in Figs. 8 and 9.

The total costs of nourishing all sites to mitigate against conditions in these two scenarios is significant. To mitigate for permanent inundation caused by a rise in sea level of one meter, the total costs of nourishment are estimated to be $436 million, or just over $4 million per year. The cost of mitigating for beach loss from a single stormy year is estimated to be $382 million. Of course, complete renourishment may not take place for all beaches.

While these cost estimates are rudimentary, there is one clear story that emerges from the analysis. The cost of adding sand to beaches to counteract the effects of sea level rise is of a similar magnitude to the costs of renourishing after an extremely stormy year. There is one important difference, however. Inundation takes place over 100 years in our analysis. That means the undiscounted average annual cost of nourishment would be approximately $4

---

[2] Data Review and Nourishment Need Assessment, prepared for Los Angeles County Dept of Beaches and Harbors by HPA Inc., September 1997, received directly from Los Angeles County. See Los Angeles County Dept of Beaches and Harbors 2007 for most recent data overview.

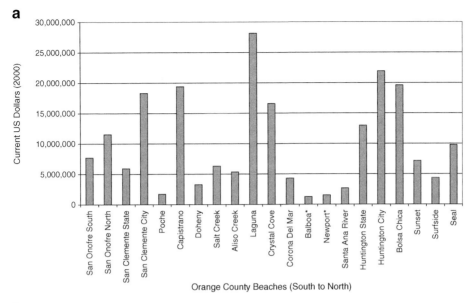

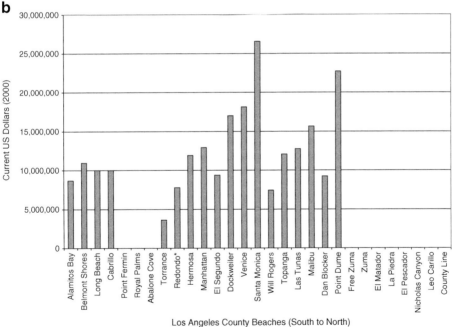

Fig. 8 Estimated costs ($2000) of beach nourishment to mitigate for permanent inundation (1 m sea level rise)

million if sea level rise resulted only in a slow flooding of beaches. This cost is just under one third the estimated loss in consumer surplus due to sea level rise and roughly equal to the average lost expenditures. This suggests that if permanent inundation were the only effect of sea level rise on beaches, then the recreational benefits from nourishment would outweigh the costs. The costs of nourishing for extremely stormy years, however, are many

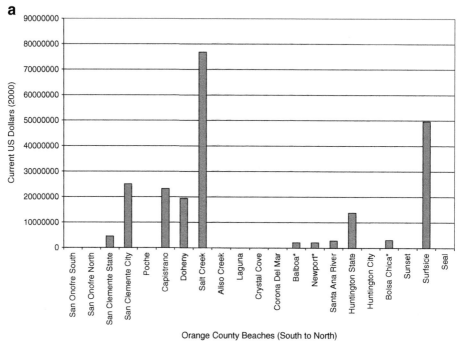

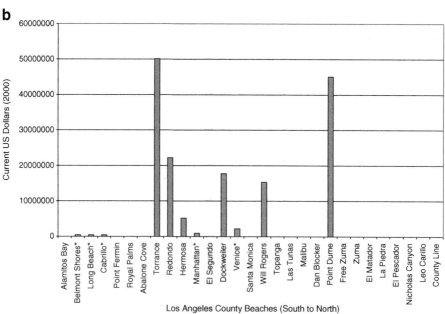

Fig. 9 Estimated costs of beach nourishment (in year 2000 dollars) to mitigate for extremely stormy year. Note: * Indicates trucking is least cost method, otherwise dredging is least cost

times the annual recreational benefit of nourishment. Recreational benefits alone are unlikely to justify the large expenditures that would be required to repeatedly replace sand

Table 7 Annual impacts caused by permanent inundation due to sea level rise of 1 m, US$(2000)

Socio-economic scenario	Total change by residents			Maximum for one beach	
	Annual attendance	Annual expenditures	Annual consumer surplus	Gain in annual expenditures	Loss in annual expenditures
No change	−588,765	−$15 million	−$63 million	$16 million	−$15 million
Low growth	−586,923	−$15 million	−$62 million	$14 million	−$13 million
High growth	−380,223	−$10 million	−$40 million	$7 million	−$8 million

lost by increasingly severe winter storms. For these beaches, the process of natural replenishment is likely to be most cost-effective.

3.4 Limitations of the models and future research

The analysis and estimates above should be considered preliminary. More analysis of the impact of severe winter storms resulting from elevated high tides and the potential impact of lengthening winter storm seasons is necessary. Future research on the probabilistic nature of large storm events could generate useful results as well (Sanstad 2008). Future research on the actual costs of nourishment and whether nourishment is politically feasible would also be helpful. We have made simplifying assumptions about the pace of beach change, sources of sediments, and the costs of transporting sediment. All of these factors need to be considered to make our predictions about potential nourishment costs more accurate. Additional research should also improve our estimates of benefits from nourishment, including buffering against storms and related property damage.[3] As noted above, nourishment costs at many sites run well into the millions of dollars, and the total estimated costs of all dredging runs into the hundreds of millions of dollars. Even with such indirect benefits, direct costs may be prohibitive for many state and local governments.

4 Conclusion

Because there are so many beaches to choose from in Southern California, many of them quite large, beach going is likely to remain an important recreational asset to the area, even in the face of sea level rise. Nevertheless, changes in sea level could reduce the number of beach visits taken in Los Angeles and Orange counties by more than half a million visits annually by 2100. In this analysis, we focus only on day-use beach visits by local residents from Los Angeles, Orange, Riverside, and San Bernardino counties. The effect on tourist visits, which could represent as much as 100 million more visitor days for the region (Pendleton and Kildow 2006), is unknown but likely to follow a similar pattern. More dramatic, however, are the uneven local effects of climate change.

Our analysis shows that even the effects of permanent beach loss due to slow and steady sea level rise would create a substantial loss in economic welfare for the region (between $40 million and almost $63 million annually), with smaller impacts on beach-related expenditures (see Table 7). Perhaps more importantly, though, the effects of the impacts of

---

[3] For example, see West et al. (2001) though they find the impacts of SLR on storm damage are likely small (5% or so) these losses could still be large in absolute terms.

Table 8 Summary of annual impacts caused by an extremely stormy year ($2000)

Socio-economic scenario	Total change by residents			Maximum for one beach	
	Annual attendance	Annual expenditures	Annual consumer surplus	Gain in annual expenditures	Loss in annual expenditures
No change	−343,447	−$9 million	−$37 million	$20 million	−$25 million

permanent beach loss due to inundation from sea level rise would be spread unevenly across the region with some beaches gaining attendance and expenditures while other beaches lose visitors and their spending.

If sea level rise proceeds at the slow pace considered in this analysis, then an opportunity to offset the losses in beach width through selective beach nourishment could exist. The costs of nourishment appear to be outweighed by the avoided potential losses in consumer surplus, a measure of beachgoer economic welfare, and avoided lost expenditures. The effects of beach change on tourist visitors are likely to show that the value of mitigating the effects of inundation are even larger than predicted here. Of course, sea level rise is unlikely to proceed slowly and gradually.

The real challenge for understanding and adapting to the effects of sea level rise on beach management could come if winter storms, combined with higher tides, lead to even more erosion than beaches experience currently (Cayan et al. 2008). A single extremely stormy year can have a temporary, but substantial impact on annual beach attendance, spending, and consumer surplus that is similar to the average annual impacts that would result from a full meter of sea level rise. We estimate the impacts that might occur during an extremely stormy year like that of the El Niño year of 1982/1983. The impacts of beach loss from these extremely stormy years are of a similar magnitude to those caused by permanent inundation (Table 8), and are likely to be highly uneven across the region. Some beaches may benefit from losses of beach size at nearby beaches and some beaches may actually grow because of future sediment accretion caused by storms. Many other beaches, however, are likely to see sharply lower attendance levels if climate change and sea level rise result in more years with high erosion impacts. As a result, local businesses at highly eroded beaches will feel the loss of beach-related expenditures.

Moving forward, our results make it clear that the real concern for beach going and the beach-related economy have to do with the impacts of wave-driven erosion and accretion. Future research needs to use the framework we provide here to take a more probabilistic approach to understanding the potential impacts that increasing sea levels may have on the erosion impacts of winter storms. Finally, this work shows that whether inundation or wave-driven erosion is the cause of beach change, the effects of climate-driven beach change are extremely uneven in their distribution throughout the region. Estimates of the impact of climate change on beaches must be conducted at a sub-regional level, preferably at the level of individual beaches.

Acknowledgments We would like to thank Guido Franco and Susi Moser for their feedback, the California Energy Commission's Public Interest Energy Research (PIER) Program for its financial support, and Kim Sterrett of the California Department of Boating and Waterways, which provided initial funding to collect and analyze the beach width data used in the modified Southern California Beach Valuation model. Help with beach width data was provided by Dr. Anthony Orme, James Zoulas, Carla Chenuault Grady, and Hongkyo Koo. Richard MacKenzie helped us with beach slope data.

References

Adams PN, Inman DL (2009) Climate change and potential hotspots of coastal erosion along the Southern California Coast, Report to the California Energy Commission, Publication Number CEC-500-2009-022-D

Adams PN, Inman DL, Graham NE (2008) Southern California deep-water wave climate: characterization and application to coastal processes. J Coast Res 24(4):1022–1035

Adams PN, Inman DL, Lovering JL (2011) Effects of climate change and wave direction on hotspots of coastal erosion in Southern California. Climatic Change

Allan JC, Komar PD (2006) Climate controls on us west coast erosion processes. J Coast Res 22(3):511–529

Bin O, Dumas C, Poulter B, Whitehead J (2007) Measuring the impacts of climate change on North Carolina coastal resources. National Commission on Energy Policy, 91 pp

Bird ECF (1985) Coastline changes: a global review. Wiley, Chichester, 219 pp

Brown AC, McLachlan A (2002) Sandy shore ecosystems and the threats facing them: some predictions for the year 2025. Environ Conserv 29(1):62–77

California Coastal Commission (2001) Overview of sea level rise and some implication for coastal California. California Coastal Commission, San Francisco, p 58

Cayan D, Luers AL, Hanemann M, Franco G, Croes B (2005) Scenarios of climate change in California: an overview. White Paper, California Climate Change Center

Cayan D, Bromirski P, Hayhoe K, Tyree M, Dettinger M, Flick R (2006) Projecting future sea level. White Paper, California Climate Change Center

Cayan D, Tyree M, Dettinger M, Hidalgo H, Das T, Maurer E (2008) Climate change scenarios and sea level rise estimates for California 2008 climate change scenarios assessment

Cayan D, Tyree M, Dettinger M, Hidalgo H, Das T, Maurer E, Bromirski P, Graham N, Flick R (2009) Climate change scenarios and sea level rise estimates for the California 2008 climate change scenarios assessment. Draft PIER-EA Discussion Paper, Sacramento, California: California Energy Commission

Changnon SA (2000) Impacts of El Niño weather. In El Niño 1997–1998: the climate event of the century. Oxford University Press, Oxford, pp 136–171

Eurosion (2004) Living with Coastal Erosion in Europe: Sediment and Space for Sustainability. Part-1 Major Findings and Policy Recommendations of the EUROSION Project. Guidelines for implementing local information systems dedicated to coastal erosion management. Service contract B4-3301/2001/329175/MAR/B3 "Coastal erosion—evaluation of the need for action." Directorate General Environment, European Commission, 54 pp

Flick RE (1998) Comparison of California tides, storm surges, and mean sea level during the El Niño winters of 1982–1983 and 1997–1998. Shore Beach 66(3):7–11

Flick RE, Bromirski PD (2008) (Draft). Sea level and coastal change. Draft PIER-EA Discussion Paper. Sacramento, California: California Energy Commission

Hanemann M, Pendleton L, Mohn C, Hilger J, Kurisawa K, Layton D, Vasquez F (2004) Using revealed preference models to estimate the affect of coastal water quality on beach choice in Southern California. Prepared for the National Ocean and Atmospheric Administration, Minerals Management Service (Department of the Interior), The California State Water Resources Control Board, and The California Department of Fish and Game

Hanemann M, Pendleton L, Mohn C (2005) Welfare estimates for five scenarios of water quality change in Southern California. National Oceanic and Atmospheric Administration

Hapke CJ, Reid D, Richmond BM, Ruggiero P, List J (2006) National assessment of shoreline change part 3: Historical shoreline change and associated coastal land loss along sandy shorelines of the California coast. U.S. Geological Survey Open-File Report 2006–1219, 72 pp

Heberger M, Cooley H, Herrera P, Gleick P, Moore E (2009)The impacts of sea level rise on the California coast. California Energy Commission paper CEC-500-2009-024-F

Inman DL, Jenkins SA (1998) Changing wave climate and littoral drift along the California coast. Proc Conf Calif World Ocean 1:538–549

King P (1999) The fiscal impact of beaches in California. Public Research Institute, San Francisco University. Report Commissioned by California Department of Boating and Waterways

King P, Symes D (2003) Potential loss in gross national product and gross state product from a failure to maintain California's beaches. Report to the California Department of Boating and Waterways

Leatherman SP (2001) Social and economic costs of sea-level rise. In: Douglas BC, Kearney MS, Leatherman SP (eds) Sea-level rise, history and consequences. Academic, New York, pp 181–223

Lew D (2005) Accounting for stochastic shadow values of time in discrete-choice recreation demand models. J Environ Econ Manag 50:341–361

Lew DK, Larson DM (2006) Valuing beach recreation and amenities in San Diego county. Sea Technol (August):39–45

Los Angeles County Dept of Beaches and Harbors (2007) Los Angeles county beach renourishment project: data review and nourishment need assessment. Prepared by HPA Inc

Nakicenovic N, Stewart R (eds) (2000) Special report on emmissions scenarios. Intergovernmental Panel on Climate Change, The Hague

Neumann JE, Hudgens DE (2006) Coastal impacts. In: Smith JB, Mendelsohn R (eds) The impact of climate change on regional systems: a comprehensive analysis of California. Edward Elgar, Northampton

NRC (1990) Managing coastal erosion. National Research Council. National Academy Press, Washington, 204 pp

Pendleton L, Kildow J (2006) The non-market value of beach recreation in California. Shore Beach 74 (2):34–37

Pendleton L, Mohn C, Vaughn RK, King P, Zoulas JG (2011) Size matters: the economic value of beach erosion and nourishment in Southern California. Contemporary Economic Policy

Rahmstorf S (2007) A semi-empirical approach to projecting future sea-level rise. Science 315(5810): 368–370

Sanstad AH (2008) (draft) Economics of mitigation and adaptation. Draft PIER-EA Discussion Paper, Sacramento, California: California Energy Commission

Seymour RJ, Strange RR III, Cayan DR, Nathan RA (1984) Influence of El Niños on California's wave climate. In: Edge BL (ed) Nineteenth coastal engineering conference: proceedings of the international conference. ASCE, New York, pp 577–592

Seymour R, Guza RT, O'Reilly W, Elgar S (2005) Rapid erosion of a small southern California beach fill. Coast Eng 52:151–158

Train KE (1998) Recreation demand models with taste differences over people. Land Econ 74(2):230–239

West JJ, Small MJ, Dowlatabadi H (2001) Storms, investor decisions, and the economic impacts of sea level rise. Clim Chang 48(2–3):317–342

Climatic Change (2011) 109 (Suppl 1)
DOI 10.1007/s10584-011-0335-y

Modifying agricultural water management to adapt to climate change in California's central valley

Brian A. Joyce · Vishal K. Mehta · David R. Purkey ·
Larry L. Dale · Michael Hanemann

Received: 16 March 2010 / Accepted: 17 October 2011 / Published online: 9 December 2011
© Springer Science+Business Media B.V. 2011

Abstract Climate change impacts and potential adaptation strategies were assessed using an application of the Water Evaluation and Planning (WEAP) system developed for the Sacramento River basin and Delta export region of the San Joaquin Valley. WEAP is an integrated rainfall/runoff, water resources systems modeling framework that can be forced directly from time series of climatic input to estimate water supplies (watershed runoff) and demands (crop evapotranspiration). We applied the model to evaluate the hydrologic implications of 12 climate change scenarios as well as the water management ramifications of the implied hydrologic changes. In addition to evaluating the impacts of climate change with current operations, the model also assessed the impacts of changing agricultural management strategies in response to a changing climate. These adaptation strategies included improvements in irrigation technology and shifts in cropping patterns towards higher valued crops. Model simulations suggested that increasing agricultural demand

---

Electronic supplementary material The online version of this article (doi:10.1007/s10584-011-0335-y) contains supplementary material, which is available to authorized users.

B. A. Joyce (✉) · V. K. Mehta · D. R. Purkey
Stockholm Environment Institute, Stockholm, Sweden
e-mail: brian.joyce@sei-us.org
URL: http://www.sei-us.org/

V. K. Mehta
e-mail: vishal.mehta@sei-us.org

D. R. Purkey
e-mail: dpurkey@sei-us.org

L. L. Dale
Lawrence Berkeley National Laboratory, Berkeley, CA, USA
e-mail: lldale@lbl.gov
URL: http://ees.ead.lbl.gov/

M. Hanemann
Economics Department, Arizona State University, Tempe, AZ, USA
e-mail: Hanemann@berkeley.edu

under climate change brought on by increasing temperature will place additional stress on the water system, such that some water users will experience a decrease in water supply reliability. The study indicated that adaptation strategies may ease the burden on the water management system. However, offsetting water demands through these approaches will not be enough to fully combat the impacts of climate change on water management. To adequately address the impacts of climate change, adaptation strategies will have to include fundamental changes in the ways in which the water management system is operated.

1 Introduction

1.1 California water resources

One of the defining features of the California landscape is the Sierra Nevada mountain range that runs along much of the eastern part of the state. The rivers that run out of the Sierra provide drinking water for the state's large urban areas and provide irrigation for the state's vast agricultural land in the Central Valley. Precipitation, however, falls mainly in the fall and winter, so flows in these rivers are sustained throughout the year by melting snow. In fact, Sierra snowpack accounts for approximately half of the surface water storage in the state. Current projections forecast that this snowpack may decline by 70% to as much as 90% over the next 100 years, threatening California's water supply (California Climate Change Center 2006).

In addition to having to manage water supplies that are unequally distributed throughout the year and, indeed, vary considerably from year to year, the state also faces the challenge of moving water from the water-rich northern part of the state to support cities and agriculture in drier areas in the south. Left to flow naturally through the state's rivers, most of the precipitation that falls in the state would flow out to the Pacific Ocean either directly through the rivers of the North Coast or through the San Francisco Bay via the Sacramento and San Joaquin rivers. This would leave the southern part of the state—which contains roughly two-thirds of the state's population—with little of the state's available fresh water supplies. To address this imbalance, several local, state, and federal water projects have been built to deliver water from the water-rich parts of the state to the arid south.

Indeed, the state has made a fairly Herculean effort to transfer water between watersheds through a complex of canals and tunnels that have been built over the last century. This has resulted in a situation where many parts of the state rely heavily upon water exports from the Sacramento River Basin, which transfers more than 5 million acre feet per year to water users in other watersheds. It is critical then for the viability of water management in California to understand how climate change may affect the sustainability of operating the water management system to deliver water throughout the state.

The importance of the Sacramento River as a source of water for the entire state led the authors, as part of the 2006 Scenarios Project reporting (Luers et al. 2006), to focus on that region when investigating the potential impacts of climate change on water management (Joyce et al 2006; Purkey et al 2008). In that work, possible changes in hydrology and water demand in the regions south of the Sacramento-San Joaquin Delta were not explicitly considered in the analysis. For this study, which was included in the 2009 Scenarios Project reporting (Moser et al. 2009), the scope of the analysis was extended to include the impact of climate change on water demand in the western San Joaquin Valley, which relies heavily upon water transfers from the Sacramento River. This expansion will allow for a more comprehensive assessment of climate change impacts, and possible management adaptations, in the California Water System, particularly as irrigated agriculture in this area constitutes a major portion of the water demand that drives water exports from the Delta. While future work would logically include bringing the rest of the system into the model, the current expansion represents an important step in developing a tool for climate change assessment in California water management.

1.2 Background

The 2006 California Climate Change Assessment (California Environmental Protection Agency 2006) included an annex report on potential impacts to Sacramento Valley agriculture (Joyce et al. 2006). This analysis was conducted using the Water Evaluation and Planning (WEAP) modeling system (Yates et al. 2005a; Yates et al. 2005b), which is designed to integrate hydrologic process into a water resources systems modeling framework such that climatic inputs can be used directly to run the model (Yates et al 2009). Thus, the hydrologic implications of a climate change scenario as well as the water management ramifications of this hydrologic change can be assessed within a single software package,.

In preparation for CalEPA's 2006 report to the governor, this WEAP application was used to assess the implications of a limited set of future climatic sequences on water demand in the various sectors and to evaluate the availability of supplies to meet these demands. These future climate scenarios were developed based on downscaling of two general circulation models (GCMs) run under two emissions scenarios (A2 and B1). The results suggested that increasing agricultural demand under climate change due to increased evapotranspiration (ET) would place additional stress on the water system in the Sacramento Valley. The model was also used to assess the effectiveness of two agricultural adaptations, increasing on-farm efficiency and crop shifts toward lower consumption/higher value crops in times of shortage. These were found to be effective at responding to supply shortfalls in agriculture and other sectors.

The completeness of this analysis was limited somewhat, however, because the water demand within the region that depends upon water deliveries from the Sacramento-San Joaquin Delta was not adjusted according to the assumed climatic sequences, and was instead a composite of historic export demands. This demand is a critical driver of water operations in the Sacramento Valley and a major factor in characterizing the status of the Delta itself, a topic of increasing urgency. The current work attempts to resolve this issue by bringing agricultural water requirements and management in the western San Joaquin Valley into the WEAP application. With this expansion, the model represents climatically driven water demand in the agricultural sector in this region along with the operations of state and federal conveyance and storage infrastructure. This expanded WEAP application, run under 12 climatic sequences using the same two adaptation strategies, will provide a much more complete assessment of the potential impact of climate change on agriculture in the Central Valley and on the other users that depend on the waters of the Sacramento River Basin.

1.3 Research objectives

This paper presents an analysis of climate change impacts on agricultural water management in California's Central Valley and is an extension of research conducted by Joyce et al. (2006) as part of the first California Climate Change Assessment (California

Environmental Protection Agency 2006). In this study, an expanded application of the WEAP model was used to evaluate changes in agricultural demand throughout the Central Valley as a result of climate change and to assess the impact of these changes on system-wide water management. The tool was also used to evaluate the relative benefits of

2 Project approach

2.1 Model description

For a complete description of the Sacramento Valley WEAP application, the reader is strongly encouraged to refer to Yates et al. (2005b, 2009). In summary, however, the WEAP application for the Sacramento Valley water system includes the major rivers; the major alluvial aquifers; the major trans-basin diversion from the Trinity River; the main reservoirs (Clair Engle, Shasta, Whiskeytown, Black Butte, Oroville, Almanor, Bullard's Bar, and Folsom); the major irrigation canals and their associated demand centers (e.g., Tehama-Colusa canal, the Glen-Colusa canal, and others); aggregated irrigation districts that draw water directly from rivers; and the principal urban water demand centers. Three flood conveyance systems included in the model are the Sacramento Weir and the Yolo and Sutter bypasses. A simplified schematic is presented in Fig. 1.

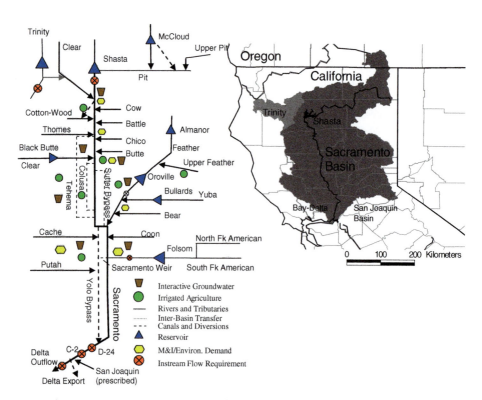

Fig. 1 Simplified schematic of the water resources elements implemented in the Sacramento River WEAP model

The expanded WEAP application developed for this study covers much of the same area and water management features that are represented in other models used in water planning in California: mainly, CalSim-II and CALVIN. The WEAP model, however, differs from these tools in a couple of important respects. First, unlike standard water resource planning tools that rely on exogenous information on water supply and demand to simulate how available water should be allocated, WEAP has embedded a watershed hydrology module into a water resources modeling framework, such that climatic inputs can be used directly to drive the model. This integration of hydrologic processes into a water resources modeling framework allows for analysis of the future climate scenarios that are unbounded by a reliance on historical hydrologic patterns. That is, analysis in the WEAP framework flows directly from the future climate scenarios and not from a perturbation of the historic hydrology as is necessary in applying standard tools to the question of potential climate change impacts in the water sector.

The other important distinction to make about the WEAP application is that it contains a rather simplified representation of the rules that guide the operations of the Central Valley Project (CVP) and State Water Project (SWP) systems. As such, we have not entered all of the sharing agreements (e.g., Coordinated Operations Agreement), regulatory guidelines (e.g., CVPIA b(2) accounting), and other rules (e.g., project allocations) that are explicitly represented in other planning models. Rather, we have attempted to capture the main features that govern the operation of the system as a whole. This choice was made in response to the main research objective which was to develop a tool that could illuminate high level implication of climate change and potential adaptive responses. This is in contrast to an objective that would focus on impacts that may be felt by individual water right and water contract holders in California.

Even though we have not focused on these individual water right and water contract impacts, we have captured enough of the details of the system to allow us, through this and other studies (Joyce et al. 2006; Yates et al. 2005b; Yates et al. 2009), to refine the representation of model features such that model simulations reliably recreate observed patterns in water supply (i.e., reservoir storage, unimpaired streamflow, groundwater elevation, snow pack), water demand (i.e., crop evapotranspiration of applied water, urban demand), and system operations (i.e., surface water deliveries, delta inflows, delta exports, delta outflows). This same type of calibration, it is argued by some, is impossible for other models that possess detailed regulations that have changed through time.

The successful calibration and validation of the model gives us confidence that WEAP can reliably simulate the water management system and, so, can be used to evaluate the impacts of changes in water management in response to changing water supply conditions. It should be understood, though, that the WEAP model is intended to complement the standard set of water planning tools. Given the simplifications made in describing project-specific operations, the WEAP model is directed toward evaluating broader-scale issues of water management. Its utility is mainly in evaluating high-level water management objectives and identifying the most promising set of strategies that may be used to optimally operate the system. Once identified, such strategies may require further investigation using standard tools, which can address management issues at a finer scale. Lastly, the integration of hydrological processes into the WEAP planning model make the tool particularly strong in evaluating proposed management alternatives in the context of climate change.

2.2 Analytical approach

The WEAP model was used to evaluate the impact of twelve future climate scenarios on agricultural water management in the region, and to investigate whether water management adaptation could reduce potential impacts. Each of the twelve climate sequences was run for three management scenarios: one in which no changes in agricultural practices occurred (Base); a second in which improvements in irrigation efficiency occurred gradually until 2050 (Increased Irrigation Efficiency, abbreviated "IE"); and a third in which annual cropping patterns changed in response to water supply conditions (Shifting Cropping Patterns, also referred to as "Logit"). All scenarios were run for an analysis period 2006–2099.

2.2.1 Future climate scenarios

The Intergovernmental Panel on Climate Change (IPCC) released a *Special Report on Emissions Scenarios* (SRES) that grouped future greenhouse gas emission scenarios into four separate "families" that depend upon the future developments in demography, economic development, and technological change (Nakicenovic and Swart 2000). Together they describe divergent futures that encompass a significant portion of the underlying uncertainties in the main driving force behind global climate change. For the purposes of this study, outputs from six general circulation models (GCMs) were used to estimate future climate conditions under two SRES scenarios: A2 and B1. By choosing six GCM and two emission scenarios that would be applied to all investigations in response to the governor's executive order (S-3-05), the Climate Action Team, which had been tasked with coordinating climate change research for the state of California, hoped to create a consistent set of output that would represent the range of future climate conditions.

The six GCMs used to generate the future climate conditions for the current investigation are summarized in Table 1. Cayan et al (2009) generated outputs from these models by applying the downscaling methodology developed by Maurer et al. (2002) to create a 1/8° gridded data set for daily climate variables. These downscaled daily data were used to derive average monthly time-series of precipitation, temperature, wind speed, and relative humidity for each of the 75 sub-catchments in the WEAP model.

2.2.2 Adaptation strategies

Adaptation to climate change within the agricultural sector is likely to occur naturally in response to economic signals that are driven by public policy, market conditions, and, in a setting like California, the availability of irrigation water supply. Understanding the evolution of this last factor under future climate conditions requires the application of a water resources systems model that tracks the management of the available hydraulic infrastructure.

Table 1 General circulation models used in study

Developer	GCM	Study Code
Center for National Weather Research, CNRM (France)	CM3	GCM1
Geophysical Fluid Dynamics Laboratory, GFDL (US)	CM2.1	GCM2
Center for Climate System Research, CCSR (Japan)	MIROC 3.2	GCM3
Max Planck Institute, MPI (Germany)	ECHAM5	GCM4
National Center for Atmospheric Research, NCAR (US)	CCSM3.0	GCM5
National Center for Atmospheric Research, NCAR (US)	PCM1	GCM6

WEAP represents dynamic changes in water management by programming in model parameters that vary over the course of a simulation. These parameter changes can be imposed as exogenous forces upon the model (e.g., as functions of the passage of time) or they can be expressed within the model as a function of the state of the system (e.g., water supply, crop yields, depth to groundwater). Both methods are used here separately to represent the adaptation strategies considered in this study.

2.2.3 Improving irrigation efficiency

With regard to irrigation efficiency, the authors believe that existing and anticipated future regulatory pressures for improved agricultural water use are likely to lead to most crops, other than rice, employing drip irrigation by the middle of the century. For this study, it is assumed that these changes occur gradually over the first half of the century and reach a maximum level by 2050.

Changes in irrigation efficiency differed among crops based upon assumptions made in the amount of land converted to low-volume (e.g., drip) irrigation systems. It was assumed that orchards, vineyards, and row crops (including tomatoes and truck crops) would be entirely irrigated with low-volume irrigation systems, while field crops (including cotton, sugar beet, alfalfa, grain, and pasture) would convert only half of the irrigated land. Rice acreage, on the other hand, will be irrigated by gravity-fed irrigation in 2050, as it is today.

2.2.4 Shifting cropping patterns

Each agricultural demand unit in WEAP possesses a characterization of how crops are distributed across the land available for irrigation. These cropping patterns were initially estimated using historical land use surveys, which show only a snapshot in time of how crops are distributed. In actuality, cropping patterns change from year to year as farmers react to water supply conditions and economic and social factors. To capture this dynamic, we have included in WEAP cropping relationships, developed by the Lawrence Berkeley National Laboratory (L. Dale, personal communication), that relate the share of various crops within a command area to water supply conditions at the time of planting. These relationships were developed based on current market conditions and, thus, reflect how cropping decisions would be made if the markets for crops remain stable.

The share of crop acreage in each demand area varies as a function of changes in the supply of surface water and depth to groundwater. The function is derived from a multinomial logit regression analysis of synthetic data of crop shares generated by the Central Valley Production Model (CVPM) for 21 regions in the Central Valley. The data were generated from CVPM model runs assuming the base water supply and groundwater depth and perturbations from these base levels. These model runs provided a suite of synthetic estimates of crop shares across a range of different regional water supply and groundwater depth assumptions. These crop share equations were then used by WEAP to show changes in crop acreage and water use over time.

How farmers respond to these changing conditions is a function of a number of factors, which change depending on the reliability of various available water sources. For example, farmers who rely solely on groundwater for irrigation base cropping decisions on the depth to groundwater, which relates directly to their operating costs. Central Valley Project settlement contractors in the Sacramento Valley, on the other hand, have guaranteed contracts for surface water deliveries that are only reduced when inflows to Lake Shasta reach a critical level (i.e., less than 3.4 million acre-feet). Their cropping choices are then more responsive to changes in surface water supplies. In the Sacramento and San Joaquin Valley there are many CVP and SWP agricultural contractors whose allocations for surface water deliveries vary from year to year based upon current storage and predicted inflows to the main project reservoirs.

The implication is that indexes of available supply must be calculated for each year in order to permit the various types of water user to make appropriate cropping decisions. Based on the value of these supply indexes, a multinomial logit model of cropping shares, estimated from historical data, is employed to determine the distribution of crops and fallow land in that year for the given user. These logit equations were programmed into WEAP so that at the start of every cropping season over the course of the twenty-first century, an adaptive simulated cropping pattern was defined.

3 Results

This section shows some results of the WEAP model simulations for each of the 12 climate change scenarios. We begin by evaluating the projected climate data for each of the scenarios used as input to the WEAP model. We then discuss the implication of these projected climate sequences by following their impacts downward through the watershed. First, we evaluate the projected changes in reservoir inflows. This includes an assessment of the changes in timing and magnitude of inflows, as well as a look at the relative magnitude and duration of future droughts. In addition to evaluating the impacts of changing climate on water supply, we also look at how climate change may affect crop water demands and how different adaptation strategies may alter these demands. We then evaluate the combined impact of these changes on water management in the Sacramento Valley and Delta export zone. Here we consider the ability of the water resources system to deliver water to satisfy future demands and evaluate the impact of water management on resources protection.

Each of the twelve climate change scenarios was run continuously over a historical period (1950–2005) and a future period (2006–2099) using downscaled GCM climate data and current operational rules. The results of these scenarios are summarized in Figs. 2, 3, 4, 5, 6, 7, 8 and 9, where climate change scenarios are compared against a historic baseline, which was generated by running the WEAP model over the period 1950–2005 using historical gridded climate data (Maurer et al. 2002).

3.1 Climatic analysis

In the following analysis, precipitation and temperature data are presented for 12 climate projections. Precipitation and temperature data are presented as averages of 56 climate locations used as inputs to WEAP, aggregated into three regions—Central Valley, Coastal Range and Sierra. These data were compared across three distinct periods: 2006–2034, 2035–2064, and 2065–2099. Figure 2 shows boxplots of simple averages annual precipitation for each period across all climate projections. These plots suggest that there is generally a decreasing trend in precipitation from the first third of the century to the latter part of the century, when considering all 12 scenarios. Comparing between emission scenarios, precipitation projections tend to be lower in the A2 scenarios compared to the B1 scenarios, with CNRM-CM3 A2 for 2006–2034 being the exception.

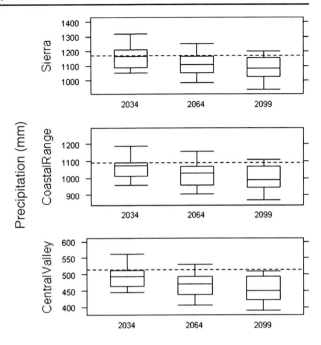

Fig. 2 Boxplots of precipitation across all projections for three periods (2006–2034, 2035–2064, and 2065–2099). The dotted horizontal line is historic (1961–1999) mean precipitation. (Units are in millimeters.) Box covers middle 50% of data, from 25th to 75th percentile. Whiskers are the 1.5*interquartile range. Outliers are not shown

Temperature projections suggested a much stronger trend than that seen with the precipitation data. Figure 3 shows a boxplot for temperature that consistently indicates warming across all projections.

3.2 Hydrologic analysis

3.2.1 Reservoir inflows

Figure 4 shows changes in monthly average inflows to the major reservoirs in the Sacramento Basin (Shasta, Folsom, and Oroville) for the end-of-century period 2065–2099. While neither emission scenario showed a statistically significant difference in annual volume of inflow to the three reservoirs as compared to the historic (1950–2005) WEAP baseline, all GCM/emission scenario combinations showed an earlier timing of streamflow. This shift in runoff timing appeared consistent for all reservoirs across models and emission scenarios. These results are consistent with the supposition that warmer temperatures lead to earlier loss of snowpack and agree with previous studies (Cayan et al. 2006; Purkey et al. 2008).

3.2.2 Occurrence of drought

Whereas some analysis approaches use historic sequences of wet and dry years for future analyses (Chung et al 2009), the WEAP model can examine evolving sequences of wet and dry years for GCM based future climate projections. Thus, WEAP can simulate conditions under different levels of drought persistence that might occur with climate change. This paper includes an estimate of possible changes in future hydrologic conditions in terms of drought persistence. Drought conditions in the Sacramento Basin were described using a

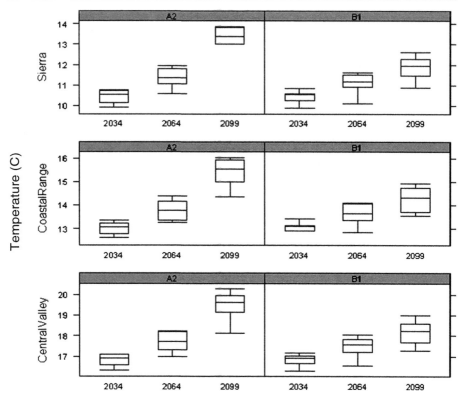

Fig. 3 Boxplots of average annual temperature (°C) for three periods (2006–2034, 2035–2064, and 2065–2099). Box covers middle 50% of data, from 25th to 75th percentile. Whiskers are the 1.5*interquartile range. Outliers are not shown. In some plots, whiskers are so close to the box as to appear missing

construction of the Sacramento Valley 40-30-30 Water Year Hydrological Classification Index (State Water Resources Control Board 1995).[1] This index is measured in million acre-feet and is composed of unimpaired runoff into Shasta, Oroville, and Folsom Reservoirs plus streamflow at the Yuba River. Based on the value of this index, a water year is classified as wet, above normal, below normal, dry, or critical. Droughts were assumed to occur during years designated as critically dry. The severity of the drought was indicated by a value called the accumulated deficit, which is calculated by subtracting the value of the 40-30-30 index for a given year for a given climate change scenario from the threshold value for the critical year designation (5.4 MAF). These deficits were accumulated in consecutive dry years and were reset to zero whenever the index exceeded the threshold for the critical year designation.

Figure 5 shows the accumulated deficits for the historic period (the 1976–1977 and early 1990s droughts are apparent) and each of the twelve climate change conditions included in this analysis. The results show much variability in drought persistence between the various climate change projections—with some GCM/emission scenario combinations replicating

---

[1] The Sacramento Valley 40-30-30 Water Year Hydrological Index is equal to 0.4 x current April to July unimpaired runoff+0.3 x current October to March unimpaired runoff+0.3 x previous year's index (if the previous year's index exceeds 10.0, then 10.0 is used).

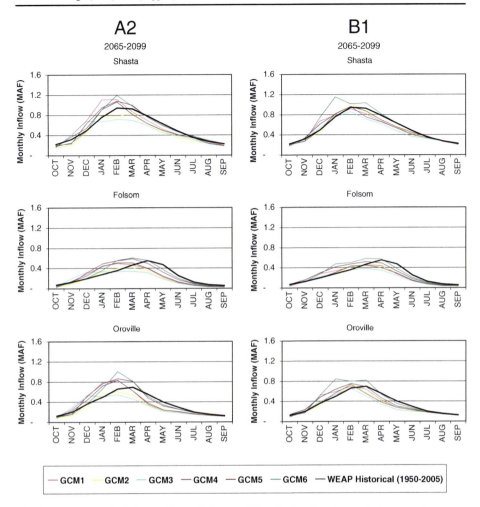

Fig. 4 Average monthly inflow to Shasta, Folsom, and Oroville for A2 and B1 emission scenarios

historic drought conditions, some showing more moderate droughts than observed, and others suggesting more severe droughts. In general, the A2 emission scenario predicted more severe droughts than the B2 scenarios, which agrees with the lower precipitation seen with these scenarios.

3.3 Demand analysis

Annual supply requirements for agricultural areas in the Sacramento and western San Joaquin valleys are summarized in Fig. 6. These are the sums of the crop water requirements for all irrigated areas calculated from the future climate time series using WEAP's internal Penman-Montieth routine, adjusted based on assumed losses in delivering water to meet these requirements.

Under the Base scenario, where the total cropped acreages remained fixed and irrigation technology and scheduling remain unchanged, both emission scenarios showed an increasing trend in water requirements with time, with the A2 scenario exhibiting a more

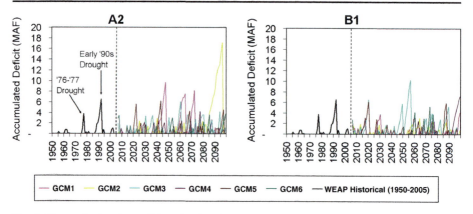

Fig. 5 Changes in drought conditions. Vertical dotted line delineates the historical period from the future climate projection period

pronounced increase than the B1 scenario. This pattern was consistent with the predicted changes in temperature.

The Base scenario also suggested that crop water requirements would experience a greater increase in the Sacramento Valley (9% under A2, 6% under B1) than in the western San Joaquin Valley (6% under A2, 4% under B1) by the end of the century. This trend was driven by differences in the mix of crops in the two regions. In particular, there is almost a 100-fold difference in the amount of rice grown—with the Sacramento Valley having just over 600,000 acres in production and the western San Joaquin Valley having only 6,600 acres in production.

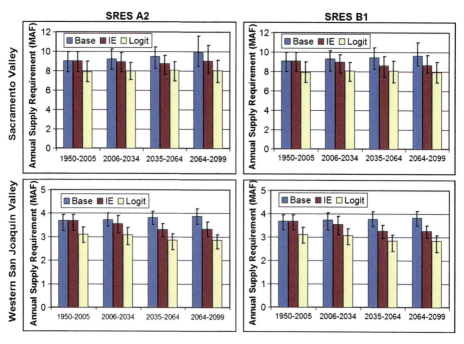

Fig. 6 Projected water supply requirements for agriculture in the Sacramento and Western San Joaquin Valleys. Hash marks indicate minimum and maximum values

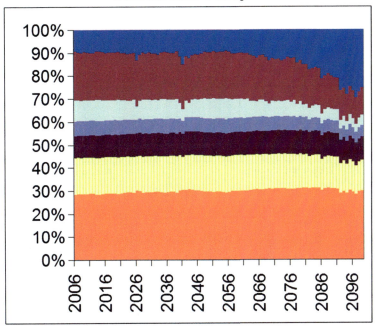

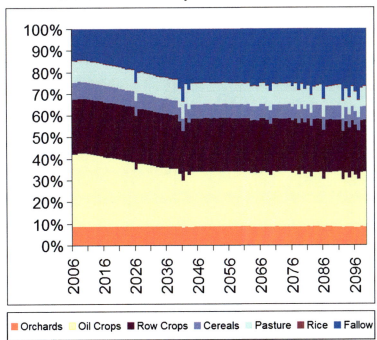

Fig. 7 Simulated changes in cropping patterns in the Sacramento and San Joaquin Valleys for A2/GFDL-CM21 scenario. Row crops include truck crops as well as process and market tomatoes. Oil crops include cotton, sugar beet, and field crops. Pasture includes alfalfa. Orchards include subtropical and vineyard. Cereals include grain

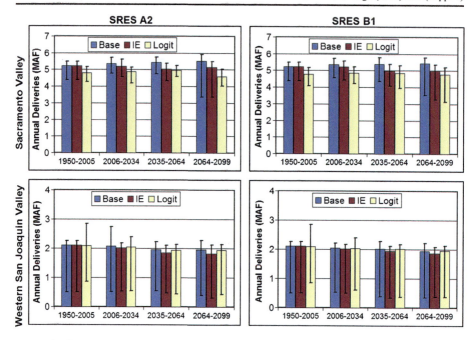

Fig. 8 Simulated surface water deliveries to agriculture in the Sacramento and Western San Joaquin Valleys. Hash marks indicate minimum and maximum values

The results of the IE scenario suggest that improvements in irrigation efficiency could largely offset increases in water demand anticipated with increasing temperatures. In fact, in

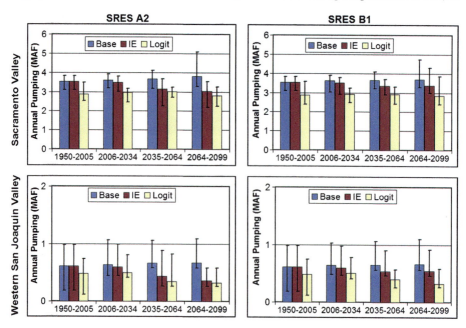

Fig. 9 Simulated annual agricultural groundwater pumping in the Sacramento and Western San Joaquin Valleys. Hash marks indicate minimum and maximum values

some cases, water demands actually decrease by the end of the century. In general, the offset in crop water demand was greatest for the B2 emission scenarios and more pronounced in the San Joaquin Valley. The difference in forecasted temperatures between emissions scenarios accounted for the greater capacity of improvements in irrigation efficiency to offset water demands in the B2 scenario. That is, changes in irrigation technology were more effective when the counteracting changes in temperature were lower. The larger impact in the San Joaquin Valley was due to the predominance of orchard, row crops, and field crops, which all have a high potential for improvements in irrigation technology. Water demands in the Sacramento Valley, on the other hand, were largely driven by rice acreage, which has little potential for improved irrigation technology because it relies on flooded fields.

An analysis of water deliveries under the changing cropping patterns is not directly comparable to the model outputs for scenarios run with no adaptation and those run with increased irrigation efficiency, because the difference in the amount of land in production between the model runs alters the baseline water demands such that the impact on the water supply system is distorted. That is, the logit model presumed an ambient presence of fallowed land that was not considered in the other scenarios. This fallow land class accounted for a minimum of 10% of irrigated land in the Sacramento Valley and 15% of irrigated land in the San Joaquin Valley. Regardless of this incongruity in model runs, it is still illuminating to consider the modeled impacts on water supply and delivery.

It is interesting to note that, unlike the previous simulations that contained either no adaptation or pre-defined changes in water usage (i.e., improvements in irrigation efficiency), the Logit scenarios exhibited similar impacts on water supply requirement for both the A2 and B1 emission scenarios. This would suggest that feedback between water supply and agricultural demands allows the model to compensate (or adapt) such that the system achieves similar water demands under different climate forcings.

The Logit scenarios also showed a very clear decrease in water supply requirements for the western San Joaquin Valley under both emission scenarios. Further, the trend in the Sacramento Valley shows more variability and, as such, is ambiguous. These general trends are again indicative of the mix of crops in the two regions. Figure 7 shows an example of how the cropping pattern changed in both regions under one climate change scenario, A2/GFDL-CM21 (or A2/GCM2). This shows that, for both regions, the decrease in water supply requirement was due to an increase in the amount of retired (or fallowed) land. In the Sacramento Valley, rice accounted for the greatest decrease in cropped acreage, while in the western San Joaquin Valley, the crop most affected was cotton.

It is interesting to observe that in this particular scenario there appear to be two different water supply conditions that lead to the large increases in fallowed lands in the two regions. In the Sacramento Valley, a prolonged drought at the end of the century led to low water supplies in several consecutive years. This prompted irrigators in this region to increase the amount of fallow land from a base of about 10% to as much as 30% in the driest years. Curiously, irrigators in the western San Joaquin Valley did not show the same type of response to the drought at the end of the century. While there was some variability from year to year, the models suggested that farmers' cropping decisions appeared to be relatively insensitive to changes in water supply. San Joaquin Valley irrigators, however, did increase the idled irrigated area by about 10% over the first half of the century, by retiring land that is currently being used to grow cotton. This trend was related to increasing pumping costs as groundwater heads declined—a trend that was at least partly due to underestimating the availability of supplemental surface water supplies from the San Joaquin and Kings Rivers.

It should be noted that these simulations reflect possible changes under future climate scenarios where the timing of planting and crop development were unchanged. It may be argued that agricultural water usage will adapt to changing climate through a combination of changes in management strategies (i.e. changing the timing of planting and harvesting) and changes in crop physiology. These changes could maintain, or even reduce, the current level of annual crop water demand. Alternatively, annual crop water demands could increase if the length of time to crop maturation shortened to a point where additional crops could be planted within a single growing season. As such, the projections presented here should be interpreted as a first-order estimate of changes in crop water demand.

3.4 Delivery analysis

The WEAP system attempts to satisfy crop water requirements by delivering water through canals and by pumping groundwater. The extent to which it is able to meet the full crop requirements depends upon surface water supplies and capacity constraints on canals and groundwater pumping. As a surrogate for contract allocations, the authors imposed limits on the amount of water that WEAP could release from reservoirs by restricting releases to a fraction of remaining active storage. This limits the amount of surface water available that can be diverted from rivers and, ultimately, pumped from the Delta.

3.4.1 Surface water deliveries

Figure 8 present the volume of surface water pumped annually from the rivers and streams of the Sacramento Valley and from the Sacramento-San Joaquin Delta. The graph suggests that under both emission scenarios higher crop water requirements resulted in increasing diversions from rivers in the Sacramento Valley as the simulation progressed into a warmer era at the end of the century. This resulted in less water flowing into the Delta and, thus, less water available to be exported to San Joaquin Valley irrigators.

This pattern of higher water deliveries within the Sacramento Valley at the expense of Delta exports underlines an important distinction in the way in which WEAP allocates water among different users. As previously mentioned, demands are given priorities, such that WEAP delivers water according to a hierarchical ordering of water users. In this scheme, lower priority water users receive surface water deliveries only after the higher priority users have received their full request for water (subject to constraints on delivery capacities). In the Sacramento-San Joaquin application, agricultural water users share the highest priority for water with environmental (i.e., in-stream flows) and indoor urban demands.

Under this configuration, Delta exports are only permissible after the environmental requirements for Delta outflow are satisfied. Because the outflow requirements are given equal priority to Sacramento Valley agricultural deliveries, it also means that the model prioritizes irrigation in the Sacramento Valley over Delta exports. This was not intended to suggest a preference for irrigators in the Sacramento Valley, but reflects a priority structure that mimics the observed historical system operations. Under the historical reference case, much of the water delivered to irrigators on the Westside of the San Joaquin Valley comes from the San Luis Reservoir, which pumps water from the Delta at a time of year when its demands are not in direct competition with those of irrigators in the Sacramento Valley.

Thus, the decline in Delta exports under future scenarios suggests that the environmental requirements within the Delta may represent the biggest constraint on Delta exports. This situation is compounded by irrigators in the Sacramento Valley using more water at the expense of inflows to the Delta. This imbalance in surface water deliveries suggests that there may be opportunities for a reallocation and/or transfer of water rights among irrigators in the Sacramento and San Joaquin Valleys.

Improving irrigation efficiencies reduced the annual surface water deliveries from the rivers of the Sacramento Basin such that they are comparable to those simulated in the historic baseline. These reductions, however, had little effect on the ability to deliver water to irrigators in the western San Joaquin Valley during dry years. This was likely due to a combination of decreasing crop water demands in the export zone and because environmental constraints in the Delta prevented the export of any additional water. Thus, the benefit of reduced water demands in the export zone materialized primarily in the form of reduced unmet demands.

Changing cropping patterns in response to water supply conditions generally led to a reduction in surface water deliveries in both the Sacramento and western San Joaquin Valleys. This trend in both regions generally follows the same trend observed for the agricultural supply requirement. In the Sacramento Valley, surface water deliveries are relatively stable throughout the simulation. In the western San Joaquin Valley, surface water deliveries decline toward the middle and end of century.

3.4.2 Groundwater pumping

In addition to changing patterns in surface water deliveries, increasing crop water requirements led to a greater usage of groundwater resources in both the Sacramento and San Joaquin Valleys (Fig. 9). The pattern of increasing groundwater pumping corresponded with the drought periods observed in Fig. 5. The higher groundwater pumping, however, was not maintained across all years, resulting in only a marginal increase in total groundwater pumping.

Increasing irrigation efficiency through improvements in technology also led to an overall stabilization of annual groundwater pumping as compared to the historical period. In fact, reductions in crop ET appeared to result in a reduction in groundwater pumping over the first half of the century, but this effect was lost as temperatures drove crop water demands higher toward the end of the century.

Shifts in cropping patterns appeared to influence annual groundwater pumping within the two regions in a similar manner. While the annual volumes were below those seen in other scenarios for reasons already discussed, the average volume of groundwater pumping in the two regions tended to follow the same pattern as changes in agricultural supply requirement.

4 Conclusions

This study used the WEAP model to evaluate the potential implications on water management of twelve climate change scenarios. The consideration of these scenarios revealed a common theme that suggested increasing agricultural demands in the Sacramento and San Joaquin valleys may lead to increased stress on the management of surface water resources and, potentially, to over-exploitation of groundwater aquifers. Further, the model results suggest that water shortages may be felt more acutely in the western San Joaquin Valley as Delta exports become more constrained. As these simulations were run using the current set of operational rules for the system, these results suggest that there may be potential to reconfigure these rules such that a more equitable allocation among water users is achieved. Nevertheless, an overall decrease in system reliability is expected in the absence of any modification of operational rules and/or changes in agricultural practices.

Two examples of how agricultural practices may change in response to changing water supply conditions brought about by climate change include improvements in irrigation efficiency through the adoption of new technology and shifts in cropping patterns to crops with higher market value and/or lower water requirements. These two examples were considered in this study and both were found to offset the increasing demands caused by rising temperatures. However, the model suggested that changing climate patterns may limit water deliveries to agriculture in the western San Joaquin Valley despite the reduced demands, because Delta exports were constrained by environmental requirements within the Delta.

References

California Climate Change Center. 2006. Our changing climate: Assessing the risks to California. A summary report from the California Climate Change Center. CEC-500-2006-077.

California Environmental Protection Agency. 2006. Climate Action Team Report to Governor Schwarzenegger and the Legislature. Sacramento, California. www.climatechange.ca.gov/climate_action_team/reports/index.html.

Cayan, D., Luers, A., Hanemann, M., Franco, G. 2006. Scenarios of climate change in California: an overview. White paper for the California Climate Change Center. CEC-500-2005-186-SF.

Cayan, D., Tyree, M., Dettinger, M., Hidalgo, H., Das, T., Maurer, E., Bromirski, P., Graham, N., and Flick, R. 2009. Climate change scenarios and sea level rise estimates for the California 2009 climate change scenarios assessment. White paper for the California Climate Change Center. CEC-500-2009-014-F.

Chung, F., Anderson, J., Aurora, S., Ejeta, M., Galef, J., Kadir, T., Kao, K., Olson, A., Quan, C., Reyes, E., Roos, M., Seneviratne, S., Wang, J., and Yin., H. 2009. Using future climate projections to support water resources decision making in California. White paper for the California Climate Change Center. CEC-500-2009-052-F

Joyce, B., S. Vicuna, L. Dale, J. Dracup, M. Hanemann, D. Purkey, and D. Yates. 2006. Climate change impacts on water for agriculture in California: A case study in the Sacramento Valley. White paper for the California Climate Change Center. CEC-500-2005-194-SD.

Luers, A.L., D.R. Cayan, G. Franco, M.W. Hanemann, and B. Croes. 2006. Our changing climate: Assessing the risks to California. White paper for the California Climate Change Center. CEC-500-2006-077.

Maurer E, Wood A, Adam J, Lettenmaier D, Nijsee B (2002) A long-term hydrologically-based data set of land surface fluxes and states for the conterminous United States. J Climate 15:3237–3251

Nakicenovic, N., and R. Swart. 2000. *Special Report on Emissions Scenarios.* Intergovernmental Panel on Climate Change.

Moser, S., G. Franco, S. Pittiglio, W. Chou, and D.R. Cayan. 2009. The future is now: An update on climate change science impacts and response options for California. White paper for the California Climate Change Center. CEC-500-2008-071.

Purkey DR, Joyce BA, Vicuna S, Hanemann MW, Dale LL, Yates D, Dracup JA (2008) Robust analysis of future climate change impacts on water for agriculture and other sectors: A case study in the Sacramento Valley. Climatic Change 87(suppl 1):S109–S122

State Water Resources Control Board. 1995. Water quality control plan for the San Francisco Bay/Sacramento-San Joaquin Delta Estuary. 95–1 WR. May 1995. California Environmental Protection Agency. Sacramento, California.

Yates D, Sieber J, Purkey D, Huber Lee A (2005a) WEAP21: A demand, priority, and preference driven water planning model: Part 1, model characteristics. Water International 30:487–500

Yates D, Sieber J, Purkey D, Huber Lee A, Galbraith H (2005b) WEAP21: A demand, priority, and preference driven water planning model: Part 2, Aiding freshwater ecosystem service evaluation. Water International 30:501–512

Yates D, Purkey D, Sieber J, Huber-Lee A, Galbraith H, West J, Herrod-Julius S, Young C, Joyce BA, MAl Raey (2009) A climate-driven water resources model of the Sacramento Basin, California USA using WEAP21. ASCE J Wat Res Management 135(5):303–313

California perennial crops in a changing climate

David B. Lobell · Christopher B. Field

Received: 9 March 2010 / Accepted: 26 September 2011 / Published online: 24 November 2011
© Springer Science+Business Media B.V. 2011

Abstract Perennial crops are among the most valuable of California's diverse agricultural products. They are also potentially the most influenced by information on future climate, since individual plants are commonly grown for more than 30 years. This study evaluated the impacts of future climate changes on the 20 most valuable perennial crops in California, using a combination of statistical crop models and downscaled climate model projections. County records on crop harvests and weather from 1980 to 2005 were used to evaluate the influence of weather on yields, with a series of cross-validation and sensitivity tests used to evaluate the robustness of perceived effects. In the end, only four models appear to have a clear weather response based on historical data, with another four presenting significant but less robust relationships. Projecting impacts of climate trends to 2050 using historical relationships reveals that cherries are the only crop unambiguously threatened by warming, with no crops clearly benefiting from warming. Another robust result is that almond yields will be harmed by winter warming, although this effect may be counteracted by beneficial warming in spring and summer. Overall, the study has advanced understanding of climate impacts on California agriculture and has highlighted the importance of measuring and tracking uncertainties due to the difficulty of uncovering crop-climate relationships.

1 Introduction

Agriculture is an important component of California's economy, landscape, and culture, and is among the human activities most vulnerable to impending climate changes. Two particularly unique and relevant features of agriculture in California are (1) the diversity of crops grown, with California the leading U.S. producer of over 80 crops, and (2) the substantial fraction of agricultural value (roughly one-third

---

D. B. Lobell (✉)
Department of Environmental Earth System Science and Program on Food Security and Environment, Stanford University, Stanford, CA 94305, USA
e-mail: dlobell@stanford.edu

C. B. Field
Department of Global Ecology, Carnegie Institution, Stanford, CA 94305, USA

according to California Agricultural Statistics Service 2006) derived from long-lived perennial crops, such as grapes and almonds. As perennials typically remain in the ground for over 20 years, climate changes over the next 20–30 years will be relevant to crops that have already been planted, and especially to those that will be planted over the next few years.

The goals of this paper are to assess the potential impacts of climate change on perennial cropping systems in California over the next 20–50 years, and to identify possible adaptation strategies to minimize the potential costs and maximize the potential benefits of climate change. We focus here on effects of changes in average monthly minimum and maximum temperature and precipitation, and therefore our results do not incorporate the potentially important additional effects of changes at sub-monthly time scales, such as increased frequency of extreme events. We consider these latter effects, which are difficult to estimate on a crop-by-crop basis because of data constraints, in a companion paper.

While perennial crops provide a unique opportunity to incorporate climate projections into decisions made today, they also present some unique challenges compared to projecting impacts and adaptation options in annual crops. First, the slow growth of perennials makes experimental warming trials difficult. Second, far fewer models exist to describe perennial crop growth compared to annual crops, in part reflecting the lack of experimental data. While annual crop studies can rely on process-based models such as EPIC or CERES, modeling of perennial crops is limited primarily to statistical models developed from historical variations in weather and crop harvests. Third, perennials can be affected by weather at all times of the year, while annual crops in California are mainly influenced by weather during the summer growing season. Identifying the particular weather variables most relevant to perennial crop growth can therefore be more difficult than with annuals.

In prior studies, we have attempted to summarize the effects of weather on perennial yields using California statewide average time series of crop harvests since 1980, combined with daily observations of weather that were spatially averaged according to the distribution of each crop throughout the state (Lobell et al. 2006b, 2007). The relatively small dataset (26 data points corresponding to 1980–2005) dictated that only two to three weather variables be considered for each crop, the selection of which relied inevitably on subjective decisions based on exploratory data analysis and physiological principles. For some crops, the relationships contained too much scatter to say anything very useful about impacts of future warming, but for others the models indicated clear negative responses to warming.

Almonds, in particular, exhibited a strong negative response to nighttime temperatures (Tmin) in February (Fig. 1a), and for projections of warming we estimated a roughly 10% loss of almond yields by 2030. We note that the importance of this variable is not likely associated with chilling hour accumulation (CHA), which is often cited as a principal control on nut tree development and growth, because most chilling hours accumulate in November-January, and not in February. Indeed, our computations of CHA, following the method of Baldocchi and Wong (2008) for individual stations and then averaging stations based on almond areas, exhibit a much weaker relationship with almond yields than February Tmin (Fig. 1b).

Instead, the importance of February Tmin likely relates to the critical period of pollination that occurs in most varieties in mid-late February. The effective period of pollination is longer when temperatures are low during the bloom season, as the stigma is receptive to pollen for longer periods of time (Polito et al. 1996). For example, 2005 had a

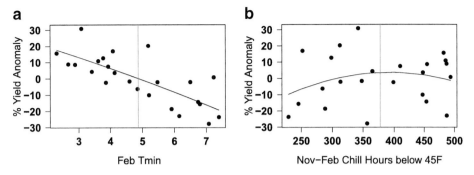

Fig. 1 The relationship between California average almond yields (% anomaly from trend) and almond area-weighted state averages of (a) average February Tmin and (b) chill hour accumulation between November–February. The best fit second-order polynomial is shown by gray line, and dashed vertical line indicates average value for 1980–2005

particularly warm February in the Central Valley of California and the United States Department of Agriculture (USDA) almond production report stated a primary reason for low yield expectations was that "bloom was rapid with an extremely poor set and numerous orchards displayed early petal fall." While poor pollination appears the most likely mechanism, it is impossible to tie a statistical relationship to any single process, which of course causes some concern when applying past empirical relationships to the future. Our perspective is that the major processes linking weather to yields are likely to be similar over the next 20–30 years, when climate is not too different than it is today, and therefore the empirical relationships should be informative in the absence of any more mechanistic predictions.

We sought to improve on the past work in three major ways. First and foremost, we analyzed county level crop and weather data to reevaluate the relationship between weather and yields for a wide range of perennial crops. There is no obvious scale at which to model weather-yield relationships. The use of county data has the main advantage that it provides additional data points, as well as access to a wider range of temperatures than when looking at statewide averages. However, there are several potential pitfalls when using county level data. First, data at the county scale are considerably "noisier" than statewide averages, because the number of fields used to estimate production is limited in each individual county. Second, factors other than climate vary between counties, so that comparing yields in a cross-section can be susceptible to omitted variable biases. (Below we evaluate this bias by repeatedly leaving some counties out of the model training and testing predictions for these counties).

The second objective was to project impacts through 2050 using down-scaled climate projections from six climate models, two downscaling methods, and two emissions scenarios (Cayan et al. 2009). For brevity, we do not consider the issue of how much recent climate trends attributable to greenhouse gases (Bonfils et al. 2008) have affected these crops. Third, we sought to assess adaptation options for almonds, which as mentioned above is a very valuable crop in California and one previously identified as susceptible to warming. Specifically, we evaluated (1) whether some almond varieties will be better suited to a warmer climate than others, and (2) whether some counties will be better suited than others. Both of these issues are relevant to decisions made when planning new almond orchards, which are expanding rapidly relative to other crops in California.

2 Methods

2.1 Statistical yield models using county data

County level area and yield data for 1980–2005 were obtained from the Agricultural Commisioners' reports.[1] Measurements of daily minimum and maximum (Tmax) temperatures and precipitation (Prec) were obtained for 382 of the National Weather Service's Cooperative Stations in California. For each county, we computed daily and monthly averages of each variable for all stations below 200 m (656 ft) elevation, to avoid inclusion of high-elevation stations removed from agricultural areas (e.g., in eastern Fresno county). We also computed daily values of chilling hour accumulation following Baldocchi and Wong (2008), equal to the total estimated hours each day below a threshold value of 45°F (7.22°C).

This study considers the 20 leading perennial crops, in terms of total state value in 2003–2005 (Table 1). For each crop we consider 72 potential weather predictor variables: monthly average Tmin, Tmax, and Prec from the September prior to the harvest year through August of the harvest year, along with their squares. We consider Tmin and Tmax separately because they are often not correlated from year-to-year with each other, particularly in winter, and often one but not the other is highly correlated with yields (Lobell et al. 2007). Combining the two into average temperature would therefore degrade model performance in these situations. We consider both the variable and its square in order to capture nonlinear relationships, as crops often possess an optimal temperature where yields are maximized relative to both cooler and warmer temperatures. A variable for harvest year is also included, to capture effects of trends in technology or other trending factors that influence yield, such as carbon dioxide (CO_2).

To develop statistical models for perennial crops, we must struggle with the fundamental problem of variable selection. Perennials are affected by weather throughout the year, with each crop potential responsive to different aspects of climate. Moreover, much less research has characterized the weather response of perennial crop growth than for annuals. A priori selection of specific months or climate variables is therefore difficult. At the same time, including all possible months and variables will result in an over-parameterized model that tends to overfit the training sample and give poor predictive performance. For this study, we adopt two statistical procedures commonly used in problems where variable selection is an important criterion.

The first is the least absolute selection and shrinkage operator (Lasso) model, which is a variation on ordinary least square (OLS) regression that "shrinks" the regression coefficient towards zero to avoid overfitting (Hastie et al. 2001). In statistical terms, the Lasso model adds a little bias to the model in return for a larger reduction in variance. In the Lasso, coefficients for many variables can be shrunk to zero, so that resulting model only uses a subset of the initial set of variables. As described in Efron et al. (2004), the Lasso can be viewed in this respect as a form of stagewise variable selection, as opposed to the more unstable method of stepwise variable selection. While the statistical details of the Lasso model are beyond the scope of this report, more information can be found in the previously cited references. Here we implement the Lasso using the "lars" package in R. An important decision in the Lasso is when to stop the stagewise process that shrinks the coefficients.

---

[1] National Agriculture Statistics Service, County Agricultural Commissioners' Data. www.nass.usda.gov/Statistics_by_State/California/Publications/AgComm/indexcac.asp

Climatic Change (2011) 109 (Suppl 1)

Table 1 Leading perennial crops in California, ranked by 2003–2005 average total statewide gross cash income, in millions of dollars

Rank	Crop	2003	2004	2005	Average
1	Almonds	1,600	2,189	2,337	2,042
2	Grapes, Wine	1,543	1,605	2,215	1,787
3	Berries, Strawberries[a]	1,173	1,206	1,110	1,163
4	Hay, All	544	609	703	618
5	Grapes, Raisin	348	616	567	510
6	Walnuts	378	452	540	457
7	Grapes, Table	407	535	384	442
8	Pistachios	145	465	577	396
9	Oranges, Navel	290	418	363	357
10	Avocados	365	375	280	340
11	Lemons	218	271	319	270
12	Berries, Bushberries	146	209	224	193
13	Oranges, Valencia	131	142	218	164
14	Peaches, Freestone	139	110	157	135
15	Peaches, Clingstone	108	141	122	124
16	Plums, Dried	132	121	81	111
17	Nectarines	119	86	120	109
18	Cherries	107	123	85	105
19	Grapefruit	69	68	130	89
20	Plums	87	74	92	85

[a] Although strawberries are perennials, they are re-planted each year in most of California

USDA

Here we use the common approach of selecting the model with the minimum value of the complexity parameter Cp, which provides an estimate of out-of-sample prediction error.

The second statistical approach we use on the county data is regression tree modeling. Regression trees work by searching for the variable and value of that variable that best splits a dataset into two subsets, where "best" is defined as the split that achieves the maximum difference between the averages of the two subsets. Each split, called a *daughter node,* is then treated as its own dataset and the process is repeated recursively. For this reason, the method is also described as recursive binary partitioning. The resulting tree model uses the mean of each node as the prediction value, so that it effectively fits a piecewise constant function to the data. Regression trees are an increasingly popular tool in data mining, as they possess many attractive features such as automated variable selection, low sensitivity to outliers and missing data, and an ability to capture interactions between variables. Here we implement regression trees using the "rpart" package in R. The tree was grown until no split improved model R^2 by more than 0.01, and then it was pruned by eliminating nodes until R^2 decreased by more than 0.05. The pruning procedure is a common technique to avoid overfitting the model to the calibration dataset.

Both the Lasso and regression tree models are imperfect. For example, the Lasso is a linear model that is incapable of capturing important interactions between weather in different months. Regression trees fit piecewise constant functions and thus provide crude approximations to linear relationships. By employing both techniques, we sought to identify for each crop where the relationships between weather and yields were robust enough that model choice had a relatively small effect on inferred impacts. In such cases, the assumptions that vary for the two methods can be viewed as having a small effect on the results.

However, comparison of the two methods does not reveal the importance of assumptions that both share. One particular concern is that differences among counties that appear due to weather are, in fact, associated with omitted variables that are correlated with weather. Possible omitted variables include soil quality, topography, and management techniques. To examine sensitivity to omitted variables, we used three approaches.

First, we simply plotted the yield data versus each climate variable identified as important in the Lasso model, with each county coded by a different color. This allowed us to visually examine whether the correlation was driven largely by differences among counties.

Second, we performed a bootstrap analysis of model performance, where for each of 100 iterations we removed one-third of the counties from the calibration procedure. The model calibrated on the other two-thirds of the data was then used to predict yields for the test subset, and the R^2 was computed between predicted and actual yields. Cases where the test R^2 was substantially lower than the training R^2 indicated the possible presence of omitted variable bias.

As a third check against omitted variables, we selected the five most important variables identified from the Lasso analysis and performed an OLS regression with and without a dummy variable for county (i.e., a county fixed-effect). Model predictions for the average statewide impact of a 2°C (3.6°F) warming in all months were compared for the two models, and when the answers diverged it indicated the presence of strong county-fixed effects.

An overview of the modeling process for the county level models study is given in Fig. 2. Only for models that appeared robust, namely with relatively high R^2 and low

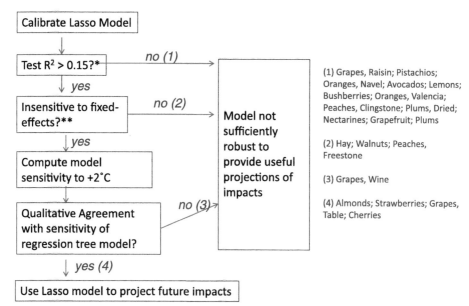

* Evaluated by calibrating model on 2/3 of counties and testing on remaining 1/3 of counties, repeating 100 times and computing average prediction R^2
** Evaluated by comparing ordinary least square regression model with top 5 Lasso variables, with and without county fixed-effects

Fig. 2 Overview of steps taken to model yield responses using county-level data

sensitivity to model structure and fixed-effects, did we then attempt to project impacts of future climate change. Figure 2 indicates which of the 20 crops were considered robust and which were eliminated for each reason (discussed in detail below). While all model evaluation was based on statistical tests, more qualitative work could be done to verify the model coefficients in the future, for example by surveying growers on their impressions of the most important weather variables.

2.2 Projections of climate change impacts through 2050

The crop models developed here were then combined with climate change projections of temperature and precipitation to assess potential impacts through 2050. We limit our projections to 2050 because temperatures beyond this date are frequently beyond the range of temperatures used to fit the statistical models. We emphasize that these projections are conditional on the assumption of no adaptation, and therefore are unlikely to represent the true future course of yield impacts. However, understanding the potential impacts in the absence of adaptation is a critical step towards planning and prioritizing adaptation options.

We used climate projections from six general circulation models and two emission scenarios (Special Report on Emissions Scenarios [SRES] A2 and B1), as described in Cayan et al. (2009). Only projections that were downscaled using the bias-corrected spatial downscaling (BCSD) were used in this study. Average monthly values of Tmin, Tmax, and Prec for 1950–2099 were computed for each of the 12 model simulations and averaged for each county over the portion of the county classified as agriculture in a California map of management landscapes.[2] This latter step was important to ensure that the climate model data were consistent with the extent of observational station data used in the crop model calibration, which were limited to low elevation areas. Their agreement was confirmed by comparing the climatology of simulated and observed temperature and precipitation averages in each county over the observation period of 1980–2005 (not shown).

The county averages of monthly climate model simulations were then fed into the crop models to project yields for each county for 1950–2099, assuming the technology of 2000 (the predictor "year" was held constant at 2000). Statewide average yields were computed by assuming the current distribution of crop area within California. The results are presented as percent changes from the 1995–2005 average yields, and as 21-year moving averages to emphasize the trend rather than year-to-year variability. We present projections only out to 2050 because (1)projections beyond this time period require substantial extrapolation of the statistical models, and are therefore less reliable, and (2) from our perspective most decisions in the agricultural sector have a timeline of 50 years or less.

Finally, to estimate the uncertainty associated with the climate projections, the yield projections were made for each of the 12 climate model simulations (six models x two emission scenarios). To estimate uncertainty associated with the crop models, the projections were repeated using crop models generated from bootstrap samples of the historical data. We present results both for climate uncertainty only and for the combination of climate and crop uncertainty.

2.3 Almond varieties

As almonds are California's single most valuable perennial and are susceptible to winter warming (see below), we investigated whether different common commercial varieties have

---

[2] California Department of Forestry and Fire Protection, FRAP. http://frap.cdf.ca.gov.

differential sensitivity to warming. If so, planting of more heat-tolerant varieties could be pursued as an adaptation strategy. Data on statewide production of individual varieties since 1980 were obtained from the Almond Board of California (courtesy of Sue Olson—Associate Director, Statistics & Compliance). Corresponding data on statewide areas of individual varieties were obtained from the USDA's National Agricultural Statistics Service, California Field Office (courtesy of Jack Rutz, Deputy Director).

The time series of each variety were then analyzed separately in an identical manner. Briefly, statewide average time series of Tmin, Tmax, and Prec were generated by averaging station data according to the fraction of 2003 statewide area for the specific almond variety that was found in the county. The almond production and yield time series were then detrended using a linear trend, and an autoregressive model was used to remove the autocorrelation that is often present in time series of alternate bearing crops such as almonds. The production and yield anomalies were then regressed against February Tmin. Here we present the results for the production data, since the relationships were slightly stronger and since production statistics are more reliable than area or yield statistics (Jack Rutz, personal communication), although results for the two variables were similar.

3 Results

3.1 County scale yield models

3.1.1 Lasso models

Twelve of the 20 perennial crops considered did not exhibit any clear relationships between weather and yields, with the models able to capture less than 15% of the variation in the yield data not used to calibrate the model. For the remaining eight crops, the models were able to explain more than 46% of the variance in training data, and, with the exception of table grapes, more than 20% of the test data (Fig. 3).

The coefficients for the eight successful models, which are useful for understanding which temperature variables most closely relate to yields, are summarized in Fig. 4.

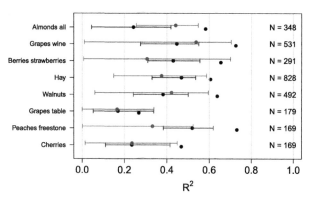

Fig. 3 Model R2 for training and test datasets for each crop with an average test R2 greater than 0.15. The twelve other crops considered in this study did not meet this criterion. Black point indicates training R2 for full dataset, red indicates test R2 when omitting one-third of years from calibration and using these to test the model, and blue indicates test R2 when omitting one-third of counties. Lines indicate 95% confidence interval based on 100 repeated tests with a different (random) one-third omitted

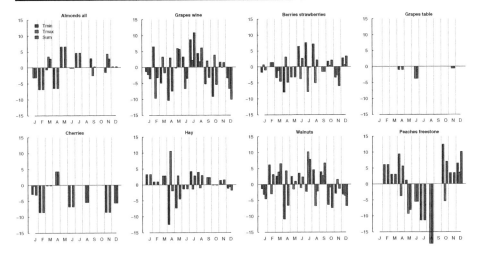

Fig. 4 Summary of temperature coefficients for Lasso model, expressed as % change in state average yields for a 2°C warming. Blue and red bars indicate Tmin and Tmax, respectively, for each month. Coefficients without bars were shrunk to zero by the Lasso model

(Precipitation coefficients are not shown, as these tended to be less important for the crops, presumably because they are all irrigated.) These coefficients were derived from a single computation of the Lasso using the full dataset (as opposed to the R^2 statistics which were derived by repeated calibrations to subsets of the data). Rather than display in units of absolute yields, these values are expressed in terms of the percent change in statewide average yields that would result from a uniform 2°C increase in each variable (Tmin and Tmax for each month.) Of course, because each variable is represented by a second-order polynomial, a positive response to 2°C warming does not necessarily imply a positive response to greater magnitudes of warming. However, we consider 2°C to be a reasonable approximation for the magnitude of warming expected by 2050 (see below).

For some crops, such as wine grapes, strawberries, and walnuts, the selected Lasso model possessed non-zero coefficients for most temperature variables. For others, such as table grapes and cherries, the majority of coefficients were shrunk to zero, indicating that interannual variations in weather for most months had insignificant effects on yields of these crops.

Several patterns emerge from these models. For some crops, there appear to be parts of the year where warming is beneficial and parts of the year where warming is harmful. For example, warming in January and February significantly reduces yields of almonds, but yields appear to be enhanced by warming in May and July. Wine grapes, strawberries, and walnuts show a qualitatively similar pattern of yield losses for warming throughout the winter but yield gains from warming in summer months. Freestone peaches exhibit an opposite pattern, with winter warming—particularly at night—beneficial, but warming during the summer extremely harmful.

Cherries and table grapes exhibit a different pattern, where warming rarely has a benefit at any time of year. The case of cherries is especially stark, with yields harmed by warming throughout November–February, the primary months in which trees accumulate chilling hours. The greater importance of Tmin than Tmax in these months supports the notion that reduced chilling (which occurs mainly at night) is the culprit for yield losses.

3.1.2 Fixed-effects

Next we consider whether these county-level relationships between weather and yields may be biased by omission of non-climatic variables that vary by county, such as soil quality. This bias was evaluated by selecting the five weather variables with the largest effect in the Lasso model (Fig. 4) and running an OLS regression with just these variables and their squares. The OLS regression was then re-run after adding a dummy variable for county.

The changes in statewide average yields for a 2°C warming were computed for both OLS regressions, using bootstrap resampling to estimate a confidence interval. The results (Fig. 5) demonstrate that five of the eight crops appear very insensitive to inclusion of county fixed-effects, indicating that the Lasso results are not biased by omitted variables. However, three crops (hay, walnuts, and freestone peaches) were significantly different between the two OLS models, with non-overlapping 5%–95% confidence intervals.

A sensitivity of results to fixed-effects indicates that much of the perceived effect of weather is obtained by comparing yields across counties, rather than by comparing across years. It does not necessarily indicate that weather is not the true reason causing yields to differ among counties. However, one cannot rule out the possibility that other differences among counties explain at least part of the yield differences. There is therefore no obvious choice between a model with and without fixed-effects, and here we simply point to the crops that are sensitive to this choice. For crops that are insensitive to this choice, such as almonds and grapes, the models utilize differences between counties but are not entirely dependent on them, as evidenced by the similar results when the model is limited to using only differences across years.

3.1.3 Comparison of Lasso and regression tree models

We next consider the importance of model structural assumptions. In particular the OLS and Lasso models assume that yield response to weather can be represented as a second-order polynomial, with no interactions among variables. In contrast, the regression tree

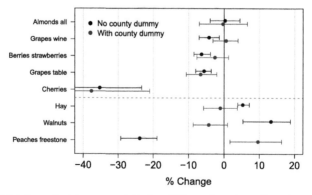

Fig. 5 The predicted change in state average yields for a 2°C warming using ordinary least square regression models without (*black*) and with (*red*) county fixed-effects. Error bars indicate 5%–95% confidence interval based on 100 bootstrap replicates. Large differences between the two indicate the potential importance of omitted variables, such as soil quality, that vary by county. The models contained the five temperature variables deemed most important in the Lasso model. Only three variables were used for table grapes since the Lasso model included only three temperature variables

models use piecewise constant fits and are capable of capturing interactions and higher order nonlinearities. For many of the crops the two models provided predictions that were qualitatively consistent, with overlapping confidence intervals (Fig. 6). An important exception was wine grapes, where the Lasso predicts on average a statewide loss of roughly 15% for 2°C warming while the regression tree predicts a 5% increase. The regression tree model can choose different variables for each bootstrap iteration, so it is difficult to describe exactly why it shows an increase. However, when fit to the entire dataset for wine grapes, the regression tree chose as predictor variables only August Tmin and Tmax, which tend to be higher in the Central Valley counties with higher yields.

Of course, for wine grapes the major concern is not total production but the quality of the grapes for winemaking. While coastal counties have varieties with lower yields, they produce wine with much greater economic value. Therefore, a shift to Central Valley yields and varieties would likely reduce, not increase, total agricultural value. While this study does not focus on climate change and wine quality, previous studies have concluded that although many factors affect wine quality, more frequent summer heat extremes could present a significant challenge to maintaining wine quality in current production zones in the future (Hayhoe et al. 2004; White et al. 2006).

To summarize, the regression tree models generally support the Lasso results, with the exception of wine grapes. We therefore do not place great confidence in the Lasso predictions of lower wine grape yields with warming. At the same time, the yields of wine grapes in a future climate will be far less important than the quality of grapes that can be grown.

3.1.4 Comparison with previous studies

Of the eight crops with significant models using county-scale data, four were also considered in previous work using statewide averages. Of these crops, the model for table grapes agreed well at the two scales, with both the county and state models indicating modest declines in statewide average yields for warming. For walnuts the model from county data was ambiguous because of the sensitivity to fixed-effects. In the state model, walnuts showed a modest decline because of sensitivity to November temperatures. The county OLS model with fixed-effects showed similar declines for warming (Fig. 5), while the OLS model without fixed-effects and the Lasso model exhibited a positive response to warming.

As shown in Fig. 6, the estimated response of wine grape yields to warming with county data was negative when using the Lasso but positive for a regression tree model. The state model exhibited a very small sensitivity to warming, falling between the predictions of the

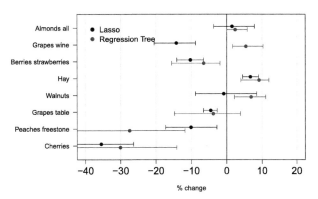

Fig. 6 The predicted change in state average yields for a 2°C warming using the Lasso (*black*) and regression tree models (*red*). Error bars indicate 5%–95% confidence interval based on 100 bootstrap replicates. Large differences between the two indicate the potential importance of structural assumptions in the models

two county level models. A reasonable estimate for wine grapes may therefore be little change in statewide average yields for warming.

The most striking difference between the state and county models was for almonds, the single most valuable perennial crop in California. The county models indicate a very low sensitivity to a uniform warming of 2°C throughout the year, with the result apparently robust to omitted variables and structural assumptions. The state model, which relied solely on February Tmin for temperature response, shows instead a negative response to warming. Importantly, the county model also shows a strong negative effect of February warming (Fig. 4). In fact, the inferred effect of February Tmin on yields is nearly identical when using the county or state average data over the range of temperatures seen in the state model (Fig. 7). The county model possesses a larger range of temperatures, and thus is also able to capture the reduction of yields at very cold temperatures.

As discussed above, the small net effect of temperature on almonds in the county model arises from a beneficial effect of spring and summer warming that cancels the effect of winter warming. To test whether the state model would also exhibit this response if summer temperatures were included in the model, we performed a stepwise OLS regression where we added variables to the original state model (Table 2). The variables added are the two that had the biggest positive effect in the Lasso model: May and July Tmin.

Table 2 displays the model R^2, as well as the predicted response to uniform 2°C warming for a model using only statewide averages and a model using county-level data. The county and state models give nearly identical results of ~13% yield loss when using the original two variables of the state model: February Tmin and January Prec. When May Tmin is added, the

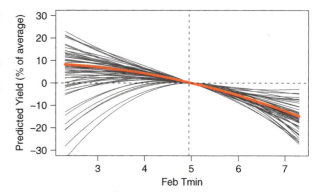
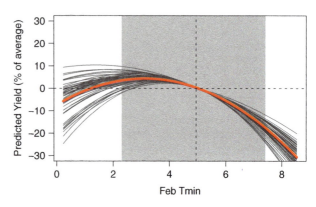

Fig. 7 The inferred relationship between February Tmin and state average almond yields using (**a**) state-wide average yields and (**b**) county-level data. The shaded area in (**b**) indicates the range of temperatures for the state model in (**a**). The red lines show the relationship when using all data, and black lines show the relationship for 50 bootstrap samples. The units of yields are the percentage above the value of yield at the average statewide temperature, which is indicated by vertical dashed line

Climatic Change (2011) 109 (Suppl 1)

Table 2 Ordinary Least Square models for almond yields using statewide average or county level data from 1980 to 2005, and using statewide average data for 1960–2006. Model R^2 and the estimated impact of 2°C warming is shown for four models of increasing complexity

Variables	State, 1980–2005				County, 1980–2005				State, 1960–2006			
	R^2	Estimated sensitivity to +2°C			R^2	Estimated sensitivity to +2°C			R^2	Estimated sensitivity to +2°C		
		mean	5th% tile	95th %tile		mean	5th% tile	95th% tile		mean	5th% tile	95th% tile
Feb Tmin	0.55	−14.9	−23.0	−5.5	0.25	−15.7	−19.6	−12.2	0.35	−14.5	−21.6	−7.9
Feb Tmin, Jan Prec	0.74	−13.5	−20.2	−7.4	0.32	−12.6	−14.8	−9.9	0.43	−12.4	−19.7	−6.2
Feb Tmin, Jan Prec, May Tmin	0.81	−14.5	−28.2	−5.4	0.41	−6.3	−9.9	−3.1	0.47	−5.7	−13.4	0.9
Feb Tmin, Jan Prec, May Tmin, Jul Tmin	0.86	−11.8	−24.0	0.3	0.46	−3.1	−9.5	2.3	0.48	−1.4	−14.4	14.8

county model R^2 improves substantially, while the projected loss is reduced by half to 6%. In the state model, the R^2 also improves but the mean projected impact changes very little. In addition, the confidence interval becomes wider because with six variables (three weather variables plus their squares) and 26 data points, the model exhibits a higher variance for bootstrap resampling that is symptomatic of overfitting.

When July Tmin is added, both the county and state model impacts are reduced by roughly 3%. Here the state model becomes very variable and the confidence interval widens further. The main difference between the two models is thus the strong beneficial effect of May Tmin warming in the county model that is simply not evident in the statewide average time series. According to Kester et al. (1996), May is the critical month of embryo growth and hardening, and "adverse conditions and stress in this period can seriously lower quality and reduce weight of the mature nut" (p. 96). Thus, it is at least plausible that warming in May does substantially benefit almond yields.

A possible explanation for the lack of this effect in statewide time series is that it is too short to accurately measure this effect. To assess this, we obtained weather records and average almond yield data back to 1960 and repeated the OLS analysis at the state scale, with results shown in the right columns of Table 2. When using this longer record, the model agrees remarkably well with the county model that showed a significant benefit of May warming. Thus, the balance of evidence leads us to believe that spring warming will, in fact, benefit almond yields enough to offset much of the losses incurred from winter warming.

3.2 Projected impacts of climate change

The climate model projections indicated similar patterns in each county, with temperatures for Fresno displayed in Fig. 8 as an example. The two emission scenarios give similar average temperature changes until roughly 2040, indicating that the next 30+ years of climate change are "locked-in" because of inertia in the climate and energy systems (Meehl et al. 2007). Temperature changes are slightly more rapid in summer months than in winter months, a result that likely reflects a simulated warming feedback from soil moisture decreases in summer months. The representation of soil moisture feedbacks in general circulation models is questionable in agricultural areas, since no models currently represent the irrigated conditions that exist in the Central Valley (Lobell et al. 2006a). Nonetheless,

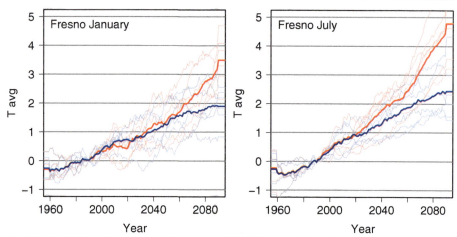

Fig. 8 Projected change in average monthly temperature for (**a**) January and (**b**) February. Each thin line shows an individual model projection, with red representing an A2 emission scenario and blue representing B1. Thick lines show the model average for each emission scenario. The results are presented as changes (°C) from the 1980–1999 climatology, and as 21-year moving averages to emphasize the trend, rather than year-to-year variability

we use these climate scenarios in the current study without adjustment for this potential bias and discuss the potential implications of the bias below.

The simulated impact of climate change on statewide average yields for the four crops with the most reliable crop models are shown in Fig. 9, assuming no shift in crop areas. For almonds, the trend is slightly positive with a projected increase of less than 5% by 2050 relative to current climate. As seen above (Fig. 6), the impact of a uniform 2°C increase was a very small net change in yield. The slight positive impact of the actual climate projections indicates the greater warming of summer months relative to winter, with the former benefiting and the latter harming almond yields in the model.

Average projections for the other three crops indicate negative trends out to 2050, ranging from less than 5% decreases for table grapes to nearly 20% average loss for cherries by 2050. The shaded areas indicate substantial uncertainties associated with these projections, arising from both the climate and crop models. For example, average statewide cherry yields may be reduced by as much as 30% or as little as 0% by 2050 relative to the climate of 2000.

The simulated impacts were fairly uniform throughout California (not shown), suggesting limited potential benefits from changing the spatial distribution of crops within their current growing areas. Slightly more negative impacts were simulated in the southern part of the state for strawberries and cherries. Again, these results have the caveat that the models used here do not consider effects of weather variables other than monthly averages.

3.3 Almond variety switching as a possible adaptation?

For almonds, the most valuable perennial crop in California, the results from the county-scale model presented above suggested that previous estimates based only on statewide data may have overestimated warming-induced losses. However, the county model agrees with the state model in predicting that winter warming, by itself, will be harmful in the absence of adaptation. An ability to adapt to this warming could thus substantially improve future yields of almond growers, even if the net effect of climate change without adaptation is small.

Different varieties of almonds have different chilling requirements and blooming periods. A reasonable hypothesis is therefore that some varieties exhibit lower sensitivity to

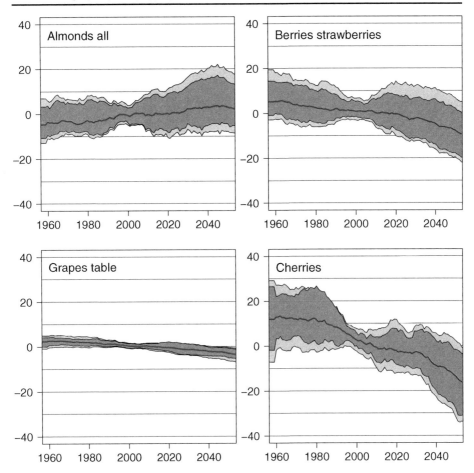

Fig. 9 Simulated change in crop yields for four crops with most reliable crop models. The thick blue line shows the average of all projections, the dark shaded area shows 5%–95% range of projections when using multiple climate models, and the light shaded area shows 5%–95% range when using multiple climate models and multiple crop models (based on bootstrap resampling). The results are presented as percent changes from the 1995–2005 average yields, and as 21-year moving averages in order to emphasize the trend rather than year-to-year variability

winter temperatures than others. Alternatively, because almond varieties are self-sterile and require other adjacent varieties for successful pollination, the response of any individual tree to weather reflects the behavior of a collection of varieties, and therefore may not exhibit unique sensitivities to warming. For example, average annual yields for common varieties exhibit strong correlations over the 1980–2005 period.

A comparison of production anomalies (in percentage of recent production) with February Tmin offers little support for the hypothesis that varieties exhibit significantly different responses to winter warming, as all of the five varieties with sufficiently long records exhibit a similar negative relationship with February Tmin (not shown). As a result, this study found little evidence that switching among the current commercial varieties offers a promising pathway towards climate adaptation.

Given the self-sterile nature of almonds, it is more likely that a holistic approach to adaptation is needed, with selection of a group of varieties that are less sensitive to warming. As discussed above, the most likely explanation for sensitivity to February Tmin is a shortening of the critical bloom period in warmer years, although this claim deserves further scrutiny. Thus, emphasis on traits controlling successful pollination may be warranted in variety development and selection.

4 Discussion and conclusions

The potential impacts discussed above considered the direct effects of temperature and rainfall on perennial crop yields. Importantly, several other factors related to climate change were not addressed. First, it was assumed that historical management practices that affect the response of yields to weather remained constant. For example, all of these crops are irrigated in the entirety of their area, and while they would not be grown without irrigation, declining water resources related to climate change may force reductions in the acreage grown or the amount of irrigation applied. Thus, the omission of water resources likely creates an overly optimistic view of net impacts on the agricultural economy, although decreased water availability will more heavily impact lower value annual crops than the higher value perennials considered here (Tanaka et al. 2006; Medellin-Azuara et al. 2011). In addition, we do not model adaptations that farmers could adopt autonomously in response to perceived shifts in climate, which would likely reduce the impacts relative to our estimates. The relevant adaptations are likely to be crop-specific, but could include shortening the period between replanting or spraying crops with bright substances that increase canopy reflectance and cool the microclimate.

Second, we did not consider the direct fertilization effect of higher CO_2 levels, which according to the SRES scenarios will reach between roughly 450–600 parts per million (ppm) by 2050. The magnitude of CO_2 fertilization for perennials is not well known. Open-top chamber experiments with sour orange trees showed substantial yield increases of up to 80% for a 300 ppm CO_2 increase, even after 13 years (Idso and Kimball 2001) but studies on other tree species show substantially lower rates. While the fertilization effect of CO_2 is relevant to projections of net yield changes, we argue that this effect is likely to be similar across the range of perennial species considered here, all of which possess the C3 photosynthetic pathway. The relative priorities for adapting California crops to climate change therefore should not depend greatly on the exact magnitude of CO_2 fertilization.

Third, we also omitted analysis of detrimental effects of high ozone levels, which may become more frequent and extreme in the next 50 years. Studies with annual crops suggest that losses from ozone may more than offset the gains from CO_2 fertilization (Long et al. 2006).

One of the main conclusions of this study is that, among the 20 most valuable perennials crops, cherries are likely be the most negatively affected by warming over the next decades. This likely relates to a loss of chilling, but again empirical models cannot unambiguously identify mechanisms. While cherries rank only eighteenth by average value for 2003–2005 (Table 1), they are rapidly increasing in popularity. For example, bearing acres of cherries hovered around 10,000 from 1920 to the early 1990s, before increasing to 20,000 in 2000 and 28,000 in 2008.[3] We do not consider here whether the economic decisions to maintain existing or establish new cherry orchards would be affected by consideration of lower yields in future climates, but this is a topic deserving of future study.

Another robust result is that almond yields will be harmed by warming of February temperatures, but likely helped by warmer springs and summers, as these effects are revealed in both state- and county-level analyses (Table 2). Adaptation of almonds to

---

[3] California Sweet Cherries, 1920–2006. www.nass.usda.gov/Statistics_by_State/California/Historical_Data/Cherries.pdf

warmer winters represents a substantial economic opportunity, as almonds are the single most valuable perennial crop and are also experiencing a surge in popularity and planted acreage.

Acknowledgments We thank Claudia Tebaldi for many helpful discussions on the statistical models, Mary Tyree for providing the observed and model climate datasets, Sue Olson and Jack Rutz for data on almond variety production and area, and three anonymous reviewers for helpful comments. This work was supported by a grant from the California Energy Commission's Public Interest Energy Research (PIER) Program.

References

Baldocchi D, Wong S (2008) Accumulated winter chill is decreasing in the fruit growing regions of California. Clim Change 87:153–166

Bonfils C et al (2008) Identification of external influences on temperatures in California. Clim Change, in press: 10.1007/s10584-007-9374-9.

California Agricultural Statistics Service (2006) California Agricultural Statistics 2005: overview. USDA-NASS, Sacramento

Cayan D, Tyree M, Dettinger M, Hidalgo M, Das T, Maurer E, Bromirski P, Graham N, Flick R (2009) Climate change scenarios and sea level rise estimates for the California 2009 climate change scenarios assessment. California Energy Commission, Sacramento, CEC-500-2009-014-F

Efron B, Hastie T, Johnstone I, Tibshirani R (2004) Least angle regression. Ann Stat 32(2):407–499

Hastie T, Tibshirani R, Friedman J (2001) The elements of statistical learning: data mining, inference, and prediction. Springer series in statistics. Springer, New York, p 533

Hayhoe K et al (2004) Emissions pathways, climate change, and impacts on California. Proc Natl Acad Sci USA 101(34):12422–12427

Idso SB, Kimball BA (2001) CO2 enrichment of sour orange trees: 13 years and counting. Environ Exp Bot 46(2):147–153

Kester DE, Martin GC, Labavitch JM (1996) Growth and development, almond production manual. University of California, pp 90–97

Lobell DB, Bala G, Bonfils C, Duffy PB (2006a) Potential bias of model projected greenhouse warming in irrigated regions. Geophys Res Lett 33:L13709. doi:10.1029/2006GL026770

Lobell DB, Cahill KN, Field C (2007) Historical effects of temperature and precipitation on California crop yields. Clim Chang 81:187–203

Lobell DB, Field CB, Cahill KN, Bonfils C (2006b) Impacts of future climate change on California perennial crop yields: model projections with climate and crop uncertainties. Agr Forest Meteorol 141(2–4):208–218

Long SP, Ainsworth EA, Leakey ADB, Nosberger J, Ort DR (2006) Food for thought: lower-than-expected crop yield stimulation with rising CO2 concentrations. Science 312(5782):1918–1921

Medellin-Azuara J, Howitt R, MacEwan D, Lund J (2011) Economic impacts of climate-related agricultural yield changes in California. Climatic Change 109 Suppl 1

Meehl GA et al (2007) Global climate projections. In: Solomon S et al (eds) Climate Change 2007: the physical science basis. Contribution of Working Group I to the Fourth Assessment Report of the Intergovernmental Panel on Climate Change. Cambridge University Press, Cambridge, United Kingdom and New York, NY, USA

Polito V, Micke W, Kester D (1996) Bud development, pollination and fertilization, Almond Production Manual. University of California Publication, pp 98–102

Tanaka SK et al (2006) Climate warming and water management adaptation for California. Clim Chang 76 (3):361–387

White MA, Diffenbaugh NS, Jones GV, Pal JS, Giorgi F (2006) Extreme heat reduces and shifts United States premium wine production in the 21st century. Proc Natl Acad Sci 103(30):11217

Effect of climate change on field crop production in California's Central Valley

Juhwan Lee · Steven De Gryze · Johan Six

Received: 12 February 2010 / Accepted: 26 September 2011 / Published online: 24 November 2011
© Springer Science+Business Media B.V. 2011

Abstract Climate change under various emission scenarios is highly uncertain but is expected to affect agricultural crop production in the 21st century. However, we know very little about future changes in specific cropping systems under climate change in California's Central Valley. Biogeochemical models are a useful tool to predict yields as it integrates crop growth, nutrient dynamics, hydrology, management and climate. For this study, we used DAYCENT to simulate changes in yield under A2 (medium-high) and B1 (low) emission scenarios. In total, 18 climate change predictions for the two scenarios were considered by applying different climate models and downscaling methods. The following crops were selected: alfalfa (hay), cotton, maize, winter wheat, tomato, and rice. Sunflower was also selected because it is commonly included in rotations with the other crops. By comparing the 11-year moving averages for the period 1956 to 2094, changes in yield were highly variable depending on the climate change scenarios across times. Furthermore, yield variance for the crops increased toward the end of the century due to the various degrees of climate model sensitivity. This shows that future climate, suggested by each of the emission scenarios, has a broad range of impacts on crop yields. Nevertheless, there was a general agreement in trends of yield changes. Under both A2 and B1, average modeled cotton, sunflower, and wheat yields decreased by approximately 2% to 9% by 2050 compared to the 2009 average yields. The other crops showed apparently no decreases in yield for the period 2010–2050. In comparison, all crop yields except for alfalfa significantly declined by 2094 under A2, but less under B1. Under A2, yields decreased in the following order: cotton (25%) > sunflower (24%) > wheat (14%) > rice (10%) > tomato and maize (9%). Under A2 compared to B1, the crop yield further decreased by a range of 2% (alfalfa) to 17% (cotton) by 2094, with more variation in yield change in the southern counties than the northern counties. The CO_2 fertilization effects were predicted to potentially offset these yield declines (>30%) but may be overestimated. Our results suggest that climate change will decrease California crop yields in the long-term, except for alfalfa, unless greenhouse gas emissions and resulting climate change is curbed and/or adaptation of new management practices and improved cultivars occurs.

J. Lee (✉) · S. De Gryze · J. Six
Department of Plant Sciences, University of California, Davis, CA 95616, USA
e-mail: ecolee@ucdavis.edu

Abbreviations

GCMs   Global circulation models
analog   Constructed analogues
bcsd   Bias correction and spatial downscaling
MSD   Mean squared deviation
SB   Squared bias
NU   Nonunity slope
LC   Lack of correlation

1 Introduction

Land use change and fossil fuel use have increased the emission of carbon dioxide (CO_2) and other greenhouse gases to the atmosphere at regional and global scales (Janzen 2004). In particular, the atmospheric CO_2 concentration, $[CO_2]$, has increased from 280 to 370 parts per million (ppm) over the past 150 years. The change in $[CO_2]$ has possibly led to an increase in both the mean and variance of the temperature anomaly at the soil surface (Jones and Mann 2004; Porter and Semenov 2005; IPCC 2007). Global variation in mean precipitation and drought has also increased (Dai et al. 1998; Jones and Mann 2004; IPCC 2007).

In agriculture, climate change will likely lead to a major spatial shift and extension of croplands as it will create a favorable or restricted environment for crop growth across different regions (Olesen and Bindi 2002; Smit et al. 1988). Agricultural crop production is fully expected to be impacted by climate change (Adams et al. 1990), but our understanding of climate change and its impacts on California cropping systems in the 21st century is limited (Lobell et al. 2006). California's Central Valley is one of the most productive agricultural regions in the world. It leads national production and sales of many crop commodities, such as almonds, cotton, grapes, hay, rice, and tomatoes (California Agricultural Statistics Service 2008). Lobell et al. (2006) investigated the impact of climate change on perennial crops (e.g., wine grapes), which are high-value commodities in California. However, long-term climate change effects have not been fully tested for major California annual crops and alfalfa (hay). Therefore, it is pertinent to further evaluate potential changes in the production of these systems in California's Central Valley under a changing climate.

Crop growth and development are simultaneously affected by numerous stress factors, which influence crop growth linearly or non-linearly (Hansen et al. 2006; Porter and Semenov 2005). Therefore, a detailed analysis of baseline climate change impacts on cropping systems should precede the development of adaptation scenarios based on alternative management practices under various climate change predictions. Complex ecosystem modeling represents a useful tool for predicting yields as it accounts for a range of interacting conditions in climate, soil, and management. However, there are several factors leading to biogeochemical model uncertainty. First, the process-based biogeochemical models have been only calibrated and validated under observed climate conditions (Adams et al. 1990). Consequently, most biogeochemical models do not predict very well the effects of extreme weather events (e.g., heat waves, floods, etc.) on crop yields. Therefore, the modeled responses of crop growth to temperature and precipitation extremes suggested by the global climate models (GCMs) are often questionable. Second, any change in climatic variables (e.g., temperature and precipitation) at different spatial resolutions and time steps is highly uncertain under any emission scenario (Hansen et al.

2006; Lobell et al. 2006). Third, the fertilization effects of rising $[CO_2]$ on crop yields currently integrated in the biogeochemical models is derived from a limited range of growing conditions and is most likely overestimated (Ainsworth et al. 2008; Long et al. 2005; Long et al. 2006; De Graaff et al. 2006). The magnitude of CO_2 fertilization effects tend to be further reduced when drought stress is minimized by irrigation (Ainsworth et al. 2008; Tubiello et al. 2002). Furthermore, belowground soil respiration (Luo et al. 1996) and weed pressure (Bunce 1995) may increase by elevated $[CO_2]$, which eventually compensate for the potential increase in crop production in the long term. However, current biogeochemical models have limited or no function to account for these factors known to affect crop production under climate change. Regardless of these limitations, process-based biogeochemical models, such as DAYCENT, can effectively integrate crop growth, nutrient dynamics, hydrology, management, and climate for diverse cropping systems and provide a best-estimate of climate change effects on crop yields.

In California, future climate change under A2 and B1 emission scenarios from the IPCC Fourth Assessment Report were evaluated extensively (Cayan et al. 2008). Briefly, the A2 emission scenario predicts medium-high emissions of CO_2 and other greenhouse gases, whereas the B1 emission scenario assumes low emissions. Under A2, $[CO_2]$ is expected to increase exponentially from 352 ppmv in 1990 to 522 ppmv by 2050 and 836 ppmv by 2100. Under B1, $[CO_2]$ is expected to be 482 ppmv by 2050 and then stabilized towards 540 ppmv until 2100. Specifically for California's Central Valley, maximum average temperatures increase from 0.9–3.9°C under B1 to 2.4–5.9°C under A2 by the end of the century relative to 1961–1990. Minimum average temperatures increase from 0.9–3.6°C under B1 to 2.0–6.7°C under A2 in the same period. More warming is expected in summer than winter with increasing frequency of heat waves. Annual precipitation shows relatively small changes (less than 10%) between the 1961–1990 and 2070–2099 periods, but the direction of changes are highly uncertain.

The objective of this study is (1) to project long-term field crop yields in California's Central Valley under the A2 and B1 emission scenarios using the DAYCENT model, and (2) to quantify uncertainties in modeled crop yields derived from uncertainties around predicted changes in climate.

2 Methods

2.1 Description of DAYCENT

To assess the impact of climate change on California cropping systems, we selected the DAYCENT model. DAYCENT is described in detail by Del Grosso et al. (2002). In short, it is the daily version of Century, a fully resolved ecosystem model simulating the major processes that affect plant productivity, such as soil organic matter, water flow, nutrient cycling, and soil temperature and water. The crop sub-model simulates crop growth, dry matter production, and yields to estimate the amount and quality of residue (i.e., C and N input) returned to the soil. It also simulates the plant's influence on the soil environment (e.g., water, nutrients). The crop sub-model simulates phenology, C to N ratios, C allocation to roots and shoots, and growth responses to soil temperature and water and nutrient availability. A variety of management options may be specified, including crop type, tillage, fertilization, organic matter (e.g., manure) addition, harvest (with variable residue removal), drainage, irrigation, burning, and grazing intensity. Specifically in DAYCENT, crop production is potentially limited by soil temperature as each crop is regulated by its specific temperature response function. Soil-water availability is also a major factor that affects crop production and is in the model controlled by current soil-water, precipitation, irrigation, and potential evapotranspiration. Nutrients (i.e., N) from the soil or fertilizer affect potential production depending on crop requirements. In particular, both soil-water and nutrient stresses affect the fraction of C allocated to roots. Germination/beginning of growing season is a function of soil temperature and harvest/end of growing season is a function of accumulated growing degree days when the growing degree day submodel is implemented. In addition, the model is able to specify the effects of elevated $[CO_2]$ on net primary production, transpiration rate, and C:N ratio for biomass. The model uses a logarithmic relationship of net primary production and transpiration rate with changes in $[CO_2]$ between 350 and 700 ppmv, and a linear relationship of C:N ratio for biomass.

2.2 Data acquisition

2.2.1 Climate data

Global climate models can simulate contrasting changes in future climate, although they reproduce the historical climate relatively well (Cayan et al. 2008). Consequently, data from various GCMs are essential for uncertainty assessments on the impacts of climate change. For this study, daily precipitation, maximum and minimum temperature, net radiation at the surface, relative humidity, and wind speed data under the A2 and B1 emission scenarios were obtained from the Climate Research Division of Scripps Institution of Oceanography, at the University of California, San Diego. Six GCMs were applied for the two emissions scenarios: (1) CNRM-CM3, (2) GFDL-CM2.1, (3) CCSR-MIROC3.2 (medium resolution), (4) ECHAM5/MPI-OM, (5) NCAR-CCSM3.0, and (6) NCAR-PCM1. A description of the GCMs can be found in Randall et al. (2007). Each climate change scenario was simulated over the time span 1950–2099.

These climate change scenarios were originally projected on a coarse resolution (hundreds of km). However, finer resolution predictions of climate change are typically required to optimize crop simulations under climate change at local and regional scales (Easterling et al. 1998; Mearns et al. 2001). Therefore, we used the downscaled climate change data to a 1/8° grid resolution (approximately 12 km) by two statistical downscaling techniques: a constructed analogues (analog) method and a bias correction and spatial downscaling (bcsd) method (Giorgi and Mearns 1991; Maurer and Hidalgo 2008). Three of the six GCMs did not provide daily data required by the analog method: CCSR-MIROC3.2 (medium resolution), MPI-OM ECHAM5, and NCAR-CCSM3.0. In total, 18 climate change predictions for the two scenarios were considered.

2.2.2 Soil data

We obtained soil data for all climate grids in California from the Soil Survey Geographic Database (SSURGO) of the Natural Resources Conservation Service (NRCS). The SSURGO database is the most detailed digital soil mapping done by the National Cooperative Soil Survey program. Estimates of soil parameters are obtained from the GIS version of the California soil survey maps, available within the SSURGO database. Specifically, soil texture class, bulk density, hydraulic properties (such as field capacity, wilting point, minimum volumetric soil-water content, and saturated hydraulic conductivity), and pH were obtained. If necessary, the hydraulic properties were estimated from texture (Saxton et al. 1986). In the simulations, soil salinity was not considered as a limiting factor for crop yield because DAYCENT does not have the capability to model soil salinity effects.

2.2.3 Crop types and parameters

The land use survey data were obtained from the California Department of Water Resources (DWR). The DWR land use data include GIS information on crop type, which was derived from exhaustive analyses of aerial photos and field surveys (www.water.ca.gov). For agriculture, nine agricultural classes were used to classify land use, such as grain and hay crops, field crops, pasture, and others. The statewide historical data were obtained from the United States Department of Agriculture (USDA)—National Agricultural Statistics Service (NASS).

Crop phenology and growing patterns were calibrated using historical crop yield data from NASS. Biomass C and N data, C allocation to shoots and roots, and N dynamics data were also calibrated from various literature sources. For field crops, these values have been validated for California conditions by De Gryze et al. (2009).

2.2.4 Management data

Details on conventional management practices in the region (e.g., planting, fertilization, irrigation, weed control, and harvesting) were obtained from the Agronomy Research and Information Center and Cost and Return Studies (2000–2005) available through the University of California Cooperative Extension. Cost and Return Studies contain details on agricultural inputs, planting and harvesting dates and other operations for crops considered in this study. In DAYCENT, the timing of planting and harvesting events, namely growing season-length, was determined by phenology for a grain-filling crop. The ability of DAYCENT to have various planting and harvesting dates is important for realistic predictions of regional crop yields (Moen et al. 1994). For non-grain crops (e.g., alfalfa and tomato), we considered the same planting and harvesting dates across the Central Valley. Data on crop rotations were derived from pesticide use reports from agricultural commissioners and survey data from Howitt et al. (2009).

2.3 Modeling strategies

We selected the six top field crops in California based on their harvested area (Fig. 1). The selected crops were alfalfa-hay (*Medicago sativa* L.), cotton (*Gossypium hirsutum* L.), maize (*Zea mays* L.), winter wheat (*Triticum aestivum* L.), tomato (*Solanum lycopersicum* L.), and rice (*Oryza sativa* L.). In addition, sunflower (*Helianthus annuus* L.) was selected because this crop is commonly included in cropping rotations. We modeled approximately 50% of California's Central Valley, currently covering 1.4×10^6 ha. A typical field crop is assumed to be grown on any soil conditions within each grid. For rice, however, the majority of soils are clayey and poor drained, which makes them generally unsuitable for other crops. Therefore, we intersected the climate, land use, and soil data on each downscaled climate grid for rice, the other field crops, or both by county. The total number of grids used was 110 for rice and 537 for the other crops.

The historical runs to initialize the size of soil organic matter pools in the model consisted of three periods: (1) temperate C_3 grassland with grazing from 0 to 1869 (Paruelo and Lauenroth 1996), (2) initiation of cropping between 1870 and 1884, and (3) pre-modern agriculture

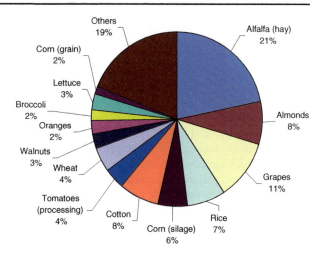

Fig. 1 Relative surface area of crops in California for 2006

between 1885 and 1949. In the first period, a medium-yield variety grass was simulated, which began growing in November and ended growing in April. Low-intensity grazing was assumed to affect 10% of the live shoots and 5% of the aboveground dead biomass. In the second period, we simulated a rain-fed low input winter wheat with minimal soil disturbance and a fallow year in 5 years. In the third period, we assumed a shift to diversified crops to include maize. For rice, we assumed continuous C_3 grass until 1911 and cropping started in 1912.

For the simulations for years 1950–2099, crop rotations were randomly selected based on the acreages of the selected crops (See De Gryze et al. 2009). The data on crop rotations were used to calculate the following conditional probabilities for each combination of crops:

$$\Pr(Cr_t) \qquad (1)$$

$$\Pr(Cr_t|Cr_{t-1}) \qquad (2)$$

$$\Pr(Cr_t|Cr_{t-1}, Cr_{t-2}) \qquad (3)$$

where $\Pr(Cr_t)$ is the probability to have a crop in the current year Cr_t; $\Pr(Cr_t|Cr_{t-1})$ is the probability of having a certain crop in the current year given that a farmer planted a certain crop the year before Cr_{t-1}; $\Pr(Cr_t|Cr_{t-1}, Cr_{t-2})$ is the probability to have a crop in the current year conditioned on previous year's crop Cr_{t-1} and the crop from 2 years before Cr_{t-2}. The survey data suggest that a farmer's decision to plant a crop only depends on the crops that were planted 2 years before. The crop following the fallow period was selected randomly according to the probabilities from Eq. (2). In all subsequent years, the crop planted was selected randomly according to the probabilities from Eq. (3). An exception was alfalfa-hay, which was typically grown in a four- to five-year rotation. Therefore, the conditional probabilities for alfalfa-hay were adjusted accordingly. For the whole simulated period 1950–2099, we considered a high intensity of soil disturbance because conservation tillage had not been extensively and will probably not be practiced in California (Mitchell et al. 2007). We also assumed the beginning of irrigation from 1950 and automatic irrigation up to field capacity was used when soil-water content dropped below 95% of available water holding capacity in the 0–1.5 m depth except for rain-fed winter wheat.

Historically, yields for major crops were relatively low until the mid 1940s. Crop production then started to significantly increase over the past 60 years due to improved mechanical, genetic, and chemical (pesticide and fertilizer) technologies (Johnson et al. 2006). In particular, commercial fertilizer nutrient inputs account for at least 30–50% of the crop yield increase since 1940 (Stewart et al. 2005). The effect of fertilization has been closely related to other improvements for most crops during the same period, such as genetic modifications (Johnson et al. 2006). To account for these effects, we simulated the increasing use of N fertilizer as indicated in NASS survey data from 1964 to 2006 that were found in Tyler (1994) and NASS' Agricultural Chemical Usage Report series. We also introduced different varieties for each period (e.g., low yielding vs. high yielding maize varieties). To maintain agricultural crop production under a changing climate, management practices and cultivars will probably have to be adjusted (Cassman 1999; Lobell et al. 2008). However, it is questionable whether or not the rate of yield improvements will continue in the 21st century (Ainsworth and Ort 2010). In this study, we did not fully consider any future adaptations for management practices (e.g., adjustment of crop variety, alternative cultivation methods, timing and amount of fertilizer and irrigation use, etc.) in response to projected climate change for years 2007–2099. We used the growing degree day submodel to control plant phenology and growing-season length. This option allowed changes in planting and harvest dates, hence providing limited adaptation to climate change. We did not directly simulate several field conditions, such as weed and pest problems.

To evaluate the effects of rising $[CO_2]$ on crop yields, we selected the climate data from the CNRM-CM3 model downscaled by the bcsd method. These climate data represent the multi-model average climate change considered in this study. We assumed that (1) the increase in $[CO_2]$ from 350 (i.e., 1990 $[CO_2]$ levels) to 700 ppmv enhanced net primary production by 10% (Long et al. 2005) and C:N ratio for biomass by 25% (Poorter et al. 1997), but decreased transpiration rate by 23% (Cure and Acock 1986); and (2) the effects of elevated $[CO_2]$ on production, transpiration, and C:N ratio for biomass were proportional to the magnitude of $[CO_2]$ changes over time. We assumed no stimulation in alfalfa and maize production by elevated $[CO_2]$ (Bunce 1995; Kim et al. 2006; Long et al. 2005).

2.4 Data analysis

All yields were expressed on a wet matter basis at harvest. The field crop moisture content at harvest was assumed to be 19% for alfalfa, 15% for cotton, 15% for maize, 14% for rice, 9% for sunflower, 94% for tomato, and 12% for wheat (Perez-Quezada et al. 2003; Wilks et al. 1993). For a combination of crops, climate models, and downscaling methods, annual average yields were calculated from 1951 to 2099, weighted by the acreage of the crop planted in each grid. A five-year moving average was computed to consider trends in yield variance. In this study, we reported 11-year moving averages from 1956 to 2094. Percent changes in yield for each year was relative to the 11-year moving averages in 2009, unless otherwise stated.

The model was validated using the observed statewide crop yield data for the period 1951–2006, except for sunflower, which was only available from 1975 to 2006. To assess model performance, we computed the mean squared deviation (MSD) between modeled and observed yield values, X and Y. The mean squared deviation was then partitioned into three components: squared bias (SB), nonunity slope (NU), and lack of correlation (LC) (Gauch et al. 2003). The SB results from two means being different, whereas the NU arises when the slope of the least-squared regression of Y on X is not equal to 1. The LC arises when the square of the correlation is not equal to 1. These MSD components are additive. The equations for these components can be found in Gauch et al. (2003). In addition, we evaluated the ability of the model to mimic crop yield response to variation in climate. At the regional scale, we described the historical climate-crop yields relationships using modeled and observed annual crop yield and annual averages for maximum/minimum temperature and precipitation. The observed data for annual mean maximum/minimum temperature and precipitation at the regional scale were obtained from Western Region Climate Center's California Climate Tracker (www.wrcc.dri.edu/monitor/cal-mon/frames_version.html). The modeled values were multi-model averages based on each emission scenario. All time series were detrended when necessary.

For this study, we focused on changes in modeled crop yields under the different climate change scenarios. We compared modeled yields by different GCMs and downscaling methods for uncertainty analysis.

3 Results and discussion

3.1 Comparison of modeled and observed yields

The modeled versus observed yields were generally clustered around the 1:1 line for the A2 and B1 emission scenarios (data not shown). By comparing the annual average yields, the observed crop yields were reproduced relatively well by DAYCENT. This is in agreement with De Gryze et al. (2009), showing that the model reliably simulated general yield trends under California conditions. In the period before 1975, the modeled yields deviated on average by 2–14% from the observed yields. Except, the maize and wheat yields averaged over this period were overestimated by $37\pm4\%$ (mean$\pm$standard error) and $31\pm3\%$, respectively. Over the period 1975 to 2006, the maize, rice, tomato and wheat yields were overestimated in the range of 5–16%, whereas the alfalfa yields were underestimated by $3\pm1\%$. Overall, the differences between the means of modeled and observed yields were in a reasonable range over the period 1951–2006 for all the selected crops.

Alfalfa and sunflower had relatively high MSD compared to the other crops. Specifically, 89–91% of errors for alfalfa resulted from LC (Table 1). However, the observed mean and trends for alfalfa were reasonably well modeled based on SB and NU, respectively. Similarly, cotton had the majority of errors resulting from LC. Sunflower had a high NU because the modeled yields slightly decreased from 1975 to 2006, while the observed yields actually increased over the same period. There was no difference between the modeled and observed means for sunflower. For the other crops, errors from LC were almost negligible, except for cotton, as indicated by the intermediate to high r^2. Yield variance for cotton and sunflower was poorly simulated, presumably due to a limited model representation of these crops (Ogle et al. 2006). Only 35–38% and 27–29% of the observed range could be simulated for cotton and sunflower, respectively. Modeled yields for the other crops accounted for 78–100% of the observed range.

At the regional scale, both modeled and observed annual mean maximum and minimum temperatures were approximately 24°C and 9°C, respectively, showing significant ($P<0.001$) increasing trends (data not shown). We found no trends in annual precipitation, which was 408 mm on average. Yield changes over time showed sigmoidal (e.g., maize) or linear (e.g., cotton) trends during the historical period. The same trends were also found in the modeled yields, except for alfalfa. In general, detrended variation in annual mean maximum/minimum temperature ranged between -1 and $1°C$. The model simulated the range of crop yield

Table 1 Components (SB = squared bias; NU = nonunity slope; LC = lack of correlation) of mean squared error (MSD) between modeled and observed annual crop yields (Mg ha^{-1}) for A2 and B1 emission scenarios

	Alfalfa		Cotton		Maize		Rice		Sunflower		Tomato		Wheat	
	A2	B1	A2	B1	A2	B1	A2	B1	A2	B1	A2	B1	A2	B1
Observed mean	13.5		1.2		7.5		6.9		1.4		58.4		3.7	
Modeled mean	13.9	13.9	1.2	1.2	8.4	8.3	7.2	7.2	1.4	1.4	66.3	66.1	4.1	4.1
b^a	0.85	0.84	1.17	1.16	1.27	1.25	0.93	0.93	−1.76	−1.92	0.86	0.86	1.04	1.02
r^2	0.31	0.30	0.35	0.38	0.90	0.90	0.98	0.97	0.17	0.18	0.89	0.89	0.90	0.90
MSD	1.74	1.72	0.02	0.02	1.09	1.03	0.09	0.09	0.04	0.03	70.88	69.02	0.19	0.19
SB (%)	9.3	6.8	0.0	0.0	68.0	69.8	75.2	76.5	1.5	0.6	87.4	86.8	87.7	87.3
NU (%)	1.8	1.9	1.7	1.8	25.3	23.6	22.7	21.1	37.2	38.5	8.4	8.8	1.4	0.6
LC (%)	88.9	91.2	98.3	98.2	6.7	6.5	2.1	2.4	61.3	60.9	4.2	4.4	10.9	12.1

[a] The slope of the least-squared regression of observed on modeled yield values

response to climate reasonably. However, there was no apparent relationship between yield deviations from the trend and climatic variation over the period 1951–2006 for both modeled and observed (Fig. 2). Historical crop yields may show increasing sensitivity to monthly changes in temperature or precipitation (Lobell et al. 2007). Nevertheless, the model would show more sensitivity to increasing temperatures above the range observed because daily production and the duration of crop growth until maturity are quite sensitive to temperature (Stehfest et al. 2007). Overall, we found no effects of precipitation changes on irrigated crop yields.

3.2 Changes in yield

3.2.1 Within each emission scenario

For a combination of emission scenarios and downscaling methods, the crop yields varied greatly between the GCMs for the period 2009 to 2094. For cotton and rice, the average differences in yield were 0.2–0.4 Mg ha^{-1} and 0.2–0.5 Mg ha^{-1} (Fig. 3), with substantial variation in projected temperature (1.0–2.1°C), precipitation (22–33 cm), and relative humidity (5–8%) between the GCMs (data not shown). The average differences in yield between the GCMs were 0.9–1.8 Mg ha^{-1} for alfalfa, 0.7–1.0 Mg ha^{-1} for maize, 0.1–0.2 Mg ha^{-1} for sunflower, 1.8–4.1 Mg ha^{-1} for tomato, and 0.2–0.4 Mg ha^{-1} for wheat (data not shown). These differences were equivalent to 3–17% changes in yield in 2009 for A2 and 2–12% for B1. Despite climate model uncertainty, the average differences in yield by GCM or downscaling method mostly showed increasing temporal trends. As a result, yield variance for the crops increased toward the end of the century due to the various degrees of climate model sensitivity. Hence, future climate change suggested by each of the emission scenarios has a broad range of impacts on crop yields. Therefore, the magnitude of changes in yield under climate change was highly uncertain for most crops.

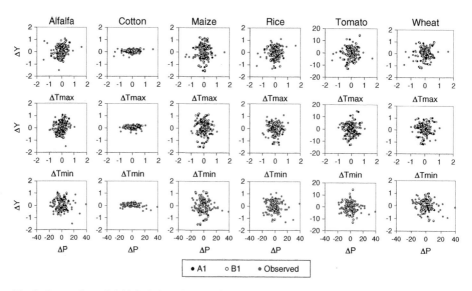

Fig. 2 Scatter plots of yield deviations from trend (ΔY) and detrended variation in annual mean maximum temperature (ΔTmax), annual mean minimum temperature (ΔTmin) and annual precipitation (ΔP). The units for yield, temperature, and precipitation are Mg ha^{-1}, °C, and cm, respectively

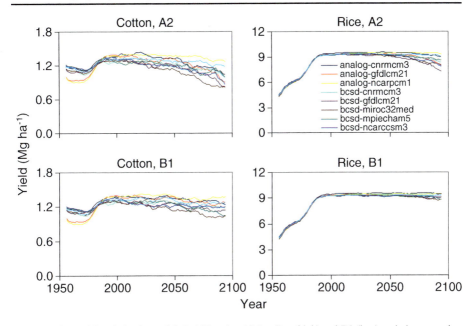

Fig. 3 Multi-model variation in modeled yield under A2 (medium-high) and B1 (low) emission scenarios. Lines are 11-year moving averages that are calculated over the period 1956 to 2094. Analog and bcsd are two methods used to downscale the original climate data from three and six climate models, respectively, for each emission scenario

Yield differences were less affected by downscaling (analog versus bcsd) method. However, we found constantly slightly higher yields for all crops with the analog method than with the bcsd method across time (2009–2094). The average yields were higher by 0.1 Mg ha^{-1} for cotton and by 0.1–0.3 Mg ha^{-1} for rice with the analog method than with the bcsd method (Fig. 3). The average differences in yield by downscaling method were 0.3–1.2 Mg ha^{-1} for alfalfa, 0.5–1.3 Mg ha^{-1} for maize, 0–0.1 Mg ha^{-1} for sunflower, 1.4–3.2 Mg ha^{-1} for tomato, and 0.2 Mg ha^{-1} for wheat (data not shown). Overall trends over time in yield were, nevertheless, similar between the downscaling methods. Maurer and Hidalgo (2008) showed that monthly and seasonal trends of temperature and precipitation produced by both downscaling methods were compatible. In addition, the observed wet and dry extremes were poorly reproduced by both methods and the difference in daily precipitation between the downscaling methods was not significant. However, when compared to observed temperatures, daily temperature extremes were effectively better reproduced by the analog method than the bcsd method (Maurer and Hidalgo 2008); partly leading to uncertainties in the model predictions of crop yields under climate change. We found lower maximum and minimum temperature (0.1–0.5°C), lower solar radiation (34–44 langleys day^{-1}), and higher relative humidity (1–2%) with the analog method than with the bcsd method (data not shown). In comparison, there was an inconsistent trend for precipitation and wind speed by downscaling method. For irrigated crops, these differences in the crops' sensitivity to climate variation were primarily due to specific crop temperature thresholds (Porter and Semenov 2005) that may affect daily production and tolerance to temperature extremes. Phenological development of a crop is strongly determined by changes in temperature. Lower solar radiation and higher relatively humidity likely resulted in less potential evapotranspiration, but changes in evapotranspiration did not affect crop yields due to the use of irrigation. Thus, the choice of the downscaling methods is a major source for uncertainties in predicting future crop yields under climate change.

3.2.2 Between the emission scenarios

Under both A2 and B1, changes in yield relative to the 2009 averages tend to increase for alfalfa (−2–+4%) and for maize (−1–+6%) in the period 2010 to 2025 (Fig. 4). Cotton (−2–+1%), sunflower (−4–0%), and wheat yields (−2–+1%) started to decrease over the same period. The exceptions were rice and tomato yields that remained closely to the 2009 crop yield level. These trends for yield variance seem to last over the next period: 2026 to 2050 (Fig. 4). Regardless of emission scenarios, cotton, sunflower, and wheat yields consistently decreased over time with expected yield losses of approximately 0–6%, 2–9%, and 1–4%, respectively. Meanwhile, our results suggest that the yield for alfalfa slightly increased toward 2050, ranging from −0.3% to 4%. In contrast, there was still no apparent change in rice and tomato yields in this period. By 2050, the average differences in yield changes between the two emission scenarios were marginal: less than 1% for all crops, except for cotton (4%) and sunflower (5%). This suggests that the differences in crop yield between the emission scenarios will generally not be noticeable until 2050.

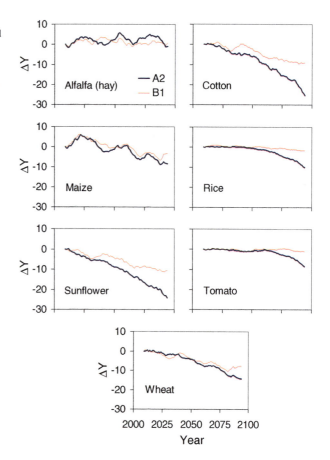

Fig. 4 Changes in yield (ΔY) under A2 (medium-high) and B1 (low) emission scenarios. 11-year moving averages are calculated for the period from 2009 to 2094. Changes in yield are then expressed as percent deviation from the 11-year moving averages in 2009

In the period 2051 to 2094, all the crop yields except for alfalfa were substantially declining under A2 (Fig. 4). Relative to the 2009 averages, the yields under A2 decreased by 2075 in the following order: sunflower (17%) > cotton (15%) > wheat (9%) > maize (6%) > rice (4%) > tomato (3%). The yields further decreased by 2094 in the following order: cotton (25%) > sunflower (24%) > wheat (14%) > rice (10%) > tomato and maize (9%). This suggests that the yields appeared to steadily decrease from 2051 on. The yields also tended to decrease under B1 in the same period, but the yield decreases were less than the ones under A2. Under A2 compared to B1, additional yield decreases were 0% to 17% depending on crop over this period. However, alfalfa yields further increased up to 5% under A2. Therefore, our results suggest that uncontrolled climate change will decrease crop yields in the long-term.

These regional yield decreases were mainly related to the increases in (maximum and minimum) temperature among the climate variables (Fig. 5). In the period 2010 to 2050, the crop yields significantly decreased with increasing temperatures for all crops, except alfalfa under A2 (data not shown) and sunflower under B1. Both emission scenarios had similar crop-specific yield responses to a range of temperature changes (0–1.1°C for A2 and 0–0.8°C for B1). In the period 2051 to 2094, we found similar temperature-crop yield relationships under A2 with the increases in temperature of 1.1–3.3°C. Under B1, all the crop yields except for alfalfa and tomato were significantly affected by increasing temperatures (0.8–1.7°C). Generally, increasing temperatures under climate change shortened the duration of phenological phases until maturity (Adams et al. 1990; Porter and Semenov 2005). As temperature increases, irrigation demand could substantially increase under climate change, although seasonal evapotranspiration was possibly reduced due to the shorter growing season (Howell et al. 1997). However, we did not consider any limitations related to water supply to irrigated croplands, which are expected to occur under climate change conditions in California (Anderson et al. 2008). Although current predictions for reductions in irrigation water supply are highly uncertain (Joyce et al. 2009), the modeled yield losses for the irrigated crops are possibly underestimated. Annual variability of crop yield could partly be affected by winter precipitation (Fig. 5) and relative humidity (data not shown), but the relationships were also highly uncertain. Daily

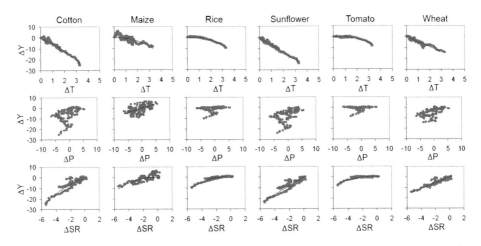

Fig. 5 Scatter plots of percent changes in crop yield (ΔY) and projected changes in annual mean temperature (ΔT), annual precipitation (ΔP), and annual mean solar radiation (ΔSR). Trend lines are shown if significant at $P=0.05$. The units for temperature, precipitation, and solar radiation are °C, cm, and langleys day^{-1}, respectively

production may decrease with decreasing solar radiation, further decreasing crop yields under climate change (Fig. 5).

3.2.3 Changes and differences in county-level yield patterns

The magnitude and direction of modeled yield changes under climate change varied considerably among counties. The county-level yield responses to climatic variation were in part associated with soil variability in the region, as has been found by Schimel et al. (1997). Across all the crops, the differences in yield changes between the A2 and B1 emission scenarios ranged from −6% to +10% in the southern counties of San Joaquin Valley and from −6% to +2% in the northern counties of Sacramento Valley by 2050 relative to the 2009 average yields (Fig. 6). Under A2, for example, alfalfa yield changes varied from +1% to +7% in the northern counties, but had a range of −10–+14% in the southern counties. Except for tomato and rice, variation in yield changes increased more progressively in the southern counties than the northern counties. This shows that the region-level changes in yield under climate change were generally more affected by the county-level changes in yield of the San Joaquin Valley than those of the Sacramento Valley. For sunflower, on the other hand, the region-level yields decreased under both emission scenarios and the changes in yield differed by −6% to −2% by emission scenario

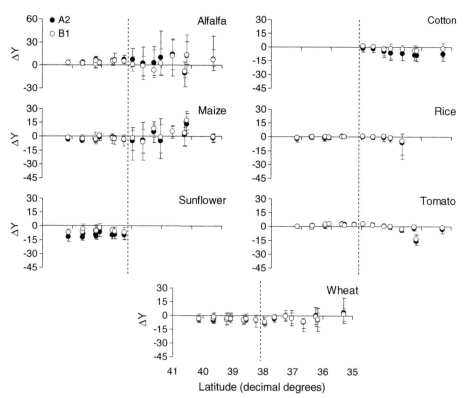

Fig. 6 Percent deviations of the 11-year moving averages in 2050 from the 2009 averages under A2 (medium-high) and B1 (low) emission scenarios. Error bars indicate the range of variation in county-level averages due to climate uncertainty. *Dotted vertical lines* indicate the boundary between the northern Sacramento and southern San Joaquin Valleys

across the counties. However, there was no obvious pattern for changes in sunflower yield by 2050 under A2 and B1. This suggests that the risk of these crops for yield losses will likely increase in response to early climate change particularly in the San Joaquin Valley, presumably due to its relatively low environmental suitability for the crops (Lobell et al. 2006).

For the period 2051 to 2094, the differences in yield changes between the emission scenarios across all the crops ranged from −16% to +6% in Sacramento Valley and from −21% to +8% in San Joaquin Valley (Fig. 7). The yields of cotton, maize, rice, sunflower, tomato, and wheat generally decreased across the counties. It has been suggested that crop production will be mostly affected by increasing climatic variation, including extreme weather events (Porter and Semenov 2005). Hence, crop production seems to be negatively affected by climate change, but the magnitude of the yield reduction was highly uncertain due to the limited model ability to simulate crop responses to extreme weather conditions. In contrast to the other crops, the effects of climate change on alfalfa seem to be not spatially consistent at the county level scale. Water delivery target in agriculture is currently approximately higher than actual delivery by 2% in the Sacramento Valley and 20% in the San Joaquin Basin (Medellín-Azuara et al. 2008). It is expected, however, to further increase by 20–25% by the year 2050 under A2 unless adaptations for water management are made. As a result, climate change will likely decrease annual water deliveries and increase water supply risk in agriculture

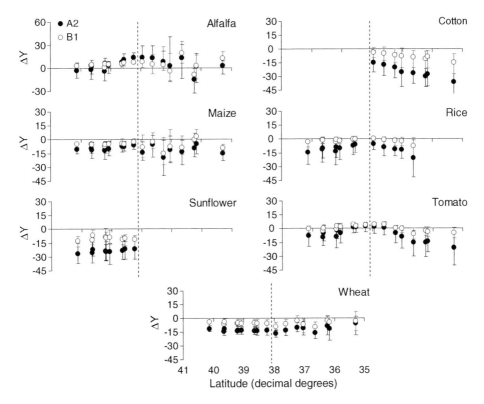

Fig. 7 Percent deviations of the 11-year moving averages in 2094 from the 2009 averages under A2 (medium-high) and B1 (low) emission scenarios. Error bars indicate the range of variation in county-level averages due to climate uncertainty. *Dotted vertical lines* indicate the boundary between the northern Sacramento and southern San Joaquin Valleys

(Anderson et al. 2008). As irrigation demand and evapotranspiration during the growing season are potentially greater in San Joaquin Valley than Sacramento Valley under climate change, the risk of the crops grown on the southern counties is expected to further increase toward the end of the century.

3.2.4 Elevated CO_2 effects on changes in yield

In theory, the negative effect of climate change on yields could be counterbalanced by the increased rate of crop production under elevated [CO_2], particularly for C3 crops (Long et al. 2006). In our simulation study, the effects of climate change and CO_2 fertilization increased crop yields by 2–16% more than the yields under climate change only under A2 in 2094 (data not shown). In comparison, there were smaller yield increases (1–8%) due to CO_2 fertilization effects under B1. The effects of elevated [CO_2] mitigated approximately 30–100% of the yield declines for all crops by 2094. However, regardless of emission scenario, this stimulation of yield seems to be overestimated in comparison to current field data from FACE experiments (Ainsworth et al. 2008). In addition, the direction and magnitude of modeled yield changes vary among climate data and crop models. Tubiello et al. (2002) showed that increasing [CO_2] up to 660 ppmv did not mitigate yield declines from baseline (1951–1994) for irrigated maize under climate change in 2090. Rain-fed winter wheat yields increased (13% to 48%) or decreased (−4% to −30%) in 2090, respectively, with increasing or decreasing precipitation (Tubiello et al. 2002). For rice, changes in [CO_2] from 330 ppmv to 660 ppmv increased yields by 9% or 15% depending on crop models (Bachelet and Gay 1993). Therefore, we have less confidence in the model to accurately predict the potential stimulation of crop yields in response to increasing [CO_2] and focused on the temperature and precipitation effects on yield in a future changed environment.

4 Conclusions

In this simulation study, we assessed the effects of future climate change on crop productivity of alfalfa (hay), cotton, maize, rice, sunflower, tomato, and wheat under current management conditions in California's Central Valley. Our study area includes 17 counties and represents approximately 50% of the crop land in California. In total 18 different climate change scenarios for both A2 (medium-high) and B1 (low) emission scenarios were used to establish a baseline for the period 1950 to 2099. We also evaluated uncertainties in modeled yields from the choice of GCMs and the downscaling methods. The model simulated the observed yields relatively well for all crops in the period 1951 to 2006, although yield variance for some crops (i.e., cotton and sunflower) was not very well reproduced. In the historical period, the range of yield deviations from the trend in response to variation in temperature and precipitation was properly simulated but did not show any trend. In the period 2010 to 2050, there were effects of climate change on changes in yield (11-yr moving average; relative to the 2009 average yields) under both emission scenarios but the differences by emission scenario were not obvious. However, in the next period (2051–2094), relative crop yield changes were different between the two emission scenarios and showed strong spatial patterns across the counties. Overall, the crop yields were negatively affected by the increase in temperature with greater precipitation variation, except alfalfa yields did not consistently respond to climate change. In part, the yields may further decrease with decreasing solar radiation over time. However, we also found that

CO$_2$ fertilization will potentially offset future yield declines under climate change but crop responses to elevated [CO$_2$] should be further validated before strong conclusions are made. Furthermore, other global change factors, such as elevated ozone and N deposition should also be considered because they could enhance or counterbalance the effects of climate change on crop yields. With respect to climate change, we conclude that California crop yields will be negatively affected in the long-term, unless there are statewide adaptation scenarios and management strategies to climate change that maintain or increase yields while mitigating emissions of greenhouse gases.

Acknowledgements This work was funded by the California Energy Commission and Kearney Foundation of Soil Science.

References

Adams RM, Rosenzweig C, Peart RM, Ritchie JT, McCarl BA, Glyer JD, Curry RB, Jones JW, Boote KJ, Allen LH Jr (1990) Global climate change and US agriculture. Nature 345:219–224

Ainsworth EA, Ort DR (2010) How do we improve crop production in a warming world? Plant Physiol 154:526–530

Ainsworth EA, Leakey ADB, Ort DR, Long SP (2008) FACE-ing the facts: inconsistencies and interdependence among field, chamber and modeling studies of elevated [CO$_2$] impacts on crop yield and food supply. New Phyto 179:5–9

Anderson J, Chung F, Anderson M, Brekke L, Easton D, Ejeta M, Peterson R, Snyder R (2008) Progress on incorporating climate change into management of California's water resources. Clim Change 89:91–108

Bachelet D, Gay CA (1993) The impacts of climate change on rice yield: a comparison of four model performances. Ecol Model 65:71–93

Bunce JA (1995) Long-term growth of alfalfa and orchard grass plots at elevated carbon dioxide. J Biogeogr 22:341–348

California Agricultural Statistics Service (2008) Agricultural overviews. In: California agricultural statistics 2007 crop year. USDA-NASS, Sacramento, CA, USA

Cassman KG (1999) Ecological intensification of cereal production systems: yield potential, soil quality, and precision agriculture. PNAS 96:5952–5959

Cayan DR, Maurer EP, Dettinger MD, Tyree M, Hayhoe K (2008) Climate change scenarios for the California Region. Clim Change 87(Supplement 1):21–42

Cure JD, Acock B (1986) Crop responses to carbon-dioxide doubling – a literature survey. Agric For Meteorol 38:127–145

Dai A, Trenberth KE, Karl TR (1998) Global variations in droughts and wet spells: 1900–1995. Geophys Res Lett 25:3367–3370

De Graaff MA, van Groenigen KJ, Six J, Hungate BA, van Kessel C (2006) Interactions between plant growth and soil nutrient cycling under elevated CO$_2$: a meta-analysis. Glob Change Biol 12:2077–2091

De Gryze S, Albarracin MV, Catala-Luque R, Howitt RE, Six J (2009) Modeling shows that alternative soil management can decrease greenhouse gases. Cal Ag 63:84–90

Del Grosso S, Ojima D, Parton W, Mosier A, Peterson G, Schimel D (2002) Simulated effects of dryland cropping intensification on soil organic matter and greenhouse gas exchanges using the DAYCENT ecosystem model. Environ Pollut 116:S75–S83

Easterling WE, Weiss A, Hays CJ, Mearns LO (1998) Spatial scales of climate information for simulating wheat and maize productivity: the case of the US Great Plains. Agr For Meteorol 90:51–63

Gauch HG Jr, Hwang JTG, Fick GW (2003) Model evaluation by comparison of model-based predictions and measured values. Agron J 95:1442–1446

Giorgi F, Mearns LO (1991) Approaches to the simulation of regional climate change: a review. Rev Geophys 29:191–216

Hansen JW, Challinor A, Ines A, Wheeler T, Moron V (2006) Translating climate forecasts into agricultural terms: advances and challenges. Clim Res 33:27–41

Howell TA, Steiner JL, Schneider AD, Evett SR, Tolk JA (1997) Seasonal and maximum daily evapotranspiration of irrigated winter wheat, sorghum, and corn - Southern High Plains. Trans ASEA 40:623–634

Howitt RE, Catala-Luque R, De Gryze S, Wicks S, Six J (2009) Realistic payments could encourage farmers to adopt practices that sequester carbon. Cal Ag 63:91–95

Intergovernmental Panel on Climate Change (IPCC) (2007) Climate change 2007: mitigation. In: Metz B, Davidson OR, Bosch PR, Dave R, Meyer LA (eds) Contribution of working group III to the fourth assessment report of the intergovernmental panel on climate change. Cambridge University Press, Cambridge, United Kingdom and New York, NY, USA

Janzen HH (2004) Carbon cycling in earth systems: a soil science perspective. Agr Ecosyst Environ 104:399–417

Johnson JMF, Allmaras RR, Reicosky DC (2006) Estimating source carbon from crop residues, roots and rhizodeposits using the national grain-yield database. Agron J 98:622–636

Jones PD, Mann ME (2004) Climate over past millennia. Rev Geophys 42:RG2002

Joyce BA, Mehta VK, Purkey DR, Dale LL, Hanemann M (2009) Climate change impacts on water supply and agricultural water management in California's Western San Joaquin Valley, and potential adaptation strategies. California Energy Commission, CEC-500-2009-051-F

Kim SH, Sicher RC, Bae H, Gitz DC, Baker JT, Timlin DJ, Reddy VR (2006) Canopy photosynthesis, evapotranspiration, leaf nitrogen, and transcription profiles of maize in response to CO_2 enrichment. Glob Change Biol 12:588–600

Lobell DB, Field CB, Cahill KN, Bonfils C (2006) Impacts of future climate change on California perennial crop yields: model projections with climate and crop uncertainties. Agr Fort Meteorol 141:208–218

Lobell DB, Cahill KN, Field CB (2007) Historical effects of temperature and precipitation on California crop yields. Clim Change 81:187–203

Lobell DB, Burke MB, Tebaldi C, Mastrandrea MD, Falcon WP, Naylor RL (2008) Prioritizing climate change adaptation needs for food security in 2030. Science 319:607–610

Long SP, Ainsworth EA, Leakey ADB, Morgan PB (2005) Global food insecurity. Treatment of major food crops with elevated carbon dioxide or ozone under large-scale fully open-air conditions suggests recent models may have overestimated future yields. Philos Trans R Soc B Biol Sci 360:2011–2020

Long SP, Ainsworth EA, Leakey ADB, Nosberger J, Ort DR (2006) Food for thought: lower-than-expected crop yield stimulation with rising CO_2 concentrations. Science 312:1918–1921

Luo YQ, Jackson RB, Field CB, Mooney HA (1996) Elevated CO_2 increases belowground respiration in California grasslands. Oecologia 108:130–137

Maurer EP, Hidalgo HG (2008) Utility of daily vs. monthly large-scale climate data: an intercomparison of two statistical downscaling methods. Hydrol Earth Syst Sci 12:551–563

Mearns LO, Easterling W, Hays C, Marx D (2001) Comparison of agricultural impacts of climate change calculated from high and low resolution climate change scenarios: Part I. The uncertainty due to spatial scale. Clim Change 51:131–172

Medellín-Azuara J, Harou JJ, Olivares MA, Madani K, Lund JR, Howitt RE, Tanaka SK, Jenkins MW, Zhu T (2008) Adaptability and adaptations of California's water supply system to dry climate warming. Clim Change 87:S75–S90

Mitchell JP, Klonsky K, Shrestha A, Fry R, DuSault A, Beyer J, Harben R (2007) Adoption of conservation tillage in California: current status and future perspectives. Aust J Exp Agric 47:1383–1388

Moen TN, Kaiser HM, Riha SJ (1994) Regional yield estimation using a crop simulation model: concepts, methods, and validation. Agr Syst 46:79–92

Ogle SM, Breidt FJ, Paustian K (2006) Bias and variance in model results associated with spatial scaling of measurements for parameterization in regional assessments. Glob Change Biol 12:516–523

Olesen JE, Bindi M (2002) Consequences of climate change for European agricultural productivity, land use and policy. Eur J Agron 16:239–262

Paruelo JM, Lauenroth WK (1996) Relative abundance of plant functional types in grasslands and shrublands of North America. Ecol Appl 6:1212–1224

Perez-Quezada JF, Pettygrove GS, Plant RE (2003) Spatial-temporal analysis of yield and soil factors in two four-crop-rotation fields in the Sacramento Valley, California. Agron J 95:676–687

Poorter H, Van Berkel Y, Baxter R, Den Hertog J, Dijkstra P, Gifford RM, Griffin KL, Roumet C, Roy J, Wong SC (1997) The effect of elevated CO_2 on the chemical composition and construction costs of leaves of 27 C_3 species. Plant Cell Environ 20:472–482

Porter JR, Semenov MA (2005) Crop responses to climatic variation. Philos T Roy Soc B 360:2021–2035

Randall DA, Wood RA, Bony S, Colman R, Fichefet T, Fyfe J, Kattsov V, Pitman A, Shukla J, Srinivasan J, Stouffer RJ, Sumi A, Taylor KE (2007) Climate models and their evaluation. In: Solomon S, Qin D, Manning M, Chen Z, Marquis M, Averyt KB, Tignor M, Miller HL (eds) Climate change 2007: the physical science basis. Contribution of working group I to the fourth assessment report of the intergovernmental panel on climate change. Cambridge University Press, Cambridge, United Kingdom and New York, NY, USA

Saxton KE, Rawls WJ, Romberger JS, Papendick RI (1986) Estimating generalized soil-water characteristics from texture. Soil Sci Soc Am J 50:1031–1036

Schimel DS, Emanuel W, Rizzo B, Smith T, Woodward FI, Fisher H, Kittel TGF, McKeown R, Painter T, Rosenbloom N, Ojima DS, Parton WJ, Kicklighter DW, McGuire AD, Melillo JM, Pan Y, Haxeltine A, Prentice C, Sitch S, Hibbard K, Nemani R, Pierce L, Running S, Borchers J, Chaney J, Neilson R, Braswell BH (1997) Continental scale variability in ecosystem processes: models, data, and the role of disturbance. Ecol Monogr 67:251–271

Smit B, Ludlow L, Brklacich M (1988) Implications of a global climate warming for agriculture: a review and appraisal. J Environ Qual 17:519–527

Stehfest E, Heistermann M, Priess JA, Ojima DS, Alcamo J (2007) Simulation of global crop production with the ecosystem model DayCent. Ecol Model 209:203–219

Stewart WM, Dibb DW, Johnston AE, Smyth TJ (2005) The contribution of commercial fertilizer nutrients to food production. Agron J 97:1–6

Tyler HH (1994) Fertilizer use and price statistics, 1960–93. Resources and Technology Division, Economic Research Service, USDA, Statistical Bull. No. 893

Tubiello FN, Rosenzweig C, Goldberg RA, Jagtap S, Jones JW (2002) Effects of climate change on US crop production: simulation results using two different GCM scenarios. Part I: wheat, potato, maize, and citrus. Clim Res 20:259–270

Wilks DS, Pitt RE, Fick GW (1993) Modeling optimal alfalfa harvest scheduling using short-range weather forecasts. Agric Syst 42:277–305

Climate extremes in California agriculture

David B. Lobell · Angela Torney · Christopher B. Field

Received: 12 February 2010 / Accepted: 26 September 2011 / Published online: 24 November 2011
© Springer Science+Business Media B.V. 2011

Abstract Changes in extreme events may represent an important component of climate change impacts on agricultural systems in California. This study considered the relative historical importance of extreme events, as measured by insurance and disaster payments. The causes for each main event for 1993–2007 were classified into general categories to compare the importance of dry vs. wet and hot vs. cold events. The study found that the most common cause of both insurance indemnity and disaster payments is excess moisture, followed by cold spells and heat waves. Climate change is likely to have different effects on the occurrence of each of these, for instance with frosts becoming less common while heat waves increase in frequency and duration. Resolving the overall net effect of changes in climate extremes will largely depend on improved understanding of future risks of excess rainfall and flooding events.

1 Introduction

California is home to a vast array of crops, all of which vary in production from year to year. These variations are driven in part by changes in average climatic conditions, such as average temperature or total rainfall in a particular month or season. However, some of the most substantial changes can be traced to singular weather events, such as freezes, floods, or hail storms. These extreme events have long been acknowledged as a potentially important aspect of climate change, given that some extremes are very likely to increase in frequency in the future (Tebaldi et al. 2006).

Two of the most rapid and significant changes associated with global warming have been, and will continue to be, a decreased occurrence of freezing nights in the winter and

---

D. B. Lobell (✉) · A. Torney
Department of Environmental Earth System Science and Program on Food Security and Environment, Stanford University, Stanford, CA 94305, USA
e-mail: dlobell@stanford.edu

C. B. Field
Department of Global Ecology, Carnegie Institution, Stanford, CA 94305, USA

increased occurrence of heat waves in the summer (Alexander et al. 2006; Meehl et al. 2007; Tebaldi et al. 2006). Changes in hydrological extremes have been more difficult to detect but are now apparent in many regions, in particular increased occurrence of heavy rainfall events (Alexander et al. 2006; Kiktev et al. 2003). Projected increases in extreme rainfall are particularly robust in extra-tropical regions like California (O'Gorman and Schneider 2009). The analysis of extreme changes in California by Mastrandrea et al. (this issue) confirm that many of these global patterns hold locally: robust and rapid decreases in freezing nights and increases in warm days and nights, significant trends in precipitation intensity, and mixed predictions of changes in the length of dry spells.

Compared to the agricultural impacts of shifting average conditions, the impacts of changes in extreme events have proven more difficult to quantify. By definition, extreme events are historically rare, and therefore few observations exist with which to calibrate and test numerical models. Some rare examples do exist in which models are modified to include effects of extremes (Rosenzweig et al. 2002; White et al. 2006), but generally our understanding of how crops respond to extremes is limited. An alternative in this circumstance is to examine specific events that have affected agriculture and estimate the likelihood of these specific events into the future. In essence, this is equivalent to an extremely simple model that has a specified loss when a specified threshold is exceeded, and zero loss otherwise. While simplistic, these can provide a first-order estimate of the direction and magnitude of change in extreme event agricultural impacts.

Here we embark on this approach for California agriculture by reviewing the extreme events that have been important over the past 15 years. Individual events are identified through insurance and disaster records, and then classified according to the type of weather event that caused crop losses. This analysis, when combined with expectations of how each type of weather event will change in the future, allows a qualitative picture of how changes in extremes may impact California agriculture.

2 Methods

To measure the impact of extreme events on agriculture, we relied on data pertaining to the two primary sources of federal aid to farmers: federal crop insurance and emergency payments, programs, and loans. A private crop insurance market has not developed in the United States because of the high risks associated with farming, such as variable weather and unpredictable price markets. The federal crop insurance system is a branch of the U.S. Department of Agriculture, administered by the Risk Management Agency (RMA). The RMA manages and oversees the Federal Crop Insurance Corporation (FCIC), 16 private firms that sell and service policies. Basic crop insurance coverage is known as Catastrophic Risk Protection, which pays farmers affected by natural disasters 50% of expected yield at 60% market price. Catastrophic level premiums are 100% subsidized. Farmers have the option of buying additional coverage for up to 75% of their crop values. To encourage participation in the program, premiums on these policies are highly subsidized by the federal government. Roughly 80% of U.S. farmers are enrolled in the program, and participation in California is currently around 60% (USDA, personal communication). The crop insurance system is supplemented by additional disaster payments and programs and emergency loans.

Crop insurance indemnity records were obtained from the website of USDA's Risk Management Agency—originally sorted by year, state, crop, county, and cause of loss.[1]

---

[1] www.rma.usda.gov/FTP/Miscellaneous_Files/cause_of_loss/

Table 1 Total number of reported claims by crop and total indemnity payments, 1980–2007

Reported crop	# Claims	Total indemnities for 1980–2007 ($M)
Almonds	1153	288.5
Prunes	639	148.9
Raisins	129	142.7
Grapes	1147	122
Navel oranges	304	107.2
Tomatoes	716	81.7
Cotton	467	76.2
Wheat	1371	73.3
Cotton ex long staple	130	47.2
Valencia oranges	232	45.8
Apples	255	40.6
Table grapes	317	39.3
Rice	327	34.1
Safflower	218	25
Sugar beets	515	24.1
Cherries	77	21.5
Lemons	99	21.3
Walnuts	515	13
Stonefruit	204	12.7
Citrus	152	12.4

The RMA provides data for individual counties and attributes losses to specific causes. The dataset covers all major crops in California, with Table 1 presenting a summary of claims and payments made for the crops with the top 20 total payments. In total, 12,093 records for 1980–2007 were obtained. These were first summed by year over counties and then causes were binned into the larger categories of Heat, Cold, Fire, Excess Moisture, Wind, Failed Irrigation Supply, and Other from smaller causes of loss.

Several sources provide estimates of disaster payments. The Environmental Working Group, a nonprofit organization that monitors and analyzes government policies related to conservation, compiles disaster payment data from USDA records. However, their figures for disaster payments include all federal payments allocated in response to natural disasters not associated with the federal crop insurance program. The USDA's Economic Research Service also provides disaster payment data, but not disaggregated by specific cause or before 1996.

We therefore relied on a third source, the Storm Event database compiled by National Oceanic and Atmospheric Administration (NOAA).[2] When storms occur, forecasters enter data about the weather event into the database, including an estimate of crop damage. While the numbers of crop damage are only a best guess made by a NOAA employee, based on a variety of sources such as the media and other government agencies, they represent a reasonable measure of damages caused by specific events. However, it is important to note that these estimates are not of actual payments, but only of total damages to agriculture. Moreover, the errors of these estimates are not well known and may vary for different types of events. For example, damages from frost events may be estimated soon after the event

---
[2] www4.ncdc.noaa.gov/cgi-win/wwcgi.dll?wwevent~storms

whereas the true magnitude of damage may not be apparent until harvest. In the database, data were originally sorted into individual events, with 237 total events for 1993–2007. For this project, data was summed by year and binned into broader categories: Heat, Cold, Fire, Excess Moisture, Wind, and Other.

3 Results

The indemnity and disaster data show similar patterns over the 1993–2007 period in terms of the average relative importance of different types of events (Fig. 1). In both datasets, excess moisture related to heavy rainfall events has been the most costly type of extreme event over this period, followed by cold events and then heat events. Damages from wind, fire, and other events account for a substantially smaller amount of damages.

A breakdown of damages by year indicates that indemnity payments are much less variable than estimated disaster losses, with the former ranging from roughly $20–$130 million per year and the latter ranging from near zero to over $1 billion, in the case of 1998 (Fig. 2). Since the disaster dataset assigns the cost to the date of the event, while the indemnity data record the date of payment, there is some mismatch between the years in which particular events show up. For example, the extreme freeze of late December 1998 shows up mainly in 1999 under the indemnity data, but mainly in 1998 in the disaster loss estimates. Nonetheless, the relative importance of different events tends to coincide between the two datasets.

A list of the top 10 events from 1993–2007, according to the NOAA dataset, indicates that the single most costly event to agriculture in California was the freeze in December 1998, which led to major losses in various crops including oranges, lemons, olives, and

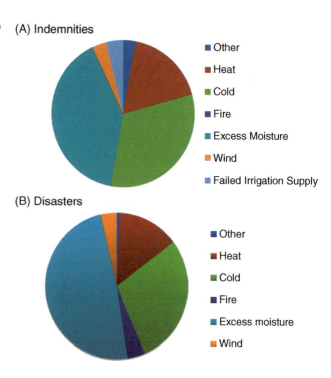

Fig. 1 Relative amount of (**a**) indemnity payments and (**b**) estimated total losses from disasters attributable to different types of extreme events

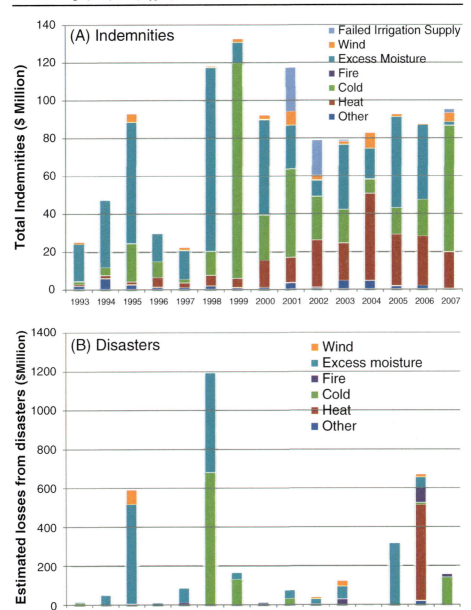

Fig. 2 Total amount of (a) indemnity payments and (b) estimated total losses from disasters for each year, by type of extreme event

cotton (Table 2). The second most important event was a heat wave in July 2006, which was especially damaging to the livestock industry. Heavy rainfall in the spring and winter months was responsible for the next three most damaging episodes in the past 15 years.

To gauge the reliability of the NOAA database, we compiled estimates of damage from these 10 events from alternate sources, such as newspaper stories that quoted federal and

Climatic Change (2011) 109 (Suppl 1)

Table 2 The top ten extreme events in California agriculture since 1993, based on the NOAA Storm Event database

Rank	Estimated crop losses ($M)	Start date	End date	Location	Event	Description
1	682	12/19/1998	12/29/1998	Sacramento Valley, San Joaquin Valley, Los Angeles, Santa Barbara, San Luis Obispo, Ventura	Extreme cold	When an arctic airmass began moving over California, the resulting cold air pool from advection and radiational cooling in the lowest levels of the atmosphere led to a devastating freeze to crops, especially citrus, and central and southern California experienced a week-long period of sub-freezing temperatures. The largest percentage of area crop losses were to lemons and oranges but several other unharvested fruit and vegetable crops were damaged, including avocados and broccoli.
2	492	07/16/2006	07/27/2006	Statewide	Excessive heat	New statewide heat records were set as temperatures soared above 100°F, and peak energy use in the state reached an all time high, causing power outages. With accompanying high humidities, consistent light or calm winds, and long durations of high temperatures, the heat negatively impacted agriculture, especially the dairy and cattle industry, although yield in produce from field crops and orchards also diminished to a slight extent.
3	342	03/10/1995	03/10/1995	Monterey, San Luis Obispo, San Benito, Napa	Flood	When the Salinas, Napa, and Pajaro Rivers overflowed due to heavy spring rains, agricultural land and crops experienced widespread flooding. Crops impacted included lettuce, broccoli, cauliflower, almonds, and strawberries.
4	310	05/01/1998	05/15/1998	Tulare, Kern, Madera, Fresno, Merced, Kings, Visalia	Heavy rain	New rainfall records were set as central California experienced heavy early spring rains and below normal temperatures. The wet, cold conditions damaged crops.
5	192	01/07/2005	01/11/2005	San Bernardino, Ventura	Heavy rain	A storm lasting 5 days dropped heavy rain across all of southern California, and flash flooding and mudslides caused millions of dollars of damage to farms, homes, businesses, vehicles, parks, roads, and bridges. Orchards were uprooted, and crops were damaged.
6	142	01/06/2007	01/24/2007	Statewide	Freeze	Many records were broken as temperatures dipped below 30°F along the coast and below 20°F in the valleys. The freeze lasted for up to a week or longer, and local farmers were hit hard by the freeze. Affected counties were declared disasters areas and made eligible to receive federal aid.

Table 2 (continued)

Rank	Estimated crop losses ($M)	Start date	End date	Location	Event	Description
7	131	04/10/1999	04/10/1999	Central and Southern San Joaquin Valley	Extreme cold	Unseasonably cool air led to minor frost episodes, which followed the disastrous freeze of December 1998. The combined effects of the two freezes caused substantial losses to agriculture, especially because during spring, deciduous trees, vineyards, and vegetable crops are vulnerable to temperatures less than 30°F.
8	113	03/01/1995	03/05/1995	Kern, Kings, Merced, Tulare, Riverside	Flood/rain/winds	Heavy rains caused extensive damage to agricultural crops due to flooding, and most field work stopped as growers waited for the soil to dry.
9	100	06/01/1998	06/30/1998	Southern San Joaquin Valley	Flood	Higher than normal water runoff from snowpack in the Southern Sierra Nevada filled reservoirs, and over 32000 acres of bottom land used for farming primarily south of Corcoran were inundated.
10	80	12/29/2005	01/03/2006	Mendocino, Sonoma, Napa, Kings	Flood	A series of strong Pacific storm systems began on December 18 and continued through the end of the month. Widespread low-land flooding occurred across Sonoma County with mainstem river gages along the Russian River remaining above flood stage for several days. An average of 4 to 6 inches of rain fell over a 24 h period, 2 days after 1 to 3 inches drenched the same area. Severe flooding occurred as the Napa River exceeded flood stage at St. Helena.

state officials. Out of ten events, alternate estimates were obtained for seven cases, and in six of these the estimates agreed well with the NOAA database. The sole exception was the January 2007 freeze, which NOAA reports as costing $142 M while most other estimates ranged from $0.8-1.3B. Even the description accompanying the estimate in the NOAA database states "crop damage was estimated at almost $1.3 billion of California's annual $32 billion agricultural production with nearly $709 million in the Interior Central California ag area". Although it remains unclear why this particular reported estimate is so low, the NOAA database appears to provide a useful if imperfect summary of recent damage from extreme events.

4 Discussion and conclusions

Data sets on crop damages from extreme events indicate that a wide variety of extremes have affected agriculture in California. Each of those events are likely to exhibit different changes in a warming climate. Cold extremes have already become less frequent throughout most of the world, and this trend will almost certainly continue into the future. Heat waves, in contrast, are very likely to become more frequent in the future. Future changes in heavy rainfall and flooding events, which have been the most costly extreme events overall in California agriculture, are less clear. Companion studies reported in this issue (Mastrandrea et al., this issue; Dettinger et al. this issue), describe projections of hydrological extremes such as precipitation intensity and flooding risk, and emphasize that projected trends from many models are relatively small compared to natural variability, and often disagree on the sign of change.

Given the importance of wet events to agricultural outcomes, refining projections of rainfall extremes is important. However, this uncertainty should not detract from the fact that much is known about temperature extremes, and that these are a formidable force in California agriculture. The ongoing trend towards less frequent and severe cold extremes, combined with the knowledge that frosts and freezes have historically been more damaging than heat waves, would suggest that climate change will bring short-term reductions in damages from extreme events. On the other hand, little is known about the nonlinearity of damage at high temperatures, and in particular how well agricultural systems will cope with heat waves of unprecedented severity. Improved understanding of how agricultural systems respond to these extremes and the specific locations most at risk will be needed to further clarify the net impact of warming and associated extremes on California agriculture.

Finally, an assessment of adaptation options and the likelihood of their adoption and effectiveness is an important remaining task in assessing future impacts of extremes and guiding adaptation policies and investments. For example, livestock losses during heat waves can be reduced by supplying more water to cattle, modifying diets, avoiding overcrowding and exposure to direct sun, adding misting and cooling systems, and changing livestock breeds (Wolfe et al. 2008). Similarly, cropping systems might be adapted by increasing frequency of irrigation during heat waves, replacing varieties, and applying foliar sprays to lower canopy temperatures. The success of any of these measures will depend on a range of factors, including cost, availability of capital for investment, farmer familiarity with technology options, attitudes toward risk, awareness of climate trends and their impacts, and institutional arrangements (Moser et al. 2008). The understanding of historical vulnerabilities to extremes presented here is therefore one small step in the overall process of adapting California agriculture to climate change.

Acknowledgments We thank the U.S. Department of Agriculture's Risk Management Agency and the National Oceanic and Atmospheric Administration for providing data used in this report. This work was supported by a grant from the California Energy Commission's Public Interest Energy Research (PIER) Program.

References

Alexander LV et al (2006) Global observed changes in daily climate extremes of temperature and precipitation. J Geophys Res 111:D05109. doi:10.1029/2005JD006290

Kiktev D, Sexton DMH, Alexander L, Folland CK (2003) Comparison of modeled and observed trends in indices of daily climate extremes. J Climate 16(22):3560–3571

Meehl GA et al (2007) Global Climate Projections. In: Solomon S et al (eds) Climate change 2007: the physical science basis. Contribution of Working Group I to the fourth assessment report of the intergovernmental panel on climate change. Cambridge University Press, Cambridge and New York

Moser S, Kasperson R, Yohe G, Agyeman J (2008) Adaptation to climate change in the Northeast United States: opportunities, processes, constraints. Mitig Adapt Strat Glob Chang 13(5):643–659

O'Gorman PA, Schneider T (2009) The physical basis for increases in precipitation extremes in simulations of 21st-century climate change. Proc Natl Acad Sci 106(35):14773

Rosenzweig C, Tubiello FN, Goldberg R, Mills E, Bloomfield J (2002) Increased crop damage in the US from excess precipitation under climate change. Global Environ Change-Hum Policy Dimensions 12 (3):197–202

Tebaldi C, Hayhoe K, Arblaster JM, Meehl GA (2006) Going to the extremes. Clim Chang V79(3):185–211

White MA, Diffenbaugh NS, Jones GV, Pal JS, Giorgi F (2006) Extreme heat reduces and shifts United States premium wine production in the 21st century. Proc Natl Acad Sci 103(30):11217

Wolfe D et al (2008) Projected change in climate thresholds in the Northeastern US: implications for crops, pests, livestock, and farmers. Mitig Adapt Strat Glob Chang 13(5):555–575

Economic impacts of climate change on California agriculture

Olivier Deschenes · Charles Kolstad

Received: 12 February 2010 / Accepted: 26 September 2011 / Published online: 24 November 2011
© Springer Science+Business Media B.V. 2011

Abstract Using county-level data from the United States Department of Agriculture's Census of Agriculture, this study evaluates the effect of weather and climate on agricultural profits in the State of California. The approach is to estimate revenue less variable production cost per acre as a function of land characteristics, weather realizations, and climate. This model is then used to evaluate the effect of two scenarios of climate change for the state of California over the coming century. The preferred estimates indicate that climate change is associated with a negative effect on aggregate agricultural profits by the end of the century. There are significant caveats to this result, including the lack of statistical precision, and keeping water supply and farm prices constant.

1 Introduction

California is at the forefront of states and even countries in having legislation on the books mandating the reduction of greenhouse gas emissions by a specified amount for specific dates. One reason for this aggressiveness is the perceived vulnerability of the state to climate change. As one example, climate change threatens to disrupt the state's precious freshwater supply, which relies heavily on snowfall and snowmelt from the Sierra Nevada Mountains. Climate change also threatens agriculture, one of the state's important industries that employed about 200,000 workers and generated $9 Billion in net value-added in 2009.

Electronic supplementary material The online version of this article (doi:10.1007/s10584-011-0322-3) contains supplementary material, which is available to authorized users.

O. Deschenes (✉) · C. Kolstad (✉)
Department of Economics, University of California, 2127 North Hall, Santa Barbara, CA 93106, USA
e-mail: olivier@econ.ucsb.edu
e-mail: kolstad@econ.ucsb.edu

O. Deschenes
IZA and NBER, Santa Barbara, CA, USA

C. Kolstad
RFF and NBER, Santa Barbara, CA, USA

This paper addresses the question of how agriculture in the state may be affected by a change in climate. To answer this question, we look at two factors influencing farm profits: (1) how farm profit differs from one county to another in the state, inferring the effect climate has on those differences; (2) how farm profit differs for a county from one year to the next, as weather turns out to be different that expected. Statistically combining these two dimensions of the problem allows us to infer how climate and weather affect profits and, in turn, how a changed California climate may affect agricultural profits.[1]

We then use our statistical model, estimated on historical data, to project agricultural profits in the state under two scenarios of global climate change: business-as-usual (A2) and a scenario with a lower projected increase in temperature (B1). These scenarios were developed for the Intergovernmental Panel on Climate Change (IPCC) and are widely used.

Our qualified conclusion is that agricultural profits and crop yields appear to be negatively affected by climate change, although most estimates lack statistical precision. The crops most negatively affected are table grapes and some citrus. In the next sections we develop our model and discuss data sources. We then turn to our results.

2 Background

A number of authors consider the effect of climate change on agriculture. Early work (i.e., from the early 1990s) focused on simulation models. Adams (1989), Adams et al. (1990), and Rosenzweig and Parry (1994) are prominent examples of the use of agricultural process models (including crop growth models) to measure the effect of climate change on crop yields. These models are typically physiological with limited scope for endogenous farmer behavioral response to climate change. Typically, adaptation and adjustment are absent or exogenous.

Some crop growth models allow a certain amount of farmer adaptation to climate change. For instance, Kaiser et al. (1993) use a simulation model to forecast a century of effects from a gradual change in the climate. Their model assumes farmers choose which variety of crop would have done best in the previous (simulated) decade. In this way, some adaptation over time is represented in their model. This results in considerably less loss from a doubling of carbon dioxide concentrations (Schimmelpfennig et al. 1996).

Hansen (1991) suggests that crop growth models (which must enumerate substitution possibilities) may miss some of the substitution opportunities available to farmers. He estimates a cross-sectional model of corn production in the United States where expected climate (July mean temperature and precipitation) as well as realized weather are used to explain cross-sectional variations in corn yield. His results for a temperature increase are mixed, showing yield increases in some climates and yield decreases in others.

Similar information can be derived statistically from observations on agricultural output. Perrin and Smith (1990) investigate the effect of weather on several crops in North Carolina and then use results of climate models to estimate the effect of climate change on crops. Hansen (1991) estimates the effect of weather and climate on corn yield and then postulates the effect of a change in climate. Some studies in agricultural economics and agronomy focus more directly on the weather effect (Kaylen et al. 2002; Thompson 1986; Kaufmann and Snell 1997). These models approach the problem after the production decisions have been made, only considering the effect of actual weather

---

[1] We use the term farm profits as the return to agricultural land revenue less non-land costs.

realizations on yield. Typically the weather data is transformed into some measure of deviation from expected weather. The underlying idea is that the effect of normal weather (represented by climatic expectations) is captured in the farmer's cropping practices, but that unusual weather will have an impact of yield. Of particular note to California is the recent work of Lobell et al. (2007), relating climate to yields of a number of specific California-relevant crops (using data from California).

Mendelsohn et al. (1994) introduced the "Ricardian" approach to econometrically estimating the effect of climate on farm output (though this approach can be traced back to Johnson and Haigh 1970). The central idea of Mendelsohn et al. (1994) is to measure the differences in land values across the United States, inferring that land value differences are due to endowed soil quality and climate. This allows the authors to infer the value of different climates. Using this approach, they infer a very small effect (possibly positive, possibly negative) on U.S. agriculture from climate change. Schlenker et al. (2005, 2006) have performed a similar analysis on a subset of the United States, focusing on non-irrigated land. In recent work, Schlenker et al. (2007) have focused on sub-county data in California and produced interesting results on long-run effects of climate on land values and water availability. Using a hedonic approach, they find that water availability is strongly capitalized in farmland values in California. Similar to our findings below, they also find that rainfall is weakly correlated with land values.

As Schlenker et al. (2005) argue, irrigation is an important issue in the West. But it is not as simple as one might expect. McFadden (1984) focuses on how a farmer (or other agent) may change behavior based on uncertainty about climate change (or even weather variability). In essence, if the farmer feels a possibility of climate change exists, he may adopt more robust practices (such as irrigation) that perform relatively well over a range of weather or climates, sacrificing a bit relative to the case of perfect knowledge about the weather. Fisher and Rubio (1997) similarly find that water storage investments should increase as the variance of precipitation increases.

The main appeal of the Ricardian approach is that if land markets are operating properly, prices will reflect the present discounted value of land rents into the infinite future. Thus the Ricardian approach promises an estimate of the effect of climate change that accounts for the adaptation behavior that undermines the earlier work based on process models. However, to successfully implement the hedonic approach, it is necessary to obtain econometrically consistent estimates of the independent influence of climate on land values, and this requires that all unobserved determinants of land values are uncorrelated with (orthogonal to) climate. Deschênes and Greenstone (2007) demonstrate that temperature and precipitation means covary with soil characteristics, population density, per capita income, and latitude. Moreover, Schlenker et al. (2005) show that the availability of irrigated water also covaries with climate (Schlenker et al. 2005). This means that functional form assumptions are important in the hedonic approach and may imply that unobserved variables are likely to covary with climate. Further, recent research has found that cross-sectional hedonic equations appear to be plagued by omitted variables bias in a variety of settings (Black 1999; Black and Kneisner 2003; Chay and Greenstone 2005; Greenstone and Gallagher 2008). Overall, it may be reasonable to assume that the cross-sectional hedonic approach confounds the effect of climate with other unmeasured factors

Kelly et al. (2005) extend the Ricardian approach by distinguishing between expected weather (climate) and the actual weather that is realized during a growing season. Expected weather determines crop choice and similar decisions; what weather is actually realized determines actual profits. They thus provide one of the first analyses of how total farm profits are affected by both climate and by weather shocks (deviations of weather from what is expected). Rather than examining a single cross-section of farm land values, they examine a pooled time-series cross-section of thirty years of farm profit data at the county level, though just for the Midwest of the United States. Their results are consistent with those of Mendelsohn et al. (1994) regarding the long-run cost of climate change. But they also explore how increased extreme weather events may increase the cost of climate change, over and above the costs associated with a change in the mean weather.

One problem with the Kelly et al. (2005) approach (in addition to its only covering the Midwest) is the treatment of unobserved farm characteristics that can play an important role in determining profits. Deschênes and Greenstone (2007) address this and other issues in their use of the profit function approach to measuring the effect of climate and weather. For one thing, their analysis involves the entire United States as well as individual states. In addition, they include climate as a county-level fixed effect, which permits them to focus on the effect of weather on farm profits. Although this provides a better estimate of profits, it is not possible to disentangle the effect of climate from other unobserved determinants of profits, since all are included in the fixed effect. Thus it is difficult to use their model to evaluate the effect of a counterfactual climate change. They suggest that focusing only on weather provides a conservative estimate of the effect of climate on profits.

3 Methodology

Our approach is to use county-level data in California over a twenty year period to understand how weather and expectations on future weather affect farm profits and crop yields. We then use this estimated relationship to simulate how a changed climate might affect farm profits and yields.

It is appropriate to clarify our distinction between weather and climate. Simply put, weather is the realization of the climate. Weather is the temperature and precipitation (or similar variables) that occur at specific locations at specific times. Climate, on the other hand, is the distribution of weather. Thus the yearly fluctuating January rainfall could be viewed as weather whereas the January rainfall averaged over long periods of time (i.e. 30 years) could be viewed as one dimension of climate.

3.1 Agricultural profits and annual weather realizations

The canonical estimating equation involving weather (W) but not climate is of the form:

$$y_{ct} = \alpha_c + \lambda_t + X_{ct}\theta + \sum_{k=1}^{K} \beta_k W_{kct} + \varepsilon_{ct} \qquad (1)$$

where the indices c and t denote county and year, respectively, and k represents different measures of weather (such as annual average temperature or annual total precipitation). In the models of agricultural profits the dependent variable is expressed in dollars per acre of farmland, while in the models for crops it is expressed in value of production per acre planted. In both cases nominal dollars are converted to 2006 dollars. The key variables of interest in Eq. 1 are the W variables for different k's, which represent degree-days and precipitation in a county c and year t. As mentioned, the "k" index simply denotes the various measures of weather we control for (e.g., winter degree-days, spring degree-days, etc.).[2] In different versions of (1), we examine weather variables from both a seasonal perspective and an annual perspective.

Equation 1 also includes a full set of county fixed effects, α_c. The appeal of including the county fixed effects is that they absorb all unobserved county-specific time invariant determinants of the dependent variable. For example, to the extent that agricultural soil quality is constant over time, the county fixed effects will account for differences in soil quality across counties. As such, the inclusion of county fixed effects will help mitigate the problem of omitted variables bias that has plagued some of the previous literature. Variants of this approach have been used in Deschênes and Greenstone (2007) and Schlenker and Roberts (2009).

The model above also includes a full set of year fixed effects, λ_t, that control for annual differences in the dependent variable that are common across counties. In particular the year fixed effects will capture the impact of changes in commodity prices on profits or value of production. An alternative is to directly control for prices, as in Kelly et al. (2005). The variables in the vector X_{ct} are the soil quality variables we described earlier. These variable change a little (but not much) from one year to another. Finally, the last term in Eq. 1, ε_{ct}, is a statistical error term.

3.2 Controlling for weather expectations

In the spirit of Kelly et al. (2005), we augment Eq. 1 to include proxies for farmer's expectations about weather. Viewing the climate as average weather, or more specifically, the distribution of weather, we control for expected weather. While expectations are not directly observed, we assume that expectations are derived from observing past weather. Specifically, we calculate the 5-year moving average of the weather variables and include them in Eq. 1, in addition to the realized degree-days and precipitation for a given year. Given the inclusion of county fixed-effects in Eq. 1, the statistical identification of the augmented equation requires that weather expectations are "changing" in the sense that the 5-year running averages must be time-varying (otherwise these variables would be perfectly collinear with the county fixed effects).[3] While the use of a 5-year moving average is arbitrary and necessarily a proxy for these expectations, the fact that this proxy is slow moving compared with the annual realizations of weather suggest that it is a reasonable proxy for expected weather in the farmer's decision making. The models for farm profits per acre as a function of weather and our proxy for expected weather are of the form:

$$y_{ct} = \alpha_c + \lambda_t + X_{ct}\theta + \sum_{k=1}^{K}\beta_k W_{kct} + \sum_{k=1}^{K}\delta_k C_{kct} + \varepsilon_{ct} \tag{2}$$

Where C_{kct} denotes the average of W_{kct} over the period t-5 to t-1. This approach follows from Kelly et al. (2005).

---

[2] We also considered models where degree-days and precipitation are modeled quadratically rather than linearly. These results were generally similar, although the statistical precision was greatly reduced.
[3] If the weather expectation variables are perfectly collinear with fixed effects, we cannot separately identify the coefficients on these variables.

3.3 Estimation

There are further issues related to Eqs. 1 and 2 that require attention. First, it is appropriate to estimate the equations using weights. Since the dependent variables are expressed in dollars per acres of farmland (or acres planted), there are two reasons to weight the models by the square root of acres of farmland (acres planted). First, the estimates of the value of farmland from counties with large agricultural operations will be more precise than the estimates from counties with small operations, and this weight corrects for the heteroskedasticity associated with the differences in precision. Second, the weighted mean of the dependent variables will equal the mean value of farmland per acre in California. It is also likely that the error terms are serially correlated over time. To account for this, we report "clustered" standard errors, where the clusters are defined by counties. This allows for arbitrary serial correlation over time within counties.[4]

3.4 Calculation of impacts

Once the gradients of the profits and yield functions are estimated, it is relatively straightforward to project the impacts of climate change. We simply combine the fixed-effect regressions estimates with the projected differences in degree-days and precipitations reported in Table 1.

We first present the calculation for the models that only control for realized weather (e.g., Eq. 1). For a given climate change model/scenario, the impact for county c is given by:

$$IMPACT_c = ACRES_c \times \left(\sum_k \hat{\beta}_k \Delta W_{kc} + \sum_k \hat{\delta}_k \Delta C_{kc} \right) \quad (3)$$

Where ΔW_{kc} is the predicted change in weather variable k in county c. These changes are specific to a climate change model, scenario, and horizon (i.e., short-run, medium-run, long-run). The variables $ACRES_c$ represent the average acres of farmland (or acres planted) during the sample period in county c. We need to "reweight" the calculations since the regression models are profits per acre (and value of production per acre). Finally, to obtain the impact for the state as a whole, we simply sum the county-specific impacts ($IMPACT_c$) across counties.

4 Data sources

4.1 Farm revenues, expenditures, and profits

The data on agricultural finances are from the 1987, 1992, 1997, and 2002 Censuses of Agriculture. Data from the Census of Agriculture are available for 1969, 1974, 1978 and 1982, as well as for years prior to 1969. However, the data from 1978 and 1982 are not comparable since the tabulation of the production expenditure variables

---

[4] Using a Moran statistic, we failed to reject the null hypothesis of no spatial correlation in the weather shocks that identify the models.

Climatic Change (2011) 109 (Suppl 1)

Table 1 County-level summary statistics on weather realizations and projections in California

	Historical:	Projected Changes: CCSM-B1		Projected Changes: CCSM-A2	
	1950–2005	2010–2039	2070–2099	2010–2039	2070–2099
	(1)	(2)	(3)	(4)	(5)
All Year					
Degree-Days (8–32)	2,601.5	387.0	668.2	309.4	1,282.8
Degree-Days (32+)	1.4	0.6	1.6	0.3	5.6
Total Precipitation (cm)	71.0	−7.8	−0.9	−6.0	−4.4
Winter					
Degree-Days (8–32)	204.6	49.3	108.1	29.1	233.4
Degree-Days (32+)	0.0	0.0	0.0	0.0	0.0
Total Precipitation	35.5	−2.4	−1.0	−3.2	−1.9
Spring					
Degree-Days (8–32)	762.1	129.0	182.6	93.8	368.3
Degree-Days (32+)	0.1	−0.1	0.0	−0.1	0.5
Total Precipitation 8.6	−0.7 -0.3	−0.2	−1.1		
Summer					
Degree-Days (8–32)	1,234.4	146.3	235.6	154.7	425.8
Degree-Days (32+)	1.3	0.7	1.6	0.4	5.1
Total Precipitation	2.0	−0.4	−0.3	−0.2	−0.8
Fall					
Degree-Days (8–32)	400.4	62.4	141.9	31.8	255.4
Degree-Days (32+)	0.0	0.0	0.0	0.0	0.0
Total Precipitation	25.0	−4.4	0.7	−2.5	−0.6

Predictions are from the National Center for Atmospheric Research's Community Climate System Model, version 3 (CCSM3), under IPCC Special Report on Emissions Scenarios (SRES) scenarios B1 and A2. Calculations are based on daily record data for the period 1950–2005 and 2010–2099. Winter is defined as the first quarter of the year; the other seasons are correspondingly defined. Degree-Days 8–32 denotes base 8°C–32°C degree-days and Degree-Days 32+ denotes base 32°C+degree-days. See the text for more details

in these years is different than in the other years.[5] All farms and ranches from which $1,000 or more of agricultural products are produced and sold, or normally would have been sold, during the census year are required to submit a census form. For confidentiality reasons, counties are the finest geographic unit of observation that is publicly available in the Census of Agriculture.[6]

From these data we construct a variable for county-level agricultural profits per acre of farmland. The numerator is constructed as the difference between the market value of agricultural products sold and total production expenses across all farms in a county. Production expenses exclude the value of or return from land. The denominator includes acres devoted to crops,

---

[5] We also estimated models pooling the 1969–1974 and 1987–2002 data, but the counterfactual results (the predicted climate change impacts) appeared highly counterintuitive. For that reason, we focus on the 1987–2002 data.

[6] We attempted to develop data at the ZIP/Postal code level but the ZIP code is for the mailing address of the farm, which may be in a totally different location from the farm. Furthermore, the United States Department of Agriculture has a policy of not releasing cost and value figures at the sub-county level, even if requested to do so on a reimbursable basis. Consequently, county-level data is the smallest geographic unit available.

pasture, and grazing. The revenues component measures the gross market value before taxes of all agricultural products sold or removed from the farm. It excludes income from participation in federal farm programs, labor earnings off the farm (e.g., income from harvesting a different field), or nonfarm sources. Thus, it is a measure of the revenue produced with the land.[7]

Total production expenses are the measure of costs. They include expenditures by landowners, contractors, and partners in the operation of the farm business. This covers all variable costs (e.g., seeds, labor, and agricultural chemicals/fertilizers). It also includes measures of interest paid on debts and the amount spent on repair and maintenance of buildings, motor vehicles, and farm equipment used for farm business. Its main limitation is that it does not account for the rental rate of the portion of the capital stock that is not secured by a loan, so it is only a partial measure of farms' cost of capital.[8] Just as with the revenue variable, the measure of expenses is limited to those that are incurred in the operation of the farm so, for example, any expenses associated with contract work for other farms is excluded.[9]

Another noteworthy issue regarding the farm profit variable is that the Census of Agriculture defines revenues as the sales of agriculture products in a calendar year, irrespective of the year of production. Since farmers may engage in inventory management decisions that allow them to smooth their income across years, it is important to derive a profit measure that accounts for this optimal storage behavior. Since it is based on revenues from the sales of agriculture products by calendar year and irrespective of the year of production, the farm profit measure derived from the Census of Agriculture embodies optimal storage decisions.

In practice, this means that, for example, a bushel of corn that is stored this year will be sold in a subsequent year. Analogously, drawing down inventories this year will reduce the crops available for sale in future years. The point is that the full impact of a weather realization on profits can only be observed over periods longer than a year because inventory decisions allow farmers to smooth the impacts over several years. This compensatory behavior suggests that measures of profits, which include revenues from sales of products regardless of their date of production, should depend on current and lagged weather variables. The inclusion of lagged weather variables controls for the possibility that the quantities sold by a farmer in any given year reflects the full history of weather realizations through his storage decisions.

More generally, the structural relationship between storage decisions and weather realizations is a complicated process. It depends on several factors including the weather realization's expected impact on current and future crop prices, storage costs, the length of time before a crop can be stored without spoiling, and the interest rate. A careful examination of this behavior would require a long panel data set with detailed information on farmer behavior. Such an analysis with lagged weather models is beyond the scope of this paper. Deschenes and Greenstone (2011) report simple dynamic estimates of the effects

---

[7] An exception is that it includes receipts from placing commodities in the Commodity Credit Corporation loan program. These receipts differ from other federal payments because farmers receive them in exchange for products.

[8] In particular, interest payments are the only measure of the rental cost of capital in the censuses. Thus, our measure understates the cost of capital by not accounting for the opportunity cost of the portion of the capital stock that is not leveraged. Further, our measure of agricultural profits does not account for labor costs that are not compensated with wages (e.g., the labor provided by the farm owner).

[9] The censuses contain separate variables for subcategories of revenue (e.g., revenues due to crops and dairy sales), but expenditures are not reported separately for these different types of operations. Consequently, we cannot provide separate measures of profits by these categories and instead focus on total agriculture profits.

of weather shocks on farm profits for the U.S. as a whole. Also see Fisher et al. (2010) for a discussion of storage in this context.

4.2 Crop production and yields

Annual county-level data on production, value of production, and acres planted for the period 1980–2005 were taken from the County Agricultural Commissioners Data.[10] This summary, which is published annually, is based on the annual Crop Reports compiled by the California County Agricultural Commissioners. These reports provide the most detailed annual data available on agricultural production by county. Basic data collected by the Agricultural Commissioners and their staffs are compiled from many sources. Sources vary from county to county. Examples of data sources include growers' surveys, regulatory and inspection data, shipment data, and industry assessments. Price data reflect the average price received by growers, except fresh market fruits and vegetables, which are on a packed and ready-to-ship basis.

4.3 Soil quality data

Like most previous analyses, we rely on the National Resource Inventory (NRI) for our measures of soil quality and characteristics. The NRI is a massive survey of soil samples and land characteristics that is conducted in census years. We follow the convention in the literature and use a number of soil quality variables as controls in the equations for profits and yields, including measures of susceptibility to floods, soil erosion (K-Factor), slope length, sand content, irrigation, and permeability. County-level measures are calculated as weighted averages across sites used for agriculture, where the weight is the amount of land the sample represents in the county. Although these data provide a rich portrait of soil quality, we suspect that they are not comprehensive. To this end, we consider models that include county fixed effects to capture these effects.[11]

4.4 Historical weather data

The weather data are drawn from the National Climatic Data Center (NCDC) Summary of the Day Data (File TD-3200). The key variables for our analysis are the daily maximum and minimum temperature, as well as the total daily precipitation for the period 1920–2005. To ensure the accuracy of the weather readings, we developed a weather station selection rule. Specifically, we dropped all weather stations that were not operating consecutively for a full year, though we allow stations to enter and exit the sample over time. The acceptable station-level data is then aggregated at the county level by taking an inverse-distance weighted average of all grid points that lie within 200 kilometers (km) of each county's centroid.[12]

With this data at hand, we now have a complete daily time-series starting January 1, 1920, and ending December 31, 2005, for every county in California, with valid measurements for daily minimum and maximum temperature, and total precipitations. We

---

[10] National Agriculture Statistics Service, County Agricultural Commissioners' Data. www.nass.usda. gov/Statistics_by_State/California/Publications/AgComm/indexcac.asp

[11] When relevant, we use the STATA command 'impute' to fill-in the county/years where some soil characteristics are missing.

[12] Ideally, the weather-station level data would be spatially interpolated in a way that recognizes the location of farmland within a county. To the best of our knowledge information on within-county farmland distribution is not available going back to the 1950s.

use the daily data to construct measures of "exposure" that follow from the agronomic literature. Agronomists have shown that plant growth depends on the cumulative exposure to heat and precipitation during the growing season. As such, monthly average temperatures may be poor predictors of agricultural outputs since they do not capture nonlinearities, and the differential impact of exposure across the temperature distribution.

The standard agronomic approach for modeling temperature is to convert daily temperatures into degree-days, which correspond to heating units (Hodges 1991; Grierson 2002). It is likely that the effect of heat accumulation is nonlinear since temperature must be above a threshold for plants to absorb heat and below a ceiling as plants cannot absorb extra heat when temperature is too high. These thresholds or bases vary across crops, but we follow Ritchie and NeSmith's (1991) suggested characterization for the entire agricultural sector and use a base of 8°Celsius (C) (46°F) and a ceiling of 32°C (90°F). Specifically, the degree-days variable is calculated so that a day with a mean temperature: below 8°C contributes 0 degree-days; days between 8°C and 32°C contributes the number of degrees C above 8 degree-days; above 32°C contributes 24 degree-days.

Ritchie and NeSmith (1991) also discuss the possibility of a temperature threshold at 34°C, above which increases in temperature are harmful. In addition, Schlenker and Roberts (2009) also propose other definitions of "harmful" degree-days. We consider this possibility by including in some models a variable for degree-days of base 32°C (90°F), without an upper limit. This variable will in effect allow the effect of degree-days on agricultural output to vary depending on its location in the temperature distribution. In that case we also truncate the 8–32 degree-days variable at 32°C.

4.5 Climate change predictions

In this paper we estimate the effect of climate change on agriculture using climate change predictions from the National Center for Atmospheric Research (NCAR) Community Climate System Model (CCSM), based on scenarios B1 and A2. These IPCC scenarios are derived from "storylines" that describe the relationships between the forces driving greenhouse gas and aerosol emissions and their evolution during the twenty-first century for large world regions and globally. Each storyline represents different demographic, social, economic, technological, and environmental developments that diverge in increasingly irreversible ways. The A2 storyline and scenario family is characterized with a very heterogeneous world with continuously increasing global population and regionally oriented economic growth ("business as usual"). The B1 storyline and scenario family features a convergent world with the same global population as in the A1 storyline but with rapid changes in economic structures toward a service and information economy, with reductions in material intensity, and the introduction of clean and resource-efficient technologies. As such, temperature increases in the B1 scenario are more moderate than in the A2 scenario.[13]

The model's prediction are then adjusted for bias correction and are spatially downscaled (BCSD) to the level of California counties. The models provide us with daily minimum and maximum temperature and precipitation predictions at several grid points throughout California for the period 1950–2099. The grid point data produced by the CCSM model are assigned to counties by taking an inverse-distance weighted average of all grid points that lie within 100 km of each county's centroid.

---

[13] A more detailed description of the SRES scenarios is available at: http://sedac.ciesin.columbia.edu/ddc/sres/

These daily predicted temperature realizations are used to develop estimates of predicted end of century climate. We utilize the historical predictions to account for the possibility of model error. In particular, we undertake the following multiple step process to correct for model-error. For simplicity, consider the end of century predicted change (2070–2099) in daily temperature. Other weather variables or prediction horizon are processed similarly:

1. Using the CCSM model data, we calculate the daily mean temperature for each of the year's 365 days during the baseline period 1950–2005, and the future period 2070–2099. These are denoted as $T_{ct,1950-2005}^{CCSM}$ and $T_{ct,2070-2099}^{CCSM}$, respectively, and where c indicates county, and t references one of the 365 days in a year (e.g., January 15).
2. For each county, we calculate the predicted change in temperature for each of the 365 days in a year as the difference in the mean from the 2070-2099 and 1950–2005 periods. This is represented as $\Delta T_{ct}^{CCSM} = T_{ct,2070-2099}^{CCSM} - T_{ct,1950-2005}^{CCSM}$.
3. Using the historical (actual) weather data, we calculate the county-specific daily mean temperature for each of the 365 days over the 1950–2005 period. This yields $T_{ct,1950-2005}^{ACTUAL}$.
4. The predicted end of century climate for each day of the year is equal to $T_{ct,1950-2005}^{ACTUAL} + \Delta T_{ct}^{CCSM}$. The resulting distribution of temperatures is the CCSM predicted end of century distribution of daily temperatures that is utilized in the subsequent analysis, and from which we calculate our various measures of degree-days

From the daily prediction data we can compute the same degree-day variables as we construct and analyze using the historical record period 1950–2005. The description of these variables is discussed above.

Before proceeding, it is important to underscore that the validity of the paper's estimates of the impacts of climate change depend on the validity of the climate change predictions. The state of climate modeling has advanced dramatically over the last several years, but there is still much to learn, especially about the role of greenhouse gases on climate (Karl and Trenberth 2003). Thus, CCSM predictions should be conceived of as realizations from a superpopulation of models and scenarios. The sources of uncertainty in these models and scenarios are unclear, so it cannot readily be incorporated into the below estimates of the impacts of climate change.

4.6 Summary statistics

Table 1 reports the averages (across counties and years) of the seasonal degree-days and precipitation variables. For convenience, the annual averages are also reported. The "Historical" column (1) shows the 1950–2005 averages of each of the listed variables for the 58 counties in California. Ideally these would be computed as weighted averages, where the weight would be given by acres of farmland. Unfortunately, acreage data are not available for all years, so the statistics reported are simply unweighted. The entries reveal that on average, the typical California county receives 71 cm (cm) of rain during the course of the year and about 2,600 degree-days between 8°C and 32°C. In addition, the typical county is exposed to 1.4 base 32°C degree-days. Clearly, the distribution of degree-days and rainfall is not uniform across the year: Most of the rainfall (e.g., 61 out of 71 cm) occurs in the winter and fall months, and half of the degree-days are in the summer months.

The "Projected Changes" columns show the predicted levels and differences from historical baselines for the degree-days and precipitation variables associated with the CCSM model, under scenarios B1 and A2. We consider two horizons: the short-run (2010–2039), and the long-run (2070–2099). The entries are simple averages over the 58 counties in California, over the 30 years in each period. As stated before, the scenarios vary in their predictions of future climates.

According to the B1 scenario, the average county in California is projected to receive an additional 668 base 8°C–32°C degree-days and 1 fewer centimeter of rain during the course of the typical year by the end of the century (column (3)). In the shorter-term, the B1 scenarios predicts a reduction of 8 cm in annual rainfall (mostly due to predicted lower Fall rainfall) and an increase in annual degree-days of about 387 (column (2)). The predictions under scenario A2 are more marked: The model predicts that the typical county will receive an additional 1,300 degree-days (8°C–32°C), 5.6 degree-days (base 32°C), and 4.4 fewer centimeters over the course of a year, on average, between 2070 and 2099 (column (5)). The increase in base 8° C–32°C degree-days is large, representing a 49% increase relative to the 1950–2005 baseline. The reduction in annual precipitation is of smaller magnitude, at 6%.

It is also informative to consider seasonal disparities in the climate change predictions. Both models show an increase in degree-days in all seasons, with the largest proportionate increases typical in the fall and winter months (although in absolute terms the summer month will gain the most degree-days). Again, the patterns for seasonal precipitations are less clear-cut. The A2 scenario predicts less precipitation in all seasons by the end of the century, with the largest proportionate declines in the summer months. The corresponding long-run prediction of the B1 scenario is a small increase in seasonal rainfall, except during the summer months. To the extent that various crops are more or less dependent on seasonal degree-days or rainfall, the impact of climate change on California's agriculture is likely to be very heterogeneous across crops.

Table 2 presents summary statistics about the financial activities of farms in California. These data are taken from the 1987–2002 Censuses of Agriculture. Averages are calculated separately for each year across the 58 counties of California. The top panel shows state totals, while the bottom panel reports county averages. The first row shows the total farmland acreage has been slightly declining over time, from 30.6 million acres to 27.6 million. Total profits (defined as revenues minus production expenditures) fluctuate over time with no clear trend, around an average of $5.5 billion (in 2006 dollars). Similar patterns are observed in the county-level averages: Profits were the highest in 1997, at $129 million and the lowest in 1992 at $73 million, on average per county. In terms of profits per acre, the average over the census years is $195 dollars per acre farmed, with a peak of $271 in 1997.

Table 2 County-level summary statistics on farm revenues, expenditures, and profits, 1969–2002

	1987	1992	1997	2002
Aggregate state totals				
Acres of Farmland (Mil.)	30.60	28.98	27.70	27.58
Total Sales ($Mil.)	21,996.0	22,884.7	28,037.5	28,794.9
Production Expenditures ($Mil.)	17,387.9	18,628.7	20,545.8	22,965.9
Total Profit ($Mil.)	4,608.1	4,255.9	7,491.8	5,829.0
County averages				
Sales ($Mil.)	379.2	394.6	483.4	496.5
Production Expenditures ($Mil.)	299.8	321.2	354.2	396.0
Profits ($Mil.)	79.4	73.4 129.2	100.5	
Profits Per Acre (1$/acre)	150.6	146.9	270.5	211.4

Averages are calculated for a balanced panel of 58 counties over 4 census years (1987, 1992, 1997, 2002)
All dollar values are in 2006 constant dollars

5 Results

5.1 In-sample estimates

Table 3 shows the basic estimation results from Eqs. 1 and 2. These are simply the estimated coefficients for the profits models based on the historical data. The dependent variable is profits per acre, while the controls are the degree-days and precipitation variables, soil characteristics, year fixed effects and county fixed effects. Columns (1a)-(1c) reports from the model with annual realizations of weather, while columns (2a)-(2c) includes annual realization of degree-days and precipitation, and the annual values of the weather expectation proxies we defined earlier. Columns (1a) and (2a) report estimates from the baseline specification that only controls for precipitation and base 8–32 ° C degree-days and underlies Table 4.[14] Columns (1b) and (2b) augment the baseline specification to account for separate effects for degree-days above 32 ° C, defined on the basis of average daily temperature. This specification underlies Table 5 below. Finally, columns (1c) and (2c) also augment the baseline specification to account for separate effects for degree-days above 32 ° C, this time defined on the basis of maximum daily temperature. Unfortunately, since the CCSM predictions that are available to us only pertain to the daily average temperature, we cannot use specifications (1c) and (2c) for our predicted climate change impacts on agricultural profits.

Foreshadowing the predicted climate change impacts reported in the coming tables, none of the coefficients on the degree-days variables are individually or jointly significant. Nevertheless the estimates are similar across specification. For example, one additional degree-day reduces profits per acre by 0.01 to 0.02 dollars per acre, depending on the specification chosen. Another key point is that it is empirically very difficult to measure the profit impact of extreme temperature degree-days, at least when it is defined on the basis of daily average temperature. The corresponding estimates in (1b) and (2b) are very imprecise. A possible solution to this is to consider measures of extreme temperature based on daily maximum temperature, since it has stronger empirical support. Nevertheless, these estimates are also statistically imprecise (see 1c and 2c). Finally, another notable result is that the effects of the weather expectation proxies for degree-days on profits per acre are generally larger in magnitude than the corresponding effects of annual weather. This is consistent with the notion that adjustment to short-run (i.e. annual) shocks is less costly to farmers than more permanent changes in the weather distribution.

5.2 Predicted impacts of climate change on aggregate farm profits in California

Tables 4 and 5 present estimates of the impact of the A2 and B1 climate change scenarios on annual agricultural profits (Eq. 3). These results are derived from the estimation of versions of Eqs. 1 and 2.

Table 4 presents our first set of impact projections for aggregate agricultural profits. The models include linear controls for annual degree-days (defined over the 8–32 ° C

---

[14] Following an insightful suggestion from an anonymous referee, we also considered models where the effect of the weather (and weather expectation) variables are allowed to vary across groups of counties in California. Specifically, we estimate separate effects across the 8 California Agricultural Statistical Districts (www.cdfa.ca.gov/phpps/fcm/pdfs/publications/2006cas-all.pdf). In the case of degree-days, we fail to reject the null hypothesis while the same hypothesis is rejected for the precipitation variables. However, incorporating agricultural district-varying weather (and weather expectation) effects in our predicted climate change impacts calculations leads to qualitatively similar impacts (that are generally smaller in magnitude).

Table 3 Estimated impact of weather and weather expectations on farm profits per acre

	Equation 1—weather only			Equation 2—weather and expected weather		
	(1a)	(1b)	(1c)	(2a)	(2b)	(2c)
Weather Variables						
Annual Degree Days (8–32 °C)	−0.0103 (0.0539)	−0.0137 (0.0523)	−0.0172 (0.0514)	−0.0039 (0.0468)	−0.0081 (0.0464)	−0.0155 (0.0432)
Annual Precipitation (cm)	−2.7950 (1.2457)	−2.7311 (1.2851)	−2.7948 (1.2137)	−2.5446 (1.2881)	−2.5016 (1.3046)	−2.6244 (1.2976)
Annual Degree Days (32 °C+)*	–	0.9988 (1.1422)	0.0084 (0.1123)	–	0.9494 (1.1834)	0.0066 (0.0976)
Weather Expectation Variables						
Annual Degree Days (8–32 °C)	–	–	–	−0.0922 (0.1133)	−0.0785 (0.1182)	−0.0550 (0.1061)
Annual Precipitation (cm)	–	–	–	−1.9000 (1.0887)	−1.8054 (1.1452)	−2.0056 (1.0536)
Annual Degree Days (32 °C+)*	–	–	–	–	0.3853 (1.5719)	−0.2000 (0.3845)
County fixed-effects	Yes	Yes	Yes	Yes	Yes	Yes
Year effects	Yes	Yes	Yes	Yes	Yes	Yes
Soil characteristics	Yes	Yes	Yes	Yes	Yes	Yes
Observations	229	229	229	229	229	229

The dependent variable is agricultural profits per acre. Degree-Days 8–32 denotes base 8°C–32°C degree-days, and Degree-Days 32+ denotes base 32°C+ degree-days as defined in the text. 32°C+Degree-days in columns (1b) and (2b) defined on basis of daily average temperature. 32°C+Degree-days in columns (1c) and (2c) defined on basis of daily maximum temperature. Standard errors in parentheses are clustered at the county level. See the text for more details

range) and annual precipitation. In addition, all models control for soil characteristics (including access to irrigation), year effects and county fixed-effects. The 4 panels in Table 4 correspond to different horizons for the predictions. Odd-numbered columns correspond to scenario A2 and even-numbered correspond to scenario B1. Columns (1)–(4) correspond to models where only the annual realizations of weather are included to predict farm profits. Columns (5)–(8) correspond to models that include both annual realizations of weather and the proxy for expected weather. In each column we can also decompose the overall impact into its component due to change in the distribution of temperature and the distribution of precipitations. In addition, the standard errors associated with each point estimate are reported in parentheses. Other specification details are noted at the bottom of the table.

Taken as a whole, the information in Table 4 reveals two keys findings. First, in the short-run (i.e. 2010–2039), aggregate profits are predicted to remain relatively on par with historical levels since the predicted changes indicate increase or decrease of 6% or less. Notably, these estimates are not statistically different from zero. Second, aggregate agricultural profits are predicted to be decreased by the end of the century (2070–2099). In all specifications, the total impact is negative, ranging from a reduction of 2% to a reduction of 58% when compared to the historical average total agricultural profits in the state. In this case, the projections from the A2 scenario are more pronounced. The decomposition of the

Table 4 Predicted impacts for aggregate agricultural profits in California (Million of 2006 dollars) under CCSM 3 B1 and A2 scenarios; models with annual realizations of weather and expected weather, degree-days 8–32°C and precipitation only

	2010–2039		2070–2099		2010–2039		2070–2099	
	(1)	(2)	(3)	(4)	(5)	(6)	(7)	(8)
	A2	B1	A2	B1	A2	B1	A2	B1
Total Impact on Farm Profits ($2006 Mil.)	249.1 (410.9)	353.3 (508.5)	−110.7 (1957.7)	−159.4 (1055.5)	−340.6 (1114.5)	−363.7 (1387.8)	−3189.8 (4940.6)	−1833.1 (2626.9)
Percent effect	4.5	6.4	−2.0	−2.9	−6.1	−6.6	−57.5	−33.1
Due to change in degree-days (8–32)								
Annual Weather	−95.4 (498.6)	−119.8 (626.4)	−390.2 (2039.6)	−204.5 (1069.0)	−36.4 (432.4)	−45.7 (543.2)	−148.9 (1768.7)	−78.0 (927.0)
Weather Expectations	–	–	–	–	−852.0 (1047.9)	−1070.4 (1316.4)	−3485.5 (4286.6)	−1826.8 (2246.6)
Due to change in precipitation								
Annual Weather	344.5 (153.6)	473.2 (210.9)	279.5 (124.6)	45.1 (20.1)	313.7 (158.8)	430.8 (218.1)	254.5 (128.8)	41.0 (20.8)
Weather Expectations	–	–	–	–	234.2 (134.2)	321.7 (184.3)	190.0 (108.9)	30.6 (17.6)
County fixed-effects	yes	yes	yes	yes	yes	yes	yes	yes
Year effects	yes	yes	yes	yes	yes	yes	yes	yes
Soil characteristics	yes	yes	yes	yes	yes	yes	yes	yes
Observations	229	229	229	229	229	229	229	229

Impacts derived from the estimates of the impact of weather/weather expectations on farm profits in Columns (1a) and (2a) in Table 3 above. Percent effect corresponds to the predicted change in aggregate farm profits divided by average historical aggregate farm profits. Standard errors in parentheses are clustered at the county level. See the text for more details

Table 5 Predicted impacts for aggregate agricultural profits in California (Million of 2006 dollars) under CCSM 3 B1 and A2 scenarios; Models with annual realizations of weather and expected weather, separating 8–32°C and 32°C+degrees-days

	2010–2039		2070–2099		2010–2039		2070–2099	
	(1) A2	(2) B1	(3) A2	(4) B1	(5) A2	(6) B1	(7) A2	(8) B1
Total Impact on Profits ($2006 Mil.)	232.2 (410.2)	339.3 (515.9)	30.6 (2011.4)	−157.1 (1081.6)	−247.2 (1140.5)	−240.7 (1439.3)	−2366.2 (5177.5)	−1560.3 (2769.9)
Percent effect	4.2	6.1	0.6	−2.8	−4.5	−4.3	−42.7	−28.1
Due to change in degree-days (8–32)								
Annual Weather	−115.2 (490.5)	−145.2 (617.9)	−448.0 (1906.8)	−247.5 (1053.3)	−73.6 (421.5)	−92.7 (530.9)	−286.2 (1638.4)	−158.1 (905.0)
Weather Expectations	–	–	–	–	−712.8 (1073.3)	−897.9 (1352.0)	−2771.0 (4172.2)	−1530.7 (2304.7)
Due to change in degree-days (32+)								
Annual Weather	6.5 (7.5)	16.1 (18.8)	201.6 (234.4)	45.7 (53.1)	5.9 (7.4)	14.8 (18.5)	185.0 (230.7)	41.9 (52.3)
Weather Expectations	–	–	–	–	2.4 (9.8)	6.0 (24.5)	75.1 (306.5)	17.0 (69.4)
Due to change in precipitation								
Annual Weather	341.0 (153.8)	468.3 (211.3)	276.7 (124.8)	44.6 (20.1)	308.4 (160.8)	423.5 (220.9)	250.2 (130.5)	40.3 (21.0)
Weather Expectations	–	–	–	–	222.5 (141.2)	305.6 (193.9)	180.6 (114.5)	29.1 (18.5)

Impacts derived from the estimates of the impact of weather/weather expectations on farm profits in Columns (1b) and (2b) in Table 3 above (N=229). Percent effect corresponds to the predicted change in aggregate farm profits divided by average historical aggregate farm profits. Standard errors in parentheses are clustered at the county level. See the text for more details

Climatic Change (2011) 109 (Suppl 1)

total impact into the degree-days/precipitation components reveals that the negative impact on farm profits is mostly attributable to increase degree-days. It is also notable that the largest impacts come from the expected weather variables (the 5-year moving average of weather). These results suggest that deviations from these farmer's expectations are costlier than year-to-year deviations in weather.

Table 5 replicates Table 4, but adds a variable for degree-days above 32°C in all models. Recall from Table 1 that the historical average of such degree-days is only 1.4 per year per county, and as such the statistical precision of models that include these variables will necessarily be lower. The main patterns in Table 4 are also observed in Table 5: Projections for the end of the century are negative and larger in magnitude than projections for the coming three decades. The A2 scenario predicts larger reductions in profits than the B1 scenario. In addition, the effects of the base 32°C and above degree-days are positive and explain the slightly smaller magnitude of the predicted impacts in Table 5. For the most part, the inclusion of harmful degree-days comes at the cost of reduced statistical precision while not leading to meaningful changes to the aggregate predictions.

5.3 Predicted impacts on specific crops

We have also explored the effect of predicted climate change on the quantity and value of crops produced in California. Based on the Census of Agriculture data for 1987–2002, crop sales represent about 70% of total farm sales in California, so it is an important sector, but sales of livestock and dairy products are also important. Another motivation for examining yields in individual crops is to assess the limitations of the profits approach we propose in the paper. Large declines in value per acre would suggest that the profit results may be biased (relative to the preferred long run measure) by short run price changes. Although farmers cannot switch crops in response to weather shocks, they are able to undertake some adaptations, although not to the same extent as it is possible in response to permanent climate change.

The analysis is based on the 15 largest-grossing crops (perennials and annuals) in California over the 1980–2005 period. For this analysis, greenhouse production was omitted as it is less clear how climate change will affect this type of production. The crops are listed alphabetically in the first row. Importantly, some of these crops are perennial (almonds, avocados, grapes, lemons, oranges, pistachios, and walnuts) and annuals (broccoli, hay, lettuce, strawberry, and tomatoes).

The results of the analysis are presented in Tables 6, 7 and 8. Table 6 reports the estimated coefficients associated with the weather (columns (2) and (3)) and weather expectation variables (columns (4) and (5)) considered in Table 4. The regression models are for the value of crop production per acre planted (i.e. dollar produced per acre) and include year fixed effects, county fixed effects and are weighted by acres planted in each crop. Table 7 presents the same information, except that the models are for crop yields per acre planted (i.e. ton produced per acre).

Column (1) in Table 6 reports value of production per acre planted. The dollar yields per acre also exhibit a wide range, from about $750 per acre planted for hay to $34,000 for strawberries. Columns (2)–(5) report the coefficients on the weather and weather expectation variables, along with their standard errors, from separate regression models for each crop. Few of the coefficients are statistically significant, although some interesting results appear. For example, table grapes revenues are positively related to degree-days while wine grapes revenues are negatively related. The same finding for grapes is reported in Table 7, where the dependent variable is physical yield (i.e. tons per acre planted).

Table 6 Estimated impact of weather and weather expectations on value of production in dollars per acre (15 Largest Crops)

Crop	Dependent Variable=Dollar Per Acre					
	Mean	Coefficients on Weather Variables:		Coefficients on Weather Expectation Variables:		Obs:
		Degree Days (8–32)	Precipitation (cm)	Degree Days (8–32)	Precipitation (cm)	
	(1)	(2)	(3)	(4)	(5)	(6)
Almonds	2,305.9	−0.20 (0.16)	−0.07 (2.29)	0.17 (0.65)	2.81 (7.16)	450
Avocados	4,806.3	−1.63 (1.24)	−22.28 (27.05)	−1.16 (0.50)	−34.28 (40.98)	259
Broccoli	4,253.2	−0.61 (0.79)	−6.23 (10.17)	1.04 (1.47)	−6.77 (12.47)	291
Cotton	1,478.7	0.06 (0.09)	−1.47 (1.98)	0.47 (0.16)	3.27 (2.77)	229
Grapes (table)	5,079.7	2.26 (0.52)	0.65 (8.53)	5.40 (2.16)	15.26 (40.31)	325
Wine Grapes	3,601.7	−0.89 (0.44)	−9.08 (5.95)	1.19 (1.50)	27.44 (9.95)	735
Hay	760.6	0.04 (0.03)	0.94 (0.51)	−0.02 (0.09)	1.99 (1.04)	1277
Lemons	8,125.3	0.98 (0.77)	−18.32 (18.52)	2.38 (2.22)	−20.69 (42.82)	279
Lettuce	6,034.9	−0.08 (1.17)	−17.32 (5.90)	0.43 (1.41)	−2.22 (16.50)	361
Oranges	5,505.4	−0.02 (0.74)	−12.61 (12.65)	−0.92 (1.51)	−40.79 (25.41)	297
Pistachios	3,402.0	−1.45 (0.43)	−28.75 (15.72)	0.62 (1.86)	18.30 (22.73)	208
Rice	1,096.8	−0.04 (0.04)	−0.10 (0.46)	−0.04 (0.16)	−2.19 (3.72)	387
Strawberries	33,659.5	2.54 (3.06)	−156.14 (46.47)	−16.64 (6.72)	127.12 (116.16)	333
Tomatoes	3,117.2	0.52 (0.59)	5.42 (10.00)	−0.60 (0.74)	30.98 (27.03)	549
Walnuts	2,040.3	−0.19 (0.07)	−0.83 (1.37)	−0.15 (0.30)	4.06 (3.87)	810

The dependent variables are crop revenues per acre planted. Degree-Days 8–32 denotes base 8°C–32°C degree-days. Standard errors in parentheses are clustered at the county level. See the text for more details

Table 8 reports the predicted climate change impacts, based on the estimation of the same models underlying Tables 6 and 7. Column (1) first reports the state total value of production for each crop, averaged over the 1980–2005 period. During this period, the highest value crops are cotton, grapes (table and wine), hay, lettuce, almonds, and oranges, each with a total value exceeding $1 billion.

Columns (2)–(3) report the climate change impact predictions for each crop according to CCSM3, scenario A2 for the period 2070–2099. Column (2) reports the predicted change in crop sales and is computed in the same manner as the predicted impacts in Tables 4 and 5. Column (3) shows the 'percentage impact' that are computed by normalizing the predicted change in value of production for each crop by the 1980–2005 average value of production for each of the crops. Column (4) also reports a 'percentage impact', this time computed by normalizing the predicted change in physical yield per acre for each crop by the 1980–2005 average yield per acre for each of the crops. For each estimate, the county-clustered standard errors are reported in parentheses.

Perhaps not surprisingly, the main message of Table 8 is that climate change is predicted to have a heterogeneous impact on value of production and yields in California. For some crops, the value of production is projected to increase by as much as 40% (i.e., cotton, table grapes and lemons), while for others (i.e., avocados and strawberries), the value of production is projected to decrease by 40% or more. However, most of these estimates are

Table 7 Estimated impact of weather and weather expectations on yields in tons per acre (15 Largest Crops)

Crop	Dependent Variable=Tons Per Acre					
	Mean	Coefficients on Weather Variables:		Coefficients on Weather Expectation Variables:		Obs:
		Degree Days (8–32)	Precipitation (cm)	Degree Days (8–32)	Precipitation (cm)	
	(1)	(2)	(3)	(4)	(5)	(6)
Almonds	0.7	0.00002 (0.00003)	0.00050 (0.00049)	−0.00009 (0.00021)	−0.00177 (0.00187)	450
Avocados	2.8	−0.00074 (0.00050)	0.00157 (0.01200)	0.00037 (0.00020)	−0.00624 (0.01536)	259
Broccoli	5.9	−0.00032 (0.00070)	−0.00121 (0.01196)	0.00202 (0.00173)	−0.02102 (0.01860)	291
Cotton	0.6	0.01514 (0.02211)	−1.15608 (0.65233)	−0.11656 (0.07284)	−1.86934 (1.77100)	229
Grapes (table)	8.3	0.00185 (0.00044)	0.05228 (0.01838)	0.00127 (0.00151)	0.02205 (0.02685)	325
Grapes (wine)	7.0	−0.00075 (0.00039)	−0.00048 (0.00383)	0.00065 (0.00079)	0.01492 (0.00647)	720
Hay	5.3	−0.00087 (0.00026)	0.00187 (0.00221)	−0.00141 (0.00080)	0.00032 (0.01210)	1277
Lemons	14.6	0.00000 (0.00093)	−0.00836 (0.03488)	0.00053 (0.00478)	−0.11295 (0.06742)	279
Lettuce	14.3	−0.00465 (0.00607)	0.02767 (0.05675)	−0.00623 (0.01174)	−0.10570 (0.15670)	361
Oranges	11.7	−0.00003 (0.00139)	−0.00561 (0.02168)	−0.00378 (0.00336)	−0.05529 (0.02676)	295
Pistachios	1.0	−0.00037 (0.00014)	−0.00633 (0.00399)	−0.00016 (0.00042)	0.00052 (0.00511)	208
Rice	3.6	−0.00014 (0.00013)	−0.00019 (0.00222)	0.00060 (0.00035)	−0.00492 (0.00357)	386
Strawberries	23.4	0.00447 (0.00281)	−0.01300 (0.03818)	0.00290 (0.00594)	0.00664 (0.20537)	333
Tomatoes	25.1	0.00001 (0.00121)	0.00677 (0.02064)	−0.00222 (0.00148)	0.05437 (0.04617)	549
Walnuts	1.3	0.00001 (0.00009)	0.00096 (0.00070)	0.00002 (0.00021)	0.00066 (0.00144)	809

The dependent variables are crop yields (ton) per acre planted. Degree-Days 8–32 denotes base 8°C–32°C degree-days. Standard errors in parentheses are clustered at the county level. See the text for more details

statistically insignificant, although there are a few exceptions (avocados, cotton, table grapes, strawberries). As such, these results must be interpreted with caution. Further, the impacts on yields per acre are also generally statistically insignificant. Nevertheless, in the case of a few crops, we observe large predicted declines in physical yields and much smaller predicted declines in value of production (lettuce, pistachios). This indicates that short-run variation in farm revenues or profits may be caused by short-run fluctuations in prices. However, given the overall lack of statistical significance of the estimates underlying Tables 4, 5, and 8, it is difficult to assess how the predicted climate change

Table 8 Predicted impacts of climate change on value of crop production and yield, under CCSM scenario A2, 2070–2099 (Percentage impacts relative to 1980–2005 baseline)

| Crop | 1980–2005 Average ($Mil.) | Projected Impacts: CCSM-A2, 2070–2099 ||| Obs: |
| | | State total value ($Mil.) | Percent Effect | Percent Effect (yield only) | |
	(1)	(2)	(3)	(4)	(5)
Almonds	1,113.2	−25.5 (407.6)	−2.3 (36.6)	−16.5 (49.3)	450
Avocados	326.9	−224.2 (92.3)	−68.6 (28.2)	−17.2 (23.2)	259
Broccoli	481.2	62.3 (281.6)	12.9 (58.5)	38.8 (46.3)	291
Cotton	1,743.8	863.3 (338.3)	49.5 (19.4)	57.1 (23.4)	199
Grapes (table)	1,665.0	3416.3 (1080.1)	205.2 (64.9)	61.0 (35.9)	325
Grapes (wine)	1,496.8	139.6 (978.6)	9.3 (65.4)	−3.7 (22.9)	735
Hay	1,180.9	28.1 (194.6)	2.4 (16.5)	−68.2 (29.8)	1277
Lemons	382.2	219.3 (169.7)	57.4 (44.4)	8.1 (49.5)	279
Lettuce	1,243.8	97.5 (407.7)	7.8 (32.8)	−101.0 (157.1)	361
Oranges	1,040.0	−221.9 (569.0)	−21.3 (54.0)	−45.5 (51.0)	297
Pistachios	225.1	−67.4 (178.1)	−29.9 (79.1)	−83.8 (86.4)	208
Rice	541.3	−45.8 (121.8)	−8.5 (22.5)	20.4 (19.3)	387
Strawberries	752.1	−385.3 (151.9)	−51.2 (20.2)	43.4 (47.8)	333
Tomatoes	943.5	−61.4 (150.2)	−6.5 (15.9)	−13.1 (12.7)	549
Walnuts	403.1	−94.2 (83.6)	−23.4 (20.7)	2.3 (33.2)	810

Regression models weighted by acres planted. Results from estimation of equation 2 separately for each of indicated crops. Weather and expected weather are annual data. Percent effect in column (3) corresponds to the predicted change in aggregate crop value divided by average historical aggregate crop value. Percent effect in column (4) is defined similarly, except that it is for physical crop yields per acre. Standard errors in parentheses are clustered at the county level. See the text for more details

impacts reported in this paper differ from the 'ideal' predicted climate change impacts that hold the prices of all agricultural inputs and outputs constant.

5.4 Caveats

There are a number of caveats to this analysis and calculations. First, the analysis ignores extreme events (e.g., droughts and floods) or the variance of climate realizations, in addition to any effects captured by degree-days and precipitation. So it is uninformative about the economic impact of these events. Similarly, it is possible that climate change will disrupt local ecosystems and/or change soil quality. Both of these factors may affect agricultural productivity. Since annual fluctuations in climate are unlikely to have the same effect on ecosystems and soil quality as permanent changes, our estimates fail to account for these effects too.

Second, global climate change will likely affect agricultural production around the world, which will cause changes in relative prices. Although our estimates are based on annual fluctuations in weather and are adjusted for year fixed effects, we are not able to account fully for future price changes, like all studies based on historical data. Third, our analysis is conditional on the existing system of government programs that impact agricultural profits and land values by affecting farmers' decisions about which crops to plant, the amount of land to use, and the level of production (Kirwan 2005). Our estimates would likely differ if they were estimated with an alternative set of subsidy policies in place.

Finally, we discuss three issues with our approach that can cause it to yield an incorrect prediction of the damage associated with climate change. First, we emphasize that these projections are conditional on the current prices and availability of water for irrigation. In the likely event that these factors change over the next century, our approach is unlikely to correspond to the true future sequence of agricultural profits. In addition, elevated carbon dioxide (CO_2) concentrations are known to increase the yield per planted acre for many plants (see e.g., Miglietta et al. 1998). Since higher CO_2 concentrations are thought to be a primary cause of climate change, it possible that carbon fertilization will lead to higher yields per acre, which in turn would affect agricultural profits, and this is not accounted for in our analysis. Also, our representation of how weather and weather expectation affect profits is admittedly simple, in part because of the limited data and sample size available to us for estimation. Finally, it may be appropriate to discount the predicted future impacts.

6 Conclusions

The question posed in this paper is: what does the historic record tell us about how agriculture in California will be affected by climate change? Although there are limitations to our analysis that prevent us from being too specific in answering this question, there are two tentative conclusions.

One conclusion is that climate change will likely lead to reduced farm profits in the state, though obviously not for all farms. A second and related conclusion is that different crops will be affected very differently. The value of production from table grapes, for instance, is expected to increase significantly, whereas the value of production from strawberries is expected to decrease.

Acknowledgments We thank Ben Hansen and Katie Kimble for research assistance and three anonymous referees for their comments. We also acknowledge financial support from the California Energy Commission, through a grant to the University of California.

References

Adams RM (1989) Global climate change and agriculture: an economic perspective. AJAE 71:1272–1279
Adams RM, Rosenzweig Cynthia, Peart RM, Ritchie JT, McCarl BA, Glyer JD, Curry RB, Jones JW, Boote KJ, Hartwell Allen L Jr (1990) Global climate change and US agriculture. Nature 345:21924
Fisher AC, Hanemann WM, Roberts MJ, Schlenker W (2010) The Economic Impacts of Climate Change: Evidence from Agricultural Output and Random Fluctuations in Weather: Comment. Am Econ Rev, forthcoming
Black DA, Kneisner TJ (2003) On the measurement of job risk in hedonic wage models. J Risk Uncertainty 27(3):205–220
Black S (1999) Do Better Schools Matter? Parental Valuation of Elementary Education. Q J Econ 114
Chay KY, Greenstone M (2005) Does air quality matter? Evidence from the housing market. J Polit Econ 113 (April):376–424
Deschênes O, Greenstone M (2007) The economic impacts of climate change: evidence from agricultural profits and random fluctuations in weather. Am Econ Rev 97(1):354–385
Deschênes O, Greenstone M (2011) The Economic Impacts of Climate Change: Evidence from Agricultural Profits and Random Fluctuations in Weather: Reply Am Econ Rev, forthcoming

Fisher AC, Rubio SJ (1997) Adjusting to climate change: implications of increased variability and asymmetric adjustment costs for investment in water reserves. J Env Econ Manag 34:207–227

Greenstone M, Gallagher J (2008) Does hazardous waste matter? Evidence from the housing market and the superfund program. Q J Econ 123:951–1003

Grierson W (2002) Role of Temperature in the Physiology of Crop Plants: Pre- and Post-Harvest. In: Pessarakli M (ed) Handbook of Plant and Crop Physiology. Marcel Dekker, New York

Hansen LeRoy (1991) Farmer response to changes in climate: the case of corn production. J Agr Econ Res 43(4):18–25

Hodges T (ed) (1991) Predicting Crop Phenology. CRC Press, Boca Raton

Johnson SR, Haigh PA (1970) Agricultural land price differentials and their relationship to potentially modifiable aspects of the climate. Rev Econ Stat 52:173–181

Kaiser HM, Riha SJ, Wilks DS, Rossiter DG, Sampath R (1993) A farm-level analysis of economic and agronomic impacts of gradual climate warming. Amer J Agr Econ 75:387–398

Karl TR, Trenberth KE (2003) Modern global climate change. Science 302:1719–1723

Kaufmann RK, Snell SE (1997) A biophysical model of corn yield: integrating climatic and social determinants. Am J Agric Econ 79:178–190

Kaylen MS, Wade JW, Frank DB (2002) Stochastic trend, weather and U.S. corn yield variability. Appl Econ 24(5):513–518

Kelly DL, Kolstad CD, Mitchell GT (2005) Adjustment costs from environmental change. J Environ Econ Manag 50(2):468–495

Kirwan BE (2005) The Incidence of U.S. Agriculture Subsidies on Farmland Rental Rates. Unpublished doctoral dissertation, MIT, Cambridge, Massachusetts

Lobell DB, Cahill KN, Field C (2007) Historical effects of temperature and precipitation on California crop yields. Clim Chang 81:187–203

McFadden D (1984) Welfare Analysis of Incomplete Adjustment to Climatic Change. In: Kerry Smith V, Witte Ann Dryden (eds) Advances in Applied Micro-Economics. 3. JAI Press, Greenwich, Connecticut

Mendelsohn R, Nordhaus WD, Shaw D (1994) The impact of global warming on agriculture: a Ricardian approach. Am Econ Rev 84:753–771

Miglietta F, Magliulo V, Bindi M, Cerio L, Vaccari FP, Loduca V, Peressotti A (1998) Free air CO2 enrichment of potato (Solanum tuberosum L.): development, growth and yield. Glob Chang Biol 4:163–172

Perrin RK, Smith VK (1990) Measuring the Potential Economic Effects of Climatic Change on North Carolina Agriculture. North Carolina State University

Ritchie JT, NeSmith DS (1991) Temperature and Crop Development. In: Hanks J, Ritchie JT (eds) Modeling Plant and Soil Systems. American Society of Agronomy, Madison, Wisconsin

Rosenzweig C, Parry ML (1994) Potential impact of climate change on world food supply. Nature 367:133–138

Schimmelpfennig D, Lewandrowski J, Reilly J, Tsigas M, Parry I (1996) Agricultural Adaptation to Climate Change: Issues of Long Run Sustainability. U.S. Department of Agriculture Report AER-740, Washington, D.C

Schlenker W, Roberts MJ (2009) Nonlinear temperature effects indicate severe damages to U.S. crop yields under climate change. Proc Natl Acad Sci 106(37):15594–15598

Schlenker W, Michael Hanemann W, Fisher AC (2005) Will U.S. agriculture really benefit from the global warming? accounting for irrigation in the Hedonic approach. Am Econ Rev 95(1):395–406

Schlenker W, Michael Hanemann W, Fisher AC (2006) The impact of global warming on U.S. agriculture: an econometric analysis of optimal growing conditions. Rev Econ Stat 88(1):113–125

Schlenker W, Michael Hanemann W, Fisher AC (2007) Water availability, degree days, and the potential impact of climate change on irrigated agriculture in California. Clim Chang 81(1):19–38

Thompson LM (1986) Climatic change, weather variability, and corn production. Agron J 78:649–653

Economic impacts of climate-related changes to California agriculture

Josué Medellín-Azuara · Richard E. Howitt · Duncan J. MacEwan · Jay R. Lund

Received: 12 February 2010 / Accepted: 26 September 2011 / Published online: 24 November 2011
© Springer Science+Business Media B.V. 2011

Abstract California agriculture is driven by the interactions between technology, resources, and market demands. Future production is a balance between the rates of change in these variables and environmental factors including climate change. With tight statewide water supplies and agriculture being an important part of the California economy, quantifying the economic consequences of changes in these variables is important for addressing related policy questions. We estimate the economic effects of climate change on California crop farming by year 2050 using the Statewide Agricultural Production Model (SWAP). With climate warming, crop yields are expected to decline, production costs to increase, and water supplies to fall. These negative effects may be partially offset by higher crop prices and technological improvements. Results indicate that gross agricultural revenues across all regions are reduced under climate change, as is water usage. However, given the climate-induced reductions in water supply and crop yields, reductions in revenue are proportionally less due to shifting crop demands, technological change, and a shift to higher value less water intensive crops. Given the long time horizon required in this study, the results should not be considered a projection or forecast, but as a probable outcome of the interaction of several uncertain driving forces.

1 Introduction

Agricultural production accounts for 1.5% of the gross state product of California and more than half a million direct jobs (AIC 2009). The 2007 Census of Agriculture estimates total

---

J. Medellín-Azuara (✉) · J. R. Lund
Department of Civil and Environmental Engineering, University of California, Davis, One shields Ave., Davis, CA 95616, USA
e-mail: jmedellin@ucdavis.edu
URL: http://cee.engr.ucdavis.edu/Medellin

R. E. Howitt
Department of Agricultural and Resource Economics, University of California, Davis, CA, USA

D. J. MacEwan
University of California, Davis, CA, USA

value of agricultural production in California at $33.9 billion dollars, of which $22.9 billion is crop farming including nurseries and greenhouses (USDA 2007). Furthermore, for some crops (including some fruits, vegetables, and tree nuts), California accounts for more than 70% of U.S. sales (AIC 2009). One challenge and uncertainty facing agriculture is climate change, so quantifying potential effects is important for informing policy discussions. However, there is a complex dynamic between agriculture and climate change. Agriculture contributes 10 to 12% of total anthropogenic greenhouse gas emissions. At the same time, agricultural production offers a possibility for designing carbon sinks to help reduce net emissions (Smith et al. 2007). Agriculture also is likely to be affected by climate change via higher global temperatures, shifting rainfall patterns, flood events, and increasing atmospheric carbon concentrations. Effects are likely to be localized due to geography and socio-economic conditions; subsistence agriculture and farming areas in lower latitudes are among the most vulnerable (Easterling et al. 2007).

In California, climate change effects on agriculture from changing crop yields and a reduced water supply are anticipated to be significant. The largest production region in California is the Central Valley which is predominately irrigated and heavily reliant on water supplies and micro-climate. Reduced water deliveries and changing micro-climates are expected to have significant effects on agriculture in the region. The net effect of climate change on agriculture in all regions will depend on the relative importance of reduced water supplies, changing yields, technology, and market conditions.

Research on the economic effects of climate change in agriculture started in the 1980s with the work of Adams et al. (1995, 1989, 1990) and Reilly et al. (1993) among others. They quantify costs in terms of U.S. economic welfare and find impacts on the order of tens of billions of dollars. Their costs estimates (and benefits in many circumstances) depend upon the underlying assumptions on CO_2 fertilization effects, temperature, and precipitation. More recently, Lobell et al. (2009) estimate economic costs of climate extremes in California at $500 million per year, based on agricultural insurance payments for disaster. The economic costs of climate change to agriculture are also modeled as a function of changing crop profits assuming stationary crop prices and constant water supply (Costello et al. 2009). Other studies emphasize with water allocation simulations, water shortages to agriculture considering current and projected cropping patterns and varying water demand scenarios (Joyce et al. 2009). Quantification of change in land cover for specific activities such as dairy and grapes under climate change has also been studied (Hayhoe et al. 2004). However, endogenous estimation of the resulting cropping patterns under different climate scenarios as a result of climate, technological, and market conditions remains largely unexplored.

We examine the economic costs of climate-related changes in crop yields and water supply to California agriculture by year 2050. We employ the Geophysical Fluid Dynamics Laboratory model, under high emissions (GFDL A2) climate scenario, with average 2.3°C increase in temperature and 3% reduction in precipitation by year 2050 in the Central Valley (Cayan et al. 2008). Reductions in precipitation and increases in temperature translate into significant reductions in surface and groundwater supplies as well as increased evaporation losses in surface reservoirs. Compared to other scenarios, the GFDL A2 scenario allow us to uniquely represent the relative importance of two extreme changes in temperature and precipitation, often connected with warmer-climate reductions and reduced water supply availability.

In the following sections we present methods, modeling assumptions and data sources for agricultural production by year 2050 with and without climate change. These include technology-related yield improvements, change in market conditions and urbanization. To investigate the potential effects of climate change on agricultural production we discuss likely changes in water availability and climate-related yield changes in our selected climate scenario. Results are quantified in terms of land use, water use, agricultural production and gross agricultural revenues. A sensitivity analysis on selected parameters and a discussion of limitations in our analysis follow the results section. Conclusions (including a comparison of our results with other studies) are presented at the end of this paper.

2 Modeling climate change and agriculture

Most crops in California's Central Valley are expected to face yield declines with rising temperatures (Adams et al. 2003; Lobell et al. 2007; Schlenker et al. 2005). Furthermore, the fertilizing effect of higher carbon dioxide (CO_2) concentrations may also inhibit photorespiration in C3 species such as tomato, wheat and barley (Bloom 2006). With temperature changes of up to 2°C, crop yields (especially for cereals) in lower latitudes will fall, whereas higher latitudes are expected to improve (Easterling et al. 2007). Some negative effects may be partially offset by technological innovations leading to increases in crop yields. For major California crops between 1960 and 2002, yields increased by an average of 1.2% per year (Brunke et al. 2004).

A second major effect of climate change will be a changing water supply. In California, 70% of yearly rainfall occurs north of the Sacramento-San Joaquin Delta, but 75% of water demands are to the south which dependens on State and Federal water projects. Some agricultural land is expected to come out of production due to urban conversion and limited water availability. To the extent that climate change shifts rainfall patterns and increase temperatures, this will have additional effects on California agriculture.

We model agricultural production in California considering the likely effects of climate related yield and water supply changes, technological yield improvement, urban land use conversion and changing market conditions by 2050. We combine estimates from agronomic, climate and hydrologic models into a statewide regional economic model of agricultural production. An advantage of explicitly modeling crop production is that one can more formally link the results from various biophysical models directly into an economic decision framework.

2.1 Model description and methods

We adopt the Statewide Agricultural Production Model (SWAP after Howitt et al. 2001) to model California agriculture under climate change. SWAP was developed by Howitt and collaborators (2001) and has been used in a wide range of policy analysis. The original use for this model was to estimate the economic scarcity costs of water for agriculture in the statewide hydro-economic optimization model for water management in California, CALVIN[1] (Draper et al. 2003). More recently, SWAP has been used to estimate economic losses due to salinity in the Central Valley (Howitt et al. 2009a), economic losses to agriculture in the Sacramento-San Joaquin Delta (Appendix to Lund et al. 2007), and economic effects of water shortage to Central Valley agriculture (Howitt et al. 2009b).

SWAP is an optimization model for major crops and regions in California, calibrated using Positive Mathematical Programming (PMP after Howitt 1995). We present a summarized version of PMP below. Examples and discussion of the mathematical derivations specific to SWAP can be found in Howitt et al. (2009c) and Medellin-Azuara et al. (2010).

---

[1] htpp://cee.engr.ucdavis.edu/CALVIN

PMP is a three-step procedure for model calibration that assumes farmers optimize input use for maximization of profits. Four inputs are defined over j, land, labor, water, supplies (chemicals and fertilizer), 26 regions over g and 12 crops over i, both defined below. In the first step, a linear program for profit maximization is solved. In addition to the traditional resource and non-negativity constraints, a set of calibration constraints is added to restrict land use to observed values. In the second step, the dual values from the calibration and resource constraints are used to derive the parameters for a constant elasticity of substitution (CES) production function (Eq. 1) as well as an exponential "PMP" cost function (Eq. 2). In the third step, the calibrated CES and PMP functions are combined into a full profit maximization program. Additional constraints and policy changes are added to account for climate change, market conditions, and other factors, as described in the sections that follow. The exponential PMP cost function captures the marginal decisions of farmers through the increasing cost of bringing additional land into production (e.g. through decreasing quality). Other input costs (supplies, land, and labor) enter linearly into the objective function in the third step.

In Eq. 1 the analytical derivation of the beta and the tau parameters in (5) is shown in Medellin-Azuara (2006). The scale parameter is τ_{gi}, and the relative use of production factors is represented by the share parameter β_{gij}. Production factor use is given by X_{gij}. The elasticity of substitution of crop i, is σ_i, such that $\rho_i = (\sigma_i - 1)/\sigma_i$, and the returns to scale coefficient is υ. SWAP uses a non-nested CES, an elasticity of substitution of 0.22 is assumed for all crops and regions which allows for limited substitution among production factors, consistent with observed production practices. In Eq. 2, TC_{gi} stands for total land PMP costs and production factor use is given by X_{gij}. Delta and gamma parameters define the slope and intercept of the exponential PMP land cost function.

$$Y_{gi} = \tau_{gi} \left(\sum_j \beta_{gij}.X_{gij}^{\rho_i} \right)^{\upsilon/\rho_i} \tag{1}$$

$$TC_{gi}(X_{gi,land}) = \delta_{gi}e^{-\gamma_{gi}X_{gi,land}} \tag{2}$$

For clarity, we present the third step of optimization in SWAP: maximize total profits (Eq. 3) which includes price time CES production less PMP land costs and other production costs, subject to resource and policy constraints (Eq. 4). Where $yred_{gi}$ accounts for reductions in crop yields in policy simulation runs, v_{gi} is the crop price (exogenous in this simplified formulation but endogenous in the true model), and ω_{igj} is the cost of production factors other than land. Finally, b_{gj} is the maximum amount of the limiting resources (often water and land) and other constraints including regional dairy herd silage requirements and stress irrigation.

$$Max_{x \geq 0} \prod = \sum_g \sum_i yred_{gi} v_{gi} Y_{gi} - \sum_g \sum_i \delta_{gi} e^{\gamma_{gi}X_{gi,land}} - \sum_g \sum_i \sum_{j,j \neq land} (\omega_{igj} x_{gij}) \tag{3}$$

$$\sum_i X_{gij} \leq b_{gj} \forall g,j \tag{4}$$

We use the SWAP model for this analysis for several reasons. First, PMP is a calibration procedure that captures the marginal production decisions of farmers, allowing for exact calibration in inputs and outputs for a range of production activities to an *observed* base year(s). This is well suited for analysis of California agriculture where there are many specialty crops. Second, SWAP is specified with a Constant Elasticity of Supply (CES) production function with water, land, labor, and supplies (fertilizer and chemicals) as inputs to production. The full CES production surface in SWAP is well suited for evaluating effects of policy changes on agricultural production at the extensive (e.g. area planted) and intensive (e.g. input use per unit land) margins; we anticipate changes in both in response to climate change. Finally, SWAP can account for yield changes, shifting crop demands and endogenous prices, and constraints for dairy and livestock feed production, to more accurately represent California agriculture and aid in economic analysis and policymaking.

2.2 Model data, regions, and crop groups

Data from geo-referenced land use surveys from the California Department of Water Resources (DWR) are combined with information from United States Department of Agriculture (USDA) statistics. Cost information is obtained from crop production budgets from the University of California (UC) Davis Cooperative Extension and the Agricultural Issues Center (AIC). Dairy herd silage requirements are estimated based on a constant California dairy herd into the future where each cow requires about 0.27 silage acres per year, based on current consumption estimates.

Agricultural regions in the SWAP model for this paper include 21 Central Valley Production Model (CVPM) regions (USBR 1997), in orange in Fig. 1, plus irrigated agriculture in the Colorado River Region and the South Coast (dark red). For the

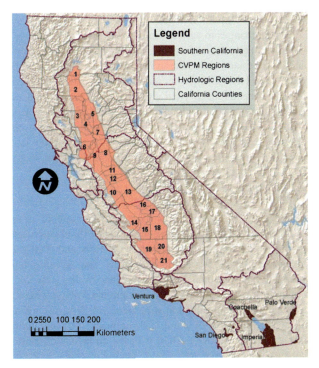

Fig. 1 Map of coverage of the Statewide Agricultural Production Model (SWAP). Areas are approximate

interpretation of results, we group SWAP regions into four larger areas: Sacramento including CVPM regions 1 thru 7; San Joaquin, CVPM regions 8 thru 13; and the Tulare Basin, CVPM regions 14 thru 21. Southern California is comprised of agriculture in the Imperial Valley, Palo Verde, Coachella, San Diego, and Ventura. The total base irrigated areas for SWAP in this study is of 3.36 million hectares (8.3 million acres) using 31.9 billion cubic meters per year [BCM/year] (26.3 million acre-feet, MAF/year) of applied water according to the California Department of Water Resources (DWR) data for 2005.

Irrigated crops in each SWAP region are classified into twelve SWAP crop groups: alfalfa, citrus, corn (including silage), cotton, field crops (e.g. safflower, dry beans, sorghum), grain crops (barley, wheat and oats), grapes (vine, table and raisin), orchards (e.g. apples, apricots, cherries, peaches, almonds, pistachios, walnuts), pasture, rice, tomato, and truck crops (vegetables and berry crops including asparagus, green beans, carrots, lettuce, melons, onion, garlic, potatoes, bush berries, spinach, strawberries, broccoli, peppers, and others). These definitions are consistent with the California DWR crop classifications.

2.3 Modeling California agriculture in 2050

We consider three different scenarios to estimate the effects of climate change: a base model for 2005, a model with historical climate in 2050 (no climate change), and a model with warm-dry climate change (GFDL A2) in 2050. The base model corresponds to conditions today and represents the base calibrated model. the historical 2050 model represents agriculture in 2050 in the absence of climate change, and the warm-dry model represents agriculture in 2050 with climate change. We model changes that occur with or without climate change, and those that only occur under climate change.

Likely changes to agriculture by 2050 that apply with or without climate change include urban and agricultural footprint, technological improvement represented as yield increases, and crop demand shifts due to increasing population and income and/or *global* commodity prices. Modeled changes in agriculture by 2050 that apply under climate change only include climate-related yield changes, water supply and availability. We employed recent empirical estimates of climate-related yield changes consistent with our selected climate change scenario. For water supply and availability we employ the CALVIN model that provides changes in water deliveries for agriculture to the SWAP model to estimate resulting agricultural production as discussed below.

2.3.1 Changes in land availability

Changes in land availability by year 2050 due to urbanization are included in both historical and climate change scenarios. Urbanization and agricultural land conversion in this study follow estimates from Landis and Reilly (2002) for prime farming land, locally important farms, unique farms, and grazing lands. Table 1 shows estimated statewide and regional land use patterns by the year 2050. Most land conversion from agriculture occurs south of the Sacramento Valley, where population growth is rapid. An average statewide reduction in agricultural land use close to 5.0% is expected between 2020 and 2050 (and of about 8.5% between 2005 and 2050). Corn silage of 323,000 acres in the Central Valley by 2050 is based on estimated dairy herd feed requirements with a constant herd size.

Table 1 Expected changes in land area between 2020 and 2050 (adapted from Landis and Reilly 2002) for CALVIN-SWAP region crop areas

	Urbanized land			Agricultural land[a]		
	Regional total area, acre (ha)		% Change	Regional total area, acre (ha)		% Change
	2020	2050		2020	2050	
Northern California and Sacramento	1,337,465 (540,737)	1,663,876 (660,576)	22.2	1,713,900 (693,590)	1,656,771 (670,495)	−3.3
San Joaquin Valley and Tulare Basin	646,381 (261,587)	1,044,333 (422,224)	61.4	5,083,100 (2,057,057)	4,797,749 (1,941,649)	−5.6
Southern California	2,550,040 (1,030,981)	3,442,696 (1,391,882)	35.0	964,360 (390,262)	905,394 (366,413)	−6.1
Statewide	4,534,516 (1,833,305)	6,120,905 (2,474,682)	35.0	7,761,360 (3,140,911)	7,363,711 (2,980,094)	−5.1

[a] For agriculture, Northern California and Sacramento includes Central Valley Production Model (CVPM) regions 1 to 7. The San Joaquin Valley and Tulare represents CVPM regions 8 to 21. Southern California considers projections of Landis and Reilly (2002) for agricultural areas in the counties that include Coachella, Imperial Valley, Palo Verde, and the counties of Ventura and San Diego

2.3.2 Technological change

Technological innovations for agricultural production are included in both historical and climate change scenarios and represented as increasing crop yields. This is calculated based on extrapolating trends detailed in Brunke et al. (2004). We model the historical (1.42%/ year) rate of yield increases until 2020, at which point an inherent limit to the rate of carbon fixation through photosynthesis will limit crop yield increases. We estimate yield increases will be lower with a log-linear growth rate of 0.25 for 2020–2050. A detailed table with yield trends by crop is presented in Howitt et al. (2009c). Total yield increases due to technological change average 29% for all SWAP crop groups by year 2050.

2.3.3 Shifts in crop demands

Shifts in demand for crop groups are included in both historical and climate change scenarios for 2050. We consider two types of crops: California specific crops and global commodities. Global commodity crops include grain, rice, and corn; all other crop groups are classified as California crops. Global commodity crops are those with no demand specific for California. For these crops, California faces a perfectly elastic demand curve, and is thus a price taker. We do not model the international trade market for these crops; we assume that California's export share will continue to remain small in the future. For California specific crops, California faces a downward sloping demand for a market that is driven by conditions in the United States. We do not model changes in tastes and preferences; we model the shift in demand for these crops that will result from increasing population and incomes.

Since California is a small proportion of global production for commodity crops, the only information we use to estimate the shift in demand by 2050 is the long run trend in real prices. Formally, we assume that California will retain its small share of the global market for these crops. A recent report by the World Bank (2009) projects price increases (in real terms) out to 2015 of 60, 48 and 40% for rice, corn and grains, respectively. Many experts in the field believe this is an overestimate as long run real prices have been historically declining for these crops. To deal with this dilema, at year 2015 we allow for the historical downward trend in real prices to resume. This translates into price drops from 2015 to 2050 of 1.45, 0.67 and 1.58%, average per year, for rice, grain, and corn, respectively. We combine the projected increases out to 2015 with the long run trend resuming in 2015 and continuing to 2050 to estimate the total percentage demand shift, or change in real price.

Demand for California specialty crops is expected to increase with increasing population and income in the United States. We estimate changes in U.S. income and population, and combine these with empirical income and population elasticities of demand, to determine the shift in demand for these crops by 2050. We estimate that 415 million people will reside in the U.S. by year 2050 (a 43% increase), and that real income in 2050 is anticipated to be about 2.5 times higher than in 2005 (Howitt et al. 2009b). Income and population increases can be directly mapped to shifts in demand through elasticity of demand with respect to income, population, and own price estimates, following the methods in Muth (1964). We assume a perfectly elastic long run supply of these crops, thus demand shifts represent a lower bound. We assume population elasticities of unity and consult the literature to select recent income and price elasticity estimates for each crop group (Green et al. 2006). The combined effect of the income induced demand shift and the population induced demand shift is reported Table 2. Details on the derivation are provided in Howitt et al. (2010).

2.3.4 Climate-related yield changes

The climate-related yield changes in Table 3 are only incorporated in the climate change analysis and are independent of the technology-related yield changes discussed above. These follow empirical estimates from Adams et al. (2003), Lobell et al. (2007), Lobell and Field (2009) and Lee et al. (2009). There is significant variation in climate-related yield changes among crops due to differences in tolerance to changes in temperature and CO_2 concentrations (Lobell and Field 2009; Lobell et al. 2006). Vegetables (truck crop group), citrus, and orchard are expected to be among the most affected crop groups under climate change. Other crops such as alfalfa, tomato (Lee et al. 2009) and almonds (Lobell and Field

Table 2 Demand shifts by crop: 2050

Crop group	Total% shift in demand intercept
Alfalfa	3.34
Citrus	3.63
Corn	17.00
Cotton	2.14
Field	3.34
Grain	−19.88
Grapes	3.83
Orchard	16.42
Pasture	5.30
Rice	−4.1
Tomato	26.86
Truck	45.45

Table 3 Expected percentage change in yields as a result of climate change

Crop groups	Sacramento River	San Joaquin River
Alfalfa[a]	4.9	7.5
Citrus[b]	1.77	−18.4
Corn[a]	−2.7	−2.5
Cotton[a]	0.0	−5.5
Field[b]	−1.9	−3.7
Grain[a]	−4.8	−1.4
Grape[c]	−9.0	−9.0
Orchard[c]	5.0	5.0
Pasture[c]	−6.0	−6.0
Rice[a]	0.8	−2.8
Tomato[a]	2.4	1.1
Truck Crops[c]	−11.0	−11.0

[a] Lee et al. (2009)
[b] Adams et al. (2003)
[c] Lobell et al. (2007) and Lobell and Field (2009)
Climate-related water changes

2009) are less affected and may even face small yield-improvements under higher carbon concentrations. Finally, changes in winter chill (Baldocchi and Wong 2008), nutrient assimilation due to carbon concentrations (Bloom 2006), and number of days within temperature thresholds (Schlenker et al. 2007; Schlenker and Roberts 2009) have also been shown to have an effect on crop yields. Effects may also differ due to differences in micro-climate between basins (Rosenzweig et al. 1996).

2.3.5 Water availability

Changes in water availability due to climate change by 2050 are only included in the climate-change model. We estimate available regional water supplies using CALVIN model runs in Medellin-Azuara et al. (2009) that optimize the statewide economic returns to all water uses. CALVIN water allocation results suggest economically worthwhile water exchanges among regions and users within the existing water supply network. We allow for these exchanges, implicitly assuming more flexible water transfer institutional arrangements by 2050. Table 4 summarizes the total change (reduction) in water availability due to climate change in 2050. While the total average reduction in water deliveries is 14% the average reduction for agricultural regions is 21%, indicating that a statewide optimal market induces voluntary sales of an additional 7% of agricultural water supplies to urban areas. Without such a market, the social costs of climate change would be significantly higher.

Table 4 Expected percentage reduction in available water from climate change (after Medellin-Azuara et al., 2009)

Region	Agriculture	Urban	Total
Sacramento	24.3	0.1	19.1
San Joaquin	22.5	0.0	17.6
Tulare	15.9	0.0	13.5
Southern California	25.9	1.12	8.9
Total	21.0	0.7	14.0

3 Results

Both 2050 scenarios model changes in land availability (Table 1), shifts in demand for crops (Table 2), and technological change. For the 2050 climate change scenario, climate-related yield (Table 3) and water availability (Table 4) are incorporated into the model. The difference between these two scenarios represents the modeled effect of GFDL A2 climate change. These results show the importance of integrating and modeling the extent of adaptations of bio-economic systems to climate change. Since such systems are primarily driven by economic incentives, they can be expected to adjust and adapt at the intensive and extensive margins simultaneously. Our results show such adaptations for irrigated agriculture in California.

We aggregate results for the 26 SWAP regions (Fig. 1) into four regions: Sacramento, San Joaquin, Tulare Basin, and Southern California as defined earlier. For the base and two 2050 climate scenarios we provide land and water use, crop prices, revenues (total harvested crop value at the farm gate) either by crop or by region as well as changes in statewide agricultural land use, water use, crop prices, production and revenues.

3.1 Total land use

Table 5 shows the land use for the 2005 base case, 2050 historical, and 2050 climate change scenarios. Comparing historical and climate change scenarios in year 2050, we estimate statewide land and water use reductions of 20% (last column of Tables 5 and 6). Water reductions follow CALVIN results and reductions in total land use occur as a result of decreased water availability. With respect to the 2005 base, we estimate reductions in agricultural land use of 7.34 and 26% for 2050 historical and climate change scenarios, respectively.

The Sacramento and San Joaquin River Basins face the largest reductions in crop area. This is a result of the functioning water markets as both regions export water to higher-value agricultural production in the Tulare Basin. Water transfers among regions in Southern California (Coachella, Imperial, Palo Verde) are limited by conveyance capacity. As such, land use falls proportionally more in this region.

3.1.1 Changes in land use by crop

Disaggregated results by crop show a shifting crop mix in response to changing market conditions and climate change. Figure 2 depicts change in acres, by crop and region, in 2050 between historical and climate change scenarios. The general trend is that under

Table 5 Land use in 2005, 2050 historical, and climate change in thousand acres (ha)

Scenario	Sacramento	San Joaquin	Tulare	Southern California	Total
2005 Base Case	2,005 (811)	2,062 (835)	3,280 (1,328)	991 (401)	8,339 (3,375)
2050 Historical	1,921 (777)	1,887 (763)	2,966 (1,200)	954 (386)	7,727 (3,127)
2050 Climate Change	1,462 (593)	1,463 (592)	2,500 (1,012)	745 (302)	6,170 (2,497)
% Change 2050 Historical vs. Climate Change	−23.89	−22.44	−15.72	−21.85	−20.15

Table 6 Applied water in 2005 base case, 2050 historical and climate change, per region in TAF/year (MCM/year)

Scenario	Sacramento	San Joaquin	Tulare	Southern California	Total
2005 Base Case	6,419 (7,917)	6,326 (7,803)	9,807 (12,097)	3,743 (4,617)	26,295 (32,434)
2050 Historical	6,150 (7,586)	5,773 (7,120)	8,856 (10,924)	3,655 (4,509)	24,434 (30,139)
2050 Climate Change	4,623 (5,703)	4,490 (5,538)	7,461 (9,203)	2,795 (3,447)	19,369 (23,891)
% Change 2050 Historical vs. Climate Change	−24.8%	−22.2%	−15.8%	−23.5%	−20.7%

climate change there is a shift to higher value, less water intensive agriculture cultivated over a smaller land area.

Acreage decreases significantly for low value agriculture such as pasture, field, and grain crops. Irrigated pasture, despite a tolerance for higher temperatures, remains an unprofitable production decision and is removed from production as water supplies fall. Sacramento, San Joaquin, and Southern California regions show irrigated pasture acreage reduced by over 90%. Field crops and grains show significant acreage decreases as well, with grain production dropping out of the Sacramento Valley entirely. Sacramento and San Joaquin, the two regions showing the largest decline in low value crops, are the two largest water export regions. Higher marginal water values in other regions induce water transfers from these regions and a reduction in low value high water use crop acreage.

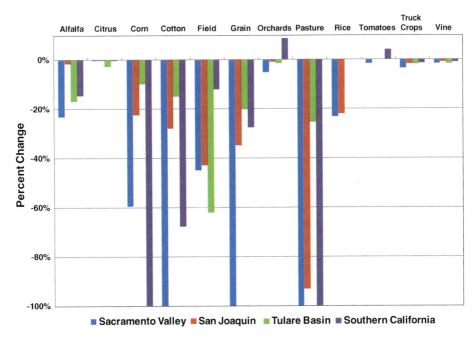

Fig. 2 Percentage change in crop area between historical and climate change scenarios for all crops and SWAP regions

Market conditions and technological innovation lead to increases in higher value crop acreage in some regions. Moderately warm-temperature resilient crops like processing tomatoes realize an increase in acreage under climate change, with some production shifting outside of the Sacramento Valley. Orchard acreage increases in Southern California in response to increasing demand and a decrease in lower value crop acreage. Grapes, vegetables (truck), and citrus show small decreases in acreage as water is shifted away from low value uses to these higher value crops.

3.2 Total water use

A breakdown of total water use, in terms of applied agricultural water, is shown in Table 6. Note that we assume crop water demand will not increase as a result of climate change. However, higher evapotranspiration rates may increase the required applied water for crops (Joyce et al. 2009).

With respect to base 2005 conditions, in 2050 with historical climate, reductions of about 7.6% in total applied water as expected. Climate change causes on additional 5,065 TAF/year (6,245 BCM/year) in water reductions which translates into a 20.1% reduction with respect to 2050 historical climate. Population growth with the accompanying water and land use conversion from agriculture to urban uses are taken into account. Changes in total average applied water per acre of crop do not change significantly (less than 3%).

3.3 Changes in crop prices and agricultural revenues

Crop prices are determined endogenously within the SWAP model and are reported in Table 7 for base, historical climate, and climate change. Most crop prices are shown to increase with climate change, although most changes are moderate. This is a result of the relative effects of shifting crop demands and regional production.

Processing tomatoes, a crop with moderate tolerance to warm climate, show a slight decrease in price under climate change. Since tomatoes have a higher tolerance for warm temperatures, yields increase due to both technological change and climate change. Additionally, the relative profitability of processing tomatoes encourages expansion of acreage in regions like Southern California, as shown in the land use section. The result is an increase in production across the state that has downward pressure on prices.

Orchards and citrus crops will likely have the largest increases in price at 8 and 12%, respectively. This is largely due to a strong demand for these crops combined with a decrease in yields due to climate change (up to 18%). Consequently, even though crop area increases in regions like Southern California, strong demand outpaces production.

Table 8 shows total gross agricultural revenues by region for base, 2050 historical and 2050 climate change scenarios. We report gross revenues instead of net revenues or other measures such as "value added" since this reflects the most comprehensive measure of economic benefits to agriculture. In contrast, value added considers returns to labor and capital.

The Sacramento Valley shows the largest reduction in gross agricultural revenues (15.5%). This is likely due to functioning water markets that allow this region to sell to other regions, and urban users, where the marginal value of water is relatively high. In contrast, the smallest decrease in gross revenues is for the Tulare basin (9%) due to the ability to import water from other regions like the Sacramento Valley.

Statewide, a 10.9% reduction in total gross agricultural revenues compared with historical projections is attributed to climate change by 2050. However, a shift from low

Table 7 Crop group prices in 2005 base, 2050 historical and 2050 climate change

	Base case 2005 price, $2008/Ton	Historical 2050 price, $/Ton	Climate change 2050 price	% Change 2050 historical vs climate change
Alfalfa	117	111	116	3.90
Citrus	436	410	460	12.26
Corn	102	119	119	0.02
Cotton	1701	1739	1759	1.16
Field Crops	295	305	305	0.06
Grain	298	238	238	0.04
Grapes	847	962	980	1.87
Orchards	1445	1257	1358	8.05
Pasture	76	80	80	0.07
Rice	290	273	277	1.55
Tomato	50	61	61	−0.38
Truck Crops	277	410	421	2.72

value to higher value crops, over less land, translates into an increase in gross revenue per unit area. Under historical climate by 2050 the average revenue per unit area is $3,671/acre ($9,072/ha), compared to gross revenues per unit area of $4,087/acre ($10,100/ha under climate change. In other words, under climate change, total gross revenues fall by 10.9%, but revenues per unit area increase by 11.1%.

3.4 Summary of statewide results

To synthesize all the results, we present statewide changes in crop prices, total production, and revenues by crop in Table 9. This table highlights the importance of the relative effects of climate change and other conditions on agriculture. For example, tomatoes are resilient to a warming climate and face strong demand in 2050, consequently a moderate decrease in price (−0.38%) is accompanied by a moderate increase in both production (1.4%) and gross revenues (0.96%). In contrast, orchards are relatively more affected by a warm climate and, even with strong future demand, show a decrease in production (−10.2) leading to a increase in price (8.05%) and a decrease in gross revenues (−2.94%).

Low value crops such as irrigated pasture and grains show a large decrease in production, with moderate increases in price, and a large decrease in gross revenues.

Table 8 Total agricultural revenues by region and scenario (in millions of 2008 dollars per year)

Scenario	Sacramento	San Joaquin	Tulare	Southern California	Total
2005 Base Case	3,637	3,872	9,807	2,730	20,046
2050 Historical	6,101	5,253	13,044	3,973	28,371
2050 Climate Change	5,157	4,709	11,863	3,488	25,217
% Change 2050vs. Climate Change	−15.5	−10.4	−9.1	−12.2	−10.9

Table 9 Percentage change price, production, revenue (historical vs. climate change scenarios) for year 2050

Crop group	% Change price	% Change production	% Change revenues
Alfalfa	3.90	−6.2	−3.20
Citrus	12.26	−18.7	−8.36
Corn	0.02	−22.9	−22.61
Cotton	1.16	−23.6	−20.59
Field	0.06	−48.9	−48.40
Grain	0.04	−31.3	−30.05
Grapes	1.87	−7.3	−5.60
Orchards	8.05	−10.2	−2.94
Pasture	0.07	−95.7	−95.70
Rice	1.55	−22.4	−21.22
Tomato	−0.38	1.4	0.96
Truck	2.72	−13.3	−10.88

However, the decrease in these crops is offset by the increase (or small decrease) in higher fruit, nut, and vegetable crops.

Changes in runoff, total water use, irrigated crop area, and agricultural revenues by 2050 are shown in Fig. 3. Comparisons are to the 2005 base case. Under climate change, total runoff is expected to decrease by 27%, relative to no change under historical climate conditions. Irrigated crop area is reduced under historical climate by 7.3% and 26% under climate change by 2050.

Total applied agricultural water follows a similar pattern to changes in land use. Total gross agricultural revenues increase by 43 and 27% under historical climate and climate change, respectively, compared to the base year of 2005. This indicates a smaller agricultural sector growing higher value crops using less land and water, with or without climate change. Under climate change, runoff is less, and land use and applied water are reduced. Consequently, gross revenues increase, but by relatively less than under historical climate conditions.

3.5 Sensitivity analysis

SWAP qualitative and quantitative results are robust for a wide range of assumptions on total available land, water and market conditions. However, uncertainties in crop prices, climate models and technological change may affect simulation results. A key parameter likely to drive results is yield changes due to climate warming and technological innovation. We conducted a sensitivity analysis on technology-related and climate-related yield changes, which ultimately affect revenues. We consider four scenarios: 1) full climate and technology yield change (main climate change scenario); 2) 50% climate yield change and full technology yield change; 3) full climate yield change and 50% technology yield change; and 4) 50% cuts in technology and climate yield change. We compare results in terms of total irrigated acres by crop with respect to the 2050 historical climate scenario. The results are summarized in Fig. 4.

The results of climate change model are shown in white bars. With half of the technology related yield change (about 14.5% average for all crops) changes in acreage are less than 5% in absolute value. For sensitivity runs where climate change yields are full and

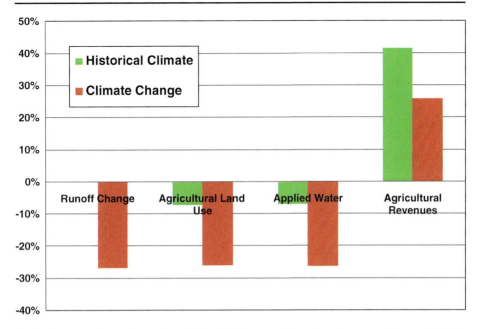

Fig. 3 Percentage change in runoff, agricultural land use, applied water and revenue with respect to 2005 base case

technology yield changes are either full or half, results are similar and changes are less than 3% in absolute value. Finally, under full technology-related yield changes but half of the climate-related yield changes, changes in crop areas remain close to the climate change results.

The most variation occurs for lower value crop groups (right on Fig. 4). Low value crops become relatively less profitable under the climate related yield changes and technological innovation is often less for these crops. Additionally, total gross revenues increase by 3% with half of the climate-related yield changes and decrease by 8% with half of the technology related yield changes.

We conclude that our results and conclusions are robust and hold over a range of plus-minus 50% of the assumed technology and climate-related yield changes.

3.6 Limitations and extensions

Limitations inherent to PMP models like SWAP have been discussed elsewhere (Howitt et al. 2009b; Medellín-Azuara et al. 2010). Three areas in our approach deserve some additional discussion: climate change uncertainties, 2050 projections, and system constraints.

Climate change uncertainties exist for many areas including yield changes, water requirements by crop, water availability, and even changes in consumer preferences towards more environmentally friendly production systems. We examined results for a reduced effect of climate change on crop yields assuming the same exogenous (CALVIN) climate change water cutbacks. Total water availability and use is driven by these likely, yet uncertain, changes. Other climate change uncertainties remain to be addressed as more information becomes available.

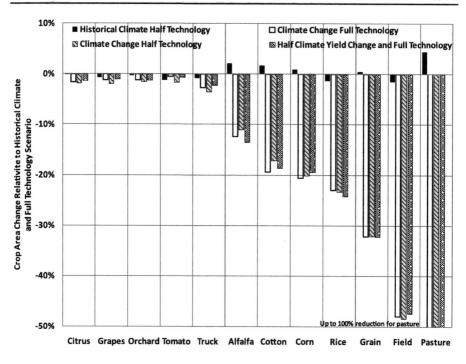

Fig. 4 Comparison of statewide changes in area by crop with respect to the historical 2050 climate scenario for full and half the assumed technology and climate related yield changes

Projections to 2050 of crop prices, urbanization, technology, and land conversion to urban uses are subject to significant uncertainty. Large crop price shocks in currently low value crops such as corn, grain, or other commodity crops are not captured in our analysis. Sustained events like the 2007–2008 commodity price spike will likely incentive additional area devoted lower value crops. This will also shift some water away from urban uses or higher value crops which will change the net effect of climate change.

Finally, physical and institutional constraints by year 2050 may prevent the economically optimal water deliveries (and exchanges) prescribed in CALVIN from happening. This will result in larger water cutbacks and an less economically efficient allocation of water across regions. However, there is a significant economic gradient for these water transfers to occur and climate change will continue to strengthen it. This, in turn, may make overcoming the institutional constraints more feasible.

4 Conclusions

We explore the effects of climate change on California agriculture in 2050 using SWAP, a mathematical programming model for agricultural production in California that calibrates exactly to *observed* base year conditions. We explicitly account for projected changes in land use due to urbanization, crop demand shifts, technological improvements in crop production, changing water supply due to climate change, and climate-related crop yield changes. Anticipated changes in climate, statewide runoff, water use, and land use generally decrease all crops. Gross agricultural revenues consequently fall, although revenues per acre increase, reflecting shifts to higher value crops. Finally, crop prices, determined endogenously and accounting for expected future demand shifts, generally increase with a warmer and drier climate.

Our results are consistent with Jackson et al. (2009) in showing a shift to more valuable and climate tolerant crops. Our cost estimates are within a similar range to studies that allow for comparison to California (Adams 1989); although, comparing our cost estimates to other studies is difficult due to the difference in locations, scope and method. In contrast to other studies (Adams et al. 1990; Costello et al. 2009), we conclude climate change is unlikely to generate significant economic gains for crop farming in California. However, technology and flexibility in water allocation may partially offset economic losses under climate change.

Our results are consistent with expectations and remarks from the fourth IPCC climate assessment of agriculture. A significant finding is that changes in water supply are more likely to have an effect on crop production than changes in temperature alone; the major effects of climate change on California agriculture are manifest through water shortages. The climate change-induced reductions in water supply (20% less), reduce irrigated crop area more than urban expansion in the Central Valley. The increasing value of water, which accompanies the increased scarcity, induces changes in crops and technology reflected in the model results.

The main effects of climate change are changing crop yields and water supply. Since climate induced yield changes differ in their intensity (and sign) among crops, we expect the comparative advantage of different crops to be changed by climate. SWAP model results show that it is possible to incorporate climate change yield data and simulate the resulting economic reactions. The results are notable for the relatively large reductions, due to climate change, of the gross revenues earned by irrigated agriculture, compared to the proportionally small changes in total land and water use shown by the model. The revenue and resource effects are shown to vary by region, with most losses in Southern California. Overall, while the effect of climate change is manifest through yield changes, after economic adaptation, the results on irrigated crop production are predominately shown in economic terms rather than in changes in aggregate land and water use.

Acknowledgements The authors are indebted for the data and comments provided by Ray Hoagland, Farhad Farnam and Tom Hawkins from the California Department of Water Resources. The authors thank Kurt Richter for providing data and Chenguang Li for her research support. Funding from the California Energy Commission's Public Interest Energy Research (PIER) is greatly appreciated.

References

Adams RM (1989) Global climate change and agriculture—an economic-perspective. Am J Agr Econ 71:1272–1279

Adams RM, Rosenzweig C, Peart RM, Ritchie JT, McCarl BA, Glyer JD, Curry RB, Jones JW, Boote KJ, Allen LH (1990) Global climate and United States agriculture. Nature 345:219–224

Adams RM, Fleming RA, Chang C-C, McCarl B, Rosenzweig C (1995) A reassessment of the economic effects of global climate change on U.S. agriculture. Clim Chang 30:147–167

Adams RM, Wu J, Houston LL (2003) Climate Change and California, Appendix IX: The effects of climate change on yields and water use of major California crops, California Energy Commission. Public Interest Energy Research (PIER). Sacramento, CA

Agricultural Issues Center (AIC) (2009) Measure of California Agriculture, University of California, Davis, Davis, CA, p. 151. Available in <http://aic.ucdavis.edu>. Access March 2011

Baldocchi D, Wong S (2008) Accumulated winter chill is decreasing in the fruit growing regions of California. Clim Chang 87:153–166

Bloom AJ (2006) Rising carbon dioxide concentrations and the future of crop production. J Sci Food Agric 86:1289–1291

Brunke H, Sumner D, Howitt RE (2004) Future food production and consumptionin California under alternative scenarios. Agricultural Issues Center, University of California, Davis

Cayan DR, Maurer EP, Dettinger MD, Tyree M, Hayhoe K (2008) Climate change scenarios for the California region. Clim Chang 87:S21–S42

Costello CJ, Deschenes O, Kolstad D (2009) Economic impacts of climate change on California Agriculture, CEC, Public Interest Energy Research (PIER), Sacramento, CA. Available in <http://www.climatechange.ca.gov/>. Access December 2010

Draper AJ, Jenkins MW, Kirby KW, Lund JR, Howitt RE (2003) Economic-engineering optimization for California water management. J Water Resour Plan Manage 129:155–164

Easterling WE, Aggarwal PK, Batima P, Brander KM, Erda L, Howden SM, Kirilenko A, Morton J, Soussana JF, Schimidhuber J, Jacob TV (2007) Food, fibre and forest products. In: Parry ML, Canziani OF, Palutikof JP, van der Lind PJ, Hanson CE (eds) Climate Change 2007: Impacts, adaptation and vulnerability. Contribution of Working Group II to the Fourth Assessment Report of the Intergovernmental Panel on Climate Change. Cambridge University Press, Cambridge, pp 273–313

Green R, Howitt R, Russo C (2006) Estimation of supply and demand elasticities of California Commodities, Working Paper. Department of Agricultural and Resource Economics. Universtity of California, Davis, Davis, California

Hayhoe K, Cayan D, Field CB, Frumhoff PC, Maurer EP, Miller NL, Moser SC, Schneider SH, Nicholas Cahill K, Cleland EE, Dale L, Drapek R, Hanemann RM, Kalkstein LS, Lenihan J, Lunch CK, Neilson RP, Sheridan SC, Verville JH (2004) Emissions pathways, climate change, and impacts on California. Proc Natl Acad Sci USA

Howitt RE (1995) Positive mathematical programming. Am J Agr Econ 77:329–342

Howitt RE, Ward KB, Msangi S (2001) Statewide Agricultural Production Model (SWAP), Department of Agricultural and Resource Economics. University of California Davis, California. Available in <http://cee.engr.ucdavis.edu/CALVIN>. Access January 2010

Howitt R, Kaplan J, Larson D, MacEwan D, Medellin-Azuara J, Horner G, Lee NS (2009a) Central Valley Salinity Report, Report for the State Water Resources Control Board. University of California, Davis, California. Available in <http://swap.ucdavis.edu/>. Access 01 November 2009

Howitt RE, Medellin-Azuara J, MacEwan D (2009b) Estimating economic impacts of agricultural yield related changes, California Energy Commission, Public Interest Energy Research (PIER), Sacramento, CA. Available in <http://www.energy.ca.gov>. Access 01 November 2009

Howitt RE, Medellin-Azuara J, MacEwan D, Lund JR (2010) Statewide agricultural production model, Davis, CA. Website. Available in <http://swap.ucdavis.edu>. Access October 2010

Jackson LE, Santos-Martin F, Hollander AD, Horwath WR, Howitt RE, Kramer JB, O'Geen AT, Orlove BS, Six JW, Sokolow SK, Sumner DA, Tomich TP, Wheeler SM (2009) Potential for adaptation to climate change in an agricultural landscape in the Central Valley of California, California Energy Commission, Public Interest Energy Research (PIER), Sacramento, CA, p. 142. Available in <http://www.climatechange.ca.gov/>. Access November 2009

Joyce BA, Methta VK, Purkey RP, Dale LL, Hanemann M (2009) Climate change impacts on water supply and agricultlural water management in California's Western San Joaquin Valley and Potential Adaptation Strategies, Sacramento, CA. Available in <http://www.climatechange.ca.gov>. Access January 2011

Landis JD, Reilly M (2002) How we will grow: Baseline projections of California's Urban Footprint Through the Year 2100. Project Completion Report., Department of City and Regional Planning, Institute of Urban and Regional Development, University of California, Berkeley, CA. Available in <http://www-iurd.ced.berkeley.edu/pub/WP-2003-04-screen.vpdf>. Access January 2009

Lee J, De Gryze S, Six (2009) Effect of climate change on field crop production in the Central Valley of California, California Energy Commission, Public Interest Energy Research (PIER), Sacramento, California, p. 26. Available in <http://www.climatechange.ca.gov>. Access 01 November 2009

Lobell DB, Field CB (2009) California Perennial crops in a changing climate, California Energy Commission, Public Interest Energy Research (PIER), Sacramento, CA, p. 37. Available in <http://www.climatechange.ca.gov>. Access 01 November 2009

Lobell DB, Field CB, Cahill KN, Bonfils C (2006) Impacts of future climate change on California perennial crop yields: model projections with climate and crop uncertainties. Agric For Meteorol 141:208–218

Lobell DB, Cahill KN, Field CB (2007) Historical effects of temperature and precipitation on California crop yields. Clim Chang 81:187–203

Lobell DB, Torney A, Field CB (2009) Climate extremes in California, California Energy Commission, Public Interest Energy Research (PIER), Sacramento, CA, p. 18. Available in <http://www.climatechange.ca.gov>. Access 01 November 2009

Lund J, Hanak E, Fleenor W, Howitt R, Mount J, Moyle P (2007) Envisioning futures for the Sacramento-San Joaquin Delta, Public Policy Institute of California, San Francisco, CA, p. 300 pp. Available in <http://www.ppic.org>. Access 01 November 2009

Medellin-Azuara J (2006) Economic-engineering analysis of water management for restoring the Colorado River Delta, Dissertation, University of California, Davis, Davis, California, p. 146

Medellín-Azuara J, Harou JJ, Howitt RE (2010) Estimating economic value of agricultural water under changing conditions and the effects of spatial aggregation. Sci Total Environ 408:5639–5648

Medellin-Azuara J, Connell CR, Madani K, Lund JR, Howitt RE (2009) Water Management Adaptation with Climate Change. California Energy Commission, Public Interest Energy Research (PIER): Sacramento, CA, p. 30

Muth RF (1964) The derived demand curve for a productive factor and the industry supply curve. Oxf Econ Pap 16:221–234

Reilly J, Hohmann N, Sally K (1993) Climate change and agriculture: global and regional effects using an economic model of international trade, Economic Research Service, U.S. Department of Agriculture. Access March 2011. Available in < http://dspace.mit.edu> Access January 2011

Rosenzweig C, Phillips J, Goldberg R, Carroll J, Hodges T (1996) Potential impacts of climate change on citrus and potato production in the US. Agr Syst 52:455–479

Schlenker W, Roberts MJ (2009) Nonlinear temperature effects indicate severe damages to U.S. crop yields under climate change, Proceedings of the National Academy of Sciences roc. Natl. Acad. Sci. U.S.A. 106, 15594

Schlenker W, Hanemann WM, Fisher AC (2005) Will U.S. agriculture really benefit from global warming? Accounting for irrigation in the hedonic approach. Am Econ Rev 95:395–406

Schlenker W, Hanemann W, Fisher A (2007) Water availability, degree days, and the potential impact of climate change on irrigated agriculture in California. Clim Chang 81:19–38

Smith P, Martino D, Cai Z, Gwary D, Janzen H, Kumar P, McCarl B, Ogle S, O'Mara F, Rice C, Scholes B, Sirotenko O (eds) (2007) Agriculture. In Climate Change 2007: Mitigation. Contribution of Working Group III Cambridge University Press, Cambridge, United Kingdom and New York, NY, USA

United States Bureau of Reclamation (USBR) (1997) Central Valley Project Improvement Act. Draft Programmatic Environmental Impact Statement. Technical Appendix Volume Eight. U.S. Department of Interior. U.S. Bureau of Reclamation: Sacramento, California

United States Department of Agriculture (USDA) (2007) Census of agriculture, NASS, Washington, D.C. Available in <http://www.agcensus.usda.gov>. Access March 2011

World Bank (2009) Double jeopardy: responding to high fuel and food prices, in G8 Hokkaido-Toyako Summit, July 2

Climatic Change (2011) 109 (Suppl 1)
DOI 10.1007/s10584-011-0306-3

Case study on potential agricultural responses to climate change in a California landscape

L. E. Jackson · S. M. Wheeler · A. D. Hollander · A. T. O'Geen · B. S. Orlove · J. Six ·
D. A. Sumner · F. Santos-Martin · J. B. Kramer · W. R. Horwath · R. E. Howitt ·
T. P. Tomich

Received: 12 February 2010 / Accepted: 26 September 2011 / Published online: 24 November 2011
© Springer Science+Business Media B.V. 2011

Abstract Agriculture in the Central Valley of California, one of the USA's main sources of fruits, nuts, and vegetables, is highly vulnerable to climate change impacts in the next 50 years. This interdisciplinary case study in Yolo County shows the urgency for building adaptation strategies to climate change. Climate change and the effects of greenhouse gas emissions are complex, and several of the county's current crops will be less viable in 2050. The study uses a variety of methods to assemble information relevant to Yolo County's agriculture, including literature reviews, models, geographic information system analysis, interviews with agency personnel, and a survey of farmers. Potential adaptation and mitigation responses by growers include changes in crop taxa, irrigation methods, fertilization practices, tillage practices, and land use. On a regional basis, planning must consider the vulnerability

---

L. E. Jackson (✉) · A. T. O'Geen · F. Santos-Martin · J. B. Kramer · W. R. Horwath
Department of Land, Air and Water Resources, University of California Davis, Davis, CA 95616, USA
e-mail: lejackson@ucdavis.edu

A. T. O'Geen
e-mail: atogeen@ucdavis.edu

W. R. Horwath
e-mail: wrhorwath@ucdavis.edu

S. M. Wheeler
Department of Environmental Design, University of California Davis, Davis, CA 95616, USA
e-mail: smwheeler@ucdavis.edu

A. D. Hollander
Information Center for the Environment, University of California Davis, Davis, CA 95616, USA
e-mail: adhollander@ucdavis.edu

B. S. Orlove
Division School of International and Public Affairs and Center for Research on Environmental Decisions, Columbia University, New York, NY 10027, USA
e-mail: bso5@columbia.edu

of agricultural production and the tradeoffs associated with diversified farmlands, drought, flooding of cropland, loss of habitat for wild species of concern, and urbanization.

1 Introduction

This paper examines climate change vulnerabilities and coping strategies for mitigation and adaptation in an agricultural landscape using a case study approach that considers agricultural sciences, economic issues, soil science, and land use change. The study focuses on Yolo County (Fig. 1) as it is representative of the diverse agricultural landscapes throughout the Central Valley of California: irrigated perennial and row crops on alluvial plains; upland grazed rangelands; small towns and cities; and a changing mixture of urban, suburban, and farming-based livelihoods through the past few decades.

Yolo County includes the Sacramento River floodplain and alluvial fans as well as terraces of the Coast Range. The area has a Mediterranean climate and rainfall mainly during the winter months. The most important crops are tomatoes, alfalfa hay, wine grapes, and almonds, but a diversity of crops can be produced which ultimately may increase resilience for future environmental changes, extreme climatic events, and market competition. Yolo County has strong local interest in agricultural preservation, but there is regional pressure for urban and suburban growth, due to its proximity to the capitol city of Sacramento.

Temperatures in this region are likely to increase 1.3°C–2°C by 2050 regardless of the postulated global emissions scenario, and 2.3°C–5.8°C by 2100 depending on the level of future emissions (Cayan et al. 2009). Warming effects are likely to be more severe in summer than winter, and precipitation will likely decline. Heatwaves will occur more frequently, last longer, experience higher peak and duration of temperature, and begin earlier in the summer than historically (Hayhoe et al. 2004; Miller et al. 2007).

Both mitigation of greenhouse gas (GHG) emissions and adaptation strategies are response options to decrease vulnerability to climate change, and they can have synergistic effects under some circumstances. This paper focuses on the potential

---

J. Six
Department of Plant Sciences, University of California Davis, Davis, CA 95616, USA
e-mail: jwsix@ucdavis.edu

D. A. Sumner
Agricultural Issues Center, University of California Davis, Davis, CA 95616, USA
e-mail: dasumner@ucdavis.edu

R. E. Howitt
Department of Agricultural and Resource Economics, University of California Davis, Davis, CA 95616, USA
e-mail: rehowitt@ucdavis.edu

T. P. Tomich
Agricultural Sustainability Institute, University of California Davis, Davis, CA 95616, USA
e-mail: tptomich@ucdavis.edu

Fig. 1 Location map of Yolo County in Northern California

modifications to farm and land use practices, rather than on changes in public policy decision-making processes (Smit and Skinner 2002). Responses to climate change are set within the larger context of other major changes in California during the next 50 years, including population growth, water availability, regulations that favor agricultural sustainability, and changes in agricultural markets. The interrelation of many of these factors is shown in Fig. 2. The goals of this research were to:

- Understand the vulnerabilities of a California agricultural landscape to climate change, based largely on review of the literature.
- Determine the key biophysical and socioeconomic uncertainties (local and regional) that will affect mitigation and adaptation to climate change in this landscape.
- Develop a template for exploring sustainable regional responses to climate change for California's agricultural counties.

2 Agricultural commodities and climate change

2.1 Crop responses to climate change

Anticipated climate change will have both positive and negative effects on the yield and quality of currently produced commodities. Increased temperatures may adversely affect yields of tomato (Sato et al. 2000), rice (Ziska et al. 1997; Moya et al. 1998), stone fruits (deJong 2005), and grapes (Hayhoe et al. 2004), but may allow for more crops of lettuce outside of the coastal regions during the winter (Wheeler et al. 1993), expansion of citrus production (Reilly and Graham 2001), and heat and drought-tolerant trees such as olives. Major physiological impacts of anticipated temperature changes include diminished yields

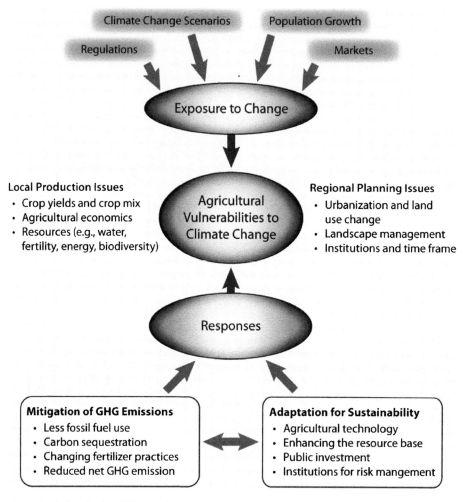

Fig. 2 Agricultural vulnerabilities to climate change, set within the context of exposure to changes driven by several societal issues, and the responses in terms of mitigation and adaptation options, including interaction and tradeoffs between them

from increased temperatures during key stages of crop development (Sato et al. 2000; West 2003; Peng et al. 2004), shorter periods of crop development (Wheeler et al. 1993; Moya et al. 1998; deJong 2005), and reduced product quality from unseasonal precipitation or adverse temperatures during fruit development (Southwick and Uyemoto 1999). Horticultural crops are more sensitive than field crops to short-term environmental stresses that affect reproductive biology, water content, visual appearance, and flavor quality, and are thus likely to be more impacted by climate change and extreme events (Backlund et al. 2008; Bazzaz and Sombroek 1996). Crop breeding is one approach to dealing with these problems, however, this requires purposeful action years in advance, especially for perennial crops.

To assess the effect of climate change on row crop productivity, alfalfa, maize, rice, sunflower, tomato, and wheat in Yolo County were modeled using DAYCENT, a process-based biogeochemical model (Del Grosso et al. 2002). We assumed current management practices (conventional management, fixed management schedules and a typical set of crop rotations, no changes in pests or diseases, and no effects of CO_2 or temperature on developmental biology) for the period 2000 to 2050 (Lee et al. 2010). Under both A2 (high emissions) and B1 (low emissions) scenarios (Cayan et al. 2009), average modeled maize, sunflower, and wheat yields decreased by 2% to 8% by 2050 relative to the 2000 to 2004 average yields (Table 1). The yields tended to decline slightly more under A2 than B1. In general, alfalfa was predicted to increase slightly under climate change in the same period, while rice and tomato were essentially unaffected by the modeled conditions.

Additional DAYCENT model runs were conducted to examine how extreme weather events may affect crop production. Early heat waves seem to have a profound effect on crop growth except for alfalfa and winter wheat (Table 1). Heat waves in May resulted in yield loss of 1–10% for maize, rice, sunflower, and tomato, whereas heat waves in June affected only maize and sunflower yields. The effects of heat waves in July on crop yields were relatively small. Repeated heat waves in May–July had the most profound effects on crop production by decreasing 3–19% of the 2000-2004 baseline yields. However, drought did not have much added effect on irrigated crop yields during heat waves.

Yolo County's horticultural row crops (e.g., tomato, cucumber, sweet corn, and pepper) are warm-season crops with a temperature optimum of 20°C to 25°C for yield, and an acceptable range of 12°C to 30°C, with a maximum tolerance of 35°C (Backlund et al. 2008). Mean mid-summer maximum temperatures by mid-century may force a shift to hot-season crops such as melon and sweet potato, with higher acceptable temperature ranges

Table 1 Effects of heat waves[a] and drought[b] on field crop yields in Yolo County under A2 and B1 emission scenarios, as determined by the DAYCENT model (Lee et al. 2010)

Commodity	Emission scenario	2000–2004			2046–2050				
			Baseline climate change		Heat waves only				Heat waves & drought
					May	June	July	May–July	May–July
		ton ha^{-1}	ton ha^{-1}	% change from 2002	Additional % change from baseline				
Alfalfa	A2	16.4	17.0	3.5	1.2	0.0	−0.4	1.0	1.2
	B1	16.5	17.8	7.3	1.1	0.4	−0.5	1.1	1.4
Maize	A2	13.9	13.5	−2.4	−4.4	−5.4	−0.2	−11.2	−11.2
	B1	13.6	13.4	−1.6	−3.5	−6.3	−0.9	−7.3	−7.3
Rice	A2	9.3	9.5	1.7	−3.8	0.0	−0.1	−6.1	−6.9
	B1	9.3	9.4	1.7	−4.1	−0.7	−1.1	−6.9	−8.0
Sunflower	A2	1.4	1.3	−7.9	−9.5	−5.2	−1.9	−18.5	−20.3
	B1	1.4	1.3	−5.4	−6.5	−7.1	−2.9	−18.7	−20.3
Tomato	A2	94.5	97.4	3.0	−1.5	−0.6	−0.8	−3.2	−4.8
	B1	95.9	97.2	1.4	−1.4	−0.3	−0.7	−2.9	−4.8
Wheat	A2	6.0	5.8	−2.4	−0.1	0.0	0.0	−0.1	−0.1
	B1	5.7	5.6	−2.6	0.0	0.0	0.0	−0.1	−0.1

[a] Temperature for heat waves (= 46°C) is the 99.9th percentile in the period 2000 to 2050. Heat waves are simulated for the last 10 days of the month of May, June, July, or all three months each year from 2000

[b] Under drought conditions, water available for irrigation is assumed to have only 75% soil water holding capacity at the time of irrigation. The baseline irrigation has 95% water holding capacity

(18°C to 35°C). Warmer winter temperatures, however, would favor cool-season crops, such as lettuce and broccoli, that are now grown in winter/early spring further south, and which have an acceptable range of 5°C to 25°C.

Stone fruits, nuts and grapes require approximately 200 to 1200 h of winter chill to flower (Backlund et al. 2008). Chill hours are computed on a daily basis relative to a reference temperature. Using climate predictions for the Central Valley, winter chill hours are expected to decrease from a baseline of 1000 h, as observed in 1950, to about 500 h by 2100 (Baldocchi and Wong 2006; Luedeling et al. 2009). Lack of winter chill hours may be more important for these perennial crops than yield losses due to higher temperatures during the growing season (Lobell and Field 2009).

For non-irrigated rangelands that are limited by cool temperatures in winter and spring, higher temperature and CO_2 enrichment could potentially stimulate productivity. But field experiments with elevated CO_2 have demonstrated decreased grassland productivity with increased temperature, precipitation and soil NO_3- compared to current ambient levels (Shaw et al. 2002). Deposition of plant residues with lower nutrient content as well as greater demand for N under elevated CO_2 may reduce grassland production over the long-term (Schneider et al. 2004; Dukes et al. 2005; de Graaff et al. 2006). Alternatively, N-fixing legumes may become more abundant in annual grasslands, partly due to warmer winter temperatures, contributing more N supply to these ecosystems. But this will require adequate soil phosphorus and micronutrients such as iron and molybdenum to support N fixation (Hungate et al. 2004; Van Groenigen et al. 2006).

For livestock, high summer temperatures, e.g., above 35°C, cause physiological stress and low consumption of feed (Conrad 1985). Dairy cows with high body temperatures have been shown to have lower milk yield (West 2003), and this is also important for cow-calf operations on rangelands. Overall, current evidence suggests that livestock forage on these rangelands may decrease due to climate change, especially in dry years and in response to N limitation, leading to lower livestock stocking rates, and earlier animal removal dates, requiring transport to irrigated, permanent pasture.

Other environmental factors contribute to the uncertainty of crop responses to climate change. Elevated CO_2 is generally considered to increase biomass production of annual crops by 10% to 20% under field conditions (Long et al. 2006). But under elevated CO_2, rates of nitrogen (N) assimilation by roots often do not keep up with shoot accumulation of carbohydrates (Reich et al. 2006; de Graaff et al. 2006). Also, nitrate (NO_3-) assimilation may decrease under elevated CO_2 due to inhibition of photorespiration, resulting in reduced growth and yield of C_3 plants (Rachmilevitch et al. 2004; Bloom 2009). Adaptation will require changes in N management. Elevated CO_2 concentration may also reduce stomatal conductance and evapotranspiration (ET), but will depend on changes in the wind environment and resulting micrometeorological resistances. If the growing seasons of certain crops shorten, then ET per crop may decrease slightly. Of course, alternative crop mixes and rotations will affect ET due to inherently different water use patterns.

2.2 Crop pests and climate change

Agricultural weeds, pests, and diseases will be impacted by climate change in uncertain ways (Field et al. 1999; Scherm 2004). Even a 2°C temperature rise can result in one to five additional generations per year for a range of invertebrates such as insects, mites and nematodes (Yamamura and Kiritani 1998). Many insect species will expand their geographical range in a warmer climate (Hill et al. 1999; Parmesan and Yohe 2003).

Climate change is likely to lead to a northern migration of weeds in the Central Valley, and disease and pest pressure will increase with earlier spring arrival and warmer winters, allowing greater proliferation and survival of pathogens and parasites (see Cavagnaro et al. 2006). Predicting these changes requires better understanding of ecophysiology, and the complexity of the trophic interactions.

Pierce's Disease has caused severe damage to grapevines in southern California, and is likely to become more prevalent northwards as the temperature warms, unless new solutions are found (Wine Institute 2002). Pierce's Disease is a bacterial disease of California grapes, caused by *Xyllela fastidiosa*, and vectored by the glassy-winged sharpshooter, a native to the southeastern USA that is more mobile than leafhoppers already present, and is now limited to climates with mild winters such as southern California (Purcell and Hopkins 1996).

Some other possible effects of higher temperature identified in discussions with the Yolo County UC Cooperative Extension farm advisors were: stripe rust on wheat (especially under wetter conditions), insect pests on nuts, medfly, corn earworm on tomato, tomato spotted wilt virus, and earlier activity of perennial weeds such as bindweed.

2.3 Crop water needs

Water supply is probably the most uncertain effect of climate change for California agriculture. Both groundwater overdraft and potentially diminished water transfers contribute to uncertainty in the quantity and sometimes the quality of irrigation water (California Department of Water Resources 2006). Periods of dry years do not permit an easy rebound for irrigated crops, especially if groundwater is not available and affordable. Perennial crops are particularly vulnerable, but growers of annual crops may need to shift crops or take land out of production. The prognosis of a drier Western USA (Barnett et al. 2008) suggests high vulnerability for crops that are abundant water users, especially if their cash value is low.

Farmers in Yolo County rely on groundwater for almost 40% of their supply in a normal water year, and this dependency is expected to increase under possible future drought and population growth conditions (Yolo County 2007). Rice, pasture, and hay have the highest applied water and ET, and are therefore the most vulnerable to water shortages. Using economic modeling, hypothetical surface water irrigation cutbacks (25% less during a normal, non-drought year with no supplemental groundwater) in Yolo County had higher impacts on high water-demanding crops including alfalfa, vegetable crops such as tomatoes, and most importantly rice (Lee et al. 2001). Crops with more return per ha and per unit of water showed less acreage reductions.

California had fewer periods of extended drought in the latter part of the last century than in the period of 1915 to 1935, when severe drought occurred, although the early 1990s also brought a serious drought. The historical runs of the global GCM models are in agreement with these records (Cayan et al. 2009). Present climate models do not predict any prolonged droughts until the end of this century. However, drought remains unpredictable and therefore planning agencies should always include extended drought events in their planning horizons. Competition for surface water supplies under drought conditions can be fierce and would impact the amount and timing of water deliveries.

Adaptation to a more uncertain water supply requires that crops be planned and managed to reduce water use by applying alternative technologies, such as using drip irrigation rather than furrow irrigation; finding ways to reduce ET, such as crop breeding for greater canopy cover; switching to crops that use less water; and/or reducing overall irrigated crop acreage.

Specific management challenges vary across the county. Eastern Yolo County relies more strongly on water originating from snowmelt from the north and east Sierra Nevada. Western Yolo County relies on rainfall runoff coming from the Coast Range, and will be more vulnerable to water shortages, although groundwater supplies are plentiful at present. Higher temperatures and increasing population and urbanization would place greater demands on water resources in Yolo County and lead to a more uncertain water supply for agriculture.

2.4 Strategies for responses to climate change

The menu of potential adaptation and mitigation responses to climate change by growers includes changes in crop diversity, irrigation methods, fertilization practices, tillage practices, and land management:

- **Crop diversity.** Growers will shift toward hot-season species, with greater winter potential for cool-season crops such as lettuce and broccoli. A shift to greater crop diversity will offset some of the risks from weather variation due to climate change. A switch is expected to higher cash value crops with greater income per amount of applied water.
- **Crop breeding.** Selection for genotypes that benefit from elevated CO_2 and are N-use efficient is essential, but breeding for water use efficiency often results in lower growth and yield (Condon and Hall 1997).
- **Irrigation.** If water supply becomes threatened, shifts toward drip irrigation, deficit irrigation, and crops that provide higher income per amount of applied water are potential adaptive responses. Subsurface drip irrigation has been shown to reduce GHG emissions (both CO_2 and N_2O), with no differences in tomato yields in Yolo County (Kallenbach et al. 2010). Reduction in water use and GHG emissions must be weighed against the need for investment, labor, energy for pressurization, and the lower potential for groundwater recharge during wet years.
- **Fertilizer use.** Reducing inputs of N-based fertilizers is a strategy to reduce emissions of N_2O, a potent GHG. Current average application rates of N fertilizers are up to 25% higher than needed for optimal yield (Krusekopf et al. 2002; De Gryze et al. 2010). Farmers may need to increase N fertilization, however, to compensate for physiological changes in N uptake and lower crop protein.
- **Cover cropping.** Cover cropping is a strategy to improve soil fertility, increase soil C sequestration, and decrease fertilizer use, but would prevent the possibility of cool-season cash crops, and potentially reduce soil water recharge for summer crops.
- **Tillage.** Low-till or no-till methods have generally not shown a soil carbon (C) benefit to the Central Valley (Veenstra et al. 2006), but can decrease fossil fuel inputs (Jackson et al. 2004). For many of Yolo County's crops, however, tillage reduction presents production constraints, such as seed establishment or efficient movement of furrow irrigation water. Also, alternative tillage practices can increase N_2O emissions due to higher soil moisture content and increased activity of anaerobic microorganisms (Kong et al. 2009). Net GHG reduction is likely only after many years of low-till practice (Six et al. 2004), which is often not feasible.
- **Manure management.** Manure management activities are important for achieving reduction in GHG (principally methane) and local air pollutants. Methane digesters are useful for dairy production (Mitloehner et al. 2009), but most livestock in Yolo County is beef cattle.
- **Farmscaping.** Use of perennial vegetation in marginal lands on farms, such as farm edges and riparian corridors, can increase C storage, reduce N availability and N_2O

emissions, and benefit water quality, habitat, and biodiversity (Young-Mathews et al. 2010). Cost-share programs would hasten widespread implementation.

- **Carbon sequestration in tree crops and vines.** Perennial woody crops are a potential opportunity for growers to receive GHG mitigation credits. But such a policy mechanism does not yet exist, and may be difficult to justify in terms of permanence of C storage.
- **Organic production.** Yolo County currently contains more than 50 organic farms, most producing a diverse mix of crops for local markets (4% of total agricultural value; Yolo County Agricultural Commissioner 2007). Organic production may hold adaptive advantages in that its diversity of crops can better respond to a changing climate. Net GHG and losses of N also may be lower than conventional production (Poudel et al. 2001; Smukler et al. 2010). New and increased pest and disease pressure may be an even greater problem for organic than conventional farms. New markets would need to be developed to support expanded organic production.
- **Biomass utilization for energy and fuel production.** Over the 40-year time horizon assumed for this study, a potential higher-value value market for agricultural commodities may be energy. Second and third generation biofuels based on ligno-cellulose (biomass) and microbial systems in both terrestrial and aquatic systems are more likely candidates than grain feedstocks for ethanol. Construction of an ethanol processing facility would not be able to increase local grain prices by a large enough margin to generate the additional acreage for local corn (Lee and Sumner 2009). Utilization of existing crop residues such as orchard and vineyard prunings, rice straw, and animal manures for bioenergy production can provide additional revenue to farming operations, or at least fix energy costs if produced and used onsite.

We conducted a survey of growers through the Yolo County Resource Conservation District and the Agricultural Commissioner's office in the summer of 2008, which found that 67% of the 36 respondents considered climate change "very important" or "somewhat important" to their investment decisions. Nearly half (43%) of growers reported that they "always" or "frequently" consider climate change in their production decisions. Growers with land in Williamson Act set-asides (a state program through which growers receive tax benefits for agreeing to keep land in production for 10 years) were more likely to be concerned about climate change. No differences were found in climate change concern between organic and non-organic producers. Ranchers were significantly more likely to be concerned about climate change than other types of farmers. The survey indicates that Yolo County farmers may be a receptive audience for strategies that could help them mitigate and adapt to a changing climate.

2.5 Agrobiodiversity as a facet of crop response to climate change

Agrobiodiversity refers to the variety of living organisms that contribute to agriculture in the broadest sense, e.g., crop and animal breeds as well as species in habitats outside of farming that affect processes such as pollination, pests and pest control, and water quality (Jackson et al. 2007). Maintenance of high levels of inter- and intra-species diversity is considered a strategy to decrease vulnerability and enhance resilience to uncertainty and to climate change (O'Farrell and Anderson 2010; Palm et al. 2010).

To assess the effect of fluctuations in crop acreage on crop diversity, we applied the Shannon-Weaver Index (Weaver and Shannon 1949), using 45 different crops found in the Yolo County Agricultural Crop reports. The index measures species richness (H) and evenness (*E*). It essentially assesses the proportion (*p*) of each crop with respect to its production category's total, and then exaggerates that relationship.

$$H = \sum p_i \ln p_i$$
$$E = H/\ln(\text{number of crops})$$

According to this Index, orchard/vineyard and grain categories share a higher average richness than other crops (Fig. 3). The Index for orchard-/vineyard rose to a two-year peak in 1992–1993 and has declined thereafter, as a consequence of increased acreage in grapes and almonds with respect to other woody crops. Grain crops are annuals such as corn and wheat, occupy a much larger amount of land overall, and are more prone to annual variation in diversity and acreage. While the diversity index for truck (i.e., vegetables and melons) and field crops (e.g., hay and alfalfa) were both initially higher, the last 25 years has brought a species-poorer crop mix. Evenness as a whole across the entire county is generally decreasing, indicating that dominant crops occupy more of the acreage with time (Fig. 4).

Several pathways exist to increase agrobiodiversity as an adaptation to climate change. One is to breed and select crop varieties resistant to heatwaves and drought, and to take advantage of the elevated CO_2 that will occur later in the century. Another is to target new crop taxa for the region, and begin the process of adaptive management. Diversified farm systems are another possibility (see Section 3.2). Long-term planning must also include processors, shippers and market opportunities.

2.6 Soil and land management options for mitigation of climate change

California's agricultural and forestry sectors contribute 8.3% of all anthropogenic GHG emissions in the state (CEC 2006), which is roughly 37.5 million metric tons of CO_2 equivalents ($MMTCO_2E$) annually, based on the fact that over 450 $MMTCO_2E$ annually is attributed to human activity in California. If California's agricultural and forestry sectors proportionately mitigate GHG emissions to maintain emissions at the 1990 level, it would be necessary to reduce emissions by 14.5 $MMTCO_2E$ by the year 2020.

Half of the California agricultural emissions is emitted as N_2O (CEC 2005) mainly due to microbial nitrification and denitrification of fertilizer N and available soil N that is mineralized from soil organic matter, breakdown of crop residues, and manure management. Methane emissions are also substantial at 37.5%, which mainly comes from

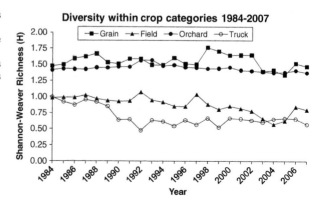

Fig. 3 Crop species richness using the Shannon-Weaver Diversity Index based on acreage 1984–2007 in Yolo County. Orchard/vineyard and grain crops maintain a higher level of species richness than other categories. The number of commodities in use continues to decline overall for grain, field and truck (i.e., vegetable) crops. Data from Yolo County Agricultural Commissioner (1984–2007)

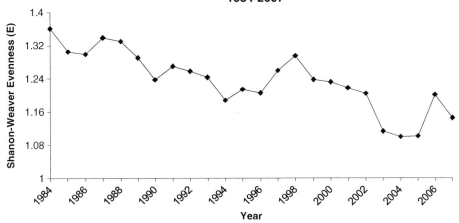

Fig. 4 Total crop species evenness using the Shannon-Weaver Diversity Index based on acreage 1984–2007 in Yolo County. Consistent total decline in the evenness between the acres of each species cropped indicates that a few commodities increasingly occupy more of the acreage with time. Data from Yolo County Agricultural Commissioner (1984–2007)

enteric fermentation of livestock. The latter source is less important in Yolo County than statewide due to the presence of very few dairies and feedlots.

Long-term research at UC Davis suggests that the use of cover crops and manure can contribute to up to 300 kg of soil C sequestration per ha under consistent management (Horwath et al. 2002). However, after five years the rate of soil C sequestration diminished significantly. Using the DAYCENT model, De Gryze et al. (2010) found that there was biophysical potential to reduce soil GHG flux up to 4577 kg CO_2E ha^{-1} yr^{-1} in agricultural soils in the Sacramento Valley, and that organic management practices had higher mitigation potential than cover crops or conservation tillage. In all cropping systems, reduction of N_2O emissions was more important for mitigation than changes in soil C.

There is uncertainty and a wide range of variability for GHG emissions estimates from agriculture in California depending on crops, management systems, and soils. Research is on-going but in the meantime, estimates should be taken with care, and management to reduce GHG emissions should be combined with the capacity to increase other ecosystem services, e.g., productivity, water quality, air quality, and erosion prevention.

3 Landscapes and land use options

Since 1850, California's agriculture has been constantly changing via growth, transition, and adjustment (Williams et al. 2005). Large changes have occurred within the last 150 years in Yolo County's alluvial plains, beginning with early attempts to raise livestock, grow grains, and develop horticulture without much irrigation. This was followed by an era of ruminants and wheat and barley production, and then intensive fruit, nut, and vegetable agriculture and large-scale cattle production began, ending with the present management-intensive, technologically-dependent agricultural industry (Mikkelsen 1983; Johnston and McCalla 2004), and expanding organic production (Broome and Worthington 2009).

Despite pressure of urbanization, Yolo County's agricultural landscapes have remained relatively stable in recent years. From 1992 until 2008, a net loss of about 12,000 ha of agricultural land occurred, and this includes a net gain of about 6,500 ha of grazing land according to the 2008 California Department of Conservation Farmland Mapping and Monitoring Program (FMMP). In 1988, several thousand ha were converted from farmland of local importance to grazing land due to sign ups for the federal Conservation Reserve Program. Urbanization has been slow, due to strong local agricultural preservation policies, and is likely to remain so unless agriculture becomes less viable. Overall, only 1% of Yolo County's total prime farmland was lost up until 2000 (Sokolow and Kuminoff 2000). Since 1998, the rate of agricultural conversion to wetlands along the Sacramento River for wildlife conservation has increased to approximately 800 ha yr^{-1} (Landon 2009), which may continue given the threats of flooding due to earlier Sierra Nevada snowmelt (see Section 3.1).

3.1 Vulnerabilities of Yolo County's agricultural landscapes to climate change

The impacts of climate change and the associated mitigations and adaptations in Yolo County will vary according to its diverse landscapes. To investigate these differences we employed a geographic information system (GIS) approach to stratify the county into the following four geographic units that represent similarities in land use, soil types and the environmental factors that formed them (Fig. 5; Appendix A):

Region 1. Flood basins, largely along the Sacramento River
Region 2. Recent alluvium (alluvial plains, fans and low terraces)
Region 3. Old alluvium (undulating dissected terraces, terraces)
Region 4. Uplands of the Coast Range

In low-lying Region 1 near the Sacramento River, farmlands are at risk of flooding due to earlier snowmelt in the Sierra Nevada. The 100- and 500-year floodplains of the Sacramento River extend westward into prime agricultural farmland (California Department of Water Resources 2006; Spencer et al. 2006). If flooding occurs late in the spring (April–June), crops planted during March–May may be damaged or destroyed. It is often too late by this time to replant fields with new crops. If soil remains wet, tillage is delayed, shortening the growing season and decreasing yields. Tomato farming in the Northern Yolo Bypass area is already subject to prolonged periods of late spring flooding that now occur more frequently than in the past (Jones and Stokes 2001). Planting beds, furrows, ditches, and other agriculture-related infrastructure (e.g., roads, canals, diversion structures, pumps, and wells) can also be damaged or destroyed by flooding.

The restoration of marginal farmlands into wetlands may have several environmental benefits including wildlife habitat, buffering from flood events, and improving water quality via filtration. Many wetlands have been shown to sequester C. These systems, however, are prone to discharge other GHGs such as N_2O and methane, which are more potent GHGs than CO_2. Thus, restored wetlands may become a net source of GHG instead of a sink (Mitsch and Gosselink 2000). More research is needed in these landscapes to document C cycling and GHG emissions. The means and outcomes for ecosystem restoration are controversial, and need to be carefully planned to maximize biodiversity and to maintain farmer livelihoods.

Landscapes of Region 2 may have the greatest potential for resilience to the effects of climate change. A variety of crops can be grown in this region, offering growers the

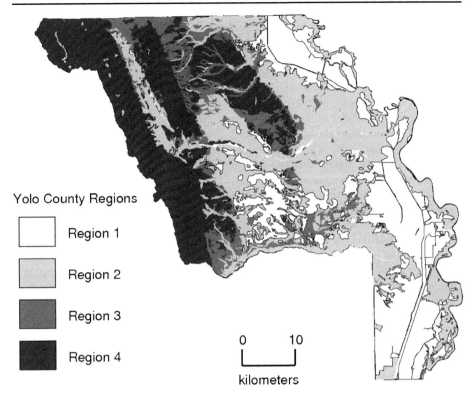

Fig. 5 Zonation of landscape regions in Yolo County, based on the aggregation of soil survey data (USDA-NRCS SSURGO database) in a GIS. Region 1 represents clay-rich soils in basin alluvium, mostly suitable for rice production. Region 2 represents other alluvial soils in their initial stages of soil development supporting a wide mix of crops. Region 3 represents marginal agricultural lands on rolling dissected fan terraces suitable for orchards and vineyards. Region 4 represents rangeland and wildlands of the Coast Range

opportunity to change commodities with lower potential losses. Best management practices to maintain or enhance C storage in soils include the use of cover crops, organic agriculture, conservation tillage, irrigation management, buffer strips, and vegetative filter strips, and many of these practices have other benefits as well, such as reduced erosion (Grismer et al. 2005; O'Geen and Schwankl 2006).

Region 3, an area of terraces and low hillslopes, can benefit from practices to increase crop diversification and practices that increase C storage. A switch to drip irrigation could conserve water and eliminate irrigation-induced soil erosion (Hanson et al. 2008). Residue management practices such as conservation tillage, cover crops, compost and mulch can improve water infiltration, increase soil C, and reduce storm water runoff (O'Geen et al. 2006). These practices have multiple positive feedbacks including enhanced productivity, improved water quality, better water use efficiency, less erosion and possible C sequestration. An added pressure in this area is the expansion of urban development, which could become more pervasive if agricultural productivity and income decline due to climate change.

Region 4 has the greatest potential to maintain its C stocks. This upland landscape has tremendous aboveground and belowground C stocks that are relatively unaffected by the present land use. Best rangeland management practices exist that can maintain this stock by reducing soil erosion and promoting forage production. These include moderate to low stocking rates, rotational grazing, seasonal use of highly erodible land, strategies to ensure oak regeneration, use of appropriate seed mixtures, and reseeding of perennial grasses. Climate variability that adversely affects forage production will, however, challenge rangeland managers. Vegetation management such as prescribed fire may be necessary to expand productive areas and avoid catastrophic fire.

To assess the viability of Region 1 for agricultural practices under changing climate conditions, we quantified the economic cost of converting the agricultural lands closest to the river to riparian habitat. This could be achieved through a vigorous restoration program to plant native trees, shrubs, and perennial graminoids. Some of the trees could then be harvested periodically as a method of storing additional C as consumer products. A simpler method would be to leave the land to successional processes, and at least initially, to ruderal, herbaceous plants. An active restoration program would protect higher elevations of Region 1 from further flooding and would simultaneously sequester more C and possibly reduce N_2O emissions. Additionally, shifting rice fields to riparian vegetation in strategic locations would provide opportunities for other sources of revenue, such as from firewood or timber, from paying the farmer to maintain high biodiversity and wildlife habitat, and from other ecosystem benefits related to water quality and recreation.

3.2 Agrobiodiversity as a source of innovation for landscape responses to climate change

Dealing with climate change can also be achieved by increasing agrobiodiversity at the landscape level. These types of changes were explored on a regional basis within the county. In one example, two hypothetical farms were created, each 1,300 ha to resemble a normal quantity of land that a Yolo County farm manager oversees. The standard farm represents a low diversity, major commodity farm with five of the most commonly grown crops in Region 2. The diverse farm contains a diversified crop mix that also includes pasture and orchards with purposefully minimized water consumption. The production costs and quantities for all crops were obtained from the Cost and Return Studies from the UC Davis Agriculture and Resource Economics Department (UC Davis 2008).

In the hypothetical diverse farm, some of the standard farm's crops are replaced with deciduous trees and pasture to simulate a diversified crop mix. Table olives and irrigated pasture replace an establishing alfalfa crop; almond and walnut orchards replace the flood-irrigated safflower crop; almonds, prunes, and high-density plantations of oil olives replace one-third of the 370-ha tomato crop. Three major components (material cost, harvest cost and non-cash overhead, and investments) contribute to the higher cost of the diverse farm (data not shown; see Jackson et al. 2009). Due to the orchard trees, the diverse farm demanded slightly more fertilizer and water than the standard farm, even with the stricter water management practices. Consequently, the mitigation benefits of diversifying with orchard trees and pasture will be related to increased C stocks and a reduction in tillage rather than a significant reduction of fertilizer, energy inputs for water deliveries, or other input requirements.

If returns are calculated with long-term intermediate commodity prices and yields, the diverse farm yields a 3–4% profit (depending on the sequence of the crop rotation), while the standard farm produces a net loss of 5%. Of course, there are additional issues associated with the increase in crop diversity, e.g., equipment needs and availability of processors and markets. But crop diversity may ensure less vulnerability to heatwaves and drought, as the different species vary in tolerance as well as timing of growth. This cost analysis suggests that a diversified farm may offer economic benefits to farmers in an era of climate change.

Farm margins offer another set of options for diversification. Only 33% of riparian zones in an irrigated cropland area in Region 2 contain stands of trees and shrubs (Young-Matthews et al. 2010). Wood C stocks, however, can increase 10-fold when native tree and shrub diversity is enhanced in restored areas along waterways, and if levee setbacks are made slightly wider than the typical steep-edged channel, then flooding potential and bank stability also improve. Planting hedgerows on field edges and installing tailwater ponds with planted edges are practiced by only a few farmers on a very small amount of land (Brodt et al. 2009). Public and private sector programs that provide assistance to farmers for restoration of farm margins are beginning to reverse these trends in Yolo County (Robins et al. 2001).

Climate change will undoubtedly impact wild species in the agricultural landscape (NCCP/HCP 2006). Swainson's Hawk is a species of special concern with a preference for nesting in old valley oak trees near alfalfa fields. Rodent-rich alfalfa crops and grain crops provide favorable foraging areas (Herzog 1996; NCCP/HCP 2006). Thus, future populations of this species will depend on the behavior of farmers with respect to: (1) fostering single, roost trees associated with field crops; (2) the economics of irrigating alfalfa as a cash crop under climate change, and (3) how the water district manages vegetation along levees (another location of roost trees). In the upland savannas and woodlands, oaks are expected to decline with climate change, due to shrinking of habitat ranges, and to increased fire frequency (Hayhoe et al. 2004; Kueppers et al. 2005). Conversion to grassland will affect C stocks, forage quality and shade for livestock, and wildlife populations (Barbour et al. 1993).

4 Mechanisms to implement climate change mitigation and adaptation

4.1 Grower decision tools and community strategies

In planning for climate change, farmers must make decisions that affect their management operations at different time scales. There will be a need for new information and tools to help make these decisions, and for merging mitigation and adaptation strategies. Decision support tools represent one of the most pressing new directions for climate change research, and effective tools will benefit from participatory input and outlook sessions with growers and other industry representations. The following are a few ideas for the types of education and decision tools that will make the agricultural community more aware and proactive in dealing with mitigation and adaptation to climate change:

- Guidelines for management practices for individual crops and cropping systems (e.g., organic vs. conventional) that mitigate GHG emissions with discussion of potential pitfalls such as associated yield or pest problems.
- Educational websites for growers to estimate their GHG footprint. For example, the Marin Carbon Project has launched a website related to rangeland C sequestration.
- Development of mechanisms to facilitate farmer participation in the California Climate Action Registry.
- Web-accessible spreadsheets and queries for individual crops and cropping systems to comply with California state government protocols to calculate, report, and verify GHG emission reductions.
- Rules and regulations that affect the adoption of GHG mitigation practices, e.g., the planting of woody, non-agricultural species on crop margins, canals, or sloughs.

- Development of different levels of participation in mitigation that are relevant to local cropping systems and land use types.
- Weather forecasting tools (e.g., AgClimate for the SE USA) to predict crop phenology and harvest after extreme events, for specific crop-location effects, and to design adaptive management.
- Designing more efficient produce distribution centers to reduce GHG emissions and encourage diversification. One example is a trucking center for small farms to reduce the miles traveled to pick up partial loads around the state.
- Programs for simplification and clarity in provisions for crop failure, e.g., insurance, subsidies etc.
- Creation of auction systems for farmers to engage in ecosystem restoration that is based on spatially explicit modeling and direct interaction with growers for best management practices on specific sites.

4.2 Scenarios for the future of agriculture and agricultural landscapes in Yolo County

Scenarios and storylines offer a way to explore possibilities and compare different outcomes. Consequently, we developed and analyzed three scenarios for the county for the period until 2050, in cooperation with farmers and personnel from governmental agencies and other organizations: high growth (IPCC A2—high emission), more sustainable (IPCC B1—lower emission), and most precautionary (AB 32-Plus, stricter than the climate change policy framework being established in California under AB 32). The narrative storylines 'downscaled' the emissions scenarios to regionally relevant situations, and were initially based on information in regional planning documents, but were then edited and augmented by stakeholders and a steering committee (Jackson et al. 2009). While the storylines are too locally-specific to be presented here, the process was essential in gaining access to new viewpoints and initiating awareness of climate change responses in the local community.

A2. "Regional enterprise." In this rapid growth and economic development scenario, the county population expands rapidly, urban land doubles, and farmland is lost. Agriculture would remain in a monoculture model with some changes in crop mix emphasizing higher value monocultures. Soil and land management and water usage would show little change, at the risk of large variation in production from year to year due to climate change-induced water shortages and flooding risks.

B1. "Global sustainability." In this more environmentally oriented scenario, the county population expands more slowly, and urban and rural residential encroachments on agricultural land are proportionally less. Growers diversify their crop mix for resilience, and reduce intensity of N-based fertilizer use and tillage. Conservation practices create wetlands in low-lying areas and vegetated corridors along waterways and farm margins. Cover cropping adds to soil fertility but reduces potential for income from cool-weather crops. Efficient water management practices are used extensively, organic-based practices increase C sequestration in soils, and these and other farming practices help reduce GHG emissions.

AB 32-Plus. "Precautionary change." In this greenest scenario, the county population stabilizes slightly above the current level, the urban footprint remains constant, farmland is strongly protected, and GHG emissions are most strongly reduced. Growers further diversify their crop mix, substantially increase heat- and drought-tolerant orchard crops, increase production by double cropping, and eliminate fossil fuel inputs to agriculture, both through fertilizers and motor vehicle fuels. Water use efficiency in dry years increases through crop mixes that reduce ET, and by alternative irrigation methods. Extensive conservation practices sequester C in wetlands and woodlands along waterways, using practices that minimize N_2O and methane emissions. Biodiversity increases in cropped and non-cropped areas of the landscape. Novel food systems encourage reduction of GHG emissions as well as new markets and greater resilience, creating the greatest long-term agricultural sustainability of all of the three scenarios.

Multiple influences in the state and region may dovetail with the types of agricultural practices and land use policies consonant with B1 and AB 32-Plus storylines, but to what extent such land use changes will actually come about is ultimately a political question. What we can say at present is that if B1 or AB 32-Plus storylines are followed, they are likely to result in substantial benefits to farmers and agricultural stakeholders in this county. These benefits include preservation of agricultural land, greater economic resilience due to a wider variety of crops and more intensive exploration of alternative farming practices, resource benefits related to reduced consumption of water and energy, and environmental benefits due to continuation of working agricultural landscapes on which some species depend as well as creation of additional habitat at farm margins and in restoration sites.

5 Conclusion

Sustainable regional responses to climate change for California's agricultural counties will undoubtedly vary from county to county, but based on our experience in Yolo County, exploration will be facilitated by including the following elements:

- Scaled-down modeling runs from GCM to determine likely temperature and precipitation impacts for the county or region in the 2050 and 2100 time frames.
- Crop-level analysis of likely impacts and adaptation strategies given the particular crops, climate, and ecology of the county or region.
- Landscape-level analysis of likely impacts and adaptation strategies given the particular crops, climate, and ecology of the area.
- Analysis of other factors potentially affecting the future viability of agriculture in the county or region: population growth and urbanization trends; economic factors affecting existing crops; potential economic feasibility of new crops including biofuels; and analysis of environmental variables including water use and availability, habitat status, and endangered species.
- Collaborative engagement throughout the research process with stakeholders, including farm organizations and local governments, to determine likely strategies and potential operational concerns.

Appendix

GIS Analyses of Agricultural Land Use and Soil Patterns

Geomorphic regions. The analysis to stratify Yolo County into four geomorphic regions used soil characteristics from the USDA SSURGO database. For each map unit, the component table in SSURGO was used to identify the name, soil order, and soil great group of its dominant soil component. The map unit was then assigned to one of the four geomorphic regions using a lookup table based upon the soil characteristics.

Crop types and soil characteristics. The relationship between cropping types and soil characteristics was analyzed by overlaying a land use map (California Department of Water Resources 1997) on the SSURGO map for Yolo County. The land use map gives 54 different crop types and 91 different land use types in total, as determined by combining values in the CLASS1 and SUBCLASS1 columns in the attribute table. The overlay of the land use map and the soils information facilitated cross-tabulating the areas and proportions of crop types with respect to the different geomorphic regions. The spatial overlays and cross-tabulations were performed using the spatial database PostGIS (http://postgis.refractions.net).

Flood frequency. Flood frequency values were taken from the dominant condition flooding frequency column in the map unit aggregated attributes table in the Yolo County SSURGO database. Four categories of flood frequency were listed in this table: Frequent (1–2 times/year), Occasional (>5 times every 50 years), Rare (once every 100 years), and None.

References

Backlund P, Janetos A, Schimel D (2008) The effects of climate change on agriculture, land resources, water resources, and biodiversity. Climate Change Science Program. Synthesis and Assessment Product 4.3

Baldocchi D, Wong S (2006) An assessment of the impacts of future CO_2 and climate on California agriculture. CEC-500-2005-187. California Energy Commission, PIER Energy-Related Environmental Research Program

Barbour M, Pavlik B, Drysdale F, Lindstron S (1993) California's changing landscapes. California Native Plant Society, Sacramento

Barnett TP, Pierce DW, Hidalgo HG, Bonfils C, Santer BD, Das T, Bala G, Wood AW, Nozawa T, Mirin AA, Cayan DR, Dettinger MD (2008) Human-induced changes in the hydrology of the western United States. Science 319:1080–1083

Bazzaz FA, Sombroek W (1996) Global climate change and agricultural production. Wiley, New York

Bloom AJ (2009) As carbon dioxide rises, food quality will decline without careful nitrogen management. Calif Agr 63:67–72

Brodt S, Klonsky K, Jackson LE, Brush SB, Smukler SM (2009) Factors affecting adoption of hedgerows and other biodiversity-enhancing features on farms in California, USA. Agroforestry Systems 76:195–206

Broome JC, Worthington M (2009) Research and extension needs assessment for organic growers in Sacramento, Solano and Yolo Counties. Univ of California Agriculture and Natural Resources. http://ceyolo.ucdavis.edu/

California Department of Water Resources (1997) Yolo County Land Use Database

California Department of Water Resources (2006) Progress on incorporating climate change into planning and management of California's water resources. http://baydeltaoffice.water.ca.gov/climatechange/reports.cfm

Cavagnaro T, Jackson LE, Scow KM (2006) Climate change: challenges and solutions for California agricultural landscapes. CEC-500-2005-189-SF. California Energy Commission, PIER Energy-Related Environmental Research Program

Cayan D, Tyree M, Dettinger M, Hidalgo H, Das T, Maurer E, Bromirski P, Graham N, Flick R (2009) Climate change scenarios and sea level rise estimates for the California Climate Change Scenarios Assessment. CEC-500-2009-014-D. California Energy Commission, PIER Energy-Related Environmental Research Program

CEC (California Energy Commission) (2005) Research roadmap for greenhouse gas inventory methods. CEC-500-2005-097. California Energy Commission, PIER Energy-Related Environmental Research Program

CEC (California Energy Commission) (2006) Inventory of California greenhouse gas emissions and sinks: 1990 to 2004. Sacramento, California

Condon AG, Hall AE (1997) Adaptation to diverse environments: variation in water-use efficiency within crop species. In: Jackson LE (ed) Ecology in agriculture. Academic, San Diego, pp 79–116, ISBN 0126312605

Conrad JH (1985) Feeding of farm animals in hot and cold environments. In: Yousef MK (ed) Stress physiology in livestock: ungulates, vol 2. CRC Press Inc, Florida, pp 205–226

de Graaff MA, van Groenigen KJ, Six J, Hungate B, van Kessel C (2006) Interactions between plant growth and soil nutrient cycling under elevated CO_2: a meta-analysis. Glob Chang Biol 12:2077–2091

De Gryze S, Wolf A, Kaffka SR, Mitchell JP, Rolston DE, Temple SR, Lee J, Six J (2010) Simulating greenhouse gas budgets of four California cropping systems under conventional and alternative management. Ecol Appl 20:1805–1819

deJong TM (2005) Using physiological concepts to understand early spring temperature effects on fruit growth and anticipating fruit size problems at harvest. Summerfruit Australia Quarterly 7:10–13

Del Grosso S, Ojima D, Parton W, Mosier A, Peterson G, Schimel D (2002) Simulated effects of dryland cropping intensification on soil organic matter and greenhouse gas exchanges using the DAYCENT ecosystem model. Environ Pollut 116:S75–S83

Dukes JS, Chiariello NR, Cleland EE, Moore LA, Shaw MR, Thayer S, Tobeck T, Mooney HA, Field CB (2005) Responses of grassland production to single and multiple global environmental changes. PLoS Biol 3:1829–1837

Field CB, Daily GC, Davis FW, Gaines S, Matson PA, Melack J, Miller NL (1999) Confronting climate change in California: ecological impacts on the Golden State. Report of the Union of Concerned Scientists and the Ecological Society of America. Cambridge, MA and Washington D.C.

Grismer ME, O'Geen AT, Lewis DJ (2005) Sediment control with vegetative filter strips (VFSs). Publ. 8195. Univ of California Department of Agriculture and Natural Resources

Hanson B, Hopmans JW, Simonek J (2008) Leaching with subsurface drip irrigation under saline, shallow groundwater conditions. Vadose Zone Journal 7:810–818

Hayhoe K, Cayan D, Field CB, Frumhoff PC, Maurer EP, Miller NL, Moser SC, Schneider SH, Cahill KN, Cleland EE, Dale L, Drapek R, Hanemann RM, Kalkstein LS, Lenihan J, Lunch CK, Neilson RP, Sheridan SC, Verville JH (2004) Emissions pathways, climate change, and impacts on California. Proc Natl Acad Sci USA 101:12422–12427

Herzog SK (1996) Wintering Swainson's Hawks in California's Sacramento-San Joaquin River Delta. The condor. 98:876–879

Hill JK, Thomas CD, Huntley B (1999) Climate and habitat availability determine 20th century changes in a butterfly's range margin. Proc R Soc Lond B Biol Sci 266:1197–1206

Horwath WR, Devêvre OC, Doane TA, Kramer AW, van Kessel C (2002) Soil C sequestration management effects on N cycling and availability. In: Lal R, Follett RF, Kimble JM (eds) Agricultural practices and policies for carbon sequestration in soil. Lewis, Florida, pp 155–164

Hungate BA, Stiling PD, Dijkstra P, Johnson DW, Ketterer ME, Hymus GJ, Hinkle CR, Drake BG (2004) CO_2 elicits long-term decline in nitrogen fixation. Science 304:1291

Jackson LE, Ramirez I, Yokota R, Fennimore SA, Koike ST, Henderson D, Chaney WE, Calderón FJ, Klonsky K (2004) On-farm assessment of organic matter and tillage management on vegetable yield, soil, weeds, pests, and economics in California. Agric Ecosyst Environ 103:443–463

Jackson LE, Pascual U, Hodgkin T (2007) Utilizing and conserving agrobiodiversity in agricultural landscapes. Agric Ecosyst Environ 121:196–210

Jackson LE, Santos-Martin F, Hollander AD, Horwath WR, Howitt RE, Kramer JB, O'Geen AT, Orlove BS, Six JW, Sokolow SK, Sumner DA, Tomich TP, Wheeler SM (2009) Potential for adaptation to climate change in an agricultural landscape in the Central Valley of California. CEC-500-2009-044-D. California Energy Commission, PIER Energy-Related Environmental Research Program

Johnston WE, McCalla AF (2004) Whither California agriculture: up, down, or out? Giannini Foundation Special Report, Univ of California

Kallenbach CM, Rolston DE, Horwath WR (2010) Cover cropping affects soil N_2O and CO_2 emissions differently depending on type of irrigation. Agric Ecosyst Environ 137:251–260

Kong AYY, Fonte SJ, van Kessel C, Six J (2009) Transitioning from standard to minimum tillage: trade-offs between soil organic matter stabilization, nitrous oxide emissions, and N availability in irrigated cropping systems. Soil Till Res 104:256–262

Krusekopf HH, Mitchell JP, Hartz TK, May DM, Miyao EM, Cahn MD (2002) Pre-sidedress soil nitrate testing identifies processing tomato fields not requiring sidedress N fertilizer. Hortscience 37:520–524

Kueppers LM, Snyder MA, Sloan LC, Zavaleta ES, Fulfrost B (2005) Modeled regional climate change and California endemic oak ranges. Proc Natl Acad Sci USA 102:16281–16286

Landon R (2009) Wetlands—problems and solutions. Yolo County Farm Bureau News. September issue, p. 12

Lee H, Sumner D (2009) Economics of biofuels feedstock in California: Projections based on policy and market conditions, Chevron-UC Davis Research Agreement Final Report (Project Number, RSO #8)

Lee H, Sumner DA, Howitt RE (2001) Potential economic impacts of irrigation-water reductions estimated for Sacramento Valley. Calif Agr 55:33–40

Lee J, De Gryze S, Six J (2010) Effect of climate change on field crop production in the Central Valley of California. Climatic Change, this issue

Lobell DB, Field CB (2009) California perennial crops in a changing climate. CEC-500-2009-039-F. California Energy Commission, PIER Energy-Related Environmental Research Program

Long SP, Ainsworth EA, Leakey ADB, Nosberger J, Ort DR (2006) Food for thought: lower-than-expected crop yield stimulation with rising CO_2 concentrations. Science 312:1918–1921

Luedeling E, Zhang M, Girvetz EH (2009) Climatic changes lead to declining winter chill for fruit and nut trees in California during 1950–2099. PLoS One 4(7):e6166. doi:10.1371/journal.pone.0006166

Mikkelsen DS (1983) Field crops. In: Scheuring AF (ed) A guidebook to California agriculture. Univ of California Press, Berkeley, pp 109–135

Miller NL, Jin J, Hayhoe K, Auffhammer M (2007) Climate change, extreme heat, and electricity demand in California. CEC-500-2007-023. California Energy Commission, PIER Energy-Related Environmental Research Program

Mitloehner FM, Sun H, Karlik JF (2009) Direct measurements improve estimates of dairy greenhouse-gas emissions. Calif Agr 63:79–83

Mitsch WJ, Gosselink JG (2000) Wetlands, 3rd edn. Wiley, New York

Moya TB, Ziska LH, Namuco OS, Olszyk D (1998) Growth dynamics and genotypic variation in tropical, field-grown paddy rice (*Oryza sativa* L.) in response to increasing carbon dioxide and temperature. Glob Chang Biol 4:645–656

NCCP/HCP (Yolo County Natural Community Conservation Plan/Habitat Conservation Plan) (2006) Yolo County natural community conservation plan/habitat conservation plan

O'Farrell PJ, Anderson PML (2010) Sustainable multifunctional landscapes: a review to implementation. Curr Opin Environ Sustain 2:59–65

O'Geen AT, Schwankl LJ (2006) Understanding soil erosion in irrigated agriculture. Publ. 8196. Univ of California Department of Agriculture and National Resources

O'Geen AT, Prichard TL, Elkins R, Pettygrove GS (2006) Orchard floor management to reduce erosion. Publ. 8202. Univ of California Department of Agriculture and Natural Resources

Palm CA, Smukler SM, Sullivan CC, Mutuo PK, Nyadzi GI, Walsh MG (2010) Identifying potential synergies and trade-offs for meeting food security and climate change objectives in sub-Saharan Africa. Proc Natl Acad Sci USA 46:19661–19666

Parmesan C, Yohe G (2003) A globally coherent fingerprint of climate change impacts across natural systems. Nature 421:37–42

Peng SB, Huang JL, Sheehy JE, Laza RC, Visperas RM, Zhong XH, Centeno GS, Khush GS, Cassman KG (2004) Rice yields decline with higher night temperature from global warming. Proc Natl Acad Sci USA 101:9971–9975

Poudel DD, Horwath WR, Mitchell JP, Temple SR (2001) Impacts of cropping systems on soil nitrogen storage and loss. Agr Syst 68:253–268

Purcell AH, Hopkins DL (1996) Fastidious xylem-limited bacterial plant pathogens. Annu Rev Phytopathol 34:131–151

Rachmilevitch S, Cousins AB, Bloom AJ (2004) Nitrate assimilation in plant shoots depends on photorespiration. Proc Natl Acad Sci USA 101:11506–11510

Reich PB, Hobbie SE, Lee T, Ellsworth DS, West JB, Tilman D, Knops JMH, Naeem S, Trost J (2006) Nitrogen limitation constrains sustainability of ecosystem response to CO_2. Nature 440:922–925

Reilly JM, Graham J (2001) Agriculture: the potential consequences of climate variability and change for the United States. Cambridge Univ Press, New York

Robins P, Holmes RB, Laddish K (2001) Bringing farm edges back to life. Yolo County Resource Conservation District. http://www.yolorcd.org/library/Farm%20Edges%20v5.pdf#search=%22Bringing%20farm%20edges%20back%22

Sato S, Peet MM, Thomas JF (2000) Physiological factors limit fruit set of tomato (*Lycopersicon esculentum* Mill.) under chronic, mild heat stress. Plant Cell Environ 23:719–726

Scherm H (2004) Climate change: can we predict the impacts on plant pathology and pest management? Can J Plant Pathol 26:267–273

Schneider MK, Lüscher A, Richter M, Aeschlimann U, Hartwig UA, Blum H, Frossard E, Nosberger J (2004) Ten years of free-air CO_2 enrichment altered the mobilization of N from soil in *Lolium perenne* L. swards. Glob Chang Biol 10:1377–1388

Shaw MR, Zavaleta ES, Chiariello NR, Cleland EE, Mooney HA, Field CB (2002) Grassland responses to global environmental changes suppressed by elevated CO_2. Science 298:1987–1990

Six J, Ogle SM, Breidt FJ, Conant RT, Mosier AR, Paustian K (2004) The potential to mitigate global warming with no-tillage is only realized when practiced in the long-term. Glob Chang Biol 10:155–160

Smit B, Skinner MW (2002) Adaptation options in agriculture to climate change: a typology. Mitig Adapt Strat Glob Chang 7:85–114

Smukler SM, Jackson LE, Moreno SS, Fonte SJ, Ferris H, Klonsky K, O'Geen AT, Scow KM, Steenwerth KL (2010) Biodiversity and multiple ecosystem functions in an organic farmscape. Agric Ecosyst Environ 139:80–97

Sokolow AD, Kuminoff NV (2000) Farmland, urbanization, and agriculture in the Sacramento region. Prepared for Capital Region Institute, Regional Futures Compendium

Southwick SM, Uyemoto J (1999) Cherry crinkle-leaf and deep suture disorders. Publ. 8007. Univ of California Department of Agriculture and National Resources

Spencer W, Noss R, Marty J, Schwartz M, Soderstrom E, Bloom P, Wylie G (2006) Report of independent science advisors for Yolo County, natural community conservation plan/habitat conservation plan (NCCP/HCP). Conservation Biology Institute

Jones and Stokes (2001) A framework for the future: Yolo bypass management strategy: (J&S 99079) August. Prepared for Yolo Basin Foundation, Davis, CA. Sacramento, CA

UC Davis Agriculture and Resource Economics Department (2008) University of California at Davis. http://coststudies.ucdavis.edu/current.php)

Van Groenigen KJ, Six J, Hungate BA, de Graaff MA, van Breemen N, van Kessel C (2006) Element interactions limit soil C storage. Proc Natl Acad Sci USA 103:6571–6574

Veenstra JJ, Horwath WR, Mitchell JP, Munk D (2006) Conservation tillage and cover crop influences soil properties in the San Joaquin Valley. Calif Agr 60:146–153

Weaver W, Shannon CE (1949) The mathematical theory of communication. Univ of Illinois, Urbana

West JW (2003) Effects of heat-stress on production in dairy cattle. J Dairy Sci 86:2131–2144

Wheeler TR, Hadley P, Morison JIL, Ellis RH (1993) Effects of temperature on the growth of lettuce (*Lactuca sativa* L.) and the implications for assessing the impacts of potential climate change. Eur J Agron 2:305–311

Williams JW, Seabloom EW, Slayback D, Stoms DM, Viers JH (2005) Anthropogenic impacts upon plant species richness and net primary productivity in California. Ecol Lett 8:127–137

Wine Institute (2002) Wine Institute web site http://www.wineinstitute.org/communications/pierces_disease/pierces_disease_update.htm

Yamamura K, Kiritani K (1998) A simple method to estimate the potential increase in the number of generations under global warming in temperate zones. Appl Entomol Zool 33:289–298

Yolo County (2007) Integrated regional water management plan. Water Resources Association of Yolo County http://www.yolowra.org/irwmp.html

Yolo County Agricultural Commissioner (1984–2007) Crop statistics. http://www.yolocounty.org/Index.aspx?page=486

Young-Matthews A, Culman S, Sanchez-Moreno S, O'Geen AT, Ferris H, Hollander AD, Jackson LE (2010) Plant-soil biodiversity relationships and nutrient retention in agricultural riparian zones of the Sacramento Valley, California. Agroforestry Systems 80:41–60

Ziska LH, Namuco O, Moya T, Quilang J (1997) Growth and yield response of field-grown tropical rice to increasing carbon dioxide and air temperature. Agron J 89:45–53

The impact of climate change on California timberlands

L. Hannah · C. Costello · C. Guo · L. Ries · C. Kolstad ·
D. Panitz · N. Snider

Received: 12 February 2010 / Accepted: 26 September 2011 / Published online: 24 November 2011
© Springer Science+Business Media B.V. 2011

Abstract California timber production has been declining in an era of warming, increased wildfires, land conversion, and growing emphasis on recreation. Climate change has the potential to further affect California timber production through changes in individual tree growth rates, forest dieback, and shifts in species ranges and ecosystem composition. Coupled with changes in global timber prices, themselves the result of productivity effects, these production impacts hold important consequences for California's private timberlands. This study uses models that project tree species productivity and movement across the landscape under climate change, coupled with economic models of landowner adaptation and returns from multiple harvest strategies. Our results show that under likely price scenarios, climate change will result in an overall decline in harvested timber value relative to no climate change, with decreases of 4.9 to 8.5% by the end of the century. The magnitude of decrease depends on climate change scenario, price scenario and management option – with dollar losses totaling up to $8.1 billion in total land value (−$2.7 billion given a 4% discount rate). There is substantial spatial variation in these changes; most areas show significant declines in timber value while some show modest increases relative to a no climate change baseline. If prices are not affected by climate change, more areas experience gains in value. We find that forestry management strategies can mitigate lost value, indicating that climate change adaptation programs can yield important economic benefits. Declining timber value corresponds disproportionately to areas already experiencing timberland conversion to housing or agriculture. Policy measures to stem conversion of devalued timberlands may warrant consideration.

Electronic supplementary material The online version of this article (doi:10.1007/s10584-011-0307-2) contains supplementary material, which is available to authorized users.

L. Hannah
Conservation International, Arlington, VA, USA

L. Hannah (✉) · C. Costello · C. Guo · L. Ries · C. Kolstad · D. Panitz · N. Snider
University of California Santa Barbara, 2400 Bren Hall, Santa Barbara, CA 93106-5131, USA
e-mail: lhannah@conservation.org

1 Introduction

California timber production has been declining in an era of increased warming, wildfires, recreational use, and land use change. Production fell 52% from 1989 to 1999 and another 62% in the subsequent decade (Fig. 1). Diminished sales from federal lands, driven by greater recreational use and protection of threatened species, has been the largest contributing factor to the decline. Climate change has the potential to further alter California timber production.

This study quantifies the likely ecological and economic consequences of climate change to California's timberlands. Warming may promote growth in certain forests, while drier conditions or earlier snowmelt may reduce growth and harvest potential. Climate change impact may also be mediated by market factors: if growth of northern softwood forests increases with warming, global timber prices may fall, depressing California profits. The combined effects of climate change, a declining timber industry, and booming land values may drive widespread conversion of timberlands to other land uses such as exurban housing and vineyard expansion. The interplay of climate change, declining production, and land use demands therefore has the potential to dramatically alter California landscapes.

There is a rich forestry literature on changes due to climate change, which incorporates dynamics in species distributions and productivity. With a focus on timber growth and productivity, that literature pays much less attention to how foresters make dynamic economic decisions in the face of changing conditions (Joyce et al. 1995, 2001; Perez-Garcia et al. 1997). Conversely, other papers introduce dynamic forest market models but rely on static climate and ecological models (Callaway et al. 1994, Alig et al. 1997). Prior studies integrating dynamic ecological models and global economic models have predicted global increases in timber production resulting in decreased timber prices (e.g., Sohngen et al. 2001). However, general patterns seen at the global scale do not necessarily reflect those projections at the state level, especially in the case of California's high climatic and landscape heterogeneity. Mendelsohn and Smith (2006) modeled climate change impacts on the timber industry in California at coarse spatial resolution and without considering species range shifts or land manager response.

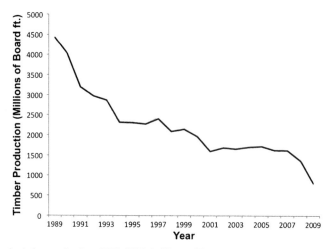

Fig. 1 California timber production, 1989–2009 (millions of board feet). Adapted from CA State Board of Equalization Timber Yield Tax & Harvest Schedules (California State Board of Equalization 2010)

It is necessary to use more specific models that incorporate downscaled general circulation model (GCM) data to predict species movement across the landscape and species-specific growth parameters to model optimization strategies for landowners. Our new structural approach to modeling dynamic decision making integrates the ecosystem response into the optimization decisions of forest landowners, and complements the hedonic approach taken by Sohngen and Mendelsohn 2003.

The goal of this study is to assess climate change impacts on California's timberlands, building on previous efforts by emphasizing value returns to land and incorporating dynamics in species distributions, productivity, and land manager behavior. Two biological models estimate climate-driven changes in productivity and species ranges, while an economic model simulates land manager behavior in three management options. This coupled approach provides more robust estimates of likely climate change-driven changes in timberland value, and can guide insights into policy measures that may influence land manager behavior and reduce losses. We restrict our analysis to California's privately held timberlands, assuming that harvest from federal lands will continue to decline and be heavily regulated, largely insulating this harvest from market forces.

2 Methods

We integrate models of biological productivity, landowner behavior, and economic value to estimate climate change impacts on California's private timberlands. The potential for carbon markets to alter commercial forestry decisions is examined with a model using fixed carbon prices, while changing timber prices are modeled based on values in the literature. While these models provide a sophisticated assessment of climate change impacts, such a broadly integrated approach requires many simplifying assumptions regarding biological, market, and industry behavior. These include omission of alternative forest biomass uses (e.g. chipping for fuel), biological effects of large scale disturbances (e.g. fire or insect attack), and industry advances that might increase land values or respond to new markets.

2.1 Domain and climatologies

Our study domain was the state of California. From a statewide grid at 1/8° (~12.5 km) resolution, we include the 791 grid cells classified as "private" or "working" by the California Department of Forestry. Douglas Fir, Ponderosa Pine, Redwood, and Western Hem-Fir together represent over 92% of harvest value on California private lands (Spero 2006). These four species are modeled, with Western Hemlock representing Western Hem-Fir (a composite category including Western Hemlock and five fir species whose wood is considered equivalent for tax purposes). Some species physiological parameters were obtained from field measurements of tree growth, stand densities, and other observable quantities, and others estimated from basic physical processes. Species parameters are taken from prior studies: Law et al. (2000) for Ponderosa Pine; Coops and Waring (2001) for Douglas Fir; and Koch et al. (2004) and Busing and Fujimori (2005) for Redwood. Growth rates were calibrated against plot data from the U.S. Forest Service Forest Inventory and Analysis (FIA) program.

Climate data provided by the California Applications Program/California Climate Change Center (CAP/CCCC) consist of monthly values for minimum temperature, maximum temperature, and precipitation for the years 1950–2099. These time series are derived from individual runs of several GCMs, downscaled to 1/8° resolution for the State of California. The present study uses two models: the National Center for Atmospheric Research Parallel Climate Model version 1 (PCM1) and the Geophysical Fluid Dynamics Laboratory Climate Model version 2.1 (GFDL). We smooth the data using a 31-year moving average filter, producing monthly values for 2055 through 2085. Emissions scenarios were Intergovernmental Panel for Climate Change (IPCC) A2 and B1 emissions scenarios. Soil texture is assumed to be sandy loam, and available water capacity [mm, depth equivalent] is derived from the STATSGO soils dataset. Soil fertility data are generally poor, with few guidelines for choosing a fertility ranking (Landsberg et al. 2002), but fertility values modeled between 0.19 and 0.53 compared well with field measurements in western Oregon (Swenson et al. 2005); soil fertility rank is assumed to be 0.4 for all sites. Monthly radiation is calculated from slope, aspect, latitude, and elevation using the method in Coops et al. (2000) with data from a U.S. Geological Survey digital elevation model (DEM) at 1-km resolution.

2.2 Ecological models

Our ecological models included models of species distribution and distribution change in response to climate change, and models of productivity. The productivity model was used to simulate the effects of climate change on productivity of individual species at locations with specified climatic conditions, both in the present and under climate change. Species Distribution Models were used to determine which species would exist at a location, both under present climate and in the future under changed climates.

The productivity model simulates climate change effects on timber growth. Growth is predicted using 3-PG (Physiological Principles for Predicting Growth), a simplified physiological process model developed by Landsberg and Waring (1997) for applying forest process models to forestry management. Growth is modelled in single-species, even-aged stands, each simulation beginning with a one-hectare stand stocked with a specified number of seedlings. Monthly climate data (temperature, precipitation, insolation) and species physiology are used to calculate carbon fixation rates, and the resulting biomass is partitioned into foliage, stems, and roots according to species-specific allometric ratios. A simplified hydrologic model tracks soil moisture and limits photosynthetic efficiency in periods of drought. Nutrient limitation is incorporated via a fertility rank parameter with low fertility values inhibiting photosynthesis and causing a greater proportion of biomass to be allocated to roots. Growth becomes less efficient as the stand ages and levels off upon reaching maturity.

As the stand matures, the population is thinned following the −3/2 power law observed in terrestrial plant populations.[1] The model tracks state variables describing the stand, including leaf area index (LAI), basal area, and tree component biomass. Stem volume is the essential metric for timber valuation, so the stem biomass is converted to volume for the economic model. The model runs in MATLAB, using code adapted from the Visual Basic implementation included with 3PGpjs (Sands 2001), a 3-PG interface. This model gave production simulations for all target species across the 791 locations modeled. Species occurring at a location were determined by species distribution models.

Species distributional changes were simulated with BIOMOD species distribution modeling (SDM) software (Thuiller 2003), which executes multiple SDMs using a common

---

[1] A self-thinning relationship has been observed in a broad range of plant species, in which a unit decrease in density will occur for every 1.5-unit increase in biomass (Li et al. 2000).

set of species presence and absence data in conjunction with environmental data. The SDM outputs were combined into a consensus model using a principal components analysis. BIOMOD input data included both climate variables (annual temperature range, mean temperature of the driest quarter, mean temperature of the coldest quarter, precipitation seasonality, precipitation in the wettest quarter, and precipitation of the warmest quarter) and soil variables from the State Soil Geographic Database (STATSGO; available water capacity [fractional volume of water per volume of soil], soil depth [in meters, m], pH, salinity [millimhos per centimeter, or mmhos cm^{-1}] and depth to water table [m]). BIOMOD output was used to calculate relative species abundance, and environmental data were extracted for each predicted presence point and subjected to a z-transformation. The transformed values were summed across all variables. The absolute values of these data were rescaled from 0 to 1 to provide a measure of relative abundance for each 1-km cell. These values were used as the metric for scaling abundance for each species across the landscape.

2.3 Economic model

2.3.1 Model description

Our approach extends prior studies on climate change impacts on forestry by incorporating forester harvesting and replanting decisions, species-specific growth rates and timber prices, land-use change, and carbon markets. We assume foresters seek to maximize net present value of timberland management by altering rotation period and species replanted after harvest, based on site characteristics such as soil, temperature, precipitation, and aspect, which dictate growth potential for any species. Each 1/8° site is independently optimized and is able to track the value and harvest decisions of multiple species over time. Where site characteristics remain constant over time, foresters calculate the value-maximizing rotation period for each viable species and replant the species with highest net present value. Climate change introduces dynamics into the problem, complicating forester decisions due to (1) changes in site suitability, (2) growth parameters that are species-specific and time-varying with climate change, (3) price changes over time, and (4) payments for carbon sequestration.

2.3.2 Management scenarios

Our approach allows forester decisions to adapt to changing site characteristics and prices. Possible landowner behaviors are consolidated into three management scenarios: "naïve," "rotation," and "optimal." Naïve foresters cannot change species or rotation period; this non-adaptive approach yields lowest profits. In the rotation scenario, landowners alter rotation period to maximize profit in response to current conditions, but cannot alter species or anticipate changing site suitability. Optimal management allows foresters to anticipate and accommodate climate change with rotation and replanting decisions. We derive optimal landowner decisions every year for each age (0 to 100 years) and species (Douglas Fir, Ponderosa Pine, Redwood, and Hemlock). The "value function" for each year is a matrix of values across ages and species, providing net present value of optimized rotation and replanting decisions for any given starting condition until the terminal period. The value function is generated in each time period, and values for the starting year are generated by iterating backwards from the terminal period to the starting period. Climate change effects on total land value are calculated by comparing the starting year value functions under climate-change and no-climate-change scenarios.

2.3.3 Integration with biological model

For integration with the economic model, biological growth model output (a series of annual volumes) is fit to a function to generate a growth curve of the form Volume (age)= exp (alpha-beta/age). The parameters alpha and beta depend upon site, species, birth year, and climate scenario. The renormalization ratio is multiplied by Mendelsohn's average volume function (Mendelsohn and Smith 2006) for a species-, site-, establishment year-, and climate change scenario-specific volume for each tree. Stated differently, we took our variation in volume relative to our average volume and rescaled it to Mendelsohn's average volume to get new volume function parameters. The end result mathematically is a change in the alpha terms while preserving Mendelsohn's original beta.

2.3.4 Global timber prices

There is a small but growing literature on global timber prices under climate change. We use the UIUC model regeneration variant found in Sohngen et al. (2001) from 2000 to 2080 for our price series, and assume that California landowners have full access to the global timber market. The UIUC regeneration variant is intermediate among the four models developed by Sohngen et al. differs no more than 10% from any one of them at any time step. We use this one price series for all GCMs and emissions scenarios because the differences between these curves are small relative to the price effect of climate change. Figure 2, panels A and B show these six price series. Global output of lumber increases in all cases, but more so under climate change, driving greater price decrease.

Assuming midsize volume per log, we apply global percent price change to each species' specific price (State Board of Equalization 2007), and calculate average harvest value across all timber value areas in California. Dollar values for statewide gains or losses were obtained by multiplying percentage gain or loss by current (2007) harvest value. Panel B shows that prices increase over time in both cases, but projected prices are lower in the climate change case relative to the baseline case.

2.3.5 Carbon payment

To illustrate the possible consequences of a carbon payment, we assume a simple fixed-price payment on the timber volume increment (proportional to amount of carbon sequestered) in a given cell. The present value of carbon payments over one rotation interval of length T is the discounted sum of all the incremental values of carbon added each year. Following van Kooten et al. (1995), present value can be represented as:

$$\int_0^T p^c \alpha v'\,(t)e^{-rt}dt$$

Where p^c is the carbon price per unit sequestered, and the coefficient α both translates volume into carbon and accounts for the fact that not all carbon is actually sequestered. Integrating by parts gives:

$$p^c \alpha v(T)e^{-rT} + rp^c \alpha \int_0^T v(t)e^{-rt}dt$$

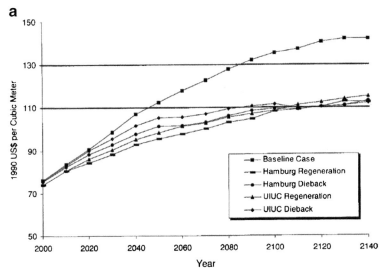

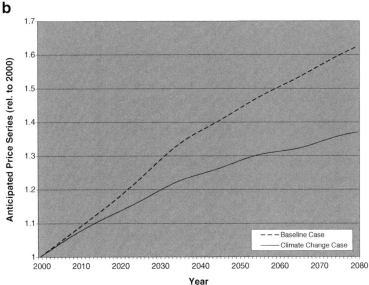

Fig. 2 Global timber prices series over time **a** Global price series from Sohngen et al. 2001; **b** price series derived for this study

Total payment is represented as the sum of two payments: a final payment at the time of harvest (T) and a payment on the stock of carbon each year. The first payment is the added value from permanent carbon storage, while the second is a rental equivalent of holding carbon out of the atmosphere for each year, priced at $r*p^c$. The forester essentially "rents out" carbon storage space each year for the price of $r*p^c$ and at the end of the harvest cycle, sells the carbon he owns at the price of p^c. Present value of carbon payments is converted to current value of carbon payments and added to the timber value at harvest. Carbon payment is thus capitalized up to the harvest date, and it is *as if* the forester receives two payments at the time of harvest: a payment for the value of timberland a payment for the value of the carbon that was sequestered during its lifetime.

Of the wood harvested, a percentage will decay and its carbon will be released back into the atmosphere; this is captured by the parameter α. We thus differentiate harvested carbon from the amount "permanently" sequestered, assuming 0.6 tons of carbon dioxide permanently sequestered per cubic meter of wood. We make the simplifying assumption in the empirical analysis that p^c is constant over our study period with a value of $20 per ton of carbon. Harvest value at any time step comprises both timber and carbon value. The value function iteration (described above) is performed with a terminal date far enough into the future as not to affect decisions made during the 2000–2080 study period. In this case it is assumed that at the terminal date of 2200, the value of a stand consists solely of harvest value, with no replanting value. Because available growth parameter and price projections end in 2080, these are assumed to stabilize at 2080 levels. Replanting costs of $988 per hectare (Mendelsohn and Smith 2006), and harvest costs of $31 per cubic meter are assumed for all species.

3 Results

The results presentation focuses on the A2 emissions scenario, which is closest to current emissions pathways. B1 scenarios were also run, but spatial patterns in B1 results were very similar to those for A2 (albeit with less dramatic change), so only A2 results are displayed here, with B1 results available in the online supplemental materials.

3.1 Baseline value change (no climate change price effect)

Climate factors affect both timber production and price, with temperature- and precipitation-driven range and productivity shifts affecting both wood volume and species mix. Productivity is expected to increase in many high latitude timberlands currently limited by low temperatures. We first examined the effect of changes in production alone, by allowing price to increase with time as if unaffected by climate change. Value change is aggregated across time into net present value, useful for tracking harvest value and comparing land value to competing land uses such as housing and agriculture. Climate change impacts are presented as percent change in net present value of land in the climate change scenario relative to the baseline scenario (Fig. 3). We focus our results assuming optimal landowner behavior; differences between management scenarios are minor relative to spatial variation within scenarios. Warmer temperatures were found to drive generally greater tree growth and harvest across a range of site conditions and species mixes. Value change in the PCM1 scenario is generally neutral or positive, while change in the GFDL scenario is more positive, with strong gains outweighing declines and moderate gains. GFDL results also show a much larger range of values than those for PCM1.

The warmer A2 scenario yielded greater value increases under both GCMs than did the cooler B2 scenario, though spatial variability within climate scenarios was much stronger than variability between scenarios. Despite general increases in value, dramatic value decreases in some areas signal possible cessation of forestry and conversion to other land uses.

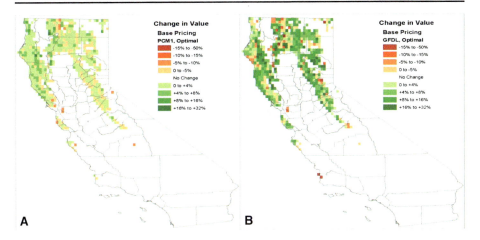

Fig. 3 Percent value change relative to no climate change for PCM1 A2 scenario **a** and GFDL A2 scenario **b**. Optimal management scenario; baseline (non-climate change) price series

3.2 Value change with climate change price effects

In order to take into account the additional effect of alteration in global timber prices due to climate change, we now apply a more realistic price scenario of global timber price trends. As seen in Fig. 2, price projections show relative price declines under climate change (prices are relatively lower in the climate change situation than the baseline situation, although they increase in both situations). The overall climate change impact on value at each location is the result of altered productivity and range changes associated with warmer temperatures interacting with climate change price effects due to increased timber supply. Figure 4 shows the spatial distribution of value change using the price series that incorporates climate change effects. In contrast to results from Section 3.1, nearly all cells show strong declines in value relative to baseline. The PCM1 scenario, showing mostly neutral change in Fig. 4a, is strongly affected by the relative decline in price. GFDL values are less impacted, but most productivity increases are nonetheless overwhelmed by relative price declines.

This price scenario, as with the last, exhibits high spatial variability, greater within than between climate scenarios. PCM1 results are dominated by value declines of 10%–15%, with declines of 15%–50% not uncommon. GFDL output shows many areas with large declines, some with little or no change, and very few increases. Areas of strong decline are found in the Santa Cruz mountains, parts of the Sierra, the northern border, and inland along the Mendocino coast. Already significant in these parts of the state, land conversion is likely to increase as declining timber values compete with rising agriculture and development land values. However, in some areas differences between climate scenarios introduce uncertainty regarding conversion threat. For example, areas north of San Francisco and bordering the Central Valley lose substantially less value under GFDL than PCM1, implying a comparatively lower incentive for conversion.

3.3 Carbon market economic impact

Modeling carbon sequestration as a function of biomass, we introduced a carbon market to the dynamic optimization problem, creating tradeoffs between carbon revenue and forestry income. The resulting impact on timber value losses can approximate the potential for

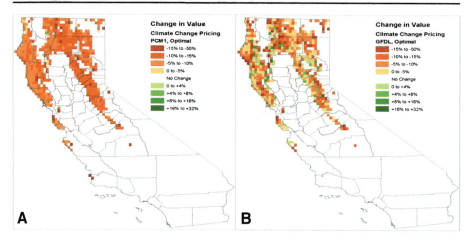

Fig. 4 Percent value change relative to climate change for PCM1 A2 scenario **a** and GFDL A2 scenario **b**. Optimal management; climate change price series

markets to mitigate economic impacts of climate change. Figure 5 illustrates these impacts with (A) and without (B) a carbon market under the GFDL A2 climate change scenario. The spatial distribution of economic impact of climate change is similar in each case, though with different magnitudes. The difference between Fig. 5a and b, shown in c, demonstrates that the carbon market diminishes climate change impact in nearly every location. The effect is greatest in the Sierra and lowest in Mendocino and Humboldt, where timber values are high.

3.4 Statewide economic impact

Results were aggregated to the state level to determine the overall economic impact of climate change to California's timberlands for the A2 emissions scenario.

The climate change price series shows substantial negative impacts relative to no climate change under all management types and climate change scenarios, with losses ranging from 4.4% to 8.5% by 2080 under A2. The baseline price series projects small gains in all scenarios, with increases between 1.5% and 5.7% by 2080. Tables 1 and 2 present gains or losses in millions of U.S. dollars with a 0% discount rate. By 2080, change in the net present value of statewide timber revenues range from losses of $8.1 billion[2] relative to no climate change to gains of $5.4 billion, depending on GCM, management, and price assumptions.

There is a strong temporal discontinuity in all A2 climate change price scenarios for all management types, with relative stability between 2020 and 2050 followed by pronounced loss in value by 2080. This loss may reflect the global price effect overwhelming climate-driven productivity gains later in the century. There are no other strong temporal discontinuities in these findings; gains and losses for other scenarios remain constant or gradually build throughout the century.

The most important finding across all scenarios is the negative trend in timberland value associated with a climate change-driven relative decline in global timber prices. This

---

[2] If we were to apply a 4% discount rate, given the value changes over time specified in Table 2, our non-discounted value change of -$8.1 billion would translate to -$2.7 billion.)

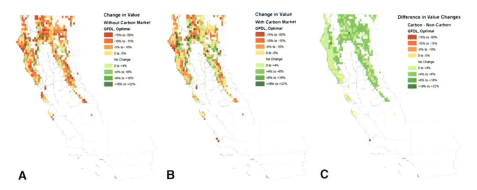

Fig. 5 a Percent value change relative to no climate change for GFDL A2 scenario without a carbon market **a**, with a carbon market **b**, and the difference between the two **c**

finding supports prior studies of climate change effects on California timber revenues. The spatial distribution of potential losses is significant for California policy, as they losses intersect strongly with areas under current threat of land conversion. In areas where such overlap occurs, climate-driven loss of timber value may gain importance among the factors shaping timberlands land-use policy.

3.5 Value of adaptation

Adaptation to climate change is increasingly critical in many settings as global policy responses to climate change lag behind emissions. However, in the case of California timberlands examined here, the role of adaptation is modest. Adaptation of harvest practices can only compensate for a small amount of the losses incurred by climate change, primarily because much of the decrease in value is driven by declines in global prices.

The losses incurred due to climate change are in the billions of dollars (0% discount rate), but this is on a base of nearly $100 billion in value. The climate change declines are therefore modest, though very significant in absolute terms. The reduction of loss achievable with adaptation of timber practices is a small fraction of this already modest amount.

What happens in global timber prices is much more important than adaptation of local timber practices. Global price effects (declines due to increasing high-latitude productivity) swamped effects of climate change on local productivity and adaptation measures (our 'Adaptation' management scenario).

But adaptation extends beyond individual sectors. Land owners will attempt to adapt to maximize income or well-being, with timber production being only one component of their overall adaptation efforts. So where adaptation of timber practices has little effect on maintaining incomes, landowners are likely to turn to other options, such as removing land from timber and converting it to more profitable uses.

3.6 Land use and policy implications

Our results predicting likely incentives favoring land use change may be useful for guiding policy interventions to maintain working timberlands or forests. Already under increasing

Table 1 Percent change in statewide timber value: A2 scenario

GCM	Scenario	Management Scenario								
		Naïve			Rotation			Optimal		
		2020	2050	2080	2020	2050	2080	2020	2050	2080
N/A	Baseline[a]	–	–	–	–	–	–	–	–	–
PCM1	Climate Change[b]	1.5%	1.4%	1.5%	1.5%	1.4%	1.6%	1.7%	1.5%	1.8%
	Climate Change & Global Prices[c]	−3.5%	−3.6%	−8.5%	−3.3%	−3.4%	−8.1%	−3.2%	−3.3%	−8.1%
GFDL	Climate Change	5.6%	5.4%	5.6%	5.7%	5.5%	5.7%	5.3%	4.9%	5.3%
	Climate Change & Global Prices	0.5%	0.4%	−4.7%	0.6%	0.5%	−4.4%	0.2%	0.0%	−4.9%

[a] No climate change effects on productivity or price

[b] Timber value change under the indicated climate change scenario

[c] Timber value change under the indicated climate change scenario and climate-driven global price series

conversion pressure from high-value alternate uses, California timberlands are likely to lose value under climate change while these alternate uses continue to appreciate in value. Climate change thus increases the need for policy interventions in areas where such conversion is undesirable. Figure 6a shows substantial overlap between timberland value decrease and projected residential growth, especially in the Sonoma-Mendocino wine country and Sierra foothills, both areas in which market dynamics already favor non-forest land uses.

Development corridors radiating from the Sacramento and San Francisco Bay regions show considerable overlap of declining timber values and projected exurban development. Strongly declining timber values in the two high-growth areas, Sonoma-Mendocino and Nevada-Placer-El Dorado (in the Sierra foothills), indicate that already-high rates of conversion in these areas may see incentives for conversion to non-timber uses increase with climate change. This trend can be seen in other areas, including Southern California and the Santa Cruz mountains. While timber value change in Southern California is modest, development pressure is intense and heavy conversion of private timberland seems likely

Table 2 Change in net present value of statewide timber revenues: A2 scenario, 0% discount rate (Millions $ US)

GCM	Scenario	Management scenario								
		Naïve			Rotation			Optimal		
		2020	2050	2080	2020	2050	2080	2020	2050	2080
N/A	Baseline	–	–	–	–	–	–	–	–	–
PCM1	Climate Change	1,392	1,310	1,443	1,441	1,356	1,492	1,618	1,433	1,685
	Climate Change & Global Prices	−3,327	−3,387	−8,057	−3,133	−3,200	−7,720	−3,020	−3,137	−7,675
GFDL	Climate Change	5,341	5,144	5,356	5,370	5,174	5,386	4,999	4,658	5,022
	Climate Change & Global Prices	438	349	−4,470	581	482	−4,212	177	16	−4,652

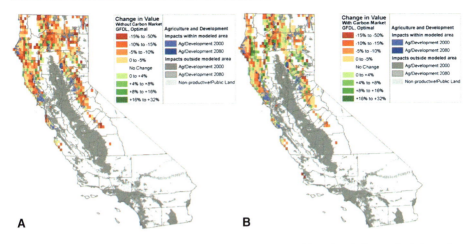

Fig. 6 Overlay of projected agriculture and development growth with projected changes in timber value, with **a** and without **b** carbon income, GFDL A2 scenario. Outputs from the CURBA urban growth model are combined with agricultural land classifications from GAP data. Existing impacts in light blue and dark gray; projected expansion into new areas in dark blue and medium gray. Note less urban expansion overlap with areas of strongly declining timber value in **b**

without policy intervention. In the Santa Cruz mountains, timber value declines are more pronounced, making conversion a concern despite less spatial overlap between timber value decline and land development.

Timberland conversion may be associated with loss of ecosystem services such as recreation value, carbon sequestration, and watershed protection. Carbon income mitigates value declines in rapidly developing areas, such as the Sierra corridor east of Sacramento and some areas of Mendocino and Humboldt counties (Fig. 6). In other locations, including some within these generally improving zones, value declines remain substantial in areas with strongly competing urbanization, indicating that other policy measures may be needed to complement carbon revenue to stem conversion of timberland in certain sites. Markets for other services may soon exist, but many are likely to remain poorly represented, underscoring policy intervention as an important consideration.

Policy tools favoring timberlands preservation include market-based solutions, smallholder tax incentives, land purchase programs, and actions incentivizing development in other areas. Policy tools to diminish the impact of climate change on land use conversion include carbon markets and climate change adaptation timber management programs. Carbon markets can generate income in areas suffering timber value declines, and adaptation programs can reduce the impact of climate change on value. Without such policy measures, it seems likely that climate change will hasten conversion of timberlands in some of the most impacted regions of the state.

4 Conclusions

Our results show that climate change will result in an overall decline in California's harvested timber value, with decreases of 4.9–8.5%, depending on climate change scenario and management option under climate change price assumptions. Dollar losses amount to up to $8.1 billion (non-discounted) by 2080 in scenarios in which prices respond to climate change in ways that have been predicted in the literature.

Optimizing timberlands management for climate change by adjusting rotation interval and/or species can conserve substantial value, and carbon markets can reduce the threat of land conversion by compensating landowners for preserving timberlands. However, declining timber value corresponds disproportionately to areas already experiencing timberland conversion to housing or agriculture. There remain large and economically significant timberland areas in which the above measures cannot offset powerful land conversion trends due to soaring property values and other development pressures. State timberlands policy is therefore likely to also assume an important role in conserving California's priority timberlands.

This study lays groundwork for further studies of greater scope and complexity. One important extension would be to expand our generic carbon valuation impact assessment to a comparison of different sets of market rules based on forester behavior and distribution of carbon payments. Implications of interacting carbon and timber markets for production and land use is a field worthy of further research. Our analyses do not investigate these issues, but we hope to inspire additional research into these important questions.

Acknowledgments C. Guo was supported by the ConvEne IGERT program of NSF.

References

Alig R, Adans D, McCarl B, Callaway JM, Winnett S (1997) Assessing effects of mitigation strategies for global climate change with an intertemporal model of the U.S. Forest and Agriculture Sectors. Environ Resource Econ 9:259–274

Busing RT, Fujimori T (2005) Biomass, production and woody detritus in an old coast redwood (*Sequoia sempervirens*) forest. Plant Ecol 177:177–188

California State Board of Equalization (2010) California timber harvest statistics. http://www.boe.ca.gov/proptaxes/pdf/harvyr2.pdf

Callaway J, Smith J, Keefe S (1994) The economic effects of climate change for U.S. forests. U.S. Environmental Protection Agency, Adaptation Branch, Climate Change Division, Office of Policy, Planning, and Evaluation, Washington

Coops NC, Waring RH, Moncrieff JB (2000) Estimating mean monthly incident solar radiation on horizontal and inclined slopes from mean monthly temperature extremes. Int J Biometeorol 44:204–211

Coops NC, Waring RH (2001) Assessing forest growth across Southwestern Oregon under a range of current and future global change scenarios using a process model, 3-PG. Glob Chang Biol 7:15–29

Joyce LA, Mills JR, Heath LS, McGuire AD, Haynes RW, Birdsey RA (1995) Forest sector impacts from changes in forest productivity under climate change. J Biogeogr 22:703–713

Joyce L, Aber J, McNulty S, Dale V, Hansen A, Irland L, Neilson1 R, Skog K (2001) Potential consequences of climate variability and change for the forests of the United States. Climate change impacts in the United States: the potential consequences of climate variability and change (Ch. 17). Cambridge University Press, Cambridge, UK

Koch GW, Sillett SC, Jennings GM, Davis SD (2004) The limits to tree height. Nature 428:851–854

Landsberg JJ, Waring RH (1997) A generalised model of forest productivity using simplified concepts of radiation-use efficiency, carbon balance and partitioning. For Ecol Manage 95:209–228

Landsberg JJ, Waring RH, Coops NC (2002) Performance of the forest productivity model 3-PG applied to a wide range of forest types. For Ecol Manage 172:199–214

Law BE, Waring RH, Anthoni PM, Aber JD (2000) Measurement of gross and net ecosystem productivity and water vapor exchange of a Pinus ponderosa ecosystem, and an evaluation of two generalized models. Glob Chang Biol 6:155–168

Li B-L, Wu H-I, Zou G (2000) Self-thinning rule: a causal interpretation from ecological field theory. Ecol Model 132(1–2):167–173

Mendelsohn R, Smith J (2006) The impact of climate change on regional systems: a comprehensive analysis of California. Elgar, Northampton

Perez-Garcia J, Joyce L, Binkley A, Clark S, McGuire A (1997) Economic Impacts of Climatic Change on the Global Forest Sector: An Integrated Ecological Economic Assessment. *Reviews in Environmental Science and Technology*. Boca Raton, FL: CRC Press.

Sands PJ (2001) 3PGPJS—a user-friendly interface to 3-PG, the Landsberg and Waring model of forest productivity. Tech. Rep. 29(2). CRC for Sustainable Production Forestry and CSIRO Forestry and Forest Products, Hobart, Tasmania

Spero J (2006) California timber yield tax volume and value trends: historical BOE timber harvest data by species, county, ownership. http://frap.cdf.ca.gov/projects/BOE/BOETimberTax.html

Sohngen B, Mendelsohn R, Sedjo R (2001) A global model of climate change impacts on timber markets. Am J Agr Econ 26(2):326–343

Sohngen B, Mendelsohn R (2003) An optimal control model of forest carbon sequestration. Am J Agr Econ 85(2):448–457

State Board of Equalization (2007) Harvest value schedule, effective July 1, 2007–December 31, 2007

Swenson JJ, Waring RH, Coops N, Fan W (2005) Estimation of forest productivity across the Pacific and Inland Northwest with a physiologically based process model, 3-PG. Can J Forest Res 35:1697–1707

Thuiller W (2003) BIOMOD- optimizing predictions of species distributions and projecting potential future shifts under global change. Glob Chang Biol 9:1353–1362

van Kooten C, Binkley C, Delcourt G (1995) Effect of carbon taxes and subsidies on optimal forest rotation age and supply of carbon services. Am J Agr Econ 77:365–374

Climatic Change (2011) 109 (Suppl 1)
DOI 10.1007/s10584-011-0329-9

Climate change and growth scenarios for California wildfire

A. L. Westerling · B. P. Bryant · H. K. Preisler ·
T. P. Holmes · H. G. Hidalgo · T. Das · S. R. Shrestha

Received: 21 September 2011 / Accepted: 17 October 2011 / Published online: 24 November 2011
© Springer Science+Business Media B.V. 2011

Abstract Large wildfire occurrence and burned area are modeled using hydroclimate and landsurface characteristics under a range of future climate and development scenarios. The range of uncertainty for future wildfire regimes is analyzed over two emissions pathways (the Special Report on Emissions Scenarios [SRES] A2 and B1 scenarios); three global climate models (Centre National de Recherches Météorologiques CM3, Geophysical Fluid Dynamics Laboratory CM2.1 and National Center for Atmospheric Research PCM1); three scenarios for future population growth and development footprint; and two thresholds for defining the wildland-urban interface relative to housing density. Results were assessed for three 30-year time periods centered on 2020, 2050, and 2085, relative to a 30-year reference period centered on 1975. Increases in wildfire burned area are anticipated for most scenarios, although the range of outcomes is large and increases with time. The increase in wildfire burned area associated with the higher emissions pathway (SRES A2) is substantial, with increases statewide ranging from 36% to 74% by 2085, and increases exceeding 100% in much of the forested areas of Northern California in every SRES A2 scenario by 2085.

Electronic supplementary material The online version of this article (doi:10.1007/s10584-011-0329-9) contains supplementary material, which is available to authorized users.

A. L. Westerling (✉) · S. R. Shrestha
University of California, Merced, 5200 N. Lake Rd, Merced, CA 95343, USA
e-mail: awesterling@ucmerced.edu

B. P. Bryant
Pardee RAND Graduate School, The RAND Corporation, Santa Monica, CA, USA

H. K. Preisler
USDA Forest Service Pacific Southwest Research Station, Albany, CA, USA

T. P. Holmes
USDA Forest Service Southern Research Station, Research Triangle Park, NC, USA

H. G. Hidalgo
School of Physics and Center for Geophysical Research, University of Costa Rica, San Jose, Costa Rica

T. Das
CH2MHILL, Inc., San Diego, CA 92101, USA

1 Introduction

The climate system interacts with various factors such as soils, topography, available plant species, and sources of ignition to give rise to both natural ecosystems and their fire regimes. Long-term patterns of temperature and precipitation determine the moisture available to grow the vegetation that fuels wildfires (Stephenson 1998). Climatic variability on interannual and shorter scales governs the flammability of these fuels (e.g., Westerling et al. 2003; Heyerdahl et al. 2001; Kipfmueller and Swetnam 2000; Veblen et al. 2000; Swetnam and Betancourt 1998; Balling et al. 1992). Flammability and fire frequency in turn affect the amount and continuity of available fuels. Consequently, long-term trends in climate can have profound implications for the location, frequency, extent, and severity of wildfires and for the character of the ecosystems that support them (Westerling 2009).

Human-induced climatic change may, over a relatively short time period (< 100 years), give rise to climates outside anything experienced in California since the establishment of an industrial civilization currently sustaining a state population that has increased approximately 41,000% since 1850. Changes in wildfire regimes driven by climate change are likely to impact ecosystem services that California citizens rely on, including carbon sequestration in California forests; quality, quantity and timing of water runoff; air quality; wildlife habitat; viewsheds and recreational opportunities. They may also impact the ability of homeowners and federal, state, and local authorities to secure homes in the wildland-urban interface from damage by wildfires (Westerling and Bryant 2008).

In addition to climate change, the continued growth of California's population and the spatial pattern of development that accompanies that growth are likely to directly affect wildfire regimes through their effects on the availability and continuity of fuels and the availability of ignitions. They are also likely to impact both wildfire and property losses due to wildfire through their effects on the extent and value of development in California's wildland-urban interface, both through their effects on the number of structures proximate to wildfire risks and their effects on fire suppression strategies and effectiveness.

The combined effects of climate change and development on California's future large wildfire occurrence and burned area are the focus of the research presented here. Our modeling projects California's fire regimes as they are currently managed onto scenarios for future climate, population, and development. The methodology we employ can incorporate the effects of spatial variations in current management strategies on average fire risks. However, the monthly and interannual variations in large wildfire occurrence and burned area that we estimate do not reflect changes in management strategies over time, although our modeling does have the capacity to reflect changes in the effectiveness of current management strategies to the extent that these changes currently tend to correlate with climate and landsurface characteristics. Thus, hypothetical effects of future changes in management in response to the impacts of climate and development on wildfire are not considered in this work.

The metrics we model—large fire occurrence and burned area—are not the only metrics we would wish to employ to assess the full range of the ecological and human impacts of wildfire. In particular, metrics of fire severity (e.g., the percent of available biomass consumed, characteristics of ecological impacts) would be highly desirable as well. These metrics are likely to change in response to climate, may also be influenced by future management decisions, and are key components for estimating many wildfire impacts due to climate change. The work reported here does not consider changes in fire severity, which are the target of multiple ongoing research efforts. However, in interpreting our results, it would be a mistake to assume a linear correspondence between increased burned area and fire severity. Fire severity is likely to increase in some ecosystems and decrease in others alongside increases in burned area.

Furthermore, severity might decrease in some ecosystems for reasons that many California residents would find undesirable (i.e., broad-scale changes in ecosystems).

This work extends an analysis by Westerling and Bryant (2008) that considered the effects of climate change on California large (>200 ha) wildfire occurrence and wildfire-related damages holding development fixed at the 2000 census. In this analysis we statistically model large (>200 and >8,500 ha) wildfire occurrence as a product of both future climate scenarios and future population and development scenarios, using nonlinear logistic regression techniques developed for seasonal wildfire forecasting in California and the western United States (Preisler and Westerling 2007). We model the expected burned area using Generalized Pareto Distributions fit to observed wildfires >200 and >8,500 ha. We assess a range of outcomes given numerous sources of uncertainty, including three global climate models (GCMs) with different sensitivities of temperature and precipitation to anthropogenic forcing, two emissions scenarios, and three population growth and development scenarios. Our goal is not to determine one most likely outcome, but rather to define a population of plausible outcomes, which can then be used in future work to assess the robustness of combined adaptation and mitigation policy choices.

2 Data and methods

2.1 Domain and resolution

The spatial domain for this analysis was a 1/8-degree lat/long grid (~12 km resolution) bounded by the political boundary for the state of California. Areas of the state outside the current combined fire protection responsibility areas of the California Department of Forestry and Fire Protection and contract counties (combined here and denoted CDF), the U.S. Department of Agriculture's Forest Service (USFS), and the U.S. Department of Interior's National Park Service (NPS), Bureau of Land Management (BLM), and Bureau of Indian Affairs (BIA) were excluded. The result was a set of 2,267 grid cells within the State. Fire was then modeled over this domain at monthly time-steps for historic and simulated climate scenarios.

Five time periods were used for this analysis. Coefficients for statistical wildfire models were estimated using monthly, gridded, historical fire and climate data available for 1980–1999. These coefficients were then applied to gridded climate scenarios derived from GCMs, and the results for three future climate periods—2005–2034, 2035–2064, and 2070–2099 (henceforth referred to as 2020, 2050, and 2085) were compared to a common reference period (1961–1990, henceforth 1975) for each scenario. These comparisons used average annual fire occurrence and burned area statistics computed for each 30-year period referenced above.

2.2 Fire history

A comprehensive wildfire history for California for 1980–1999[1] was assembled from digital fire records obtained from CDF, USFS, NPS, BLM, and BIA. The CDF records included perimeters for large fires under both direct CDF and contract counties' fire protection responsibility (obtained online at http://frap.cdf.ca.gov/). Federal fires were

---

[1] Comprehensive data prior to 1980 are not available from some of these sources. Fire data after 1999 are available, but the hydrologic simulations forced with historic climate data used here were developed for the California Scenarios Project, of which this research is a component. These data ended in 1999, so the common period of overlap between the available fire history and the hydroclimatic data was 1980–1999.

sourced from point data records compiled from individual fire reports. The methods used in compiling these data are described in Westerling et al. (2006, online supplement) and Westerling and Bryant (2008). Westerling et al. (2002) describe in detail the federal data in this sample and their response to climate variability.

Our wildfire history was aggregated to a 1/8-degree gridded monthly data set of frequencies of fires >200 ha in burned area and the total burned area in these large wildfires (544,080 data points=2,267 grid points × 20 years × 12 months/year). Wherever we refer to fire occurrence in the text, we are referring to the occurrence of wildfires greater than 200 ha, unless otherwise specified. For federal-sourced wildfire data, fires were allocated to the grid cell in which they were reported to have ignited. For CDF fires reported as polygon perimeters, fires were allocated to the grid cell corresponding to their centroid. Fires were assigned to the month in which they were discovered. In many cases fires continued to burn for additional months, but we did not have the means to apportion burned area by month.

Wildfires managed by the Fisheries and Wildlife Service (FWS), the Department of Defense (DOD), and the Bureau of Reclamation (BOR) were not included because they were not available with sufficient comprehensiveness and quality. The FWS and BOR lands are relatively small in area and—particularly in the case of FWS—located in areas (e.g., California's Central Valley) that would likely have been excluded from this analysis for other reasons. The Department of Defense lands in California are significant, similar in scale to those of NPS. We could extend our analysis spatially to DOD lands by applying model coefficients estimated using other agencies' fire histories to DOD lands, for which we have the explanatory variables. However, the vast majority of DOD lands in California lie in southeastern desert areas of the state, which show negligible changes in fire risks by the end of the twenty-first century under many, though not all, of our scenarios.

2.3 Land surface characteristics

2.3.1 Vegetation

Coarse vegetation characteristics used here were compiled from the Land Data Assimilation System (LDAS) for North America's 1/8-degree gridded vegetation layers that use the University of Maryland vegetation classification scheme with fractional vegetation adjustment (UMDvf) (Mitchell et al. 2004; Hansen et al. 2000) (For an analysis of how our modeling relates to the LDAS vegetation categories, see the supplemental materials.)

In our modeling, the primary variable we use from LDAS is the fraction of each grid cell that is vegetated and not used for agriculture (V). In our model for estimating fire probabilities, V plays an important role. In the limit of complete urbanization, it is clear that this variable is affected by encroaching human development, because a grid cell entirely covered by dense population would lack any sufficiently large vegetated space in which wildfires could exist. However, vegetation cover may be reduced by encroaching human development at intermediate scales as well, depending on how new growth is allocated. In Section 2.4 and the supplemental materials we describe how scenarios for allocating projected growth within a grid cell affect the vegetation fraction.

2.3.2 Topography

Topographic data on a 1/8-degree grid were also obtained from LDAS. The LDAS topographic layers are derived from the GTOPO30 Global 30 Arc Second (~1 km) Elevation Data Set (Mitchell et al. 2004; Gesch and Larson 1996; Verdin and Greenlee

1996). We tested mean and standard deviation of elevation, slope, and aspect as explanatory variables in our wildfire model specification.

2.3.3 Protection responsibility

Logistic regression techniques attempt to estimate probabilities of wildfires occurring at a given time and location as a function of predictor variables. The predictand is always valued zero or one, and in this case most data points are zero, since fires are relatively rare occurrences at ~12 km × monthly scales. It is important to distinguish between data points that are zero because no fire occurred, and data points that are zero because the fires that may have occurred there are not reported by the agencies included in our sample. The practical effect of including in our modeling areas for which we do not have good fire reports is to underestimate the probability of fire occurrence everywhere on average. We used the fraction of each grid cell's area covered by the protection responsibility of the agencies in our sample to estimate the area over which fires could occur.

A GIS layer of Local, State, and Federal protection responsibility areas in California was obtained online from the CDF Fire and Resource Assessment Program (http://frap.cdf.ca.gov/). The polygons in this layer were extracted and intersected with our 1/8-degree grid to obtain the fraction of each grid cell in Local (LRA), State (SRA), and Federal (FRA) protection responsibility areas. Since federal protection responsibility includes agencies excluded from our sample, we obtained GIS layers of federal and Indian lands online from the United States National Atlas (http://nationalatlas.gov/). These were also extracted and intersected with our grid in order to determine what fraction of the federal responsibility area in each grid cell corresponded to the federal agencies in our sample.

Our analysis was limited to the SRA (for which we have CDF wildfire histories) and those parts of the FRA administered by the federal agencies for which we have wildfire histories (USFS, NPS, BLM, BIA). The LRA was excluded both because we do not have wildfire histories for those lands and because they are predominately developed for urban and agricultural uses. We limited our analysis to grid cells with a combined protection responsibility (SRA, USFS, NPS, BLM, BIA) exceeding 15% of the total grid cell area.

2.4 Growth and development scenarios

We employed spatially explicit 100 m-resolution twenty-first century population growth and development scenarios based on work by Theobald (2005) and developed by the U.S. EPA (2008) as the Integrated Climate and Land Use Scenarios (ICLUS) for the United States. These scenarios incorporate assumptions about future trajectories with regards to sprawl as well as population growth rates based on the Intergovernmental Panel on Climate Change (IPCC) Special Report on Emissions Scenarios (SRES) social, economic, and demographic storylines, downscaled to the United States. We report results here for ICLUS scenarios with growth and sprawl that would be consistent with the global A1 and B2 SRES scenarios, and an intermediate base case trajectory (Table 1). While we do not model spatial variation in wildfire below the 1/8-degree grid threshold, we do consider effects of assumptions about the allocation of new development below that level (see supplemental materials). Our scenarios for allocating development consider the effects of expanding the wildland-urban interface into areas that are currently vegetated versus siting new development in areas that are currently bare or agricultural land. For the intermediate case, we allocate new development proportionately across baseline vegetated, bare and agricultural areas. For the high growth and sprawl scenario (A2), we allocate new

Table 1 ICLUS growth and development scenarios for California

Scenario	2100 CA Population	Change vs 2000	Description (identifier)
A2	81 million	+154%	high growth & sprawl (HIH,HIH)
Base Case	62 million	+84%	medium growth & sprawl (MID,MID)
B1	50 million	+54%	Low growth & sprawl (LOW,LOW)

development to vegetated areas, and for the low growth and sprawl scenario (B1) we allocate new development within bare and agricultural areas. Additionally, we use two different thresholds for describing when a 100 m pixel is fully "urbanized," such that a wildfire cannot occur there: 147 and 1000 households per square kilometer (the limits used in the ICLUS scenarios to define suburban areas).

2.5 Climate and hydrologic data

2.5.1 Historical data

A common set of gridded historical (1950–1999) climate data including maximum and minimum temperature, precipitation (PCP), radiation, and wind were designated for the Second California Assessment (see Maurer et al. 2002; Hamlet and Lettenmaier 2005). We used these data with LDAS vegetation and topography to drive the Variable Infiltration Capacity (VIC) hydrologic model at a daily time step in full energy mode, generating a full suite of gridded hydroclimatic variables such as actual evapotranspiration (AET), soil moisture, relative humidity (RH), surface temperature (TMP) and snow-water equivalent (SWE) (Liang et al. 1994). Because Potential Evapotranspiration (PET) was not easily extracted from the version of the VIC model used here, we used the Penman-Monteith equation to estimate PET directly (Penman 1948; Monteith 1965). Moisture deficit (D) was then calculated from PET and AET (D = PET - AET).

As indicators of drought stress, we calculated the cumulative water-year moisture deficit for each of the preceding 2 years (D01 and D02). Water year (October - September), rather than calendar year, moisture deficit is used because in California most of the precipitation that affects fuel availability and flammability falls between October and April. We also calculated the standardized cumulative moisture deficit from October 1 through the current month (CD0). The 30-year means for 1961–1990 for moisture deficit (D30), AET (AET30), PCP (PCP30), and TMP (TMP30) were also used below to analyze the spatial distribution of climate-vegetation-wildfire interactions to be represented in the wildfire model specification (Electronic Supplementary Material).

2.5.2 Climate scenarios

As described by Cayan et al. (2009), several GCMs and emissions scenarios were selected for the Second California Assessment. In this report we describe results across three models —CNRM CM3, GFDL CM2.1, NCAR PCM—and two emissions scenarios—SRES A2 (a medium-high emissions trajectory) and SRES B1 (a low emissions trajectory) comprising six realizations for future climate. Daily climate data from these models were downscaled to the 1/8-degree grid using the Constructed Analogues method (see Hidalgo et al. 2008; Maurer and Hidalgo 2008). After downscaling, the VIC model was used to simulate the same suite of hydrologic variables for each target time period (1975, 2020, 2050, 2085) in each climate scenario as described for the historical period (Section 2.5).

As described in Cayan et al. (2009), the models were selected for their fidelity in representing historical California temperature and precipitation in particular, with appropriate seasonality. Also, the selected models were required to simulate tropical Pacific sea surface temperature variability consistent with observed ENSO variability, and to have appropriate spatial resolution over California for the downscaling methodologies employed here. That left a set of six models. Daily precipitation and daily maximum and minimum temperature had to be included in the models' historical and projection saved sets in order to use the Constructed Analogs downscaling methodology, which narrowed the set to three models.

While we only use the three models downscaled with the Constructed Analogs methodology, the larger set of six models show a consensus toward drier conditions in southern and central California, but more scattered results in the region from Lake Shasta to the northernmost parts of California. An unpublished review of a larger set of 12 models showed similar results, increasing our confidence that the models used here are representative of what a larger set of models project for California. So while it is true that our models do not encompass the full range of model uncertainty of the AR4, the GCMs used here appear to cover a range of projected temperature and precipitation that is representative of the state of the art modeling guidance for California. A possibly more relevant concern is that the estimated emissions growth for 2000–2007 exceeded the most fossil fuel-intensive scenario of Intergovernmental Panel on Climate Change (Le Quéré et al 2009). The longer this trend in emissions growth rates persists, the harder it may be to reach a future scenario like those modeled here, and the more likely it is our results may underestimate future impacts.

Each of the GCMs evidences different sensitivities to anthropogenic forcing, with the CNRM CM3 and GFDL CM2.1 models generally showing warmer temperatures than the NCAR PCM, particularly in summer. The other three models assessed by Cayan et al (2009) that are not used here were also warmer than the NCAR PCM. The GFDL CM2.1 model tended to be drier than the others in both northern and southern California in most scenarios, while CNRM CM3 was typically drier than NCAR PCM. Compared to the larger set of models, NCAR PCM is also the least dry by the end of the century under the A2 scenario (Cayan et al 2009). It is not clear how significant variation in precipitation across the models is, since that variation is small compared to the potential natural variability in precipitation in the region. Notably, the tendency towards dryness across all the models is much more pronounced and uniform for California in the A2 emissions scenarios, when anthropogenic forcing is higher, than for the B1 scenarios, and it seems reasonable to speculate that the forcing is overwhelming the natural variation in the A2 scenarios.

The six climate scenarios used here encompass three separate sources of uncertainty for our modeling: the degree of anthropogenic forcing (represented by the A2 and B1 emissions paths), the extent of climate system sensitivity to anthropogenic forcing (greater in GFDL CM2.1 and CNRM CM3 than NCAR PCM), and the range of random variation in climate across 30-year windows in six scenarios (unknowable for a sample of this size).

3 Modeling

3.1 Fire modeling

Because large wildfires are relatively rare events, statistical models for wildfire must aggregate fire occurrence over space and/or time in order to avoid fitting a model of zeros.

Logistic regressions allow us to model wildfire occurrence at arbitrarily fine spatial and temporal resolutions (limited only by the available data) while statistically aggregating across locations with similar characteristics. We estimated the monthly probability of large (>200 ha and >8500 ha) wildfires occurring as a function of land-surface characteristics, human population, and climate on a 1/8-degree grid using generalized linear models in R. Candidate model specifications were tested by comparing the Akaike Information Criterion (AIC) estimated for each model. The best model specification was then tested using leave-one-out cross-validation. That is, for each of the 20 years in the model estimation period, a separate set of model coefficients were estimated using the other 19 years of data.

The climatic variables used here as predictors for fire occurrence (Table 2) were selected to describe variation in moisture available for the growth and wetting of fuels over a range of time scales, from long term averages over 30 years (D30, AET30), to interannual variations in the 3 years through the current year (D01, D02), to seasonal variations and conditions at the month a fire could potentially burn (CD0, PCP, RH, TMP). Westerling and Bryant (2008) and Westerling (2009) show that long-term climate averages can serve as a proxy for coarse vegetation and fire regime types, distinguishing between diverse fire regime responses to antecedent and concurrent climate. Preisler et al (2011), Westerling (2009) and Westerling et al. (2006) show that cumulative annual moisture deficits in the current and 1–2 preceding years are strongly associated with variability in large wildfire occurrence, and the latter work also links large wildfire occurrence to temperature and variations in spring snowmelt timing. Finally, relative humidity (RH) is a key component of many fire weather indices, and a good predictor of fuel moisture (e.g., Schlobohm and Brain 2002).

Vegetation fraction and population are indicators of burnable area, availability of human-origin ignitions, and accessibility. Other variables we tested—such as total protection responsibility area, and mean and standard deviation of elevation—were not included in the model specification because they did not substantially improve model fit as measured by the AIC. The climatic and land surface variables we selected are not independent of topography and protection responsibility area. Federal protection responsibility and U.S. Forest Service protection responsibility area were highly significant, and may be indicative of differences in management or resources across protected areas, or they may be proxies for other spatial variables such as accessibility, vegetation type, sources of ignitions, etc.

The model specification employed here for estimating fire occurrence builds on Preisler and Westerling (2007) and Westerling and Bryant (2008), and adopts methodologies presented in Brillinger et al. (2003), and Preisler et al. (2004):

$$\text{Logit}(\Pi_{200}) = \beta \times [1 + D30 + D01 + D02 + PCP + X(D30, AET30)$$
$$\times (1 + TMP + CD0) + P(TMP) + P(RH)$$
$$+ P(POP) \times (1 + D30) + X(V) + FRA]$$

where Π_{200} is the probability of a fire >200 ha occurring, $\text{Logit}(\Pi_{200})$ is the logarithm of the odds ratio $(\Pi_{200}/(1 - \Pi_{200}))$, β is a vector of parameters estimated from the data, and $X(\bullet)$ and $P(\bullet)$ are matrices describing basis spline and polynomial transformations of the data. The interaction terms:

$$X(D30, AET30) \cdot (1 + TMP + CD0)$$

result in estimation of a set of constants and a set of coefficients on average monthly temperatures and cumulative moisture deficits that are associated with the distribution of vegetation types and patterns of fire regime response to climate variability.

Climatic Change (2011) 109 (Suppl 1)

Table 2 Predictor variables

Variable	Description	>200 ha[1]	>8,500 ha[2]
D30	30-year average cumulative Oct.–Sep. moisture deficit	√	
AET30	30-year average cumulative Oct.–Sep. actual evapotranspiration	√	
D02	Cumulative Oct.–Sep. moisture deficit, 2 years previous	√	
D01	Cumulative Oct.–Sep. moisture deficit, 1 year previous	√	
CD0	Current water-year cumulative moisture deficit, Oct. through current month	√	
PCP	2-month cumulative precipitation, through current month	√	
RH	Monthly average Relative Humidity	√	√
TMP	Monthly average surface air Temperature	√	
V[3]	Vegetation Fraction	√	
POP	Total population (2000)	√	
FRA[3]	Federal protection responsibility area as percent of total area	√	
USFS[3]	USFS protection responsibility area as percent of total area		√
Aspect[4]	Average north/south facing		√

[1] Predictors for logistic regression estimating probability of a fire exceeding 200 ha

[2] Predictors for logistic regression estimating probability total burned area exceeds 8,500 ha conditional on one fire having exceeded 200 ha

[3] Fractions are transformed using $\log((\text{fraction} + .002)/(1 - \text{fraction} + .002))$ to generate a continuous variable centered around zero

[4] The transformation $\cos(\text{pi}/2 + \text{aspect} \cdot \text{pi}/180)$ yields the north/south component of aspect

The interaction terms:

$$P(POP) \cdot (1 + D30)$$

capture direct effects of local population on the probability of large fire occurrence, which vary in this specification with the average summer moisture deficit in any given location. These factors may be indicative of several effects, including ignitions, accessibility, suppression resources, and suppression strategies.

We also estimate the probability that one or more very large fires burning in excess of 8,500 ha in aggregate may occur, conditional on there being at least one fire exceeding 200 ha:

$$\text{Logit}(\Pi_{8500|200}) = \beta \times [1 + RH + \text{Aspect} + USFS]$$

This model says that whether or not a large fire becomes a very large fire or not depends on the relative humidity, the north/south facing, and whether the fire is on Forest Service land or not. The latter variable might be related to management strategies, or it could simply be that so much of the forested area of the state, especially in less accessible locations, is managed by the Forest Service.

The probabilities estimated in this way can be summed to get an expected number of fires. For example, the number of wildfires >200 ha expected in California in a given year is just the sum of the probabilities Π_{200} for each of 12 months and 2,267 grid cells. The probabilities can also be used to simulate fire histories, drawing [0,1] randomly from binomial distributions with probabilities Π_{200} for each grid cell and month.

We also estimated generalized Pareto distributions (GPDs) for the logarithm of burned area for fires >200 ha and for fires >8,500 ha (for a discussion of methodology as it applies to wildfire and evidence that fire size regimes are consistent with heavy-tailed Pareto distributions, see Straus et al. 1989; Malamud et al. 1998; Ricotta et al. 1999; Cumming 2001; Song et al. 2001; Zhang et al. 2003; Schoenberg et al. 2003; Holmes et al. 2008). From these we estimated the expected burned area for a fire given that its area is greater than 200 ha and less than 8,500 ha, and the expected burned area for fires given that they are >8,500 ha. Expected burned area in any grid cell and month is thus:

$$\textbf{Expected Burned Area} = \Pi_{200} \times \widehat{A}\big|>_{200,<8500} + \Pi_{200} \times \Pi_{8500|200} \cdot \times \widehat{A}\big|>_{8500}$$

where the probabilities Π are as above, $\widehat{A}\big|_{>200,<8500}$ is the expected burned area less than 8,500 ha given that there is at least one fire greater than 200 ha, and $\widehat{A}\big|_{>8500}$ is the expected burned area given that at least 8,500 ha burned.

We use a two-step process like this because the empirical distribution of fire sizes in California appears to be non-stationary. Extremely large fires are governed by different processes than are more frequent large fires. As with the logistic regression results above, we can also draw randomly from the GPDs to simulate fire histories, generating a random burned area for each fire simulated with the binomial distributions.

3.2 Summary of scenarios

To summarize, we estimate 108 future scenarios, considering two emissions scenarios, three GCMs, two thresholds for defining urbanization, three scenarios for growth rates and the allocation of development, and three future time periods (Table 3).

4 Results and discussion

4.1 Model fit

The statistical significance of coefficients for the wildfire model specification was very stable across sub-samples, using cross-validation. The logistic regression for fires greater than 200 ha fit the observations well (Fig. 1), and the maximum probability of predicted fire occurrence was over 33%, which compares well with earlier fire models for climate change impact assessments in Westerling and Bryant (2008) and for seasonal forecasting in Preisler and Westerling (2007). The scale of interannual variation in expected fires aggregated statewide was comparable to that of observed fire. Forty-two percent of interannual variability in statewide large (>200 ha) wildfire occurrence was explained by the predicted probabilities, and the correlation between annual observed fire occurrence and predicted fire occurrence was highly significant (Spearman's $\rho=0.73$, p-value$=0.0003$). For monthly fire

Table 3 Summary of scenarios

Emissions	Model	Urban Threshold (households/km^2)	Growth Rate & Allocation	Period
SRES A2	CNRM CM3	147	LOW, LOW	2020
SRES B1	GFDL CM2.1	1,000	MID, MID	2050
	NCAR PCM1		HIH, HIH	2085

Fig. 1 Logistic regression model fit for fires>200 ha: Observed fire frequency (*vertical axis*) versus predicted probabilities (*horizontal axis*), binomial 95% confidence interval (*upper and lower lines*)

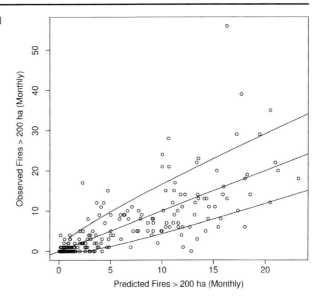

occurrence aggregated statewide, 58% of month-to-month variability was explained, and the correlation was also highly significant (Spearman's $\rho=0.84$, p-value=2e-16). The additional skill at the monthly time scale is due to the model fitting the seasonal cycle in the incidence of fires >200 ha, as well as the interannual variability.

Results for our estimation of the frequency of fires greater than 8,500 ha were also significant, but the skill was lower. This is likely due to both the small number of observed fires of this magnitude, and also it is likely that the size of these fires is highly sensitive to factors we are not able to consider here, such as the management strategy on individual fires, meteorological conditions on hourly to daily timescales during the fire, and landsurface characteristics at finer scales than those practical here (~12 km). Twenty percent of interannual variability in statewide very large (>8,500 ha) wildfire occurrence was explained by the predicted probabilities, and the correlation between annual observed and predicted very large fire occurrence was significant (Spearman's $\rho=0.46$, p-value= 0.04). Similarly, monthly predicted very large fire occurrence explained 23% of the variation in observed monthly values, and the correlation was highly significant (Spearman's $\rho=0.46$, p-value<7e-14).

While the Generalized Pareto Distribution fit the logarithm of monthly burned areas exceeding 200 ha and 8,500 ha very well (Fig. 2), the empirical fire size distributions still have a "fatter" right tail (i.e., higher probability of extremely large fires) than estimated by the modeled distributions, as indicated by the values to the far right in the Quantile plots sitting above the 1:1 ratio line (Fig. 2). This under-prediction of the extremes in fire size appears to be associated with two factors that are not captured in our modeling—a high degree of clustering in lightning-caused ignitions in some years, and high wind events (e.g. Santa Ana winds in coastal Southern California). Management factors may also play a role, such as the use of backfires, variation in the availability or effectiveness of suppression resources, and also a lack of consistency in reporting as fires both individual ignitions versus fire "complexes" resulting from multiple, coincident ignitions. The correlation between predicted and observed burned area was, however, highly significant (Spearman's $\rho=0.82$, p-value<2e-16).

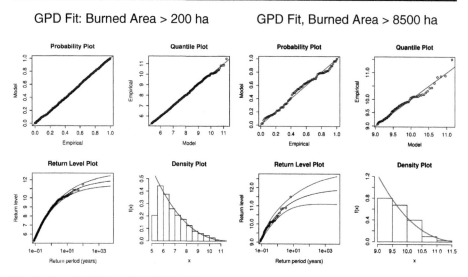

Fig. 2 Generalized Pareto Distributions fit to occurrence of burned area >200 ha (*left*) and burned area >8,500 ha (*right*). Distributions are fit to the logarithm of burned area and assume stationarity

Generalized Pareto Distributions can be fit with covariates such as climate and land surface characteristics, such that the parameters describing the distribution vary over time and space (see e.g., Holmes et al. 2008). We tested the full suite of predictor variables described in the preceding sections. While some of them were highly significant statistically, in practice including covariates had a trivial impact on our ability to predict year-to-year variations in burned area. Consequently, the results reported here use GPD fits assuming stationarity (i.e., that the fire size distribution does not change over time). Thus, the interannual variation in our predicted burned area is due entirely to variation in estimated probabilities for burned area to exceed the two specified thresholds (200 ha and 8,500 ha).

Together these four probabilistic models (logistic models of the probability of fire occurrence >200 ha and >8,500 ha, and Generalized Pareto Distributions for burned area >200 ha and >8,500 ha) can be used to estimate both the expected burned area at any given time and place in our modeling domain, and the associated variability implied by the four models around those estimates. Using random draws from binomial distributions with the probabilities predicted by logistic regressions, and random draws from Generalized Pareto Distributions, we simulated 1,000 fire histories over our gridded monthly model domain for 1980–1999, and aggregated them statewide by year (Fig. 3). Observed annual burned area was more variable than median simulated values, but within the range of variability across the simulations.

While for aggregate annual burned area for the state our combined models only explain about a quarter of interannual variability, the extremes in the residuals are accounted for by essentially two factors. Clusters of lightning-caused ignitions in a few outlying years in northern California account for a large part of the unexplained variability in large fire occurrence. Unexplained variability in burned area is in large part due to a combination of the effects of these lightning ignitions, and of Santa Ana-wind driven fire events in coastal southern California. Our climate-driven modeling is not able to predict the timing of lightning ignitions or Santa Ana winds. However, this range of variability is represented in

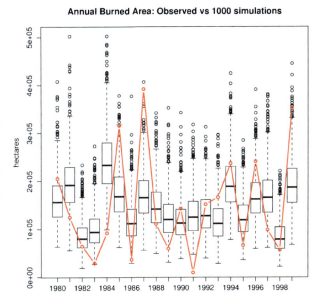

Fig. 3 Median (*horizontal lines*), interquartile range (*box*), extremes within 1.5 × interquartile range (*whiskers*) and extremes outside 1.5 × interquartile range are shown for annual burned area aggregated statewide for 1,000 simulations versus observed (*red line*) burned area for large fires in California. Predicted probabilities of large fires from logistic regressions, and Generalized Pareto distributions of fire size, were used to generate 1,000 simulations of burned area for each grid cell and month and then aggregated by year for the state for each simulation

the probabilities associated with fire occurrence and size in our models, and thus simulations using these probabilities do encompass the variability observed in burned area (Fig. 3).

4.2 Changes in California wildfire

Predicted large fire occurrence and total burned area increase over time for both emissions scenarios (Fig. 4).[2] Initial increases for burned area are relatively modest, with little difference between emissions scenarios—by 2020 the increases range from 6% to 23%, with median increases between 15% and 19%. By 2050 the spread in modeled outcomes widens, with predicted increases in burned area ranging from 7% to 41%, and median increases between 21% and 23%, but again differences due to emissions scenarios are relatively small compared to other factors. By 2085, the range of modeled outcomes is very large, with total burned area increasing anywhere from 12% to 74%. On average, the largest increases occur in 2085 for SRES A2 scenarios, with a median statewide increase in burned area of 44%, and the biggest increases occurring for the warmer, drier GFDL CM2.1 and CNRM CM3 model runs (range: 38%–74%, median 56%).

The SRES A2 scenarios in 2085 seem to be qualitatively different from either earlier periods or SRES B1 in 2085 (Fig. 4), implying that the most important policy implication of this study may be that moving to an emissions pathway more like that in SRES B1 (or lower) could be highly advantageous.

A robust result of this study is that forest burned area increases substantially—exceeding increases of 100% throughout much of the forested areas of Northern California—across all three of the GCMs analyzed here for the SRES A2 emissions scenario by 2085 (Fig. 5). To highlight the effects of potential increases in forest burned area, we analyzed a set of transects running from the edge of California's Central Valley near Merced northwest through the Sierra Foothills and Yosemite National Park to Mono Lake and Lee Vining on

---
[2] We show results for expected total area burned only, which are similar to predicted changes in large fire occurrence.

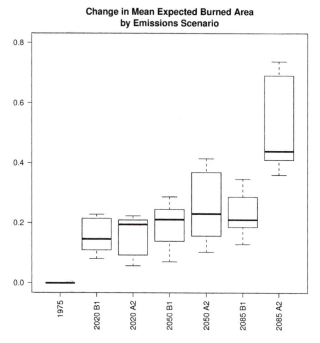

Fig. 4 Percentage change in expected burned area for large California fires from a 1961–1990 reference period for 108 scenarios, estimated for 30-year periods centered on the indicated dates, by emissions scenario. Bold horizontal lines indicate median scenario, boxes indicate middle 50% of values, *whiskers* indicate extremes

the eastern side of the Sierra Nevada range for a GFDL CM2.1 SRES A2 scenario with intermediate population growth and sprawl (MIDMID) and a high threshold for urbanization (Fig. 6). While increased temperatures and modest variations in precipitation generally resulted in similar increases in moisture deficit across the transects, the highest elevation sites were somewhat buffered by their greater available moisture. The largest

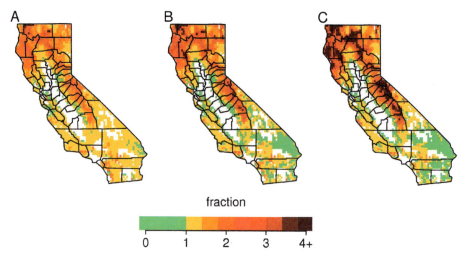

Fig. 5 2085 Predicted burned area as a multiple of reference period predicted burned area for three SRES A2 climate scenarios: **a** NCAR PCM1, **b** CNRM CM3, and **c** GFDL CM2.1, with high population growth, high sprawl, and a high threshold housing density for defining the limit to the wildland urban interface. The location of fire regimes is assumed to be fixed. A value of "1" indicates burned area is unchanged, while 4+ indicates that burned area is 400% or more of the reference period (i.e., a 300% increase)

increases in burned area tended to be mid-elevation sites on the west side of the Sierras. Much of the lower portion of these mid-elevation sites lies outside of federal reserves, and is comprised of private land holdings with very low-density development. In consequence, a significant portion of private households in this region of the state would be exposed to a substantially increased threat of wildfire under the SRES A2 emissions scenarios.

To highlight the effects of growth and sprawl on fire risks in the wildland urban interface in areas with a higher density of households, we analyzed model results for several SRES A2 scenarios on California's central coast, running from San Mateo County in the north through Santa Cruz to Monterey in the south, and east to San Benito and Santa Clara counties (Fig. 7). CNRM CM3 and GFDL CM2.1 model results were quite similar, though with higher burned area for GFDL CM2.1 (Fig. 7a&b). For both models, the largest increases were in Big Sur and the Ventana Wilderness, and in the redwood forests at the northern end of Santa Cruz county. High-growth, high-sprawl scenarios in this region tended to reduce burned area over a large region, as increased development reduced the vegetated area available to burn (Fig. 7c&d). Notably, results for rural northern Santa Cruz county were sensitive to the threshold set for determining when development becomes too dense for wildfires to occur (Fig. 7c&d).

The decrease in burned area in the southeastern deserts of California in Fig. 5 is the effect of greater drying across the three SRES A2 scenarios, which reduces fuel availability in these fuel-limited fire regimes.

5 Conclusion

We examine a range of future scenarios, considering multiple GCMs, emissions scenarios, and a range of growth and development scenarios. While the range of outcomes for these scenarios was large by the end of century, a majority of the scenarios indicated significant increases in large wildfire occurrence and burned area are likely to occur by mid-century. By 2085, substantial increases in large wildfire occurrence and burned area seem likely, particularly under the SRES A2 emissions scenarios, and particularly in forested areas of the state. This is mainly due to the effects of projected temperature increases on evapotranspiration in this scenario, compounded by reduced precipitation. The middle 50% of scenarios for SRES A2 in 2085 ranged from average statewide burned area increases of 41% to 69% compared to the reference period centered around 1975 (Fig. 4).

The spatial pattern of increased fire occurrence and burned area in the forests of Northern California was robust across a wide range of scenarios. Wildfire burned area increased dramatically throughout the mountain forest areas of Northern California across most of the various 2085 SRES A2 scenarios examined here. The SRES A2 scenarios by 2085 were qualitatively different from the B1 scenarios and the earlier period A2 scenarios in terms of the scale and spatial extent of increased fire occurrence and burned area. Given that we do, as a species, face choices regarding future emissions pathways, it would seem desirable to avoid persisting on a high greenhouse gas emissions pathway like A2.

Projected increases in burned area in the Sierra Nevada appear to be greatest for mid-elevation sites on the west side of the range. These are locations with significant areas outside of federally managed forest and park lands—and low-density development patterns—potentially exposing private landowners to substantially increased wildfire threats. To the extent that future low-density development in this region is vulnerable to

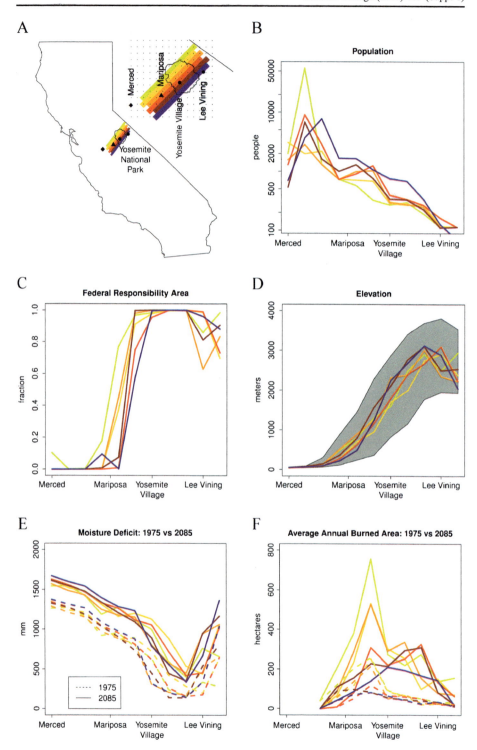

◄ **Fig. 6 a** Six transects through the Sierra Nevada at Yosemite National Park with four reference communities; Color-coded by transect: **b** Population (*log scale*); **c** Federal area as a fraction of total area; **d** Mean elevation, and range (*shaded*); **e** 30-year Mean Annual Deficit and (**f**) Predicted Average Annual Burned Area

losses from fire, continued growth could increase the state's economic vulnerability to increased wildfire due to climate change.

Forest areas in coastal California also faced increased risk of wildfires under the warmer CNRM CM3 and GFDL CM2.1 SRES A2 scenarios, including iconic coastal redwoods and the Ventana wilderness area in central California. In contrast to communities in the Sierra foothills, continued sprawling growth around more densely settled communities in coastal California could actually reduce fire risks in some areas while increasing them in others. These results are sensitive to model assumptions that define the demarcation between areas where increased development is threatened by wildfire, and areas where urbanization has increased to the point where wildfires can no longer occur. This highlights an important area for further research.

The results presented here reflect a set of illustrative models and their underlying assumptions that together result in a cascading series of cumulative uncertainties, such that results for any one time and location cannot be considered a reliable prediction, even contingent on the scenario represented. While aggregating results over time and space and comparing outcomes against a common reference period estimated with the same methods and data sets can reduce the impact of some types of systematic error, these measures are not foolproof. Nonlinear effects of errors, or qualitative systematic changes over time that are not captured by our models, can lead to significant errors in future projections that are

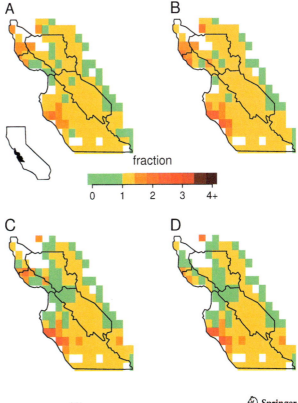

Fig. 7 Projected wildfire burned area around Monterey Bay communities for four scenarios **a** CNRM CM3 SRES A2, low growth and sprawl, high threshold for defining urbanized areas (1,000 households/km^2); **b** GFDL CM2.1 SRES A2, low growth and sprawl, high threshold for defining urbanized areas; **c** GFDL CM2.1 SRES A2, high growth and sprawl, low threshold for defining urbanized areas (147 households/km^2); (**D**) GFDL CM2.1 SRES A2, high growth and sprawl, high threshold for defining urbanized areas

not encompassed by the range of uncertainty represented in our results. That is, the results are conditional not only on the storylines of the SRES A2 and B1 scenarios and choice of global climate models, but also on the specifications of the statistical models of fire activity that we have estimated from historic data. To the extent that these data reflect processes that will no longer operate in the future because of qualitative changes to the systems we are modeling, our results will be in error.

Our results project the current managed fire regimes of California onto future scenarios for climate and for population and development footprint. We do not consider hypothetical effects of future changes in management strategies, technology, or resources that might be adopted with the intent of mitigating or adapting to the effects of climate and development on wildfire. Our models' implicit assumptions that such management effects are fixed may prove untenable under some future scenarios. Explicitly including management factors that can vary over the long term might significantly affect the outcomes modeled here in a systematic fashion. Such an exercise is left to a future study.

While we do not model metrics of fire severity (e.g., percent of vegetation consumed, ecological impact of burning) here, we expect that fire severity may be correlated with increases in fire occurrence and spatial extent in some ecosystems, particularly in some mountain forest areas of the state. It seems likely that outcomes such as those described here would have important implications for ecosystem services such as carbon sequestration in California forests, air pollution and public health, forest products and recreation industries, and the quality and timing of runoff from precipitation and snowmelt. All of these merit intensive further study.

References

Balling RC, Meyer GA, Wells SG (1992) Relation of surface climate and burned area in yellowstone national park. Agric For Meteorol 60:285–293
Brillinger DR, Preisler HK, Benoit JW (2003) Risk assessment: a forest fire example. In Science and Statistics, Institute of Mathematical Statistics Lecture Notes. Monograph Series
Cayan D, Tyree M et al (2009) Climate Change scenarios and sea level rise estimates for the California 2008 Climate Change Scenarios Assessment, Public Interest Energy Research, California Energy Commision, Sacramento, CA
Cumming SG (2001) A parametric model of the fire-size distribution. Can J For Res 31:1297–1303
Gesch DB, Larson KS (1996) Techniques for development of global 1-kilometer digital elevation models. In Pecora Thirteen, Human Interactions with the Environment—Perspectives from Space. Sioux Falls, South Dakota, August 20–22, 1996
Hamlet AF, Lettenmaier DP (2005) Production of temporally consistent gridded precipitation and temperature fields for the continental U.S. J Hydrometeorol 6(3):330–336
Hansen MC, DeFries RS, Townshend JRG, Sohlberg R (2000) Global land cover classification at 1 km spatial resolution using a classification tree approach. Int J Remote Sens 21:1331–1364
Heyerdahl EK, Brubaker LB, Agee JK (2001) Factors controlling spatial variation in historical fire regimes: a multiscale example from the interior West, USA. Ecology 82(3):660–678
Hidalgo HG, Dettinger MD, Cayan DR (2008) Downscaling with constructed analogues: Daily precipitation and temperature fields over the United States. CEC Report CEC-500-2007-123. January 2008
Holmes TP, Hugget RJ, Westerling AL (2008) Statistical analysis of large wildfires. Chapter 4 of economics of forest disturbance: Wildfires, storms, and pests, Series: Forestry Sciences, Vol. 79. In Holmes TP, Prestemon JP, Abt KL (Eds), XIV, p 422. Springer. ISBN: 978-1-4020-4369-7
Kipfmueller KF, Swetnam TW (2000) Fire-climate interactions in the Selway-Bitterroot wilderness area. USDA Forest Service Proceedings RMRS-P-15-vol-5
Le Quéré C, Raupach MR, Canadell JG, Marland G (2009) Trends in the sources and sinks of carbon dioxide. Nat Geosci 2:831–836

Liang X, Lettenmaier DP, Wood EF, Burges SJ (1994) A simple hydrologically based model of land surface water and energy fluxes for general circulation models. J Geophys Res 99(D7):14,415–14,428

Malamud BD, Morein G, Turcotte DL (1998) Forest fires: an example of self-organized critical behavior. Science 281:1840–1842

Maurer EP, Hidalgo HG (2008) Utility of daily vs. monthly large-scale climate data: an intercomparison of two statistical downscaling methods. Hydrology and Earth System Science 12:551–563

Maurer EP, Wood AW, Adam JC, Lettenmaier DP, Nijssen B (2002) A long-term hydrologically-based data set of land surface fluxes and states for the conterminous United States. J Climate 15:3237–3251

Mitchell KE et al (2004) The multi-institution North American Land Data Assimilation System (NLDAS): Utilizing multiple GCIP products and partners in a continental distributed hydrological modeling system. J Geophys Res 109:D07S90. doi:10.1029/2003JD003823

Monteith JL (1965) Evaporation and the environment. Symp Soc Expl Biol 19:205–234

Penman HL (1948) Natural evaporation from open-water, bare soil, and grass. Proc R Soc Lond A193 (1032):120–146

Preisler HK, Westerling AL (2007) Statistical model for forecasting monthly large wildfire events in the Western United States. J Appl Meteorol Climatol 46(7):1020–1030. doi:10.1175/JAM2513.1

Preisler HK, Westerling AL, Gebert KM, Munoz-Arriola F, Holmes T (2011) Spatially explicit forecasts of large wildland fire probability and suppression costs for California. International Journal of Wildland Fire 20:508–517

Preisler HK, Brillinger DR, Burgan RE, Benoit JW (2004) Probability based models for estimating wildfire risk. Int J Wildland Fire 13:133–142

Ricotta C, Avena G, Marchetti M (1999) The flaming sandpile: self-organized criticality and wildfires. Ecol Model 119:73–77

Schlobohm P, Brain J (2002) Gaining an understanding of the national fire danger rating system. National Wildfire Coordinating Group Publication: NFES # 2665. www.nwcg.gov

Schoenberg FP, Peng R, Woods J (2003) On the distribution of wildfire sizes. Environmetrics 14:583–592

Song W, Weicheng F, Binghong W, Jianjun Z (2001) Self-organized criticality of forest fire in China. Ecol Model 145:61–68

Stephenson NL (1998) Actual evapotranspiration and deficit: Biologically meaningful correlates of vegetation distribution across spatial scales. J Biogeog 25:855–870

Strauss D, Bednar L, Mees R (1989) Do one percent of the forest fires cause ninety-nine percent of the damage? Forest Sci 35:319–328

Swetnam TW, Betancourt JL (1998) Mesoscale disturbance and ecological response to decadal climatic variability in the American Southwest. J Clim 11:3128–3147

Theobald D (2005) Landscape patterns of exurban growth in the USA from 1980 to 2020. Ecol Soc 10(1):32

U.S. EPA (2008) Preliminary steps towards integrating climate and land use (ICLUS): The Development of Land-Use Scenarios Consistent with Climate Change Emissions Storylines (External Review Draft). U. S. Environmental Protection Agency, Washington, D.C. EPA/600/R-08/076A

Veblen TT, Kitzberger T, Donnegan J (2000) Climatic and human influences on fire regimes in ponderosa pine forests in the Colorado Front Range. Ecol Appl 10:1178–1195

Verdin KL, Greenlee SK (1996) Development of continental scale digital elevation models and extraction of hydrographic features. In: Proceedings, Third International Conference/Workshop on Integrating GIS and Environmental Modeling, Santa Fe, New Mexico, January 21–26, 1996. National Center for Geographic Information and Analysis, Santa Barbara, California

Westerling AL (2009) "Wildfires." Chapter 8 in Climate Change Science and Policy. Schneider, Mastrandrea, Rosencranz and Kuntz-Duriseti (Eds), Island Press"Washington DC, USA

Westerling AL, Bryant BP (2008) Climate change and wildfire in California. Clim Chang 87:s231–249. doi:10.1007/s10584-007-9363-z

Westerling AL, Gershunov A, Cayan DR, Barnett TP (2002) Long lead statistical forecasts of Western U.S. Wildfire area burned. Int J Wildland Fire 11(3,4):257–266. doi:10.1071/WF02009

Westerling AL, Brown TJ, Gershunov A, Cayan DR, Dettinger MD (2003) Climate and wildfire in the Western United States. Bull Am Meteorol Soc 84(5):595–604

Westerling AL, Hidalgo HG, Cayan DR, Swetnam TW (2006) Warming and earlier spring increases Western U.S. forest wildfire activity. Science 313:940–943. doi:10.1126/science.1128834, Online supplement

Zhang Y-H, Wooster MJ, Tutubalina O, Perry GLW (2003) Monthly burned area and forest fire carbon emission estimates for the Russian Federation from SPOT VGT. Remote Sens Environ 87:1–15

The impact of climate change on California's ecosystem services

M. Rebecca Shaw · Linwood Pendleton · D. Richard Cameron · Belinda Morris · Dominique Bachelet · Kirk Klausmeyer · Jason MacKenzie · David R. Conklin · Gregrory N. Bratman · James Lenihan · Erik Haunreiter · Christopher Daly · Patrick R. Roehrdanz

Received: 12 February 2010 / Accepted: 26 September 2011 / Published online: 24 November 2011
© The Author(s) 2011. This article is published with open access at Springerlink.com

Abstract Ecosystem services play a crucial role in sustaining human well-being and economic viability. People benefit substantially from the delivery of ecosystem services, for which substitutes usually are costly or unavailable. Climate change will substantially alter or eliminate certain ecosystem services in the future. To better understand the consequences of climate change and to develop effective means of adapting to them, it is critical that we improve our understanding of the links between climate, ecosystem service production, and the economy. This study examines the impact of climate change on the terrestrial distribution and the subsequent production and value of two key ecosystem services in California: (1) carbon sequestration and (2) natural (i.e. non-irrigated) forage production for livestock. Under various scenarios of future climate change,

M. R. Shaw (✉) · D. R. Cameron · K. Klausmeyer · J. MacKenzie · E. Haunreiter
The Environmental Defense Fund, 123 Mission Street 28th Floor, San Francisco, CA 94105, USA
e-mail: rshaw@edf.org

L. Pendleton
The Nicholas Institute, Duke University, P.O. Box 90335, Durham, NC 27708, USA
e-mail: linwood.pendleton@duke.edu

B. Morris
Environmental Defense Fund, 1107 9th Street, suite 1070, Sacramento, CA 95814, USA

D. Bachelet · J. Lenihan
USDA Forest Service, Oregon State University, Corvallis, USA

D. R. Conklin · C. Daly
Conservation Biology Institute, Corvallis, USA

D. R. Conklin
e-mail: david.conklin@lifeboatearth.org

G. N. Bratman
Stanford University, Stanford, USA

P. R. Roehrdanz
University of California at Santa Barbara, Santa Barbara, USA

we predict   that the provision and value of ecosystem services decline under most, but not all, future greenhouse gas trajectories. The predicted changes would result in decreases in the economic output for the state and global economy and illustrate some of the hidden costs of climate change. Since existing information is insufficient to conduct impact analysis across most ecosystem services, a comprehensive research program focused on estimating the impacts of climate change on ecosystem services will be important for understanding, mitigating and adapting to future losses in ecosystem service production and the economic value they provide.

1 Introduction

Ecosystems generate a variety of goods and services important for human well-being, collectively called *ecosystem services*. Ecosystem services are components of nature, directly enjoyed, consumed, or used to yield human well-being (Boyd and Banzhaf 2007). These goods and services play a crucial role in sustaining economic viability in ways that may or may not be immediately understood (MA 2005a, b). Ecosystem services are categorized into four types in an effort to clarify the broad spectrum of their value and significance: provisional services (e.g., food, timber, water and fuels), regulating services (e.g., water purification and carbon sequestration), supporting services (e.g., climate regulation), and cultural services (e.g., aesthetic values and sense of place) (MA 2005a, b). It is the provisioning services that are most familiar to the general public, but the whole suite of services contributes to human well-being and generates economic value. These ecosystem goods and services generate value when they are enjoyed directly by people (e.g., eating fish) or indirectly when they support the production and quality of other things people enjoy (e.g., instream flows support recreational fishing).

The importance of valuing ecosystem services in decisions affecting natural resources and their management is highlighted in Bingham et al. (1995). A significant body of literature exists that assesses various means of quantifying the relationship of ecosystem functions to human well-being in economic terms. Mendelsohn and Olmstead (2009) provide a critical synthesis of ecosystem service valuation studies and Costanza, et al. (1997) offer an early attempt to provide an aggregate estimate of the total value of global ecosystem services. The effect climate change will have on ecosystem function has also received significant attention (see Schmitz et al. 2003; Shaver 2000, and MA 2005a, b), but it is not yet well understood how climate change will impact ecosystem services and their associated economic value in the 21st century.

If ecosystems change dramatically as a result of climate change, the direct value we enjoy from the ecosystem services they produce also will change, in some cases dramatically. Climate change is likely to affect the abundance, production, distribution, and quality of terrestrial ecosystems. Therefore, ecosystem services such as climate stabilization through carbon sequestration, the provision of non-irrigated forage for livestock and wildlife species, the delivery of water which supports fish for commercial and recreational sport fishing, the provision of critical habitat for biodiversity, and many other types of ecosystem services are likely to be impacted by a changing climate. For example, the ability of forests to sequester carbon and thus limit climate change could be hindered as forest extent and productivity decrease and fire frequency and/or intensity increase with rising atmospheric temperatures. The distributions of grasslands suitable for forage production to support livestock grazing could shift along with changing patterns of precipitation. Because ecosystem services have economic value, that value is likely to change under a changing climate.

Ecosystem services contribute to state, national and global economies, which are thus vulnerable to perturbations in these services and in their net economic value. As a case study to explore this vulnerability, we focus on two ecosystem services that depend on the type and quantity of vegetation cover in California. By focusing on these two concrete examples for which we have in-depth knowledge, we hope to demonstrate how climate change can lead to changes in the provision and value of ecosystem services and in doing so shed light on the potential magnitude of economic effects that could result from the impacts of climate change on ecosystems.

Specifically, we focus on the direct effects climate change may have on (1) ecosystem attributes that influence the timing, magnitude and geographic distribution of ecosystem service production (e.g. ecosystem extent and distribution) and (2) the subsequent impacts on the production of ecosystem services including *carbon sequestration*, or the ability of terrestrial ecosystems to store carbon, and *forage production*, or the production of native or naturalized, non-irrigated vegetation by woodlands and grasslands, for cattle. We then (3) establish the link between these climate-induced changes in ecosystem services and potential impacts to economic values by estimating changes to the future net economic value of *carbon sequestration and forage production*. We chose these two ecosystem services for analysis because climate change impacts to the provision of each service can be directly projected and quantified using existing models and the subsequent economic value can be directly quantified through current markets or rigorous estimates of social, net economic value. Finally, we discuss the importance of future research to consider the economic impacts of climate change, and adaptation to climate change, on other ecosystem services. By bringing together climate, biophysical and economic models in a spatial analysis, it is our intent to highlight the integrated and interdisciplinary nature of examining the impact of climate change on ecosystem services and their value. We also underscore the need for a more directed and concerted research effort for accurately projecting potential changes to the future delivery of these vital services that underpin human well-being.

2 Methods

This section describes the methods used to (1) generate climate change projection data for California under two emissions scenarios, (2) model the impact climate change will have on ecosystems and provision of two ecosystem services, and (3) assess the value of the projected changes on those services (carbon sequestration and forage production).

2.1 Biophysical models

2.1.1 Climate data and projections

To explore the range of impacts on California ecosystem services projected, we consider the Intergovernmental Panel for Climate Change's (IPCC) high (A2) versus low (B1) greenhouse gas emissions scenarios (IPCC 2007); and three atmospheric-oceanic general circulation models (AOGCMs): GFDL-CM2.1 (Delworth et al. 2006), NCAR-CCSM3 (Collins et al. 2006, data only shown for carbon sequestration), and NCAR-PCM1 (Washington et al. 2000). The AOGCM data were statistically downscaled to 12 km resolution using the bias correction and spatial disaggregation (BCSD) method (Wood et al.

2004; Maurer and Hidalgo 2008; Hugo et al. 2008). Each AOGCM was selected based upon strong regional performance in California and were selected to bracket future projected extremes ranging from a *warm, wet future* (NCAR-PCM1) to a *hot, dry future* (GFDL-CM2.1, NCAR-CCSM3).

We used three sets of climatic data for terrestrial California in this analysis: (1) historical climate data generated from interpolating weather station data from 1895 to 2006 across the state (PRISM Group, Oregon State University); (2) constructed climate neutral future based on historical trends from 2005 to 2099; (3) projected future climate from the downscaled results of the AOGCMs from 2005 to 2099. We then summarized our results based on 43-year time periods; one historical time period (1961 to 1990) and three future time periods (2005–2034, 2035–2064, and 2070–2099).

2.1.2 Ecosystem extent and distribution

To project changes in vegetation distribution throughout California, we used the MC1 Dynamic Global Vegetation Model (MC1-DGVM) developed by the U.S. Forest Service (USFS) and Oregon State University at the Forestry Sciences Laboratory, Corvallis, Oregon. MC1 is a dynamic vegetation model that estimates the distribution of vegetation and associated carbon, nutrients, and water fluxes and pools. The biogeochemistry module is a modified version of the CENTURY model (Parton et al. 1994), which simulates plant productivity, organic matter decomposition, and water and nutrient cycling (Bachelet et al. 2004). The direct effect of an increase in atmospheric carbon dioxide (CO_2) is simulated using a beta factor (Friedlingstein et al. 1995) that increases maximum potential productivity (about 10% increase in net primary productivity at 550 ppm) and reduces proportionally the moisture constraint on productivity. The fire module (Lenihan et al. 2008b) simulates the occurrence, behavior, and effects of fire using several mechanistic fire behavior functions (Rothermel 1972; Cohen and Deeming 1985; Peterson and Ryan 1986; van Wagner 1993; Keane et al. 1997) that provide interaction between the biogeography and biogeochemistry modules (Lenihan et al. 2003). The current life form mixture is used to select factors that allocate live and dead biomass into different classes of live and dead fuels. The moisture content of the live fuel classes is estimated from soil moisture provided by the biogeochemical module. Dead fuel moisture content is estimated directly from climatic inputs (Cohen and Deeming 1985). Fire events are triggered in the model when the Build Up Index and the Fine Fuel Moisture Code (BUI and FFMC of the Canadian Forest Fire Weather Index System) meet set thresholds. Sources of ignition (e.g., lightning or anthropogenic) are assumed to always be available. The fraction of a cell burned by a fire event is estimated as a function of the prescribed mean fire return interval for each vegetation type, and of the number of years since a simulated fire event. Fire effects include the consumption and the mortality of dead and live vegetation carbon (functions of fire line intensity and tree canopy structure, Peterson and Ryan 1986), which is removed from (or transferred to) the appropriate carbon pools in the biogeochemistry module. Dead biomass consumption is simulated using functions of fuel moisture that are fuel-class specific (Anderson et al. 2005).

2.1.3 Carbon sequestration

The MC1 model generates the monthly estimates of carbon stored or lost in each grid cell under historical climate, neutral future climate, and future climate change scenarios. We generate results for all ecosystem carbon pools but we focus the analysis on above-ground live tree carbon to be consistent with the existing protocol within the Climate Action

Reserve (CAR) a protocol that focuses on securing carbon offsets under a cap and trade program capitalizing on the carbon stored as live tree biomass (tree bole [trunk], roots, branches, leaves/needles) and dead standing wood. (www.climateregistry.org/tools/protocols/project-protocols/forests.html). This provides a conservative estimate of the carbon stored by terrestrial forest systems since it does not take into account below-ground carbon storage pools.

We ran MC1 for both historical and future climate conditions and documented changes in (1) carbon pools (stems, leaves, branches, roots) and (2) soil carbon and soil moisture; (3) wildfire occurrence and impacts to estimate carbon losses and changes in the resilience of ecosystems if/when the fire regime changes; and (4) vegetation cover that will affect species range and extent, and ecosystem service production. We then averaged the summed annual carbon pool values for each of the four 30-year time periods and subtracted the carbon pools generated for the neutral climate future dataset from the projected carbon pools for each of six model-emission combinations to describe potential change due to climate change.

We accounted for urban expansion impacts on carbon sequestration potential by including estimates of future urban growth. For the baseline urban and agricultural land cover, we used the Multi-source Land Cover data to represent the extent statewide. We held agricultural land cover constant in all time periods. The agricultural extent represents row crops and other intensive agriculture, not rangelands or timberlands. For the future extent of urban lands in 2035, 2065, and 2100, we used the methods described in Sanstead et al. (2009). We calculated urban extent for each 1/8° grid cell using the mid-range projections for household density and set a threshold of 1 unit per hectare as the minimum density for "urban". We combined this urban extent data with the agricultural extent to generate a combined "converted land" extent. The percentage of the remaining natural cover in each cell was multiplied by the carbon and forage production values to account for the additional impact of future urbanization on these services.

2.1.4 Forage production

Annual natural (non-irrigated) forage production for livestock in California is determined primarily by the amount and timing of precipitation. We analyzed the projected changes in distribution of the production of forage vegetation within grassland and oak woodland habitat under future climate scenarios using monthly precipitation data from two AOGCMs for each emissions scenario, the projected vegetation production from MC1, and the Natural Resources Conservation Service (NRCS) State Soil Geographic (STATSGO) soil data. We use the following formula to estimate monthly forage production/precipitation relationship for grid cell x:

$$Monthly\ Forage\ Per\ Precipitation = \frac{(Forage_{ann}*Productivity\ Pct_{monthly})}{Precipitation_{monthly}}$$

For each emission/AOGCM combination, we used the average monthly precipitation for each grid cell to generate the estimates of production. Because the timing of precipitation greatly affects the forage production, we used the growth curve for rangeland sites available through the NRCS Ecological Site Description website (http://esis.sc.egov.usda.gov) to proportionally allocate the annual range production into growing season months. We summed the resultant monthly production across the six growing season months to generate the annual production for each cell (pounds/acre, kg/ha). To account for the effects of current and projected anthropogenic land cover on forage availability, we multiplied the annual production value for each grid by the percentage of each grid cell in agriculture or urban for each time period and subtracted those from the estimate of available rangeland forage. We define rangelands as areas in either herbaceous or hardwood woodland land cover including non-irrigated grasslands and woodlands with shrubs that produce forage for livestock and wildlife. Modeling other factors that may affect the carrying capacity of rangeland for livestock such as the nutrient content of the forage, management costs, or adjacent land ownership is beyond the scope of this study.

2.2 Economic valuation

2.2.1 Carbon sequestration valuation

We estimate the net economic effects of carbon sequestration using the social cost of carbon (SCC). The SCC measures the full global net economic impact today of emitting an incremental unit of carbon at some point of time in the future, and it includes the sum of the global cost of the damage it imposes the entire time it is in the atmosphere (Price et al. 2007; Pearce 2003). The SCC attempts to capture how much society would be willing to pay to avoid damage from climate change in the future and still be as well off as they would be in the absence of climate change.

In 2005, Richard Tol published a meta-analysis of 103 estimates from 28 public studies of the marginal net economic costs of CO_2 emissions. In 2008, Tol further refined his analysis. With conservative assumptions, Tol determined the mean for peer-reviewed estimates is $23/metric ton of carbon (MTC). Watkiss and Downing (2008) provided further updates of Tol and reported that in 2002, the UK Government recommended a marginal global SCC estimate of $185/MTC (USD 2007), with a range of $93 to $371 with all three estimates increasing $1.50/MTC per year from the year 2000. We conservatively consider a central value from Watkiss and Downing of $185/MTC noting that the authors expect significant increases in this value over time.

Nordhaus (2008) used the DICE-2007 model, to estimate the optimal carbon taxes that would accurately reflect the cost of carbon emissions and showed that the trajectory of optimal carbon prices (e.g. carbon taxes or the SCC) should rise to reflect the increasing damage caused by climate change and the need for increasingly tight constraints. In the model, the optimal price rises steadily over time, at between 2% and 3% per year in real terms, to reflect the rising damages from climate change. In this trajectory, Nordhaus' carbon price rises from $34/MTC to $113/MTC by 2050 and $251 per MTC in 2100 all of which reflect the SCC as predicted by the DICE model.

To account for the range of accepted SCC values, we present the social value of carbon sequestration with estimates of SCC from Tol (2008), Watkiss and Downing (2008), and Nordhaus (2008).

2.2.2 Forage valuation

For the purposes of this study, we examine the economic value of natural, non-irrigated forage as an input to the livestock market. In our analysis, we estimate the value of non-irrigated forage production using two methods for valuation–replacement of feed and as an intermediate good in the production of livestock. To be conservative, we do not value quality of life, landscape generated by grazing lands nor do we estimate the costs associated with fire and invasive species management. We also do not attempt to calculate what, if any, additional costs to society would be created by changes in greenhouse gases caused by more or fewer head of cattle.

We consider the final ecosystem service (or end product) of forage production to be livestock. We identify the following two mechanisms for valuing forage production: (1) the market in livestock and its products (see Chan et al. 2006); and (2) the price of the least cost replacement for forage as a livestock feed.

In our land cover model, each cell of rangeland generates associated forage dry matter (DM) in units of annual tons, which in turn supports livestock production. We measure animal production from forage as an Animal Unit Month (AUM). AUM values can range up to 1,000 lbs (454.5 kg)DM, depending on forage quality. Recent studies support an average of 791 lbs (359.5 kg) of DM as equivalent to one AUM (Brown et al. 2004; Thorne et al. 2007), the value we use in our calculations, which represents high quality forage and likely results in a high estimate for total available forage. We also assume only 50% of the forage produced on an acre of land is available for livestock production; the rest must be used for management of land productivity, or it is lost due to trampling and contamination from animal waste. This 50% utilization is based on perennial grasslands as opposed to annual grasslands. By using this conversion rate, we may underestimate potential forage production and err on the side of conservatism. If climate change dramatically reduces production, the 50% rule may overestimate utilization.

The value of the marginal product of forage (change in market value due to change in forage), expressed in profits per AUM, provides a lower bound estimate for the economic value of forage production. Brown et al. (2004) consider costs and revenues on ranches in each county in California. They report the following as average statewide values (USD 2007).

Average per cow profitability :	**\$110.00(2007 USD)**
Annual average DM requirements per cow :	**9,492 lbs(12 months @ 791 lbs(359.5 kg)/AUM))**
\$ profit per pound of DM :	**\$0.011553(.025495 per kg)in profit**

To calculate the climate change impacts on the economic value of change in forage production due to climate change, we assume only 50% of forage is available for livestock (see above biophysical model description) in order to arrive at the quantity of forage available for livestock production in each month. To find the market value of climate change on forage production, we multiply the predicted change in DM (from the vegetation model) by the value of its marginal product in livestock production as measured through average state livestock prices (\$0.011553 per pound or 0.025495 per kg of DM).

As an upper bound for the value of forage, we also consider the marginal replacement cost of forage. Following on the common practice of substituting hay for forage, we assume that the lowest grade hay available from each county in California in 2003 (USDA 2004) is an equivalent substitute for natural forage. We use the market price for low grade hay, averaged across all counties in California that provide the same hay type (USDA 2008) to calculate a maximum bound for the potential change in value of forage production resulting from climate change. Assuming that forage and hay are equally nutritious on a one-to-one basis, a simple, direct cost of forage substitution is calculated at approximately \$78/t (\$70.91/tonne).

3 Results

3.1 Changes to ecosystem attributes, services and values

3.1.1 Ecosystem extent and distribution

The MC1 DGVM projects widespread changes in the distribution of vegetation across the state by the end of the century that will impact carbon sequestration and forage production. The most pronounced change that occurs across a majority of model-emissions scenarios is a 15% to 70% increase in shrublands when compared to the neutral climate future scenario (Fig. 1). In addition, there is a consistent decline, through the end of the century, in conifer woodland and forest, as well as in herbaceous cover across all model-emissions scenarios. Under the hot, dry GFDL scenario, the model projects an increase in shrubland and hardwood forest as well as a decrease in grassland, conifer woodland and forest, under both emissions scenarios. Changes predicted under the warmer, wetter PCM1 model projections are less pronounced and vary by emissions scenario, with the exception of a 10–20% increase in hardwood woodland and a decrease in conifer woodland and for conifer forest (~10%). Shrublands are projected to decrease (<10%) under the B1 emissions scenario but to increase (~30%) under the A2 emissions scenario. While this study projects an increase in shrubland extent at the expense of grasslands, previous studies using the MC1 model for California have shown an increase in grasslands at the expense of woody dominated vegetation types (Lenihan et al. 2003, 2008a, b). This difference is due to two changes in the model. First, the MC1 model was calibrated to more accurately represent current actual land cover types, especially in Southern California, which is not the case with previous studies (Lenihan et al. 2003, 2006, 2008a, b). Second, we used an advanced fire model, which resulted in simulations of more frequent but lower intensity fires, which

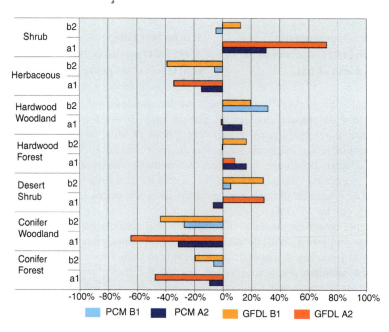

Fig. 1 Percent change in extent of major vegetation types projected by 2070–2099. The chart shows the difference between the areal extents of vegetation types in 2070–2099 as compared to the base scenario for that time period

resulted in fire events that did not kill woody lifeforms. In addition, the increase in atmospheric CO_2 throughout the 21st century enhances shrub water use efficiency allowing them to better survive simulated drought conditions.

The spatial distribution of the projected vegetation changes by 2070 to 2099 is presented in Fig. 2. The map labeled "Historical" reflects the modeled potential natural vegetation simulated for the period spanning 1961 to 1990. The expansion of the hardwood forest up in altitude into the interior Sierra Mountain Range and up in latitude to Northern California (Modoc, Klamath, and North Coast ecoregions) is evident in all future scenarios, but it is most pronounced with the

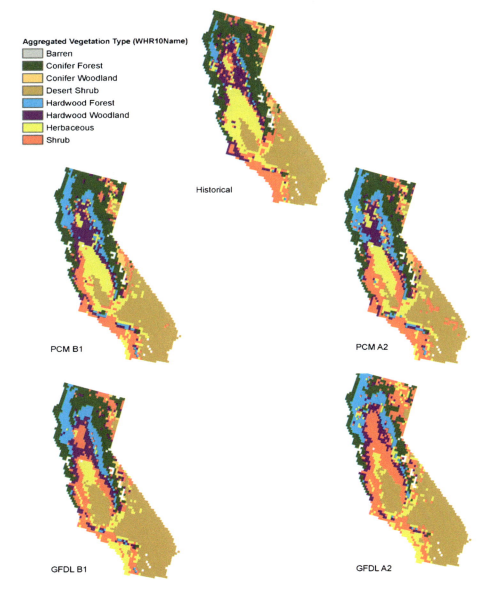

Fig. 2 Distribution of major vegetation types during historical time period (top of figure) and at the end of the 21st century (modal values for the period 2070–2099)

hot, dry GFDL climate under the high A2 scenario (Fig. 2). In a majority of model-emissions scenario combinations, shrublands expanded. Desert shrubland expands into the California's interior central valley in the hotter, drier GFDL model, but retreats in the warmer, wetter PCM1 climate model along the coast. Our analysis shows that land use change increases aboveground carbon loss, on average across all model-emissions scenarios, by an additional 1.5% by the end of the century.

3.1.2 Carbon sequestration

Projected changes in vegetation distribution will affect the carbon sequestration services provided by terrestrial ecosystems in California. The warmer, wetter climate model (PCM1) projects an increase in aboveground carbon storage relative to the neutral climate future scenario under both emissions scenarios (Fig. 3). In contrast, under the hotter, drier model (GFDL), MC1 projects much lower carbon pools than it does under neutral climate future scenario, with steep declines by the end of the century under the high A2 emissions scenario. The future climate generated by hotter, drier CCSM3 causes an even greater loss in carbon storage over the next century, with the largest loss simulated under the high A2 emissions scenario.

The spatial distribution of carbon storage in aboveground live carbon pools changes dramatically across the state by the end of the century, depending on the model-emissions scenario used. Under both the low (B1) and high (A2) emissions scenarios, there is a large increase in carbon stored by terrestrial ecosystems in the northwest of the state under the warmer, wetter climate conditions projected by PCM1. As a result, total carbon storage in live trees increases statewide, outweighing the losses in carbon in the central valley of California and along the coast. Under both

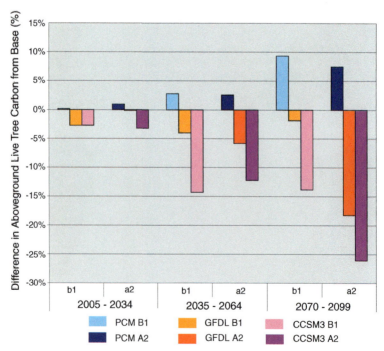

Fig. 3 Percent change from the neutral climate future in carbon storage in aboveground live tree biomass under low and high emissions scenarios for three AOGCMs (PCM1, GFDL, and CCSM3) simulated climate conditions

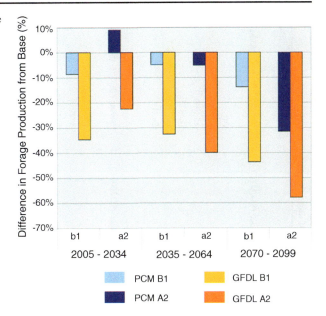

Fig. 4 Percent change from the neutral climate future in forage production under low and high emissions scenarios for three AOGCMs (PCM1, GFDL, and CCSM3) simulated climate conditions

the B1 and A2 emissions scenarios, large losses in aboveground live carbon pools are projected in the eastern side of the Sierra mountain range under the future conditions simulated by the hotter, drier GFDL model, and in northern California (Klamath Mountains and Modoc Plateau) under a future simulated by the hot and dry CCSM3 model. Under the A2 emissions scenario, there are relatively few areas projected to have increased carbon storage under the hotter, drier conditions simulated by GFDL and CCSM3.

3.1.3 Carbon sequestration value

We estimated the net economic value due to expected climate-driven changes in the carbon stored by natural ecosystems in California. We based our estimates on the carbon storage estimates described above combined with the estimates of the social cost of carbon (SCC) released into the atmosphere (or the social benefit of carbon not released). The estimates we use for SCC are described earlier in the paper and are summarized as: a) mean values (Tol 2008), b) existing UK SCC (Watkiss and Downing 2008) and c) the globally aggregated DICE-2007 model (Nordhaus 2008) (Table 1). Driven by the warmer, wetter PCM1 climate model, the MC1 vegetation model consistently projects a higher capacity to store carbon. Consequently, the effect of climate change on natural carbon storage in California would result in a net benefit to society of $38 million annually during the period 2005–2034 and as high as $22 billion annually by 2070. Driven by the hotter, drier climate model, however, the MC1 vegetation model projects a sharp loss in carbon storage capacity in natural areas leading to social costs of −$646 million to −$5.2 billion annually for the period 2005–2034 (under scenario B1 using the hotter, drier CCSM3 model of climate change) to as high as −$62 billion annually by the period 2070–2099, under scenario high A2 using the Nordhaus' DICE-2007 model predictions.

3.1.4 Forage production

Forage production declines dramatically by the end of the century (2070–2099) in all future projections, ranging from a 14% decline in annual mean (Tg) production under the warm, wet

Table 1 Projected change in live aboveground carbon sequestered and the economic value including social cost of carbon of these changes

2005–2034

Scenario	Model	Carbon		Change in value (2007$ million)		
		Total (Tg)	% Change from base	Tol 2008 mean ($23/MTC)	DICE-2007 optimal price ($34/MTC)	Existing UK SCC ($185/MTC)
Base		1,025				
B1	PCM1	1,027	0%	$38	$56	$303
	GFDL	997	−3%	($651)	($962)	($5,236)
	CCSM3	997	−3%	($646)	($955)	($5,194)
A2	PCM1	1,035	1%	$230	$340	$1,847
	GFDL	1,024	0%	($31)	($45)	($245)
	CCSM3	992	−3%	($761)	($1,125)	($6,119)

2035–2064

Scenario	Model	Carbon		Change in value (2007$ million)		
		Total (Tg)	% Change from base	Tol 2008 mean ($23/MTC)	DICE-2007 optimal price ($113/MTC)	Existing UK SCC ($185/MTC)
Base		1,028				
B1	PCM1	1,057	3%	$655	$3,220	$5,271
	GFDL	987	−4%	($950)	($4,669)	($7,644)
	CCSM3	881	−14%	($3,390)	($16,656)	($27,269)
A2	PCM1	1,055	3%	$608	$2,987	$4,890
	GFDL	968	−6%	($1,381)	($6,786)	($11,109)
	CCSM3	902	−12%	($2,894)	($14,220)	($23,281)

2070–2099

Scenario	Model	Carbon		Change in value (2007$ million)		
		Total (Tg)	% Change from base	Tol 2008 mean ($23/MTC)	DICE-2007 optimal price ($251/MTC)	Existing UK SCC ($185/MTC)
Base		952				
B1	PCM1	1,041	9%	$2,044	$22,309	$16,443
	GFDL	935	−2%	($399)	($4,350)	($3,207)
	CCSM3	820	−14%	($3,035)	($33,123)	($24,413)
A2	PCM1	1,023	7%	$1,632	$17,807	$13,125
	GFDL	778	−18%	($3,992)	($43,570)	($32,113)
	CCSM3	704	−26%	($5,707)	($62,281)	($45,904)

PCM1-B1 scenario to a 58% decline under the hotter, drier, GFDL-A2 scenario. The geographic pattern of the changes in forage production projected by the end of century (2070–2099) differs dramatically among model-emissions scenarios (Fig. 4). Many of the largest losses in forage production are due to conversion of rangeland to shrublands in highly suitable climates. The spatial pattern of change in forage production driven by the warmer, wetter PCM1 model under both B1 and A2 scenarios is heterogeneous. In contrast, using the drier and warmer GFDL climate model and both B1 and A2 scenarios, the vegetation model projects extensive losses in forage production, concentrated in the inner central coastal region and along the foothills of the interior mountain range, the Sierra Nevada. The hottest and driest model-emission scenario combination, GFDL-A2, projects extensive and consistent losses in production over virtually all of the current extent of rangelands.

3.1.5 Forage value

Table 2 shows the potential impacts of climate change as estimated in changes of the value of natural forage for livestock. While the neutral climate future scenario for 2005–2034 projects an estimated positive increase in the value of forage (estimated at an average annual increase in profits of $15 million using the marginal product for livestock approach or an average annual savings in hay purchases of $50 million using the forage substitution approach), most projections show losses in forage production for all three time periods (with losses ranging between $14 million and $570 million). The choice of valuing forage using livestock profits or hay prices clearly makes a large difference (a factor of five) in the estimates provided here. This disparity illustrates the need for a more robust and county-specific calculation of profit

Table 2 Absolute and percent changes in forage production (Tg) from the neutral climate future and changes in forage production value under low and high scenarios for two AOGCMs (PCM1 and GFDL) across three time periods in the future. Difference in value is millions of dollars per year averaged over the time period

Scenario	Model	Forage		Difference in value ($ million)	
		Total (Tg)	% change from base	Profits from livestock	Cost of replacement for hay
2005–2034					
Base		13			
B1	PCM1	11.9	−8%	($14)	($47)
	GFDL	8.54	−34%	($57)	($192)
A2	PCM1	14.19	9%	$15	$51
	GFDL	10.14	−22%	($36)	($123)
2035–2064					
Base		12.24			
B1	PCM1	11.65	−5%	($8)	($25)
	GFDL	8.28	−32%	($50)	($170)
A2	PCM1	11.63	−5%	($8)	($26)
	GFDL	7.39	−40%	($62)	($209)
2070–2099					
Base		12.52			
B1	PCM1	10.97	−14%	($22)	($74)
	GFDL	7.05	−44%	($70)	($235)
A2	PCM1	8.56	−32%	($50)	($170)
	GFDL	5.26	−58%	($92)	($312)

per AUM as well as a more thorough investigation of the true cost of a substitute for lost forage production.

4 Discussion

Climate change will alter the fundamental character, production, and distribution of the ecosystems that contribute to the well-being of California citizens. In this study, we show how climate change will impact terrestrial ecosystem services, especially those associated with vegetative land cover change, and their economic values (measured as net economic value). Although there have been attempts to systematically model the link between ecosystem change and ecosystem service value (e.g. de Groot et al. 2002), our scientific understanding of the link between ecosystem service production and the value to society is still in the early stages of development for most ecosystems services (Turner et al. 2003).

4.1 Carbon sequestration by terrestrial ecosystems

Although the exact magnitude of the economic effect of increased carbon in the atmosphere has not been defined, the literature is clear that more atmospheric carbon will lead to more climate change, which will, in turn, have economic impacts around the globe (IPCC 2007). Indeed, the effects of climate change on the ability of California ecosystems to store carbon could result in more carbon being released into the atmosphere. The impact of these changes differs substantially depending upon the climate change models and scenarios employed. Models that predict a wetter future climate indicate that California terrestrial ecosystems could increase in their carbon sequestering capabilities and could generate additional value to the world's economy of over to $300 million annually in the near future and as much as $22 billion annually by 2070 (Table 1). Other models of climate change, however, are far more pessimistic, predicting social costs from climate change of −$650 million to more than −$5 billion annually for the period 2005–2034 under the low (B1) emission scenario using the hotter, drier CCSM3 model of climate change to as high as −$62 billion annually by the period 2070–2099 (Table 1).

4.2 The value of natural, non-irrigated forage

Despite some localized variation and spatial heterogeneity in the impacts of climate change, our projections of the ecosystem services associated with land cover vegetation show that climate change will lead to a reduction in the statewide provision of forage under most model-emissions scenarios by the end of the century. As a result, the economic value of forage production in California (measured as lost profits or increased costs of feeding cattle) will be substantially lower under climate change. In the near term, annual changes in profits are predicted to range from a slight increase in profits ($15 million) to losses of up to $36 million (Table 2). By 2070, the average annual profits of cattle ranching could be between $22 million and $92 million lower due to climate change. To put these figures in context, we consider what steps ranchers may take to offset losses of natural forage. We estimate the least-cost option of replacing natural forage with hay and find that the cost of replacing lost forage with hay could be as high as $235 million per year. As a result of the loss in forage production, most model-emissions scenarios predict that climate change will result in significant losses in the economic value of cattle and forage-reliant livestock industries in California.

4.3 The market value of natural carbon sequestration

In this paper we estimate the net economic effects of carbon sequestration using the social cost of carbon. The market value of carbon could also be used to estimate the value of carbon sequestration as long as the carbon sequestered is 'additional' to an established baseline, meaning that project-based greenhouse gas emissions reductions are additional to what would have happened under a business as usual scenario (VCS 2008). The market value of carbon reflects the least-cost method for reducing carbon emissions in the atmosphere, as revealed by the market. Generally, market price is determined by the cost of meeting a cap on total carbon permitted to be released into the atmosphere through reductions in carbon emission or through the sequestration of carbon. Currently, carbon trading occurs through a number of allowance-based markets and project based transactions. The two main markets are the European Union Emissions Trading Scheme (EU ETS) and the New South Whales GHG Reduction Scheme, which are both regulated markets. Additionally, a compliance-based cap and trade program is planned to begin in 2012 under California's Global Warming Solutions Act of 2006 (Assembly Bill 32). The program allows for 8% of emissions to be achieved through offsets, and offsets achieved through forest carbon sequestration (aboveground live biomass in trees) are included in the program. The actual market value of forest carbon will depend on the development of this program. If emissions from forests were included in the cap, then the market price could be used to estimate the change in value of forest carbon resulting from climate change impacts. As forests are not included in the cap, we only estimate the change in value to landowners who may be eligible to participate in offset markets.

4.4 Other economic impacts of climate change

Changes in precipitation, temperature and ecosystem distribution with climate change will alter the timing and distribution of water availability in the future which will impact an array ecosystem services on which we depend including the delivery of clean water for drinking and food production, the generation of electricity through hydropower, the transportation, recreation, and the mitigation of floods. In California, all climate change projections show a shift in the proportion of precipitation falling as rain rather than snow in the mountains, a shift to higher river flows earlier in the spring and a shift in river flow from northern rivers to southern rivers. Under all emissions scenarios and all climate models, all rivers show an increase in average flow from January to April by the end of the century (2070–2099) compared to the historical period and a decrease in average flow from April to October with the greatest drops in June and July (Shaw et al. 2009). This temporal shift in flow will be compounded by spatial shifts in precipitation that increase instream flow in the southern rivers at the expense of the northern rivers.

These shifts will cause a disruption in the delivery of many ecosystem services and their values, but we have yet to fully comprehend the impact to our economy. The economic values associated with the provision of surface water differ depending on the ultimate use of water. The values, but arise from the direct use of water by residences, municipalities and industry, irrigation for farming, and hydropower. Indirectly, surface water is an intermediate input to commercial fisheries, recreational fishing, recreational boating, and snow-related recreation. These intermediate services, in turn, affect the production of end users or final services. Change in instream flow impacts commercial fisheries, recreational fishing, recreational boating, municipal and industrial use, irrigation, hydropower and flood mitigation. For example, the primary effects of climate change on recreational fishing for salmon is through its impact on stream temperature and precipitation-related changes to the quantity and timing of stream flow, especially the timing of spring runoff and average flow (Flemming and Jensen 2002; Anderson et al. 1993). Change in snowpack can affect related recreation (skiing/snowmobiling), flood mitigation, municipal and industrial use, and hydropower. Excessive surface water can also provide an economic disservice or economic cost by creating flooding and causing coastal and freshwater pollution that can affect beach and other recreation.

Another important attribute of ecosystem change that contributes significantly to ecosystem service production is biodiversity (Edwards and Abivardi 1998). Loss of species, especially iconic species around which tourism industries are developed, will have an impact on the economic well-being of California residents as the economy in many parts of California is dependent on tourists who come to California, in part or specifically, to see such species as otters, redwoods, sequoias, condors, and many more plants and animals (USFWS 2007). Projected species' response to climate change varies considerably between alternative emissions scenarios but we expect species migrations to track shifting climates poleward, higher in altitude, toward the coast and toward stable water sources with large contractions of suitable areas for persistence across the state (Shaw et al. 2009). Under scenarios of hotter, drier future climate, the magnitude of the simulated species' responses and loss is greater. Because of the complex interactions between plants, animals, and humans under a changed climate, we do not yet have a comprehensive understanding of the impacts of climate change on biodiversity, ecosystem function, ecosystem and species distribution, nor ecosystem resilience.

In addition, ecosystems generate value beyond those that appear in organized markets. In some cases, especially recreational services that depend on ecosystems, Californians enjoy economic benefits that exceed what they have to pay. These non-market values are important and changes in these values due to climate change can represent real losses in the economic well-being of Californians. Indirectly, some changes in non-market values can eventually reveal themselves in the hedonic value of homes near recreation sites, the cost of hotels, and other premiums that can be charged to recreationists. Much of this value, however, resides with the user. The economic well-being of homeowners, land owners, outdoor workers, and even motorists who choose to drive on scenic byways depend on ecosystem conditions. All of these non-market values could change substantially due to climate change. Future research is needed to understand how recreational behavior, home values, and other non-market economic behavior will likely change due to climate change.

4.5 The challenges with estimating the economic effects of climate change

The results of this study demonstrate that there remains a great deal of uncertainty in our understanding of the potential economic effects of climate change, even on services like carbon sequestration and forage for which we have relatively more data than other types of ecosystem services. We find that climate change could have positive or negative impacts on the ability of natural systems to sequester carbon and that these differences can vary spatially and temporally by several orders of magnitude depending on the levels of anthropogenic emissions. We are hampered in our ability to provide precise estimates of the value of changes in natural forage due to climate change because the quality and value of forage differs significantly across the state. Observations and models indicate that these systems are sensitive to climatic conditions, but the precise impacts of change are greatly uncertain.

The challenges in estimating the potential economic impacts of climate change include uncertainty in a) the amount of anthropogenic carbon that will be emitted into the atmosphere, b) the effects of atmospheric carbon on atmospheric temperature and precipitation, c) our understanding of how ecosystems and ecosystem outputs will change due to these changes, and d) how human beings and economic activity may change as a result of ecosystem change.

Part of the uncertainty associated with our ability to predict the future impacts of climate change is linked to our inability to predict long-term weather patterns and teleconnections which are complex systems and we are only beginning to understand the complexity needed for modelling. More fundamental, however, is the fact that we have yet to collect long-term, spatially explicit data on enough types of ecosystems, ecosystem services, and how people use and benefit from these ecosystem services. We know these ecosystems are sensitive to climate and weather, we know people value the goods and services produced by these ecosystems, but our knowledge of just how climate affects ecosystems and how ecosystems affect people is still rudimentary. Without good, coordinated data on ecosystems, ecosystem services, and human uses and benefits of ecosystem services we are unable to accurately predict the future effects of climate change on people and economies, nor are we able to predict how these effects differ across states, regions, or the world.

If we are to plan to adapt to future impacts of climate change, it is essential that we begin now to collect rigorous and consistent data on ecosystems, ecosystem services, and human uses and values for ecosystem services. These data should meet the same criteria already in place for the data we collect to monitor and model climate and important economic sectors: the data must be rigorous, collected systematically across the nation, and collected at relevant spatial and temporal intervals that are sufficient to allow scientists to understand the effects of change in these systems. Most importantly the data collected across atmospheric, biophysical, and human systems should integrate well to facilitate statistical analyses, modelling and decision making under uncertainty.

5 Conclusion

Our research reveals that climate change will affect the distribution and production of terrestrial ecosystems and the services they provide. Yet, our understanding of how ecosystem services contribute to economic well-being and productivity is still rudimentary. Beyond a general knowledge of the overall importance of ecosystem services, we have only a few concrete examples of the value of these ecosystem services. Even more rudimentary is our understanding of how these ecosystem services will change due to climate change, how these changes will affect people and the economy, and how the economy will respond to these changes. The problem isn't just uncertainty about the future, but our general lack of understanding of how to deal with risk and make decision under this uncertainty.

We highlight two ecosystem services for which we have some knowledge. Our findings show that even small changes in terrestrial ecosystem productivity can cause large changes in the value of the ecosystem service, but the uncertainty of any given outcome is huge. In the case of the economic value of carbon sequestration—a service that helps mitigate climate change— this value is large and shared globally. In the case of natural forage, its global impact is smaller, but its proportional impact looms large on the sectors affected. It is important to remember that these examples were chosen not because of the size of their expected change, but largely because of the availability of data, the feasibility of modeling change in the ecosystem and the subsequent change in the ecosystem service production, and availability of methods to value the change in ecosystem service. There are likely to be many other ecosystem services for which the effects of climate change will be larger and proportionately more important. For instance, consider the potential effect of climate change on: the ability of forests and natural vegetation to moderate urban and suburban temperatures; the ability of the ocean and coastal habitats to sequester carbon, cool coastal areas (important to people and to agriculture), and provide seafood and recreation; the ability of montane and riparian forests to recharge groundwater and protect against flooding; the contribution of natural pollinators to agriculture, horticulture, and even home gardens; the list goes on.

To better understand, avoid, and adapt to the impacts of climate change on our economy, it is critical that we develop a better quantitative understanding of the links between climate change, ecosystems, and economic activity. While we have begun to develop a literature on the economic value of many of these ecosystem services we are largely ignorant of the value of ecosystem services to the economy and even less knowledgeable about the ways in which climate change will affect these services and how we can best adapt to these changes. Our approach has not been systematic and has not been designed to specifically address those ecosystem services that are most likely to change due to changes in climate. We recommend a research agenda that employs a strategic approach to prioritizing, understanding and modeling the impacts of climate change (and other environmental change) on ecosystem services. We also urge the development of new approaches and guidance for decision-makers and managers to deal with the uncertainty of model outcomes and the risks associated the array of decisions before them in the context of those outcomes. Until we close this gap in our understanding, we will be unable to fully comprehend or begin to mitigate and adapt to the effects of climate change on California's economy and the well-being of California residents.

Open Access This article is distributed under the terms of the Creative Commons Attribution Noncommercial License which permits any noncommercial use, distribution, and reproduction in any medium, provided the original author(s) and source are credited.

References

Anderson D, Shangle S, Scott M, Neitzel D, Chatters J (1993) "Valuing effects of climate change and fishery enhancement on Chinook Salmon." Contemporary Policy Issues. Vol. XI, October 1993

Anderson GK, Ottmar RD, Prichard SJ (2005) CONSUME 3.0 user's guide. Pacific Wildland Fire Sciences Laboratory. USDA Foreset Service Pacific Northwest Research Station, Seattle, p 183

Bachelet D, Neilson RP, Lenihan JM, Drapek RJ (2004) Regional differences in the carbon source-sink potential of natural vegetation in the U.S. Ecol Manag 33(Supp#1):S23–S43. doi:10.1007/s00267-003-9115-4

Bingham G, Richard B, Brody M, Bromely D, Clark E, Cooper W, Costanza R, Hale T, Hayden G, Kellert S, Norgaard R, Norton B, Payne J, Russell C, Suter G (1995) Issues in ecosystem valuation: improving information for decision making. Ecol Econ 14:73–90

Boyd J, Banzhaf S (2007) What are ecosystem services? The need for standardized environmental accounting units. Ecol Econ 63:616–626

Brown S, Dushku A, Pearson T, Shoch D, Winsten J, Sweet S (2004) "Carbon supply from changes in management of forest, range, and agricultural lands of California." Winrock International for California Energy Commission 144

Chan KMA, Shaw MR, Cameron DR, Underwood EC, Daily GC (2006) Conservation planning for ecosystem services. PLoS Biol 4:e379. doi:10.1371/journal.pbio.0040379

Cohen JD, Deeming JE (1985) The National Fire Danger Rating System: basic equations. USDA Forest Service Pacific Southwest Forest and Range Experimental StationGeneral Technical Report PSW-82. 16 pp

Collins WD, Bitz CM, Blackmon ML, Bonan GB, Bretherton CS, Carton JA, Chang P, Doney SC, Hack JJ, Henderson TB, Kiehl JT, Large WG, McKenna DS, Santer BD, Smith RD (2006) The Community Climate System Model Version 3 (CCSM3). J Clim 19:2122–2143

Costanza R, d'Arge R, de Groot R, Farber S, Grasso M, Hannon B, Limburg K, Naeem S, O'Niell RV, Paruelo J, Raskin RG, Sutton P, van den Belt M (1997) The value of the world's ecosystem services and natural capital. Nature 387:253–260

de Groot R, Wilson MA, Boumans RMJ (2002) A typology for the classification, description and valuation of ecosystem functions, goods and services. Ecol Econ 41:393–408

Delworth TL, Rosati A, Stouffer RJ, Dixon KW, Dunne J, Findell K, Ginoux P, Gnanadesikan A, Gordon CT, Griffies SM, Gudgel R, Harrison MJ, Held IM, Hemler RS, Horowitz LW, Klein SA, Knutson TR, Lin S-J, Milly PCD, Ramaswamy V, Schwarzkopf MD, Sirutis JJ, Stern WF, Spelman MJ, Winton M,

Wittenberg AT, Wyman B (2006) GFDL's CM2 global coupled climate models. Part I: formulation and simulation characteristics. J Clim 19:643–674

Edwards PJ, Abivardi C (1998) Valuing biodiversity: where ecology and economy blend. Biol Conserv 83:239–246

Flemming IA, Jensen AJ (2002) Fisheries: effects of climate change on the life cycles of Salmon. Causes and Consequences of Global Environmental Change, 3:309–312. Edited by Ian Douglas in Encyclopedia of Global Environmental Change. John Wiley & Sons, Ltd, Chichester, UK

Friedlingstein P, Fung I, Holland E, John J, Brasseur G, Erickson D, Schimel D (1995) On the contribution of CO2 fertilization to the missing biospheric sink. Glob Biogeochem Cycles 9:541–556

Hugo GH, Dettinger MD, Cayan DR (2008) Downscaling with constructed analogues: daily precipitation and temperature fields over the United States. California Energy Commission Report. CEC-500-2007-123. January

IPCC (Intergovernmental Panel on Climate Change) (2007) Impacts, adaptation and vulnerability. Contribution of Working Group II to the Fourth Assessment Report of the Intergovernmental Panel on Climate Change. Cambridge University Press, Cambridge

Keane RE, Long D, Basford D, Levesque BA (1997) Simulating vegetation dynamics across multiple scales to assess alternative management strategies. In: Conference Proceedings—GIS 97, 11th Annual Symposium on Geographic Information Systems—Integrating Spatial Information Technologies for Tomorrow. GIS World, Inc. Vancouver, British Columbia, Canada. 310–315

Lenihan JM, Drapek RJ, Bachelet D, Neilson RP (2003) Climate changes effects on vegetation distribution, carbon, and fire in California. Ecol Appl 13:1667–1681

Lenihan JM, Bachelet D, Drapek RJ, Neilson RP (2008a) The response of vegetation distribution, ecosystem productivity, and fire in California to future climate scenarios simulated by the MC1 dynamic vegetation model. Clim Chang 87(Supp):S215–S230

Lenihan JM, Bachelet D, Neilson RP, Drapek RJ (2008b) Simulated response of conterminous United States ecosystems to climate change at different levels of fire suppression, CO_2 emissions rate, and growth response to CO_2. Glob Planet Chang 64:16–25

MA (Millennium Ecosystem Assessment) (2005a) Ecosystems and human well-being. Synthesis. Island, Washington

MA (Millennium Ecosystem Assessment) (2005b) Ecosystems and human well-being. Biodiversity synthesis. Island, Washington

Maurer EP, Hidalgo HG (2008) Utility of daily vs. monthly large-scale climate data: an intercomparison of two statistical downscaling methods. Hydrology Earth Syst Sci 12:551–563

Mendelsohn R, Olmstead S (2009) The economic valuation of environmental amenities and disamenities: methods and applications. Annu Rev Environ Resour 34:325–347

Nordhaus W (2008) A question of balance: weighing the options on global warming policies. Yale University Press, New Haven

Parton W, Schimel D, Ojima D, Cole C (1994) A general study model for soil organic model dynamics, sensitivity to litter chemistry, texture, and management. Soil Sci Soc America Spec Publ 39:147–167

Pearce D (2003) The social cost of carbon and its policy implications. Oxford Rev Econ Policy 19:362–384

Peterson DL, Ryan KC (1986) Modeling postfire conifer mortality for long-range planning. Environ Manag 10:797–808

Price R, Thornton S, Nelson S (2007) The social cost of carbon and the shadow price of carbon: what they are, and how to use them in economic appraisal in the UK. Department for Environment Food and Rural Affairs, UK

Rothermel RC (1972) A mathematical model for predicting fire spread in wildland fuels. Res. Pap. INT-115. Ogden, UT: U.S. Department of Agriculture, Forest Service, Intermountain Forest and Range Experiment Station

Sanstead AH, Johnson H, Goldstein N, Franco G (2009) Long-run socioeconomic and demographic scenarios for California. California Energy Commission Scenarios Report. CEC-500-2009-013-F-1

Schmitz OJ, Post E, Burns CE, Johnston KM (2003) Ecosystem responses to global climate change: moving beyond color mapping. Bioscience 53:1199–1206

Shaver GR, Canadell J, Chapin FS III, Gurevitch J, Harte J, Henry G, Ineson P, Jonasson S, Melillo J, Pitelka L, Rustad L (2002) Global warming and terrestrial ecosystems: a conceptual framework for analysis. Bioscience 50:871–883

Shaw MR, Pendleton L, Cameron R, Morris B, Bratman G, Bachelet D, Klausmeyer K, MacKenzie J, Conklin D, Lenihan J, Haunreiter E, Daly C (2009) The impact of climate change on California's ecosystem services. California Energy Commission Scenarios Report. CEC-500-2009-025-D

Thorne M, Cox LJ, Stevenson MH (2007) Calculating minimum grazing lease rates for Hawai'i. Published by the College of Tropical Agriculture and Human Resources, Pasture and Range Management

Tol RSJ (2005) The marginal damage cost of carbon dioxide emissions: an assessment of the uncertainties. Energy Policy 33:2064–2074

Tol RSJ (2008) The social cost of carbon: trends, outliers and catastrophes. Economics: The Open-Access, Open-Assessment E-Journal, 2:2008-25

Turner RK, Paavola J, Cooper P, Farber S, Jessamy V, Georgiou S (2003) Valuing nature: lessons learned and future directions. Ecol Econ 46:493–510

United States Department of Agriculture (2008) Agricultural Marketing Service, Livestock & Grain Market News, March, 2008. Alfalfa Hay, 2008 Year-to-date cumulative: California Market Summary

United States Fish and Wildlife Service (USFWS) (2007) 2006 National survey of fishing, hunting and wildlife-associated recreation activities: California. December, 2007

USDA (United States Department of Agriculture, National Agricultural Statistics Service) (2004) Livestock county estimates United States Department of Agriculture, Agricultural Marketing Service, Alfalfa Hay 2008 Year-to-Date cumulative, California Market Summary, March 2004

USDA (United States Department of Agriculture, National Agricultural Statistics Service) (2007) Livestock county estimates United States Department of Agriculture, Agricultural Marketing Service, Alfalfa Hay 2008 Year-to-Date cumulative, California Market Summary, March 2008

van Wagner CE (1993) Prediction of crown fire behavior in two stands of jack pine. Can J For Res 23:442–449

Voluntary Carbon Standard (VCS) (2008) Voluntary carbon standard program guidelines. 18 November 2008

Washington WM, Weatherly JW, Meehl GA, Semtner AJ Jr, Bettge TW, Craig AP, Strand WG Jr, Arblaster J, Wayland VB, James R, Zhang Y (2000) Parallel climate model (PCM) control and transient simulations. Clim Dyn 16:755–774

Watkiss P, Downing TE (2008) The social cost of carbon: valuation estimates and their use in UK policy. Integr Assess J 8:85–105

Wood AW, Leung LR, Sridhar V, Lettemaier DP (2004) Hydrological implications of dynamical and statistical approaches to downscaling climate model outputs. Clim Chang 62:189–216

Climatic Change (2012) 110:1067
DOI 10.1007/s10584-011-0384-2

ERRATUM

Erratum to: The impact of climate change on California's ecosystem services

M. Rebecca Shaw · Linwood Pendleton · D. Richard Cameron ·
Belinda Morris · Dominique Bachelet · Kirk Klausmeyer ·
Jason MacKenzie · David R. Conklin · Gregory N. Bratman ·
James Lenihan · Erik Haunreiter · Christopher Daly ·
Patrick R. Roehrdanz

Published online: 3 January 2012
© Springer Science+Business Media B.V. 2012

**Erratum to: Climatic Change (2011) 109 (Suppl 1):S465–484
DOI 10.1007/s10584-011-0313-4**

Unfortunately some authors' affiliations were captured incorrectly on the first page of the article. Please find the correct affiliation information below. Also, Gregory N. Bratman's first name was wrongly listed as Gregrory in the original publication.

The online version of the original article can be found at http://dx.doi.org/10.1007/s10584-011-0313-4.

M. R. Shaw (✉)
The Environmental Defense Fund, 123 Mission Street 28th Floor, San Francisco, CA 94105, USA
e-mail: rshaw@edf.org

D. R. Cameron · K. Klausmeyer · J. MacKenzie · E. Haunreiter
The Nature Conservancy, 201 Mission St. 4th Floor, San Francisco, CA 94105, USA

L. Pendleton
The Nicholas Institute, Duke University, P.O. Box 90335, Durham, NC 27708, USA
e-mail: linwood.pendleton@duke.edu

B. Morris
The Environmental Defense Fund, 1107 9th Street, Suite 1070, Sacramento, CA 95814, USA

J. Lenihan
USDA Forest Service, Oregon State University, Corvallis, OR, USA

D. Bachelet · D. R. Conklin
Conservation Biology Institute, Corvallis, OR, USA

C. Daly
Oregon State University, Corvallis, OR, USA

G. N. Bratman
Stanford University, Stanford, CA, USA

P. R. Roehrdanz
University of California at Santa Barbara, Santa Barbara, CA, USA

Climatic Change (2011) 109 (Suppl 1)
DOI 10.1007/s10584-011-0310-7

The climate gap: environmental health and equity implications of climate change and mitigation policies in California—a review of the literature

Seth B. Shonkoff · Rachel Morello-Frosch · Manuel Pastor · James Sadd

Received: 12 February 2010 / Accepted: 26 September 2011 / Published online: 24 November 2011
© Springer Science+Business Media B.V. 2011

Abstract Climate change is an issue of great importance for human rights, public health, and socioeconomic equity because of its diverse consequences overall as well as its disproportionate impact on vulnerable and socially marginalized populations. Vulnerability to climate change is determined by a community's ability to anticipate, cope with, resist, and recover from the impact of major weather events. Climate change will affect industrial and agricultural sectors, as well as transportation, health, and energy infrastructure. These shifts will have significant health and economic consequences for diverse communities throughout California. Without proactive policies to address these equity concerns, climate change will likely reinforce and amplify current as well as future socioeconomic disparities, leaving low-income, minority, and politically marginalized groups with fewer economic opportunities and more environmental and health burdens. This review explores the disproportionate impacts of climate change on vulnerable groups in California and investigates the costs and benefits of the climate change mitigation strategies specified for implementation in the California Global Warming Solutions Act of 2006 (AB 32). Lastly, knowledge gaps, future research priorities, and policy implications are identified.

S. B. Shonkoff
Department of Environmental Science, Policy, and Management, Division of Society and Environment, University of California, Berkeley, 137 Mulford Hall, MC 3144, Berkeley, CA 94720, USA

R. Morello-Frosch (✉)
Department of Environmental Science, Policy and Management & School of Public Health, University of California, Berkeley, 137 Mulford Hall, MC 3114, Berkeley, CA 94720, USA
e-mail: rmf@berkeley.edu

M. Pastor
Departments of Geography and American Studies and Ethnicity, University of Southern California, 3620 S. Vermont Ave, KAP-462, Los Angeles, CA 90089-0255, USA

J. Sadd
Department of Environmental Science and Geology, Occidental College, 1600 Campus Rd., Los Angeles, CA 90041, USA

1 Introduction—the climate gap: environmental health and economic equity

The *Climate Gap* refers to the disproportionate and unequal implications that climate change and climate change mitigation hold for people of color and the poor (Pastor et al. 2010b; Morello-Frosch et al. 2009; Shonkoff et al. 2009b). Vulnerability to climate change is determined by the ability of a community or household to anticipate, cope with, resist, and recover from the direct and indirect impacts of extreme weather events and geophysical shifts such as sea level rise (Pacific Institute 2009), hurricanes (Greenough and Kirsch 2005), floods (Greenough et al. 2001), heat waves (Knowlton et al. 2009), air pollution (O'Neill et al. 2008), and infectious diseases (Gage et al. 2009). Therefore, to understand concerns regarding the climate gap, it is critical to explore disparities in the costs and benefits of climate change, the abilities of different groups to adapt to it, and the equity dimensions of the mitigation strategies imposed to attenuate it in order to better inform policy and regulatory action.

This review focuses on the current and projected disparate impacts of climate change and climate change mitigation policies on groups of lower socioeconomic status (SES)[1] in California. We begin with a review of the disproportionate health and economic impacts of climate change itself and examine differences in the capacity of certain groups to adapt to its direct and indirect effects, such as extreme weather events, increased or re-located air pollution, infrastructure impacts, and major economic shifts. Second, we review a subset of the health and economic equity implications of different climate change mitigation strategies, with an emphasis on those included in The Global Warming Solutions Act of 2006 (AB 32) in California. We end with a discussion of the implications of this wide-ranging body of literature for future policy-relevant research on the climate gap.

2 Environmental health inequities and climate change

Globally, climate change and climate change mitigation strategies hold a variety of implications for differential environmental health outcomes across socio-demographic strata. In the California context, the primary climate change exposures that pose risks for population health are increases in the incidence and duration of extreme weather events, such as heat waves and the exacerbation and changing patterns of outdoor air pollution. We thus focus our review of health implications of climate change in California on these two factors.

2.1 Extreme weather events—heat

Extreme weather events, such as heat-waves and floods are expected to increase in frequency and intensity over the next hundred years (Solomon et al. 2007). This could amplify the risk of associated morbidity and mortality for populations that are not able to adapt to, or protect themselves against, such events.

Regarding heat wave mortality, in a study of nine California counties from May through September of 1999–2003, Basu and colleagues (2008) found that for each 10°F (5.6°C)

---

[1] The term *socioeconomic status* or *socioeconomic position* (used synonymously) will refer to the position of an individual or group along the spectrum of access to the resources necessary to maintain their health and economic livelihoods. Socioeconomic status thus encompasses variables such as income level, inherited wealth, educational status, beneficial social networks, and race/ethnicity.

increase in ambient temperature, there is a 2.6% (95% confidence interval [CI]: 1.3, 3.9) increase in cardiovascular mortality with ischemic heart disease being the most dominant of these outcomes (Basu et al. 2008). In this analysis mortality risks were highest for African Americans at 4.9% (95% CI: 2.0, 7.9) (Basu and Ostro 2008).

In terms of heat-wave morbidity, a study on the 2006 California heat wave (July 15–August 1, 2006) estimated an excess of 16,166 emergency department visits and 1,182 excess hospitalizations statewide, compared with a temporally-proximate summer referent period (July 8–14 to August 12–22, 2006) (Knowlton et al. 2009). Emergency department visits for heat-related causes (i.e., acute renal failure, diabetes, cardiovascular diseases, electrolyte imbalance, and nephritis) increased across the state (relative risk [RR] 6.30; 95% CI 5.67–7.01), especially in the Central Coast, which includes San Francisco. Elevated rate ratios of emergency department visits of 1.05 (95% CI: 1.04–1.07) and 1.03 (95% CI: 1.02–1.04) were found for children (0–4 years of age) and the elderly (≥65 years of age) respectively (Knowlton et al. 2009).

2.2 Intrinsic and extrinsic risk factors for heat-associated mortality and morbidity

Although heat exposure alone is implicated in increased morbidity and mortality, physiological, social and economic factors are also fundamental to understanding the uneven distribution of these adverse heat-specific health outcomes across diverse populations (Klinenberg 2002). Thus, risk factors for heat-associated mortality and morbidity can be categorized as intrinsic (i.e., age, disability, medical status) or extrinsic (e.g., housing, access to cooling centers, transportation) and low SES groups are disparately affected by both of these risk categories.

In terms of intrinsic factors, people suffering from chronic medical conditions have an elevated risk of death during heat waves (Kilbourne 1997; Kovats and Hajat 2008) compared with those who are healthy. In fact, a study on heat specific mortality during the 2003 heat wave in France reported that over 70% of the victims found at home had pre-existing medical conditions, particularly cardiovascular and/or psychological illnesses (Poumadere et al. 2005). Because low SES groups are disproportionately affected by medical conditions partially due to their lack of access to technological, informational, and social resources to cope with these conditions (Phelan et al. 2004), they tend to be most adversely affected by extreme heat events.

In terms of extrinsic risk factors, low-income urban communities and communities of color are particularly vulnerable to increased frequency of heat waves and higher temperatures because they are often segregated in the inner city (Schulz et al. 2002; Williams and Collins 2001), which is more likely to experience "heat-island" effects (Harlan et al. 2008). Heat-island effects occur in urban areas when lighter-colored (higher albedo) materials such as grass, trees, and soil are replaced by darker-colored (lower albedo) materials such as roads, buildings, and other surfaces, leading to increased absorption of sunlight. This increased absorption of sunlight decreases the dissipation of heat, thus warming the local area (Oke 1973). A recent land cover analysis (Shonkoff et al. 2009a) shows a positive relationship between the proportion of impervious land cover in neighborhoods and an increasing proportion of residents living in poverty, as well as a negative relationship between the amount of tree canopy coverage and the proportion of residents living in poverty in four California urban areas (Fig. 1). Further, there is a positive relationship between the proportion of neighborhood residents of color and the proportion of impervious land cover and a negative relationship between the proportion of people of color and the amount of tree cover (Fig. 2). These data suggest a

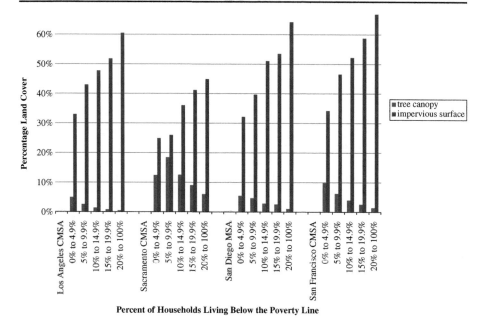

Fig. 1 Land cover characteristics across comparable neighborhood poverty groups. Cited From: Shonkoff et al. (2009a, b)

disproportionate exposure to heat-island risk factors on communities of color and low income.

From home building materials to devices used to heat and cool living environments, technologies are employed by all communities to protect themselves from exposure to weather and climate fluctuations. In terms of technological adaptation as an extrinsic factor in heat-associated health outcomes, studies have documented that lack of access to air conditioning is correlated with risks of heat-related morbidity and mortality among urban elderly of low SES in the United States (Kovats and Hajat 2008; Semenza et al. 1996; Knowlton et al. 2009). In the Los Angeles-Long Beach Metropolitan Area, for example, a higher proportion of African-Americans do not have access to home air conditioning compared to the general population (59% vs. 40%, respectively). Similar trends hold for Latinos (55%) and communities living below the poverty line (52%) (USCB 2004) (Table 1). Although these data do not fully explain the drivers of observed racial and SES disparities in air conditioner ownership, the differential proportions of ownership of these technologies is important because some households may rely on air conditioning during extreme heat events and days when communities are instructed to stay indoors and avoid outdoor pollution exposures.

Further, nearly 84% of residents in the Los Angeles metropolitan area rely on cars to commute to work compared to 7% of residents who rely on public transportation (ACS 2007). The paucity of public transit options makes residents extremely reliant on car ownership to meet basic transportation needs.[2] During extreme heat events, households

---

[2] Since the 1930s when National City Lines, a holding company run by corporate partners in the automotive industry, bought and dismantled a considerable portion of the public transit infrastructure in Los Angeles, residents without a personal automobile in the Los Angeles-Long Beach Metropolitan Area have been at a severe disadvantage (Kunzli et al. 2003).

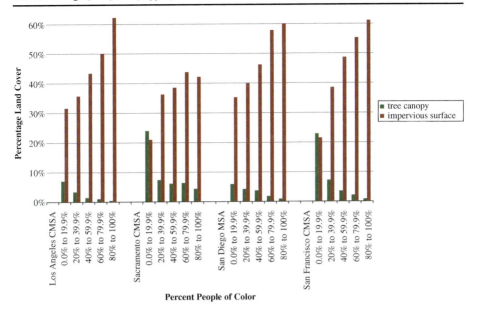

Fig. 2 Land cover characteristics across comparable neighborhood racial/ethnic minority groups. Cited From: Shonkoff et al. (2009)

without air conditioning may need to relocate to cooler areas and government-sponsored cooling stations, which can be a logistical challenge for those without access to a car or adequate public transportation. In the Los Angeles-Long Beach Metropolitan Area, elevated proportions of African-American (20%), Latino (17.1%), and Asian (9.8%) households do not have access to a car, compared to white households (7.9%) (USCB 2004).

Using heat-wave data from Chicago, Detroit, Minneapolis, and Pittsburgh, O'Neill et al. (2005) found that African Americans had a 5.3% higher prevalence of heat-related mortality than Whites and 64% of this disparity is potentially attributable to disparities in prevalence of central air conditioner technologies (O'Neill et al. 2005). These results are bolstered by other studies that found associations between being African American and lack of air conditioning as an indicator for vulnerability to heat-related poor health outcomes (Curriero et al. 2002; Greenberg et al. 1983; O'Neill et al. 2003; Rogot et al. 1992; Semenza et al. 1996; Whitman et al. 1997). It should be noted that this same analysis found that both Asian and Hispanic groups suffer a lower mortality burden than that of white groups in Los Angeles.

Material and socioeconomic deprivation, especially in the inner city, is highly correlated with heat-wave and heat-stroke mortality risk in the United States, including California (English et al. 2007; Kovats and Hajat 2008; Klinenberg 2002). For example, the heat wave in Phoenix, Arizona, in 2006 was responsible for 13 heat-stroke-related deaths, 11 of which were homeless people who tend to lack access to material and social resources (Kovats and Hajat 2008).

This is in line with other findings that African American Los Angeles residents have a projected heat-wave-mortality rate that is nearly twice that of the Los Angeles average under different GHG emission scenarios (Fig. 3) (Cordova et al. 2006).

Because SES is fundamentally associated with occupation, it is important to note that California's agricultural and construction workers experience severe heat-related morbidity and mortality with data pointing towards possible increasing trends in recent years (English et al. 2007; MMWR 2008). Mexican and Central American immigrants who come to

Table 1 Proportion of households without access to any air conditioning by race and SES—Los Angeles-Long Beach Metropolitan Area, California (2003)

	Total number of households (General Los Angeles population)	Total occupied units (General Los Angeles population)	Black (not Hispanic)	Hispanic	Elderly (65 years or older)	Below poverty level
All Occupied units	3,131,000	39.7%	58.5%	54.6%	37.5%	51.5%
Renters	1,608,900	48.1%	59.1%	58.4%	38.7%	56.3%
Homeowners	1,522,100	30.9%	57.4%	48.9%	36.8%	38.8%

Percentages are likely an underestimate of the true value due to the fact that more than one category may apply to a single unit in the dataset

Adapted from: American Housing Survey for the Los Angeles-Long Beach Metropolitan Area 2004 (USCB 2004)

California to work in the agricultural and construction sectors are particularly vulnerable because of the cumulative impacts of their long workdays under strenuous conditions, low capacity to protect themselves on the job and assert labor rights, and exposure to chemicals such as pesticides. Between the years 2003–2006, 71% of the crop-workers that died due to heat-associated complications were identified as Mexican, Central or South American and 72% of these deaths were in adults aged 20–54 years, a population typically considered at low-risk for heat illnesses (MMWR 2008). As heat-wave incidence and intensity increases, disparities will persist among those with high levels of material and social deprivation that characterize the context within which low-SES groups live and work.

2.3 Air pollution

The literature on outdoor (ambient) air pollution in California has primarily focused on tropospheric ozone (O_3), nitrogen oxides (NO_x), and chemically undifferentiated particulate matter (PM). Hence, we focus our review on these three pollutants. It is nonetheless

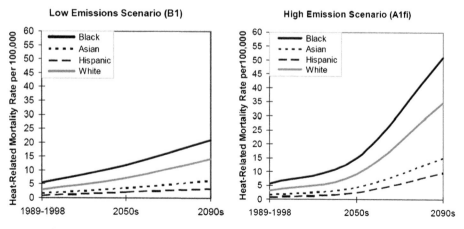

Fig. 3 Relative heat-wave mortality rates by race/ethnicity for Los Angeles. Actual historical values (1989–1998) and projected future values (2050s and 2090s) for high-emissions (A1fi) and low-emissions (B1) scenarios. (HadCM3 projections only). Cited from: Cordova et al. (2006)

important to mention that other greenhouse gas (GHG) co-pollutants such as sulfur dioxide (SO_x), black carbon (BC) (Smith et al. 2009) and carbon monoxide (Kaur and Nieuwenhuijsen 2009) are associated with: population health disease burdens as well as additional climate forcing.

Five of the ten most ozone-polluted metropolitan areas in the United States are in California (Los Angeles, Bakersfield, Visalia, Fresno, and Sacramento) (ALA 2011). Because of this, Californians suffer a relatively high air pollution associated disease burden, including 18,000 (95% CI: 5,600–32,000) premature deaths each year and tens of thousands of other illnesses (CARB 2008a). Primarily due to the combustion of fossil fuel among mobile sources and the stationary energy sectors, California's levels of NO_x, PM, O_3, and a myriad of other health damaging air pollutants are very high, particularly in California's Central Valley and South Coast Region where ambient levels frequently exceed National Ambient Air Quality Standards (US EPA 2010). As mentioned above, these sectors not only emit criteria air pollutants but also climate altering pollutants (i.e., CO_2, NO_x, BC, O_3) that contribute to additional climate forcing on local, regional and global scales (Smith et al. 2009). In turn, elevated temperatures interact with NO_x and sunlight and lead to increases in ambient O_3 concentrations in urban and suburban areas. This contributes to both respiratory health effects (Jerrett et al. 2009) as well as even more climate forcing (Meleux et al. 2007; Stathopoulou et al. 2008; Smith et al. 2009).

In California, the five smoggiest cities are also the locations with the highest projections of climate change induced ambient ozone increases as well as the highest densities of people of color and low-income residents (Cordova et al. 2006). A recent study projects a dose–response relationship in which for each 1 degree Celsius (1°C) rise in temperature in the United States, there is an estimated 1,000 (CI: 350–1800) excess air-pollution-associated deaths (Jacobson 2008). About 40% of the additional deaths may be due to the exacerbation of ozone production due to increased temperatures and the rest to particulate matter—which increases due to CO_2-enhanced stability, humidity, and biogenic particle mass—annually (Jacobson 2008).

3 Disproportionate economic impacts of climate change on low SES groups

On scales from the global to the local, climate change holds direct and indirect implications for the strength of economic systems as well as the distribution of their impacts and benefits. This section reviews how climate change could have negative implications for the economic prospects of lower SES groups in California compared with their higher SES counterparts. We focus our equity analysis on three issues: 1) Increased prices of basic necessities such as water, food, and household energy; 2) Downturns in productivity and employability in the agricultural sector; and 3) Increased infrastructure damage from extreme weather events, sea level rise, and wildfires.

3.1 Price of basic necessities

Under a business-as-usual scenario it is estimated that between the years 2025 and 2100, the cost of providing water to the western United States will increase from $200 billion to $950 billion per year, representing an estimated 0.93–1% of the United States' gross domestic product (GDP) (Ackerman and Stanton 2008). Under the same scenario annual U.S. energy expenditures (excluding transportation) could be $141 billion higher in 2100 than they would be if today's climate conditions continued throughout the century

(Ackerman and Stanton 2008). This increase is approximately equal to 0.14% of US GDP, an economically significant figure (Ackerman and Stanton 2008). Four climate change impacts—hurricane damage, energy costs, real estate losses, and water costs—alone are projected to cost 1.8% of the GDP of the United States, or, just under $1.9 trillion (2008 U.S. dollars [USD]) by the year 2100 (Ackerman and Stanton 2008).

Low-income groups spend the highest proportion of their income on basic necessities such as water, household energy (electricity), and food (BLS 2002) and price increases due to these climatic shifts may lead to increases in necessity prices. There is nearly a three-fold difference in the proportion of the sum of expenses allocated to water between the lowest- and the highest-expenditure quintiles of the US population. Households in the lowest economic quintile use more than twice the proportion of their total expenditures on electricity than do those households in the highest economic quintile. Similarly, food, the commodity that represents the largest portion of total spending out of all the basic necessities in the expenditure quintiles, shows a two-fold discrepancy between the lowest and the highest economic quintiles (Fig. 4) (BLS 2002).

3.2 Disproportionate impact of climate change on agricultural employment in California

The majority of jobs in sectors that will likely be significantly affected by climate change in California, such as agriculture are held by low-income people of color (EDD 2004; USCB 2005). In the event of climatic shifts that impact the productivity or location of agriculture, these workers would be the first to lose their jobs. The literature suggests that climate change will affect employment within the agricultural sector in two main ways: (1) Increases in the frequency and the intensity of extreme weather events will expose agriculture to greater productivity risks and possible revenue losses that could lead to abrupt layoffs (Costello et al. 2009; Howitt et al. 2009); and (2) Changing weather and precipitation patterns could require expensive adaptation measures such as the relocation of crop cultivation, the modification of the composition or type of crops (Jackson et al. 2009), and the increase in agricultural inputs, such as pesticides, to adapt to changes in ecological composition.

Latinos comprise 77% of the U.S. agricultural workforce and the majority of these men and women are also categorized as low-income (EDD 2004). In California, as of 2003, agriculture provided approximately 500,000 jobs with 315,000 of them being held by Latinos (EDD 2004). The majority of these jobs are seasonal, pay very low wages, and do not provide health insurance or job security. Because of the low wages and the seasonality of the work, agricultural counties are among the poorest in the state. As climate change affects agricultural

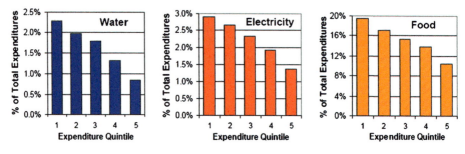

Fig. 4 Household expenditures on water, electricity, and food by income group (as percentage of total expenditures). Expenditure quintile is a proxy for income with quintile 1 representing the lowest-income households and quintile 5 representing the highest-income households. Adapted From: BLS 2002; Cited from: Cordova et al. (2006)

productivity in California, agricultural laborers could be increasingly affected by job losses. For example, the two highest-value agricultural products in the $30 billion California agriculture sector are dairy products (milk and cream, valued at $3.8 billion annually) and grapes ($3.2 billion annually) (CASS 2002). Climate change is expected to decrease dairy production by as much as 7–10% under the IPCC B1 scenario and 11–22% under the A1fi scenario by the end of the century (Pittock et al. 2001). It is also expected to adversely impact the ripening of wine grapes, substantially reducing their market value (Hayhoe et al. 2004). These data suggest that communities in the Central Valley and other crop growing areas, where agriculture is most concentrated and low-income Latino communities are most common, would be the hardest hit by these climate change impacts.

3.3 Infrastructure, SES, and insurance access

As extreme weather events such as sea level rise, wildfires, and storms become more frequent and severe, California's infrastructure will be increasingly threatened and damaged. The literature suggests that the capacity of households and communities to prepare for and adapt to these risks is a function of their income, access to infrastructure insurance and other SES-related factors. For example, sea level rise, due to climate change is expected to put the California coast at increased risk of property damage (Pacific Institute 2009). An analysis by the Pacific Institute (2009) indicates that there is likely to be disproportionate impacts on low-income households in 13 of the 20 counties that lie along the California coast because of disparities in abilities to purchase emergency preparedness materials, buy insurance policies, and obtain needed building reinforcements.

Increased risk of property and financial damage from wildfires are also correlated with socioeconomic position. For example, Ojerio and colleagues (2010) found that in Arizona, low income households are less likely to have control over ignitability of their property, are more likely to be located in districts with less wildfire suppression capability, and have less access to federal government resources to mitigate wildfire risks than their more wealthy community counterparts.

Although the issue of insurance is a large question and a detailed analysis is beyond the scope of this review, the literature indicates that those in low socioeconomic positions are consistently underinsured (Blaikie et al. 1994; Fothergill and Peek 2004). Households that have home or renters' insurance can, relatively rapidly, recuperate and resume living much in the same way as prior to the disaster. In contrast, low-income households—who are often under-insured—may spend the rest of their lives struggling to recover from property damage related to an extreme weather event (Blaikie et al. 1994; Fothergill and Peek 2004; Thomalla et al. 2006). As the frequency and intensity of extreme weather events increase, the price of disaster insurance will also likely increase. This could make disaster insurance even more prohibitively expensive for low-income people thus decreasing the ability of this group to cope with infrastructure and property losses. Swiss Re (2006) indicates that insurance losses have been on an upward trend since 1985. During the years 1987–2004 property insurance losses due to natural disasters averaged USD 23 billion per year and in 2005, losses rose to USD 83 billion, of which USD 60 billion was due to hurricanes (Katrina, Rita, and Wilma) alone (Swiss Re 2006). Increases in the price of disaster insurance will add insult to injury to those who are already disproportionately affected by these events.

Lastly, disproportionate impacts of extreme weather events on low SES households have the potential to exacerbate homelessness, especially in urban areas. This would be largely due to the lack of access to insurance and emergency credit, lower amounts of savings, fewer personal resources, and the accumulated suffering from previous economic stresses of low-income groups (Bolin and Bolton 1986; Fothergill and Peek 2004; Tierney 1988). It is also possible that increased government spending on infrastructure protection could hold deleterious consequences for low-income communities due to a diversion of funds away from educational and social programs, public transportation projects, population health initiatives, and other services (CRAG 2002).

4 Implications of climate change policies for low-SES groups

Because low SES groups are disproportionately affected by climate change, they could significantly benefit from sound climate change policies that are sensitive to demographic distribution of economic and health vulnerabilities. This section examines the equity dimensions embedded in the most prominent climate change mitigation strategies included in California's climate change law to be implemented by the California Air Resources Board (CARB) (AB 32) (CARB 2008b). We discuss two overarching themes: (1) The economic implications of different climate change policies on low-SES households; and (2) Positive and negative human health implications of different mitigation strategies for low-SES communities and households.

4.1 Economic costs and benefits of different climate change mitigation strategies

A major concern with regard to policies to reduce emissions is that they will be regressive; the burden of costs that arise from mitigation will fall disproportionately on lower-income households (Hassett et al. 2008; Walls and Hanson 1996). For example, the Congressional Budget Office projects that a United States-wide cap-and-trade scenario aimed to cut carbon dioxide emissions by 15% could cost 3.3% of the average income of households in the lowest income quintile as opposed to only 1.7% of the average income of households in the top income quintile (CBO 2007).

Substantial equity issues are raised by how pollution credits are allocated to facilities as well as—in the case of policies that include fees on emissions or the auctioning of emission credits—how revenues generated from these programs are redistributed to society and individual consumers. Under cap-and-auction or fee-based strategies, the sale of emission credits to polluters could generate sizable revenues that could help to offset the potentially regressive qualities of the emission cap program (Hepburn et al. 2006). These funds could be distributed to the public through tax cuts, tax-shifting, investments in clean energy, education, or through direct periodic dividends to consumers (CBO 2007), assuaging the regressive impacts that could accrue if the prices of necessities increase. Other investments generated from cap-and-auction or fee-based revenues could include investments in public transportation that could both reduce the emissions of greenhouse pollutants while simultaneously adding the co-benefit of reductions in emissions of health damaging co-pollutants due to lowering the numbers of mobile sources on the road. These types of programs should, however, be geographically targeted to reduce the pollution from the most air pollution-impacted areas—the majority of which are found in areas with low SES populations and people of color (Morello-Frosch and Jesdale 2006).

4.2 Health concerns of cap-and-trade policies

While cap-and-trade, under certain circumstances, is efficient at reducing GHGs and their associated co-pollutants on a regional basis, the strategy makes no guarantee about the reduction of these emissions from any one source (O'Neill 2004). Hence, low-SES communities geographically situated in highly polluted areas are concerned about the persistence and potential exacerbation of co-pollutant hotspots at the local community level. The bundles of measures that CARB has already begun to implement to reduce GHG emissions could also contribute to notable reductions in co-pollutants such as SO_x, PM, ozone, and other health damaging contaminants and toxic air pollutant precursors (CARB 2008b). These measures could hold the most notable benefits for low-income groups and people of color who are disproportionately segregated in neighborhoods in close proximity to highways, ports, and other sections of transportation and goods-movement corridors where air quality has been noted as poor (CARB 2006b, 2008c; Morello-Frosch and Jesdale 2006; Morello-Frosch and Lopez 2006).

Cap-and-auction (assuming that fewer than 100% of permits are auctioned) reduces, and fees eliminate the need for emissions trading in comparison to free-allocation programs because industry is likely to buy only what it needs (Hepburn et al. 2006). Auctioning credits also decreases financial incentives to keep old polluting facilities open by eliminating the grandfathering in of old facilities. It also decreases the problem of over-allocation and excessive banking and trading of emission credits. An over-allocation of credits paired with excessive emission credit banking and trading could possibly lead firms to not reduce local GHG emissions. This could lead to the under-achievement of significant co-pollutant benefits in communities that are currently highly impacted by multiple pollution sources (Ellerman and Buchner 2007).

An example of such an emission reduction underachievement is the Regional Clean Air Incentives Market (RECLAIM), an emission trading system employed to lower NO_x emissions in Southern California. Data suggests that this program may have increased NO_x emissions in Wilmington, California, while region-wide emission levels declined (Lejano and Hirose 2005). Further, under Rule 1610, licensed car scrappers could purchase old, polluting vehicles and destroy them, and in return receive emission credits through the South Coast Air Quality Monitoring District (SCAQMD) that could be sold to oil refineries (Drury et al. 1999). The majority of the emission credits were purchased by four oil companies: Unocal, Chevron, Ultramar, and GATX to avoid the cost of installing pollution-reduction technologies that would capture volatile organic compound (VOC) gases forced out of oil tankers into the air when being loaded. These refineries are all located in close proximity to one another in the City of Wilmington and San Pedro except for the Chevron facility located in El Sugundo (Drury et al. 1999). In their analysis, Drury et al. (1999) indicate that this mobile-to-stationary trading program led to a situation where workers and local community residents were unnecessarily exposed to benzene, a known human carcinogen, and other VOCs that were contained in the emissions. These emissions could have been remediated by pollution reduction technologies that were already in widespread use in similar operations along the West Coast.

4.2.1 Co-benefits of AB 32 measures

As mentioned, GHG reduction measures under AB 32 are predicted to greatly reduce health damaging co-pollutant emissions (Bailey et al. 2008). For example, NO_x emissions, a precursor of ozone formation and a group of health damaging pollutants in their own right, are expected to be reduced by 86,000 t by 2020, more than three-quarters of which will be achieved through regulatory requirements for cleaner cars and trucks (Bailey et al. 2008).

Under AB 32, projected PM and NO_x reductions together are estimated to prevent approximately 400 premature deaths, 11,000 fewer cases of asthma-related and other lower respiratory symptoms, 910 fewer cases of acute bronchitis, and 67,000 fewer work days lost in California (CARB 2008b). These health benefits are projected to be valued at $1.4 billion to $2.3 billion in 2020 alone (Bailey et al. 2008). A review by CARB (2008a) indicates that there is a 10% (CI: 3% to 20%) increase in the number of premature deaths per 10 $\mu g/m^3$ increase in $PM_{2.5}$ exposure (CARB 2008a). CARB (2008a) also estimates that diesel PM contributes to 3,500 (CI: 1,000–6,400) premature deaths statewide on an annual basis. These projections could be an underestimate of the actual health and economic impacts of diesel PM because many emission reduction measures and public health benefits such as reduced cancer risks have not been accounted for in their calculation (Bailey et al. 2008). Some known carcinogens, that were not included in these analyses, such as benzene, formaldehyde, and toluene may be reduced by the implementation of GHG reduction measures because they are predominantly produced, directly and indirectly, by mobile sources and by the refinement and combustion of fossil fuels (EPA 2005). From an environmental equity perspective, the reduction of air toxics may be important as several studies indicate that communities of color and the poor bear a disproportionate burden of health risks associated with air toxics exposures (CARB 2008b, c; Morello-Frosch and Jesdale 2006; Morello-Frosch and Lopez 2006; Morello-Frosch et al. 2002).

4.2.2 Co-benefits of AB 32 early action and other mitigation measures

AB 32 also includes Early Action Measures (EAMs) (CARB 2008b) that regulate the inputs and functions of landfills, types of motor vehicle fuels, varieties of refrigerants in cars, types of port operations, and many other processes that are involved in emissions of climate altering pollutants. It is estimated that if all EAMs are adopted, 52,000 tons of NOx and PM pollution would be removed from the air. This would avert approximately 250 additional premature deaths, thousands of cases of asthma and other lower respiratory symptoms, and could save an estimated $1.1 to $1.8 billion in health costs in 2020 alone (Bailey et al. 2008). Table 2 shows the CARB analysis of the health co-benefits of these actions. These measures could benefit low-SES groups that tend to be segregated in neighborhoods that may be host to significant industrial and transportation emission sources.

4.2.3 Fuel switching

The Low Carbon Fuel Standard (LCFS) was adopted as an EAM under AB 32 (CARB 2008b). The goal of the LCFS is to reduce lifecycle GHG emissions from transportation by at least 10% (CARB 2008b). One of the primary foci of this EAM is to transition from mobile source reliance on pure gasoline to partial or pure biofuels such as ethanol (CARB 2008b). However, some studies suggest that biofuel refineries could negatively impact the health of adjacent communities by exposing them to chemical as well as microbial byproducts of the distillation processes necessary for fuel production (Madsen 2006).

Widespread use of biofuels may also hold implications for outdoor air pollution concentrations. For instance, Jacobson (2007) predicts that E85 (85% ethanol, 15% gasoline) may increase ozone-related mortality, hospitalization, and asthma by 9% in Los Angeles and 4% nationwide if used to power vehicles. In fact, E85 may prove to have as much or more of a public health impact than the use of 100% gasoline (Jacobson 2007). This suggests that low-income and minority communities that live closest to highways and goods transport corridors could bear disproportionate health burdens if these fuels prove to be more toxic than gasoline.

Table 2 Estimates of statewide air quality-related health benefits in 2020 Cited from: (CARB 2008b)

Health endpoint	Health benefits of existing measures and 2007 SIP mean	Health benefits of recommendation in the proposed scoping plan mean
Avoided premature death	3,700	400
Avoided hospital admissions for respiratory causes	770	84
Avoided hospital admissions for cardiovascular causes	1,400	150
Avoided asthma and lower respiratory symptoms	110,000	11,000
Avoided acute bronchitis	8,700	910
Avoided work loss days	620,000	67,000
Avoided minor restricted activity days	3,600,000	380,000

Lastly, it should be noted that growing crops for use as fuel will likely raise prices of food crops (Tenenbaum 2008). This could prove to be regressive, damaging socioeconomic prospects of low-income consumers and low-income agricultural laborers who are most vulnerable to job loss and hunger (Tenenbaum 2008).

5 Future research needs and implications for policy development

Research on climate equity ranging from health effect estimates to economic impacts remains a nascent field with substantial knowledge deficits. Empirical and theoretical approaches in the fields of climate science, industrial ecology, epidemiology, environmental health, sociology, economics, geographic information system (GIS) spatial analysis, and statistics are key to understanding and predicting the socioeconomic, cultural, and health implications of complex ecological, meteorological, and air pollution phenomena. Moreover, these diverse analyses will also be integral to the determination of which policies and mitigation practices could most effectively narrow the climate gap.

5.1 Expanding climate gap research: climate change

Research that sheds light on the state of the climate gap, as it pertains to climate change directly is in high demand. Substantial arguments ensue over the scale at which measurements of localized impacts and co-pollutants should be evaluated in order to meet the intent and requirements of AB 32. In order to design effective policies and to monitor the efficacy of those policies in regards to localized impacts, future research should: (1) explore how to characterize, quantify, and maximize co-benefits of pollution reductions in existing or new "toxic hotspots"; (2) determine the geographic scale at which these evaluations should take place given the data available; and (3) identify the data necessary to improve future evaluations.

More research is needed to investigate the rates and impacts of climate change events that are projected to occur specifically in California. The identification of possible adaptation strategies that could be used to evade morbidity and mortality burdens from climate change impacts specifically in California is also important foci for future analyses.

Although much research has been done to characterize the geographic and demographic characteristics that increase the risk of heat-associated health impacts of communities (Knowlton et al. 2009), fewer studies have shed light on how to best deliver targeted messages about extreme heat exposure. As the literature suggests, heat-related mortality and morbidity is borne disproportionately by groups of older residents, children, and those of low SES (Basu and Ostro 2008; English et al. 2007; Knowlton et al. 2009). Strategies to prevent heat-related illness should include messages targeted toward parents and caregivers of young children, the elderly and, most importantly, to socially isolated populations.

Differential exposures to the health-damaging impacts of climate change, such as excessive heat, extreme weather events, and increases in air pollution could be examined from a geo-equity perspective by using GIS maps overlaid with vulnerability models and current socioeconomic, racial/ethnicity, and cultural group distributions in California. Interaction between these data layers should be taken into account when developing climate change policies so as to reduce the likelihood that future policies would create or amplify disproportionate burdens on vulnerable populations.

5.2 Expanding climate gap research: climate mitigation policies

Important foci for climate change mitigation strategy development are: (1) To conduct multi-level policy scenario comparisons to evaluate combinations of regulations and mechanisms that produce the most efficient, effective, and equitable outcomes with the most health and economic co-benefits on the local level (Shonkoff et al. 2009b); (2) To investigate the ways that impacted communities could play a role in climate change mitigation policy and regulatory deliberations; (3) To develop tools to measure the socioeconomic, environmental, and population health benefits and impacts of an expanded green economy; (4) To identify which climate altering pollutant source sectors will, most cost-effectively, be able to reduce pollution with the least amount of socioeconomic disruption and health impact; (5) To develop robust methods to characterize and quantify the co-benefits of health damaging pollution reductions in new or pre-existing air pollution hotspots.

Under cap-and-trade policies, it is essential to develop analytical tools to track where carbon credits are traded in order to assess the subsequent burden of co-pollutant emissions that may increase or decrease on local and regional levels. Building on these analyses, climate gap research should characterize patterns of human population exposure that results from local sources of pollution in a variety of settings, especially in population dense urban areas.

Health and economic risks of fuel switching and fuel innovations (i.e., ethanol) as specified in the LCFS should be characterized and presented in policy-relevant formats. For example, epidemiologic studies should better assess the effects of exposure to new fuels and their externalities during combustion (Jacobson 2007) as well as during production and distillation—for which there are no studies available. More research must also focus on the dangers of food shortages and food price increases associated with the production of ethanol and other biofuels (Tenenbaum 2008).

5.3 Cumulative impacts screening to guide decision-making

AB 32 requires that, prior to implementation, there be consideration and prevention of cumulative or additional impacts on already disproportionately impacted communities (CARB 2006a). However, no established method for identifying these communities currently exists. Researchers continue to develop environmental justice or cumulative impact screening methods that employ GIS-based mapping to consider risks from criteria and toxic air pollutants, proximity to sources of pollution, and socioeconomic factors (Sadd et al. 2011). Such tools could be useful to evaluate community-level cumulative impacts from climate change itself as well as the implications of mitigation policies. Research to expand upon this work could develop a screening method that provides consistent monitoring and evaluation across air districts and cities to insure that all communities are assessed using similar metrics (Pastor et al. 2010a; Su et al. 2009). Such screening tools could be valuable for the evaluation of permitting, land-use change, and growth pattern decisions that are made at multiple scales; such data can also assist decision-makers to more accurately assess the local implications of regional planning strategies that address climate change.

6 Conclusions

Climate change is not only an environmental issue; it also has human rights, public health, and social equity dimensions. This review indicates that climate change is likely to disproportionally impact the health and economic stability of Californian communities that are least likely to cope with, resist, and recover from the impacts of climate change. This review also finds that low-income and minority communities could be disparately affected by the economic shocks associated with climate change both in price increases for basic necessities (i.e., water, energy, and food) and by threats of job loss due to economic and climatic shifts that affect important industries in California such as agriculture. Without proactive climate change mitigation policies that are sensitive to their economically regressive potential and their distribution of benefits, these strategies could potentially reinforce and amplify current as well as future socioeconomic and racial disparities in California. The consistency of racial and SES disparities as they relate to climate change has made these issues of mounting concern to regulators, policy-makers, researchers, and environmental justice advocates.

As California moves closer to a full implementation of AB 32, it will become a national and international leader in the development of aggressive strategies to reduce greenhouse gas emissions. Ensuring that climate equity is part of the equation will be critical to this implementation process. Research on climate equity—ranging from health effects to economic impacts—remains in its infancy. Interdisciplinary approaches are key to understanding the drivers of the climate gap and to specify which policies and mitigation practices would best address equity concerns. To proactively attenuate disproportionate environmental health burdens borne by the poor and people of color, agency officials and policy makers should ensure that vulnerable communities play a significant role in the development of future solutions to climate change. Non-technical knowledge, such as local expertise, community experience, and other contextual information is important to supplement technical knowledge as policy formation is underway (Minkler and Wallerstein 2003). In other words, researchers who hope to generate climate change-impact information that is sensitive to community-specific concerns should employ community-engaged approaches in their study designs (Minkler and Wallerstein 2003; Corburn 2005, 2009).

Although this paper is a comprehensive review of the environmental health and equity implications of climate change and climate change mitigation policies in California, limitations in the data exist. More extensive research on the mechanisms that underlie associations between inequities and climate change as well as mitigation policies should be undertaken as other competing risk factors that could confound relationships between race, SES, and climate change impacts may exist.

References

Ackerman F, Stanton E (2008) The cost of climate change: what we'll pay if global warming continues unchecked. NRDC, New York

ACS (2007) 2007 American community survey 1-year estimates. American Community Survey. http://factfinder.census.gov/servlet/ADPTable?_bm=y&-geo_id=05000US06037&-qr_name=ACS_2007_1YR_G00_DP3&-context=adp&-ds_name=&-tree_id=307&-_lang=en&-redoLog=false&-format=. Accessed July 10 2010

ALA (2011) State of the air: 2011. American Lung Association, New York, NY

Bailey D, Knowlton K, Rotkin-Ellman M, NRDC (2008) Boosting the benefits: improving air quality and health by reducing global warming pollution in California. Natural Recources Defense Council, New York

Basu R, Ostro BD (2008) A multicounty analysis identifying the populations vulnerable to mortality associated with high ambient temperature in California. Am J Epidemiol 168(6):632–637

Basu R, Feng WY, Ostro BD (2008) Characterizing temperature and mortality in nine California counties. Epidemiology 19(1):138–145

Blaikie P, Cannon T, Davis I, Wisner B (1994) At risk: natural hazards, people's vulnerability, and disasters. Routledge, New York

BLS (2002) Consumer expenditure survey. Bureau of Labor Statistics, Washington

Bolin R, Bolton P (1986) Race, religion, and ethnicity in disaster recovery, program on environment and behavior. University of Colorado, Institute of Behavioral Science, Natural Hazards Research and Applications Information Center, Boulder

CARB (2006a) Assembly Bill No. 32: California Global Warming Solutions Act of 2006

CARB (2006b) Emission reduction plan for ports and goods movement in California. Sacramento

CARB (2008a) Methodology for estimating premature deaths associated with long-term exposures to fine airborne particulate matter in California. California Air Resources Board, Sacramento

CARB (2008b) AB 32 Climate change scoping plan: a framework for change. California Air Resources Board, Sacramento, CA

CARB (2008c) Diesel particulate matter health risk assessment for the West Oakland community: preliminary summary of results. California Environmental Protection Agency, Air Resources Board, Sacramento

CASS (2002) California agriculture statistical review. California Agriculture Statistics Service, Sacramento

CBO (2007) Trade-offs in allocating allowances for CO2 emissions. A series of issue summaries from the Congressional Budget Office. Congressional Budget Office, Washington

Corburn J (2005) Street science: community knowledge and environmental health justice. MIT Press, Cambridge

Corburn J (2009) Cities, climate change and urban heat island mitigation: localising global environmental science. Urban Stud 46(2):413–427

Cordova R, Gelobter M, Hoerner A, Love J, Miller A, Saenger C, Zaidi D (2006) Climate change in California: health, economic and equity impacts. Redefining Progress, Oakland

Costello C, Deschênes O, Kolstad C (2009) Economic impacts of climate change on California agriculture. Climate action team report. California Energy Commission; California Air Resources Board, Sacramento

CRAG (2002) Preparing for a changing climate: the potential consequences of climate variability and change for California. California Regional Assessment Group, US Global Change Research Program, Washington

Curriero FC, Heiner KS, Samet JM, Zeger SL, Strug L, Patz JA (2002) Temperature and mortality in 11 cities of the eastern United States. Am J Epidemiol 155(1):80–87

Drury R, Belliveau M, Kuhn J, Bansal S (1999) Pollution trading and environmental injustice: Los Angeles' failed experiment in air quality. Duke Environ Law Pol Forum 9:231–289

EDD (2004) Occupational employment (2002) and wage (2003) data, occupational employment statistics survey results. California Employment Development Department. http://www.calmis.cahwnet.gov/file/occup$/oeswages/Cal$oes2003.htm. 2004

Ellerman A, Buchner B (2007) The European Union Emissions Trading Scheme: origins, allocation, and early results. Rev Environ Econ Pol 1(1):66–87

English P, Fitzsimmons K, Hoshiko S, Kim T, Margolis H, McKone T, Rotkin-Ellman M, Solomon G, Trent R, Ross Z (2007) Public health impacts of climate change in California: community vulnerability assessments and adaptation strategies. Climate Change Public Health Impacts Assessment and Response Collaborative, California Department of Public Health Institute, Richmond

EPA (2005) Toxics release inventory (TRI) basis of OSHA carcinogens. United States Environemental Protection Agency, Washington

Fothergill A, Peek L (2004) Poverty and disasters in the United States: a review of recent sociological findings. Nat Hazards J 32(1):89–110

Gage KL, Burkot TR, Eisen RJ, Hayes EB (2008) Climate and vectorborne diseases. Am J Prev Med 35(5):436–450

Greenberg JH, Bromberg J, Reed CM, Gustafson TL, Beauchamp RA (1983) The epidemiology of heat-related deaths, Texas—1950, 1970–79, and 1980. Am J Public Health 73(7):805–807

Greenough PG, Kirsch TD (2005) Hurricane Katrina. Public health response—assessing needs. N Engl J Med 353(15):1544–1546

Greenough G, McGeehin M, Bernard SM, Trtanj J, Riad J, Engelberg D (2001) The potential impacts of climate variability and change on health impacts of extreme weather events in the United States. Environ Health Perspect 109(Suppl 2):191–198

Harlan S, Brazel A, Jenerette G, Jones N, Larsen L, Prashad L, Stefanov WItSoATIDotHIEatE, Research in Social Problems and Public Policy V, 173–202. (2008) In the shade of affluence: the inequitable distribution of the heat island. equity and the environment. Research in Social Problems and Public Policy 15 (173–202)

Hassett K, Mathur A, Metcalf G (2008) The incidence of a U.S. carbon tax: a lifetime and regional analysis. American Enterprise Institute for Public Policy Research, Cambridge

Hayhoe K, Cayan D, Field CB, Frumhoff PC, Maurer EP, Miller NL, Moser SC, Schneider SH, Cahill KN, Cleland EE, Dale L, Drapek R, Hanemann RM, Kalkstein LS, Lenihan J, Lunch CK, Neilson RP, Sheridan SC, Verville JH (2004) Emissions pathways, climate change, and impacts on California. Proc Natl Acad Sci USA 101(34):12422–12427

Hepburn C, Grubb M, Neuhoff K, Matthes F, Tse M (2006) Auctioning of EU ETS phase II allowances: how and why? Clim Policy 6(1):137–160

Howitt R, Medellín-Azuara J, MacEwan D (2009) Estimating the economic impact of agricultural yield related changes for California climate action team report. California Energy Commission; California Air Resources Board, Sacramento

Pacific Institute (2009) The impacts of sea-level rise on the California coast. The Pacific Institute, Oakland

Jackson L, Santos-Martin F, Hollander A, Horwath W, Howitt R, Kramer J, O'Geen A, Orlove B, Six J, Sokolow S, Sumner D, Tomich T, Wheeler S (2009) Potential for adaptation to climate change in an agricultural landscape in the Central Valley of California of California. Climate Action Team Report. California Energy Commission; California Air Resources Board, Sacramento

Jacobson MZ (2007) Effects of ethanol (E85) versus gasoline vehicles on cancer and mortality in the United States. Environ Sci Technol 41(11):4150–4157

Jacobson M (2008) On the causal link between carbon dioxide and air pollution mortality. Geophys Res Let 35 (L03809)

Jerrett M, Burnett RT, Pope CA 3rd, Ito K, Thurston G, Krewski D, Shi Y, Calle E, Thun M (2009) Long-term ozone exposure and mortality. N Engl J Med 360(11):1085–1095. doi:10.1056/NEJMoa0803894

Kaur S, Nieuwenhuijsen MJ (2009) Determinants of personal exposure to PM2.5, ultrafine particle counts, and CO in a transport microenvironment. Environ Sci Technol 43(13):4737–4743

Kilbourne E (1997) Heat waves and hot environments. In: Noji E (ed) The public health consequences of disasters. Oxford University Press, New York

Klinenberg E (2002) Race, place, and vulnerability. In: Heat Wave

Knowlton K, Rotkin-Ellman M, King G, Margolis H, Smith D, Solomon G, Trent R, English P (2009) The 2006 California heat wave: impacts on hospitalizations and emergency department visits. Environ Health Perspect 117(1):61–67

Kovats RS, Hajat S (2008) Heat stress and public health: a critical review. Annu Rev Public Health 29:41–55

Kunzli N, McConnell R, Bates D, Bastain T, Hricko A, Lurmann F, Avol E, Gilliland F, Peters J (2003) Breathless in Los Angeles: the exhausting search for clean air. Am J Public Health 93 (9):1494–1499.

Lejano R, Hirose R (2005) Testing the assumptions behind emissions trading in non-market goods: the RECLAIM program in Southern California. Environ Sci Pol 8:367–377

MMWR (2008) Heat-related deaths among crop workers - United States, 1992–2006. Mortality and Morbidity Weekly Report 57(24):649–653

Madsen AM (2006) Exposure to airborne microbial components in autumn and spring during work at Danish biofuel plants. Annals of Occupational Hygiene 50 (8):821–831. doi:10.1093/annhyg/mel052

Meleux F, Solmon F, Giorgi F (2007) Increase in summer European ozone amounts due to climate change. Atmos Environ 41(35):7577–7587. doi:10.1016/j.atmosenv.2007.05.048

Minkler M, Wallerstein N (2003) Community-based participatory research for health. Jossey-Bass, San Francisco

Morello-Frosch R, Jesdale B (2006) Separate and unequal: residential segregation and estimated cancer risks associated with ambient air toxics in U.S. metropolitan areas. Environ Health Perspect 114(3):386–393

Morello-Frosch R, Lopez R (2006) The riskscape and the color line: examining the role of segregation in environmental health disparities. Environ Res 102(2):181–196. doi:10.1016/j.envres.2006.05.007

Morello-Frosch R, Pastor M Jr, Porras C, Sadd J (2002) Environmental justice and regional inequality in southern California: implications for future research. Environ Health Perspect 110(Suppl 2):149–154

Morello-Frosch R, Pastor M, Sadd J, Shonkoff S (2009) The climate gap: inequalities in how climate change hurts Americans & how to close the gap. The Program for Environmental and Regional Equity (PERE), University of Southern California

Ojerio R, Moseley C, Bania N, Lynn K (2010) The limited involvement of socially vulnerable populations in federal programs to mitigate wildfire risk in Arizona. Natural Hazards Review Posted ahead of print 10 June 2010

Oke T (1973) City size and the urban heat island. Atmos Environ 7:769–779

O'Neill C (2004) Mercury, risk and justice. ELR News and Analysis 34:11070–11115

O'Neill MS, Zanobetti A, Schwartz J (2003) Modifiers of the temperature and mortality association in seven US cities. Am J Epidemiol 157(12):1074–1082

O'Neill MS, Zanobetti A, Schwartz J (2005) Disparities by race in heat-related mortality in four US cities: the role of air conditioning prevalence. J Urban Health 82(2):191–197

O'Neill MS, Kinney PL, Cohen AJ (2008) Environmental equity in air quality management: local and international implications for human health and climate change. J Toxicol Environ Health A 71(9–10):570–577

Pastor M, Morello-Frosch R, Sadd J (2010a) Air pollution and environmental justice: integrating indicators of cumulative impact and socio-economic vulnerability into regulatory decision-making. California Air Resources Board

Pastor M, Morello-Frosch R, Sadd J, Scoggins J (2010b) Minding the climate gap: what's at stake if California's climate law isn't done right and right away. Program for Environmental and Regional Equity, University of Southern California, Los Angeles

Phelan JC, Link BG, Diez-Roux A, Kawachi I, Levin B (2004) "Fundamental causes" of social inequalities in mortality: a test of the theory. J Health Soc Behav 45(3):265–285

Pittock B, Wratt D, Basher R, Bates B, Finlayson M, Gitay H, Woodward A, Arthington A, Beets P, Biggs B (2001) Climate change 2001: impacts, adaptation, and vulnerability. Cambridge Univ Press, Cambridge

Poumadere M, Mays C, Le Mer S, Blong R (2005) The 2003 heat wave in France: dangerous climate change here and now. Risk Anal 25(6):1483–1494

Rogot E, Sorlie PD, Backlund E (1992) Air-conditioning and mortality in hot weather. Am J Epidemiol 136 (1):106–116

Sadd J, Pastor M, Morello-Frosch R, Scoggins J, Jesdale B (2011) Playing it safe: Assessing Cumulative Impact and Social Vulnerability through an Environmental Justice Screening Method in the South Coast Air Basin, California. International Journal of Environmental Research and Public Health. 8:1441–1459

Schulz A, Williams D, Israel BA, Lempert LB (2002) Racial and spatial relations as fundamental determinants of health in Detroit. Milbank Q 80(4):677–707

Semenza JC, Rubin CH, Falter KH, Selanikio JD, Flanders WD, Howe HL, Wilhelm JL (1996) Heat-related deaths during the July 1995 heat wave in Chicago. N Engl J Med 335(2):84–90

Shonkoff S, Morello-Frosch R, Pastor M, Sadd J (2009a) Environmental health and equity impacts from climate change and mitigation policies in California: a review of the literature. Climate Action Team Report, CARB, Sacramento

Shonkoff S, Morello-Frosch R, Pastor M, Sadd J (2009b) Minding the climate gap: environmental health and equity implications of climate change mitigation policies in California. Environ Justice 2(4):173–177

Smith K, Jerrett M, Anderson H, Burnett R, Stone V, Derwent R, Atkinson R, Cohen A, Shonkoff S, Krewski D, Pope C III, Thun M, Thurston G (2009) Public health benefits of strategies to reduce greenhouse-gas emissions: health implications of short-lived greenhouse pollutants. Lancet 6736(9):56–69

Solomon S, Qin D, Manning M, Chen Z, Marquis M, Averyt KT, M, Miller H, (Eds.) (2007) Working Group I Report: "The Physical Science Basis". Intergovernmental Panel on Climate Change, Cambridge, United Kingdom and New York, NY, US

Stathopoulou E, Mihalakakou G, Santamouris M, Bagiorgas HS (2008) On the impact of temperature on tropospheric ozone concentration levels in urban environments. J Earth Syst Sci 117(3):227–236

Su J, Morello-Frosch R, Jesdale B, Kyle A, Shamasunder B, Jerrett M (2009) An index for assessing demographic inequalities in cumulative environmental hazards with application to Los Angeles, California. Environ Sci Technol 43(20):7626–7634

Swiss Re (2006) Natural catastrophes and man-made disasters 2005: high earthquake casualties, new dimension in windstorm losses

Tenenbaum D (2008) Food vs. fuel: diversion of crops could cause more hunger. Environ Health Perspect 116(6):A254–A257

Thomalla F, Downing T, Spanger-Siegfried E, Han G, Rockström J (2006) Reducing hazard vulnerability: towards a common approach between disaster risk reduction and climate adaptation. Disasters 30(1):39–48

Tierney K (1988) The Whittier Narrows, California earthquake of October 1, 1987—social aspects. Earthquake Spectra 4(1):11–23

US EPA (2010) Air trends. http://www.epa.gov/air/airtrends/values.html

USCB (2004) Current housing reports, American Housing Survey for the Los Angeles-Long Beach Metropolitan Area: 2003. U.S. Census Bureau

USCB (2005) California—County. Percent of people below poverty level in the past 12 months (for whom poverty status is determined): 2005. United States Census Bureau. http://factfinder.census.gov/servlet/GCTTable?_bm=y&-geo_id=04000US06&-_box_head_nbr=GCT1701&-ds_name=ACS_2005_EST_G00_&-_lang=en&-mt_name=ACS_2005_EST_G00_GCT1701_ST2&-format=ST-2. Accessed July 2008 2008

Walls M, Hanson J (1996) Distributional impacts of an environmental tax shift: the case of motor vehicle emissions taxes. Resources for the Future, Washington

Whitman S, Good G, Donoghue ER, Benbow N, Shou W, Mou S (1997) Mortality in Chicago attributed to the July 1995 heat wave. Am J Public Health 87(9):1515–1518

Williams D, Collins C (2001) Racial residential segregation: a fundamental cause of racial disparities in health. Publ Health Rep 116:404–416

Climatic Change (2011) 109 (Suppl 1)
DOI 10.1007/s10584-011-0316-1

Climate change-related impacts in the San Diego region by 2050

Steven Messner · Sandra C. Miranda · Emily Young · Nicola Hedge

Received: 12 February 2010 / Accepted: 26 September 2011 / Published online: 24 November 2011
© Springer Science+Business Media B.V. 2011

Abstract This paper explores what the San Diego region may look like in the year 2050 as projected changes in regional climate conditions take place. Focusing on interrelated issues of climate change, sea level rise, population growth, land use, and changes in water, energy, public health, wildfires, biodiversity, and habitat, the paper reviews the potential impacts of a changing climate by 2050 and makes recommendations for changes in planning processes at the local and regional levels to prepare for these impacts. The original research for this study was completed in 2008 by a team of 40 experts from the region including universities, nonprofit organizations, local governments, public sector agencies and private sector entities. This paper has now been updated with more recent research regarding climate change adaptation while preserving the integrity of the original research team's work. The simulated impacts discussed in this study are based on regional projections of climate change generated by scientists at Scripps Institution of Oceanography, employing three climate models and two emissions scenarios used by the Intergovernmental Panel on Climate Change. The impacts are discussed in the context of significant regional growth expected during the period as well as an aging population base. Key issues explored in the report include potential inundation of six selected low-lying coastal areas in San Diego due to sea level rise, potential shortfalls in water deliveries, peak energy demand increases due to higher temperatures, growing risk of devastating wildfires, migrations of species in response to higher temperatures in an increasingly fragmented natural habitat, and public health issues associated with extreme temperature events.

S. Messner (✉)
ENVIRON Corporation, Novato, CA, USA
e-mail: smessner@environcorp.com

S. C. Miranda
Energized Solutions, San Francisco, CA, USA

E. Young · N. Hedge
The San Diego Foundation, San Diego, CA, USA

E. Young
e-mail: eyoung@sdfoundation.org

N. Hedge
e-mail: nicola@sdfoundation.org

1 Introduction

The San Diego region is renowned worldwide for its unique combination of mild climate, low rainfall, breathtaking shorelines, mountains, and deserts—all in close proximity. Not surprisingly then, the region has been one of the fastest-growing areas in the country. This unique set of climate and population characteristics also creates a unique fragility. The complex and fragile interrelationship of urban and natural systems here has been dramatically highlighted by devastating wildfires, as well as by more gradual changes in the region's natural ecosystems. Higher temperatures, changing precipitation patterns, and a rising sea level will create a number of issues that will require new planning and coordination activities, as well as exacerbate existing stresses due to regional growth.

This study considers the regional impacts due to climate change that can be expected by 2050 based on projections of climate change using three climate models and two emissions scenarios drawn from those used by the Intergovernmental Panel on Climate Change (IPCC). A number of analytical models were developed and used for this study to provide quantitative estimates of the impacts where possible. For example, wave and sea level modeling was used to develop a range of impacts on six low-lying coastal areas in the region. Also, temperature data from the IPCC scenarios were applied to regional ecosystems models to provide information on the migration patterns of species trying to adapt to higher temperatures. These temperature data were also used to extrapolate forecasts of peak electricity demand in the region, which will be exacerbated by higher temperatures as well as the faster inland population growth where the region is hottest.

According to the San Diego Association of Governments (SANDAG) the population of San Diego County in January 2010 was 3,224,432 (2010a).[1] SANDAG's most recent Regional Growth Forecast projects that between 2004 and 2030, the region will add about one million more people (2010b). The region's population is projected to reach 4.5 million by 2050 (State of California, Department of Finance 2007), an increase of 524,000 persons beyond the 2030 projections. This growing population will not only affect the way in which the San Diego region adapts to climate change, but exacerbate the overall regional impacts of climate change as well.

As the region's population grows, it will also become older. Approximately one quarter of the region's current population is baby-boomers, the large cohort born between 1946 and 1964. Their presence helps increase the median age in the region from 33.7 years in 2004 to 39 years in 2030. By 2050, almost one quarter of the region's residents (over 1,000,000) will be age 65 and older, with over half being older than age 41. The aging population of San Diego will be more vulnerable to the public health impacts of climate change, including increased heat waves and air pollution.

The aim of this was to provide a scientific basis for local governments and public agencies to develop climate-preparedness plans, which include strategies for mitigating the damage from, and adapting to, climate change. A key message of this study is that there is not any single "silver bullet" to solve these projected impacts, rather that there is a serious need for coordinated actions among local, regional and state authorities to begin or advance planning activities in all of these areas. Already, local governments in the San Diego region are beginning to integrate this science into regional and local plans, including the City of Chula Vista, the City of San Diego, SANDAG, as well as a sea level rise and climate

---

[1] SANDAG's estimates available at www.sandag.org/index.asp?subclassid=84&fuseaction=home.subclasshome

adaptation project for the Unified Port of San Diego and its five member cities. To continue growing public awareness beyond this study, The San Diego Foundation is now funding follow-on research being undertaken to deepen understanding of risks posed to wetlands and regional infrastructure from sea level rise, as well as climate impacts to regional water supply and demand, and key native wildlife.

The original research for this study was conducted in 2007–2008 by a team of 40 experts from regional universities, nonprofit organizations, local governments, public and private sector entities. Since then, additional research reports and adaptation planning guidelines have been issued from a variety of sources that have advanced the body of knowledge of this emerging issue. Recognizing this, this study has been updated for more recent research regarding climate change adaptation while preserving the integrity of the original research team's work. This update also addresses one of the objectives of the original research and report – namely that this effort would be part of a "living" process to consider new climate change research and new projections to inform better future regional planning.

2 Project approach and outcomes

2.1 Climate change in the San Diego area

The studies presented here are based on analysis using three climate models,[2] and two scenarios of energy use and greenhouse-gas (GHG) emissions.[3] The models and scenarios were among those used in the IPCC's 2007 climate assessment, and they are included in the set of models Scripps Institution of Oceanography prepared for the 2008 California Climate Change Scenarios Assessment. More details on the scenarios and modeling work, including the approaches to downscaling used for this study can be obtained in Cayan et al. 2008b. Because there is considerable uncertainty about future greenhouse-gas emissions scenarios, specific probabilities to any of these simulations were not assigned. However, the analysis provides strong and clear indications that the climate that the San Diego region must plan for will not be the climate to which it has been accustomed.

2.1.1 Warming

Figure 1 represents warming trends in San Diego from the three climate models for scenarios A2 and B1 from 1900 to 2100. All six of the simulations warm over the next five decades. It can be observed that the temperatures began to warm more substantially in the 1970s. This is likely a response to effects of GHG accumulation, which began to increase significantly during this time period (Bonfils et al. 2007; Barnett and Pierce 2008).

All of the climate model simulations exhibit warming across San Diego County – ranging from about 1.5°F to 4.5°F (0.8°C to 2.5°C). The warming becomes progressively greater through the decades of the twenty-first century. There is greater warming in summer than in winter, with surface air temperatures in summers warming from 0.7°F to more

---

[2] The three models are the National Center for Atmospheric Research's Parallel Climate Model (PCM), the National Oceanic and Atmospheric Administration's Geophysical Fluids Dynamics Laboratory (GFDL) version 2.1, and the French Centre National de Recherches Météorologiques (CNRM).

[3] The IPCC's Special Report on Emissions Scenarios (SRES) A2 and B1 scenarios.

Fig. 1 Change in annual mean temperature, San Diego region from the three GCMs, for the historical period (*blue*) and for the A2 (*red*) and B1 (*brown*) emission scenarios

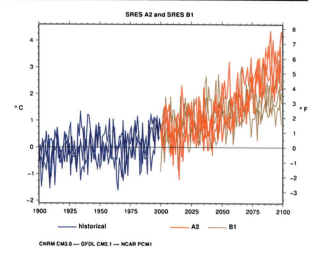

than 2°F (0.4°C to 1.1°C) over that found in winter. There is a distinct Pacific Ocean influence wherein warming is more moderate in the zone within approximately 50 km from the coast, but rises considerably, as much as 2°F (1.1°C) higher, in the interior areas of San Diego County as compared to the warming that occurs right along the coast.

2.1.2 Heat waves

Extreme warm temperatures in the San Diego region mostly occur in July and August, but as climate change takes hold, the occurrences of these events will likely begin in June and could continue to take place into September. All simulations indicate that hot daytime and nighttime temperatures (heat waves) increase in frequency, magnitude and duration. For instance, in one inland portion of San Diego County there is a predicted threefold increase in heat wave events.

2.1.3 Precipitation

The simulations indicate that the San Diego region will retain its strong Mediterranean climate with relatively wetter winters and dry summers. Projections of future precipitation have mixed results; three of the simulations become drier (12%–35% drier than historical annual average) and three are wetter (12%–17% wetter than historical annual average) overall. This reflects the reality that precipitation cannot yet be modeled with the same degree of consistency as other climate change parameters. The models vary in their projections of storminess[4] but none show a significant change from past patterns. One important aspect of all of the climate model projected simulations is that the high degree of variability of annual precipitation that the region has historically experienced will prevail during the next five decades. This suggests that the region will remain highly vulnerable to drought, with significant implications for water management planning.

---

[4] Indicated by the number of days per year when sea level pressure equals or falls below 1005 mb.

Recent research (MacDonald et al. 2008) has suggested that there is an increased likelihood of a so-called 'perfect drought' in California's future – i.e., a prolonged drought that simultaneously affects southern California, the Sacramento River basin and the upper Colorado River basin. Over the past century such perfect droughts have occurred, but lasted for less than 5 years. Tree-ring research of past drought events indicate that prolonged perfect droughts (30–60 years) developed in the mid-11th century and the mid-12th century during the period of the 'Medieval Climate Anomaly' (MCA), also called the Medieval Warm Period (MWP). This period is discussed in more detail in the IPCC 4th assessment report.[5] This prolonged aridity appears to have been associated with the differential thermal response of the western and eastern Pacific to increased tropical radiative forcing at that time. This potential linkage between positive radiative forcing and prolonged perfect droughts led to a conclusion by MacDonald et al. that future droughts in California have the potential of far exceeding the duration of any droughts experienced during the past 100 years.

2.1.4 El Niño/Southern Oscillation

Historically, the El Niño/Southern Oscillation (ENSO) has been an important influence on weather conditions in southern California. Each of the climate models contains ENSO within its historical simulations. Although there is no evidence for an increase in the frequency of ENSO, each of the simulations exhibits continued ENSO activity within the twenty-first century. There is already a modest tendency for the San Diego region to experience higher than normal precipitation during El Niño winters and lower than normal precipitation during La Niña winters. This pattern is expected to continue under climate change conditions in the future. Regarding ENSO intensity, the IPCC scenarios show little change over the century, but this conclusion has come under recent criticism from some scientists who conclude that there is a "significant probability of a future increase in ENSO amplitude" in this century (Lenton et al. 2008). Changes in ENSO intensity will result in stormier years and drier years, which have implications for public health planning, as well as coastal impacts.

2.2 Sea level scenarios and coastal impacts in 2050

To better understand the combined effects and possible impacts of sea level rise and wave activity on San Diego's coastline by 2050, models have been employed that take into account both of these key issues. Predicted impacts are mapped for six low-lying coastal areas to highlight the severity and frequency of shoreline inundation in areas of San Diego County prone to flooding. The predictions also illustrate how impacts may vary along the San Diego coastline in areas with cliffs, estuaries, sea walls, rock jetties, and other man-made structures. The analytical approach used to predict climate change impacts associated with sea level rise and wave activity is further described below.

2.2.1 Modeling sea level rise in the San Diego region

A well-known approach for forecasting sea level rise is the Rahmstorf (2007) semi-empirical method. This method links sea level rise to observed global mean temperatures. The method

---

[5] http://www.ipcc.ch/publications_and_data/ar4/wg1/en/ch6s6-6.html#6-6-1

also assumes that sea level rise along the southern California coast will be the same as global estimates. The sea level projections developed here also include a lower range of estimates which account for the worldwide growth of dams and reservoirs, which have changed, and will continue to change surface runoff into the oceans (Chao et al. 2008).

Results of three simulation scenarios indicate sea level increases of 12–18 in. by 2050. As the decades proceed, the simulations show an increasing tendency for heightened sea level events to persist for more hours, which will likely cause greater coastal erosion and related damage (Cayan et al. 2008a). Sea level rise was projected based on application of the Rahmstorf (2007) method with and without adjustment for the effects of dams and was compared with observed values between 1900 and 2000. Dams have retained significant amounts of water in the past and have skewed past trends. The model assumes that dams will not be able to continue this trend and a new equilibrium is being established.

2.2.2 Combining effects of sea level rise and wave activity

The results from the sea level rise projections were then applied to existing wave models used in the San Diego region to better understand future wave activity on lower-lying areas. The wave model forecasts, derived from a global climate model simulation, were transformed to 10 m depth using the Coastal Data Information Program[6] (CDIP) spectral refraction model developed by the Scripps Institution of Oceanography (SIO). The CDIP model was revised to look at offshore wave conditions for a coastline that is slowly progressing landward along with sea level rise. The CDIP model accounts for coastline variations that affect wave height and energy including island sheltering, wave refraction, and shoaling of waves in the southern California Bight. The increased elevation of the shoreline water level owing to wave run-up (called super-elevation) is estimated from the wave conditions using an empirical engineering formula.[7] Wave-induced super-elevation is then combined with tides, weather effects (e.g., storms), El Niño effects, and longer-term sea level changes (Cayan et al. 2008a) to develop a time series profile of shoreline water levels at each site. Finally, digital elevation data from LIDAR coastline surveys were combined with an analysis of the shoreline water level time series to create the maps of potential inundation.

Figures 2 and 3 show the projected impacts by 2050 in two of the six already flood-prone areas analyzed with the revised CDIP wave model, with a brief explanation of the specific impacts at each site. The colored zones represent new flooding areas. The Figures depict predicted wave event frequencies using the following definitions[8]:

- Very Likely: predicted high tide range in 2050.
- Moderately Common: estimated sea level + tide + wave run-up elevation[9] recurrence, on average, every 5 years in the 50-year simulation.
- Moderately Rare: estimated sea level + tide + wave run-up elevation recurrence, on average, every 10 years in the 50-year simulation; but expected in most years when El Niño conditions are present.

---

[6] Model documentation is available at: http://cdip.ucsd.edu/?sub=faq&nav=documents&xitem=future
[7] Run-up elevation=0.4 * Wave Height @ 10 m depth.
[8] These definitions are unique to this report. As a point of reference, all of the definitions here except "Very Rare" would be associated with a FEMA "high risk" coastal flooding area.
[9] Wave run-up is the maximum vertical extent of wave uprush on a beach or structure above the still water level

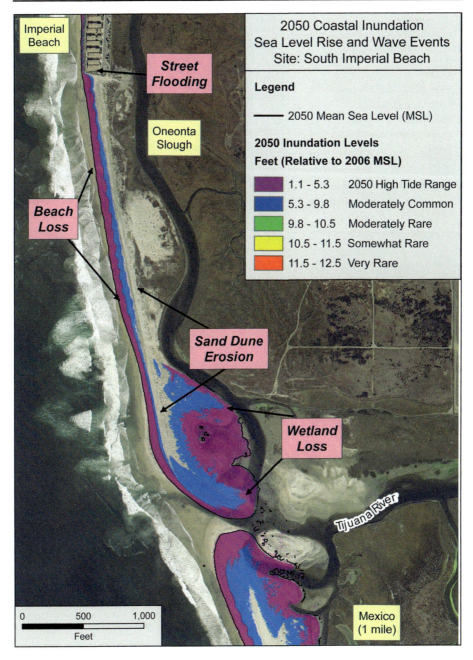

Fig. 2 South Imperial Beach

- Somewhat Rare: estimated sea level + tide + wave run-up elevation recurrence on average every 25 years, based on the 50-year simulation.
- Very Rare: highest combination of sea level + tides + wave run-up elevation in the 50-year simulation.

Fig. 3 Mission Beach

As shown in Fig. 2, tidal fluctuations alone (shown in purple) appear to inundate sandy beaches and the Tijuana River mouth. Run-up from moderately common wave events (shown in blue) together with tidal fluctuations flood the majority of sandy beach. Very rare wave events (shown in red) flood sandy beaches, areas of sensitive sand dune habitat and streets in south Imperial Beach. The dune line shown north of the river mouth would likely be eroded by even a moderately rare inundation event.

As shown in Fig. 3, tidal fluctuations alone (shown in the purple) appear to inundate portions of sandy beach and streets from bayside flooding. Run-up from moderately common wave events combined with tidal fluctuations (shown in blue) flood the majority of sandy beaches, streets and parts of Mission Beach Park. Moderately rare wave events (shown in green) appear to breach the seawall and inundate streets and sidewalks. Very rare wave events (shown in red) flood sandy beaches, surface streets and the heavily used boardwalk in Mission Beach.

2.3 Climate change impacts on water

A key objective of this part of the study was to extrapolate existing water demand and supply forecasts for 2030 out to 2050 to highlight the pressures of population growth and climate change on the regional water situation. The data used to project water demand and supply from 2005 to 2030 were primarily from the San Diego County Water Authority (Water Authority) 2005 Urban Water Management Plan, along with additional information provided by the San Diego County Water Resources Department Staff. The projected water demands and supplies in 2005, 2030, and 2050 are illustrated in Fig. 4. The height of the bars represents the water demand, and the expected sources of water are illustrated by the different colors of the bar segments. The two bars labeled "2050" and "2050*" show expected water sources for San Diego in the year 2050 under historical climate and climate change scenarios, respectively.

2.3.1 Demand impacts

Water demand was projected based on assessments of agency-by-agency trends for 2030. Projections from 2030 to 2050 are straight-line extensions of per capita trends leading up to 2030 scaled down to reflect the expectation that the population growth rate in San Diego is

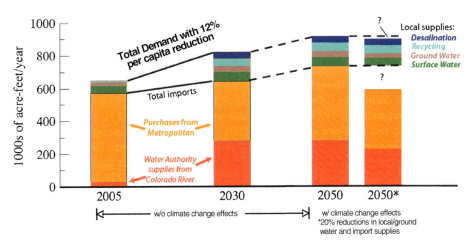

Fig. 4 Projected water demand and supply in 2005, 2030, and 2050, under "normal year" and climate change conditions

projected to decline by 25% (from the pre-2030 rate) after 2030. It is important to note that the Water Authority projections assume a 12% per capita demand reduction by 2030 to reflect planned conservation and efficiency measures. These conservation efforts are assumed to continue from 2030 to 2050. It is expected that the Water Authority's future conservation efforts will focus more on landscape irrigation and commercial, institutional, and industrial savings as well as new residential construction standards. Residential surveys conducted in the region by the Water Authority indicate that more than 50% of urban water use in San Diego is spent on landscape irrigation and other outdoor uses (e.g., pools, water features, etc.).[10]

As shown in Fig. 4, the Water Authority predicts an increase in water demand from 668,000 acre-feet/yr (the 2001–2005 average) to about 830,000 acre-feet/yr in 2030. Seventy percent of future demand in 2030 is expected to be met by imported water sources. In 2050, estimated demand is 915,000 acre-feet/yr. Under current demand and planned local supply projections, about 80% or 730,000 acre-feet/yr of the water supply would need to be imported. Production from groundwater supplies is anticipated to increase by 75% to about 31,000 acre-feet/yr by 2015. After that, local surface and groundwater supplies will have reached their foreseeable limit. Figure 4 also illustrates that increased demand from 2005 to 2030 is expected to be accommodated by increased supplies from the Colorado River as a result of recent Federal agreements pertaining to Colorado River water management,[11] as well as from desalination and recycling. From 2030 to 2050, increased demand will then have to be accommodated by increased purchases from the Metropolitan Water District (MWD) which imports much of its water from the Sacramento Valley (the source region of the California State Water Project or SWP).

Future water demand is expected to further increase as a result of climate change and increased drought tendency. For this study, future soil moisture conditions for western San Diego County were simulated using the Variable Infiltration Capacity (VIC) water-balance model,[12] with temperature and precipitation data supplied by the Geophysical Fluid Dynamics Laboratory's CM2.1 climate-model simulations for the emissions scenarios used here. The future soil moisture conditions were then compared to the 1989 (Water Authority representative drought year) drought conditions as an index of climate effects on water supplies. In order to calibrate the model the VIC simulation was run and compared against the recent historical period (1950–1999).

As illustrated in Fig. 5, future soil moisture conditions are expected to drop below the 1989 drought threshold with an increasing frequency and greater severity. The model results shows droughts becoming 50% more common during the 2000–2049 period than during the 1950–1999 period. The VIC model projects that in 2030, a drought comparable to 1989 would increase water demand by 6.5%, in large part due to decreased soil moisture content.[13] In 2050, the demand would increase from 915,000 acre-feet/yr in a normal year to 980,000 acre-feet/yr in a drought year. Drought years also appear to occur as much as twice as often (Fig. 5). Drought years in the future might also yield larger demand than the 6.5% increase associated with 1989 because they are projected to be considerably drier than 1989.

---

[10] http://www.20gallonchallenge.com/about.html
[11] http://www.usbr.gov/lc/region/programs/strategies.html
[12] http://www.hydro.washington.edu/Lettenmaier/Models/VIC/
[13] Lower soil moisture content increases agricultural and landscaping water demands.

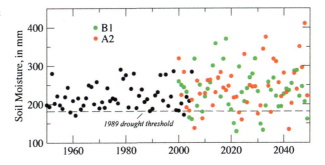

Fig. 5 Simulated annual-mean soil moisture, Western San Diego County

2.3.2 Supply impacts

There are two supply scenarios for 2050; the first was projected assuming existing climate conditions (climate change effects were not taken into account), and the second scenario makes the assumption that climate change could result in 20% reductions in the availability of imported water and local surface and ground water. The 20% reduction in water availability is based on the results of studies that predict Colorado River flows will decrease in response to climate changes by 18% to 20%. This is a middle-of-the-road choice, given the range of published estimates of reductions in Colorado River flows by 2050 ranging from 6% to as much as a 45% (Christensen et al. 2004; Milly et al. 2005; Christensen and Lettenmaier 2007; Hoerling and Eischeid 2007).[14] Although California-specific downscaled projections are somewhat divergent in their future estimations of precipitation, broader regional projections of precipitation affecting the Colorado River watershed as a whole are consistently drier across the various climate models in the cited studies. In addition, zero to 40% is the range of water supply declines suggested by recent studies that estimated the reliability of SWP deliveries under different climate change scenarios (Department of Water Resources 2006; Vicuna 2006; Zhu et al. 2005).

The potential linkage between positive radiative forcing and prolonged perfect droughts cited earlier (MacDonald et al. 2008) also raises serious concerns regarding the hydrological impacts of future climate warming in southern California and the western United States. There will be an increased likelihood of the SWP and the Colorado River simultaneously providing significantly reduced supplies during these prolonged extended droughts.

In recent years, the states that draw water from the Colorado River negotiated a shortage-sharing agreement that specifies how supply shortfalls from the river of as much as 8% might be shared by water users.[15] Barnett and Pierce (2008) estimate that, without this agreement, the major reservoirs of the Colorado River could be emptied within a few decades by a combination of increasing demand and climate change. Specifically, there is a 10% chance active storage in Lakes Mead and Powell will be gone by about 2013 and a 50% chance it will be gone by 2021 with no changes in water allocation from the Colorado River.

Blank areas with question marks in the far right bar in Fig. 4 (climate change scenario) indicate shortfalls in water supplies by 2050 due to the projected 20% reductions in the volumes of available imported water and local surface and ground water. This potential shortfall in water supplies combined with risks of prolonged droughts represents a significant concern to the San Diego region.

---

[14] Christensen et al. 2004, −18%; Milly et al. 2005, −20%; Christensen and Lettenmaier 2007, −6%; Hoerling and Eischeid 2007, −45%.

[15] The agreement can be viewed at: http://www.usbr.gov/lc/region/programs/strategies.html

2.4 Wildfires

Extended drought conditions forecasted in the coming decades are expected to increase the likelihood of large wildfires. A past study of the western United States (Westerling et al. 2006) showed that the frequency and duration of large wildfires began increasing in the mid-1980s when there was a marked increase in spring temperatures, a decrease in summer precipitation, drier vegetation and longer fire seasons. A more recent study (Spracklen et al. 2009) explores these relationships to 2050 using temperature and precipitation data from a global climate model (GISS). Their study suggests that 34% more California Coastal Shrub acreage will burn in the decade around 2050 (2046–2055) as compared to the baseline comparison decade (1996–2005) and that overall, 54% more acreage in the western United States will burn compared to the baseline comparison decade.

Climate change models yield somewhat different predictions about the frequency, timing and severity of future Santa Ana wind conditions (dry hot winds blowing down the mountains from the deserts in the east), leading to uncertainty regarding how Santa Ana winds will affect San Diego regional fire regimes in the future. A recent study on anthropogenic reduction of Santa Ana winds (Hughes et al. 2011) indicates that these events are expected to decrease by the mid-twenty-first century in both frequency and intensity due to higher temperatures in the desert during wintertime creating less pressure gradient between the mountains and the ocean. An earlier analysis (Miller and Schlegel 2006) for the period 2005–2034 suggested that Santa Ana conditions may significantly increase earlier in the fire season (especially September), while they may decrease somewhat later in the season (in particular, December). This predicted shift to earlier Santa Ana occurrences would increase the frequency of Santa Ana fires, as severe fire weather would coincide more closely with the period of most frequent fire ignitions.

Wildfires in the San Diego region occur throughout the year, but most strongly during late summer and early fall. Over the twentieth century, the area burned by wildfires has undergone substantial fluctuations, but in the last 10 years the extent of these wildfires has been unprecedented, greatly exceeding that during any past decade. Specifically, in 2003 and 2007, wildfires burned a combined total of nearly 760,000 acres (The San Diego Foundation 2003, 2007). The question of whether the region is now in a higher state of vulnerability to such fires due to climate change merits further research to better inform disaster preparedness efforts.

2.5 Ecosystems

Located at the heart of a global biodiversity hotspot, the biological richness of the San Diego region is difficult to overstate, with high densities of both endangered plants and animals (Rodriguez et al. 1997, Myers et al. 2000). Past and present land use changes have brought significant, often cascading impacts to biodiversity across the San Diego region County. The starkness of the fragmentation[16] pattern in San Diego reveals how the size, shape, and isolation of habitat fragments affect their ability to support native species. When habitat is fragmented by human land uses, it can trigger ecological cascades that result in the loss of species.

A changing climate will add to the stress on ecological systems in ways that may create feedback cycles with significant consequences. With climate change, the "climatic envelopes"[17] that species need will move due to increasing temperatures and more frequent

---

[16] Fragmentation is the emergence of discontinuities in an organism's preferred environment (habitat). Habitat fragmentation can be caused by geological processes that alter the layout of the physical environment or by human activity such as land conversion, which can alter the environment on a much faster scale.

[17] Climate envelops are defined as locations where the temperature, moisture and other environmental conditions are suitable for the persistence of particular species.

fires. For many species, a changing climate is not the problem per se. The problem is the rapid rate of climate change: the envelope will shift faster than species are able to follow. To put the rate of temperature change for species survival into context, a 1°F–5°F (0.56°C–2.8°C) increase by 2050 predicted by the three climate change models is 10–50 times faster than the temperature changes (2°F, or 1.1°C per 1000 years) that occured when ice sheets receded from review of the Vostok climate records.[18]

Future trends of San Diego ecosystems in response to climate change are evaluated in this report based on relevant literature associated with the growing body of research in paleoclimatology, related studies of paleorecords and assessments of ecosystem changes in correspondence to major climate regimes, and use of new tools and models to refine local predictions. Shrubland and distribution models were used in this study to predict how climate change can affect terrestrial ecosystems and distribution of species.

2.5.1 Shrubland models

The Center for Conservation Biology (CCB) at the University of California, Riverside has developed models predicting potential habitat for a variety of plant and animal species in different ecosystems in southern California (Preston and Rotenberry 2007) with a particular focus on shrubland communities that support a diversity of sensitive plant and animal species in the region (Preston et al. 2008). To understand how changing climate conditions might affect these natural communities in the San Diego region, the CCB conducted climate sensitivity analyses for coastal sage scrub and chaparral vegetation as well as for plant and animal species found in these shrublands (Preston et al. 2008). To assess the sensitivity of the species and the vegetation types to climate change, the models used different temperature and precipitation values associated with climate futures. The CCB also developed models predicting suitable habitat for the federally endangered Quino Checkerspot butterfly (*Euphydryas editha quino*) and threatened California Gnatcatcher (*Polioptila Californica*). The intent was to investigate whether associations between species, such as an animal species' dependence on a particular type of vegetation or specific plant species for food or shelter, might affect their potential distribution in a changing climate. The models developed included associations between animal and plant species under the 2050 climate scenarios.

The results of the CCB modeling showed that in response to rising temperatures and reduced precipitation, each vegetation type moves to higher elevations where conditions are cooler and there is greater precipitation. The suitable environmental conditions for coastal sage scrub were predicted to decrease between 10% and 100% under altered climate conditions, with the greatest reductions at higher temperatures and extremes in precipitation. Chaparral responded in a similar manner as coastal sage scrub, although higher percentages of suitable habitat remain at the elevated temperatures with current or reduced levels of precipitation.

Plant and animal species will each differ in their sensitivity to a changing climate, but the fact that they depend on each other increases the overall effects. The CCB models predicting suitable habitat for the Quino Checkerspot butterfly and California Gnatcatcher, when in association with plant species, were compared with predictions from models that included only climate variables and did not consider species associations. It was found that when vegetation, shrub or host plant species were included in the animal models, potential

---

[18] http://www.ncdc.noaa.gov/paleo/icecore/antarctica/vostok/vostok.html

habitat for the butterfly and songbird were reduced by 68%–100% relative to the climate change-only models.

2.5.2 Coastal ocean

The intertidal[19] and subtidal[20] habitats along the coast of San Diego contain a large diversity of marine algae, invertebrates and fish. Marine ecosystem productivity is strongly influenced by climate regime shifts (Chelton et al. 1982; MacCall et al. 2005). Human-induced impacts associated with discharges and harvesting also affect populations on local to regional scales.

The predicted increase in global temperatures over the next century will cause an increase in the temperature of the sea and will also have other effects on coastal oceanography, such as changes in the intensity of winds along the coast that can lead to major changes in upwelling patterns (Rykaczewski and Checkley 2008), which, in turn, influence nutrient supply and coastal ecosystem dynamics. However, the relationship between climate change and wind-driven upwelling is complex (Harley et al. 2006) and specific predictions of how upwelling patterns in San Diego and in adjacent regions will change are not yet available. Predicted sea level rise can also impact the marine communities in San Diego County, in particular intertidal species (Harley et al. 2006), because as sea level rises, the boundary between land and sea moves landwards. Whether intertidal habitat is lost would depend on the coastal topography. When intertidal habitats are bordered by high cliffs or anthropogenic structures such as seawalls and breakwaters, existing intertidal habitats (beaches, rocky shores) are prevented from migrating landwards, which results in a net loss (drowning) of these habitats (Galbraith et al. 2002).

Loss of rocky beach habitat is of particular concern because the two main intertidal marine reserves in San Diego, Cabrillo National Monument and Scripps Coastal Reserve, are bordered by steep cliffs and will almost certainly lose much of their intertidal habitats. Predicting which species will persist or not and how changes in species composition and abundance may affect local productivity and fisheries remains a complex challenge.

Increasing ocean temperature is expected to impact several colder water species in intertidal and shallower subtidal areas in that their reproduction ability is diminished as water temperature increases. The intertidal area contains many of the most recognizable flora and fauna for beachgoers such as abalone, anemones, brown seaweed, crabs, green algae, mussels, sea cucumber, starfish, sea urchins, shrimp, snails, sponges, and some coral species. Rogers-Bennet et al. (2010) cited significant loss in reproduction ability for red abalone when exposed to elevated (18°C or 64°F) water for extended (6 month or more) periods. When combined with temperature's other known impacts on abalone growth, kelp abundance, and disease proliferation, the authors conclude that ecosystem-wide consequences are likely.

2.6 Public health

Given the importance of air pollution issues in the San Diego region today, and increased significance of air quality issues in the future due to the aging population, the study team

---

[19] Intertidal refers to the area along an ocean coastline that is exposed to air during low tide and submerged at high tide; organisms in the intertidal zone are adapted to harsh conditions.
[20] Subtidal refers to the area along an ocean coastline below the intertidal zone; the subtidal zone is always covered by water.

conducted a literature review and summary of public health issues associated with climate change including heat stress illness, respiratory illness due to higher ground level ozone concentrations resulting from higher temperatures, respiratory effects from wildfires, as well as infectious disease implications. The study team also performed quantitative modeling analysis of aerosol fine particulate matter emissions and ambient concentrations trends in the San Diego air basins to evaluate the possible impacts of changes in this key air quality parameter on public health.

2.6.1 Air quality modeling

Past research has identified links between fine particulate matter ($PM_{2.5}$) and numerous health problems including asthma, bronchitis, acute and chronic respiratory symptoms such as shortness of breath and painful breathing (Hodgkin et al. 1984; Ferris et al. 1973), and premature death (Dockery et al. 1993; Pope et al. 1995; Krewski et al. 2000). Public health risks from $PM_{2.5}$ are highest among young children (Romieu et al. 2004; Kaiser et al. 2004; Woodruff et al. 1997) and the elderly (Dockery et al. 1993; Pope et al. 1995; Krewski et al. 2000). On an annual basis, San Diego County meets the U.S. Environmental Protection Agency's (EPA) Annual National Ambient Air Quality Standards (NAAQS) for $PM_{2.5}$ but exceeds the 24-hr NAAQS for $PM_{2.5}$ a few times during the cooler months of the year (County of San Diego Air Pollution Control District 2009).

The mathematical model developed in this study is a first attempt to evaluate the influence of climate change on air pollutant levels in San Diego County. The model projects changes in emissions through population expansion, the application of emission control programs, and interaction between future temperature and future emissions through 2020, and extrapolates these trends through 2050. An air quality box model approach was used to project fine particle ($PM_{2.5}$) concentrations for San Diego County. A detailed discussion of this model is provided in the previous research paper of this study provided to the California Energy Commission.[21] Key parameters were incorporated into the model, including current and projected air pollution emissions patterns within the county, current and projected local meteorology, and atmospheric chemical transformation and removal processes (wet and dry deposition) for air pollutants. San Diego County was divided into five sub-regions in the model and the vertical layer was divided into five levels. Concentrations of chemical species were assumed to be well-mixed within each sub-region and vertical layer, essentially using 25 separate "boxes" to simulate the county. The PM contribution from outside San Diego was considered by using the California Air Resources Board (CARB) emissions data for 2006 and San Diego Air Pollution Control District's Del Mar, Camp Pendleton, and Otay Mesa monitoring stations data for 2006 since these monitors lie towards the boundary of the model.[22]

One challenge to this work was the development of projections for emission inventories to 2050. For the A2 scenario, 2050 emissions were assumed to be equal to the base year 2006. Since California regulations actually require close to a 30% reduction from 2006 levels by 2020, the A2 scenario reflects a conservative case that no progress will actually be made towards this goal. For the B1 scenario, CARB emissions projection for the years 2010, 2015, and 2020 were used and emissions were assumed to be constant at the 2020 level for subsequent years until 2050. This

---

[21] http://www.energy.ca.gov/2009publications/CEC-500-2009-027/CEC-500-2009-027-F.PDF
[22] Data available at: http://www.arb.ca.gov/html/ds.htm and http://www.sdapcd.org/air/air_quality.html.

is a conservative assumption since California policy beyond 2020 calls for continued emission reductions until 2050.

To model the effects from particulate matter ($PM_{2.5}$) on human mortality in the region, data for mortality rates of diseases associated with fine particulate aerosols were collected from the California Department of Health Statistics Website.[23] These data were then analyzed using the projected $PM_{2.5}$ concentrations from our model to obtain projected mortality rates, using parameters of Pope et al. (1995). According to Pope et al. (1995) each 10 microgram per cubic meter (g/m3)(over a 16 year span) elevation in long-term average $PM_{2.5}$ ambient concentrations was associated with approximately a 4%, 6%, and 8% increased risk of all-cause, cardiopulmonary, and lung cancer mortality, respectively. These data were used to generate the mortality values in Table 1. In this mortality model, it was assumed that the population size, demographic characteristics, and long-term ecological effects were constant throughout the period under study.

The $PM_{2.5}$ model is predicting an overall decrease in $PM_{2.5}$ concentrations until approximately 2015, which explains the decrease in mortality in 2015. This decrease of $PM_{2.5}$ is driven by the expected decrease over the next decade in emissions in California for many of the chemical species (NOx, Reactive Organic Gases) that lead to the formation of secondary particulate matter (nitrates and secondary organic compounds). However, in 2035 there is a slight increase in mortality compared to 2006, and a significant increase by 2050 – as many as 45 additional deaths from lung cancer and 258 from cardiopulmonary causes for the A2 scenario as emissions rise again. For the B1 scenario, there appears to be a decrease in mortality – associated with the reduced level of emissions.

This model's predictions should be viewed as conservative estimates, given predicted future increase in both the absolute size of the population in the San Diego region and the relative proportion of this population that is at high risk from life-threatening respiratory illness from air pollution and heat stress.

2.6.2 Ozone air pollution

San Diego County is currently out of compliance with the federal ozone standard (County of San Diego Air Pollution Control District 2009). The effect of hot, sunny days on the generation of ozone air pollution can be demonstrated by relating ozone pollution data in San Diego with temperature. Ozone levels exceeded the state 8-hour standard in San Diego 8% of the time for days with temperatures between 85°F–89°F (29°C–32°C) (Environment California 2007). For days over 90°, the state ozone standard is exceeded 16% of the time. An increase in these hot, sunny days due to climate change has been projected for San Diego in the year 2050 as discussed above and shown in Fig. 6.

2.6.3 Extreme heat events

Heat waves, by far, claim more lives than all other weather related events and disasters.[24] Heat waves are expected to increase in frequency, magnitude and duration in San Diego over the next 50 years. As shown in Fig. 6, the number of days over 97.3°F (36.3°C) in the Miramar area is projected to increase six-fold, accompanied by a projected four-fold increase in the number of days over 93.8°F (34.3°C) for the years 2041–2050. Days over 84°F (28.9°C) are projected to increase from the recorded current average of 78 days to

---

[23] Mortality data available at: http://www.dhcs.ca.gov/dataandstats/Pages/default.aspx.

[24] http://www.noaawatch.gov/themes/heat.php

Table 1 Expected mortality change in 2015, 2035, and 2050 from base year 2004

IPCC SRES	Cause of Mortality	Mortality in 2004	2015			2035			2050		
			NCARPCM1	CRNMCM3	GFDLCM21	NCARPCM1	CRNMCM3	GFDLCM21	NCARPCM1	CRNMCM3	GFDLCM21
SRES A2	Cardiopulmonary	9110	−0.06%	−0.24%	−0.30%	0.96%	0.72%	1.49%	2.53%	1.93%	2.83%
	Lung Cancer	1,187	−0.08%	−0.32%	−.040%	1.28%	0.96%	1.99%	3.39	2.57%	3.79%
	All Cause	19104	−0.04%	−0.16%	−0.20%	0.64%	0.48%	1.00%	1.69%	1.28%	1.89%
SRES B1	Cardiopulmonary	9110	−1.08%	−1.26%	−1.08%	−1.55%	−1.67%	−1.26%	−1.38%	−1.56%	−1.08%
	Lung Cancer	1,187	−1.44%	−1.68%	−1.44%	−2.07%	−2.23%	−1.68%	−1.84%	−2.07%	−1.44%
	All Cause	19104	−0.72%	−0.84%	−0.72%	−1.04%	−1.12%	−0.84%	−0.92%	−1.04%	−0.72%

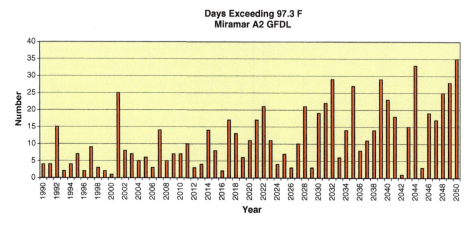

Fig. 6 Days exceeding 97.3°F (36.3°C), Miramar A2 GFDL

129 days during the period 2041–2050 and these hot days are expected to occur from April to December. These results were obtained analyzing the A2 scenario for the GFDL (moderate warming) model. The B1 scenario shows approximately 15% fewer hot days during this period. The resulting public health risks do not affect populations equally, however; certain individuals, populations and communities are at greater risk than others (Knowlton et al. 2009; Kunzli et al. 2006; Iñiguez et al. 2010). A recent analysis of temperatures during summers with no heat waves (1999–2003) found a 3% increase in deaths in any given day for a 10°F (5.6°C) increase in temperature (including humidity) (Basu et al. 2008).

Factors that should be considered when identifying community-level risk include the incidence of relatively high percentages of (1) children under 5 years of age and elderly people 65 and over; and (2) chronically ill persons (especially those suffering cardiovascular or respiratory conditions) (Knowlton et al. 2009; Iñiguez et al. 2010). As stated in the Introduction, in 2050, there will be one million seniors 65 years and older in the San Diego region, roughly equal to nearly one-quarter of the region's total population. San Diego's aging population will likely face more mortality events associated with an increase in temperature due to climate change.

2.6.4 Wildfire impacts on public health

Wildfires can be a significant contributor to air pollution in both urban and rural areas, and have the potential to significantly impact public health through particulates and volatile organic compounds in smoke plumes. Wildfire smoke contains numerous primary and secondary pollutants, including particulates, polycyclic aromatic hydrocarbons, carbon monoxide, aldehydes, organic compounds, gases, and inorganic materials with toxicological hazard potentials (Künzli et al. 2006). Future land use and climate change will exacerbate the risk of wildfires as a result of the alteration of fire regimes in the county. Fires also create secondary effects on morbidity as the result of increased air particulates that can worsen lung disease and other respiratory conditions. People most at risk of experiencing adverse effects related to wildfires are children and individuals with existing cardiopulmonary disease, and that risk increases with advancing age. Two studies conducted after the 2003 southern California wildfires link wildfire-related $PM_{2.5}$ with increased respiratory hospital admissions, especially asthma, in the general population (Delfino et al. 2009) and increased eye and respiratory symptoms, medication use and physician visits among children (Kunzli et al. 2006)

2.6.5 Infectious disease

Climate change in San Diego County could increase the risk of certain vector-borne diseases, while decreasing the risk of others. The occurrence of vector-borne disease is influenced by a variety of factors. Prevailing temperature influences the rate of development of larvae of some vectors, as well as the rate of development of the infectious agent in the vector (NRC 2001; Colwell and Patz 1998). Humidity and rainfall patterns affect both the composition and abundance of arthropod vectors (mosquitoes, fleas, ticks, etc.), as well as animal hosts (Lang 2004). Behavior patterns of hosts, such as indoor living, and vector preferences for particular hosts and periods of peak activity also influence transmission opportunities.

The San Diego region is expected to experience increased public health risks from mosquito-transmitted West Nile Virus (Dudley et al. 2009), especially if there are more intense El Niño cycles in the future (Anyamba et al. 2006) and from rodent-transmitted hantavirus (Yates et al. 2002). Higher temperatures predicted for the region could also facilitate the local establishment of tropical vector-borne diseases such as malaria and dengue fever, while reducing public health risks from the endemic mosquito-transmitted diseases Western Equine Encephalitis and St. Louis Encephalitis (Gubler et al. 2001). Climate warming effects on the geographic and altitudinal ranges and population densities of rodent hosts and flea vectors will also alter the distribution of high-risk areas for plague (*Yersinia pestis*) in the San Diego region (Lang 2004).

Projected surface temperature increases of 1.5°F to 4.5°F (0.8°C to 2.5°C) in 2050 in local lagoons and waterways as well as the nearshore ocean could increase the risk of disease from exposure to harmful algal blooms (red tides), microbes (*Vibrio spp., Listeria monocytogenes, Clostridium botulinum, Aeromonas hydrophila*), and other waterborne agents (Feldhusen 2000; Tamplin 2001).

2.7 Electricity: meeting growing power demand

Electricity consumption in San Diego County has increased steadily since 1982 with the exception of 2000–2001 due to the California energy crisis.[25] Voluntary efforts to reduce consumption have helped San Diego avoid extensive outages since 2001, but more recently consumption increases have resumed and even exceeded pre-crisis levels (see footnote 27). The peak demand in 2006 was the highest on record in the San Diego Gas and Electric (SDG&E) territory, driven largely by cooling loads as a result of high summertime temperatures. This section provides an analysis of future peak temperature data and electricity demand as well as a quantitative analysis of future Cooling Degree Day (CDD) trends and annual electricity consumption. The section also provides an analysis of future high temperature days against two high temperature thresholds that SDG&E uses for critical peak pricing determinations to curtail demand. Lastly, it considers other energy supply (wind energy) and demand (desalination) issues that are relevant to climate change.

---

[25] http://sdenergydata.com/ElectricitySD.html

Fig. 7 Change in peak demand temperatures by 2050

2.7.1 Peak temperature and cooling degree day trends

Cooling degree days (CDD) are the number of days during the year above a reference temperature of 65°F (18°C) – and so are indicative of annual temperature trends rather than daily peak trends. Temperature data from the three climate models were analyzed to generate maps for peak temperature and CDD trends (see Fig. 7). The maps divide the San Diego region into four climate zones,[26] using data from four commonly used temperature station locations (Lindbergh Field, Miramar, El Cajon, and Borrego Springs). For San Diego energy forecasting, the California Energy Commission (CEC) uses temperature data from Lindbergh Field, Miramar, and El Cajon to simulate future demand. Figure 7 shows CDD increases by 2050 for one model (CNRM) and the two emissions scenarios.

2.7.2 Peak demand trends for electricity

Peak summertime temperatures have a well-established relationship to peak electricity demand that utilities use for the purpose of load planning. The study team reviewed actual and predicted peak temperatures and electricity demand data from 1980 through 2050 for the three climate models and two growth scenarios. The analysis was performed using the moderate peak temperature model (GFDL) with the A2 scenario) to correlate historical electricity demand with regional population and the four climate zone temperature trends and compared with a CEC 10-year peak demand forecast through 2018. This comparison showed very good agreement and gives confidence in using the same technique to project demand through 2050.

The forecast shows a dramatic increase of 60%–75% in peak electricity demand by 2050 (see Fig. 8) –an increase of more than 2,500 megawatts (MW) from present levels. The differences between the models account for roughly 7% of the total, or approximately 400 MW. The "base case" on the graph shows what peak demand would be if temperatures did not increase (i.e., demand based on population growth alone).

2.7.3 Annual consumption trends for electricity

Annual electricity consumption forecasts can be quite complex, with many variables influencing the outcome (e.g., economic growth, population, temperature, and efficiency). In order to simplify the analysis, the only variables taken into account were annual temperature and population. The study team converted the annual temperature data into CDDs by averaging the daily data for maximum and minimum temperature included in the climate models to determine the daily average temperature. The daily CDD values were calculated on a Base 65 basis, meaning that a day with a mean temperature of 85°F would correspond to 20 CDDs with a reference temperature of 65°F (85°F–65°F).[27] The daily CDDs were then summed for each year to get the annual values that were used for the analysis. These values were verified through a regression analysis that confirmed that this

---

[26] California has 16 climate zones as defined by the California Energy Commission. These zones represent regions with similar weather characteristics and are used in Title 24 energy analysis and compliance. A map of the climate zones is presented in Appendix M. http://www.energy.ca.gov/maps/building_climate_zones.html

[27] The National Weather Service describes calculating CDDs with a 65°F reference temperature here: http://www.cpc.ncep.noaa.gov/products/analysis_monitoring/cdus/degree_days/ddayexp.shtml

a

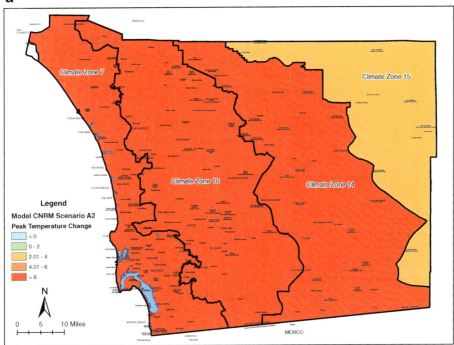

Change in Peak Temperatures by 2050 Model CNRM Scenario A2

b

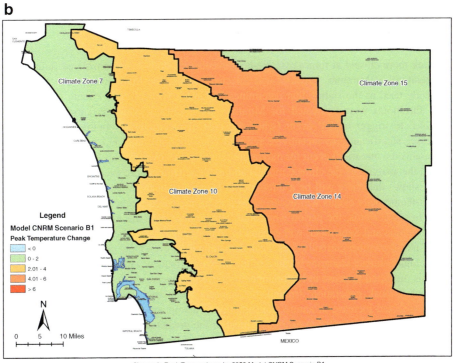

Change in Peak Temperatures by 2050 Model CNRM Scenario B1

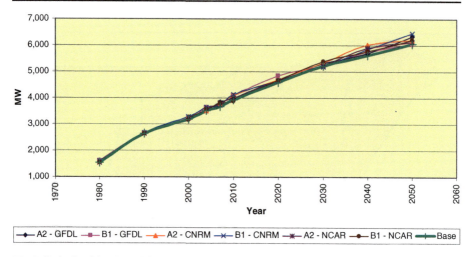

Fig. 8 Peak electricity demand forecast

correlation of population, CDDs, and energy consumption matches closely the CEC's current 10-year forecast. There is a nominal difference in the forecasts based on the model and scenario. This means that assumptions about annual electricity consumption in the forecasting model are primarily population dependent and only marginally temperature-dependent for estimating annual electric consumption. Overall annual electricity consumption is expected to increase of 60%–62% by 2050 compared to current demand. Rising temperatures account for only approximately 2% of the increase in consumption.

2.7.4 Extreme temperature events and impact on system reliability

To look more closely at future extreme-heat events, the study team considered future peak temperatures at Miramar. Miramar was selected because the Marine Corps Air Station is currently used as the station that determines SDG&E's Critical Peak Pricing (CPP) tariff. Two thresholds were evaluated for Miramar: 84°F (29°C), which is currently used by SDG&E to trigger a CPP event (when the cost of electricity to commercial customers increases significantly to incentivize reduced consumption); and 93.8°F (34.3°C), which represents the one-in-10 event for maximum temperatures over the last ten years. In general terms, the analysis of the climate models and extreme temperature events conducted in this study shows that there will be a three-month expansion of the season during which these higher-temperature events will occur, as well as an increased frequency of such events per year. In other words, the period when high temperature days are most frequent, currently between June and September, will expand to May through November. Early November will "feel" like September currently does. Peak demand will be even more challenging to deal with under future climate scenarios because of the increased frequency of extreme-heat events.

This predicted increase in high temperature events appears consistent with the predicted increase in peak generation demand, but appears inconsistent with the seemingly small (2%) predicted increase in annual consumption. This apparent contradiction is partially a result of predicted population patterns – most of the predicted population increase is in the coastal regions of the County which will experience fewer high temperature events than at Miramar. It is also a result of the simplified correlation between population and temperature used to develop the annual consumption forecast. Data from 1980–2004 was used to develop historic correlations, but this period also experienced significant improvements in energy efficiency in California which would have the effect of suppressing the relationship between increasing temperature and consumption.

2.7.5 Wind energy

The availability of wind power may also be affected by climate change, although projected climate impacts on wind are highly uncertain at this time. A 2005 study estimated that there could potentially be as much as 1,500 MW of wind power generated in or near eastern San Diego County (San Diego Regional Renewable Energy Group 2005). This would make up a substantial portion of the increased demand projected (2,500 MW) with non-fossil fuel energy resources. Further research in this area is needed because changes in wind resources are not currently modeled in the global climate scenarios. One such study (Young et al. 2011) indicates a general global trend of increasing values of wind speed and, to a lesser degree, wave height, over the past 23 years. The U.S. Climate Science Program predicted that overall wind-power generation would decrease in the mountain areas of the West, but could increase in California (U.S. Climate Change Science Program 2007). In turn, the impact of large-scale wind power generation on regional climate also still needs examination (Keith et al. 2004).

2.7.6 Desalination demand

The region is exploring seawater desalination as a means of diversifying its supply of water resources to better prepare for a water-constrained future exacerbated by climate change and droughts. A 50 million gallon per day (56,000 acre-feet per year (AFY) reverse osmosis (RO)) seawater desalination plant and associated water delivery pipelines was approved in 2008 and is under construction in Carlsbad. As an energy-intensive process, the development of desalination facilities will bring with it an increase in regional energy demand. Assuming an energy intensity of 4,000 kilowatt-hours per acre-foot (kWh/af) of water produced (California Department of Water Resources 2003), the rise in energy demand attributable to meeting the Water Authority's desalination goals are summarized in Table 2.

Some other regional options such as recycling and local groundwater are less energy intensive, requiring 400 kWh/af and 570 kWh/af respectively, according to a Pacific Institute study (Cohen et al. 2004). That study concluded that satisfying future growth in water demand in the San Diego region via conservation would reduce the overall energy intensity of the Water Authority water supply by 13%. In comparison, satisfying growth in water demand via recycling would reduce overall energy intensity by 4%, while using seawater desalination to satisfy growth would increase overall energy intensity by 5% (California Department of Water Resources 2003 and Table 2). This emphasizes the significance of conservation versus

Table 2 Increases in annual power consumption attributable to saltwater desalination throughout the San Diego region

Year	Scenario	Desalination capacity added to region (acre-feet/year)	Resulting increase in annual power consumption (MWh)
2020	low case	40,000	160,000
	high case	56,000	224,000
2030	low case	56,000	224,000
	High case	89,000	358,400

other regional alternatives for saving energy. The energy savings will also result in less overall GHG emissions from the utility sources providing the power.

3 Conclusions and recommendations

A key message of this study is that the San Diego region is threatened by climate change through a unique combination of impacts on its climate-sensitive sectors. The San Diego region by 2050 will have to concurrently deal with the major challenges of protection against sea level rise, increased risk of large wildfires, increasingly uncertain water supplies from the Sacramento Delta and Colorado River imports, increased energy demands, and public health issues associated with heat waves and an increase in some infectious diseases. Important to note is that while this study is confined to the period between now and 2050, the levels of warming, the amount of sea level rise, and other impacts from climate change are not expected to reach their peaks by mid-century due to the long-lasting accumulation of GHG in the atmosphere (Solomon et al. 2009; IPCC 2007; Hansen 2005; Meehl et al. 2005) as well as the potential for substantial man-made GHG emissions to continue beyond 2050.

An overarching recommendation is that public decision makers and agencies keep moving in a common direction on understanding the climate projections for the region, which in turn should facilitate better joint planning. For example, fire protection agencies, utility planners, and public health planners should have a common understanding of temperature increase expected for the region. Likewise, water agencies and fire prevention agencies should have a uniform understanding about the likelihood of droughts and precipitation patterns. Land use planning agencies will have to deal with the combined challenges of sea level rise in coastal areas, increasing fragmentation of ecosystems, as well as mitigation measures to address local emissions that could require increasing future population centers around transportation corridors. Although this study has focused on adaptation needs, it is important to also recognize the importance of local GHG emission reduction measures as they can create positive health effects as well as provide a local economic stimulus.

An additional overarching recommendation is that a process of quantitatively and qualitatively assessing risks and vulnerabilities from climate change should take place to aid in the development of better public planning and prioritization of resources. This process has already begun for sea level rise issues affecting San Diego Bay with 2011 studies being carried out by the Port of San Diego, five communities around San Diego Bay, The San Diego Foundation,[28] and the U.S. Navy.[29]

Acknowledgments This paper relies heavily on the research conducted in The San Diego Foundation's Regional Focus 2050 Study (Focus 2050 Study) in 2007–2008, which was conceived of and commissioned by The Foundation's Environment Program. The Foundation contracted with the University of California, San Diego's Environment and Sustainability Initiative (ESI) during this period to serve as the project manager for the Focus 2050 Study and The Foundation was the project manager for the subsequent Public Interest Energy Research (PIER) Program study that was incorporated into the California Climate Change Center's Second Biannual Assessment of the implications of climate change for the State of California. A team of 40 experts from the region including universities, nonprofit organizations, local governments, public sector agencies and private sector entities collaborated to produce this document. It drew upon current scientific analyses from an array of experts in climate science, demography and urban/regional planning, water, energy, public health and ecology. The Focus 2050 Study for the San Diego region was modeled, in part, on the Focus 2050 study undertaken by King County, Washington.

---

[28] http://www.signonsandiego.com/news/2011/feb/16/king-tides-offer-look-rising-sea-levels/

[29] http://meso.spawar.navy.mil/Projects/SL_rise/index.html

References

Anyamba A, Chretien JP, Small J, Tucker CJ, Linthicum KJ (2006) Developing global climate anomalies suggest potential disease risks for 2006–07. Int J Health Geogr 5:60

Barnett T, Pierce D (2008) When will lake mead go dry? J Water Resour Res 44:W03201

Basu R, Feng W, Ostro B (2008) Characterizing temperature and mortality in nine California counties. Epidemiology 19:138e45

Bonfils C, Duffy PB, Santer BD, Lobell DB, Phillips TJ, Wigley TML, Doutriaux C (2007) Identification of external influences on temperatures in California. Clim Chang 87(Suppl 1):S43–S55. doi:10.1007/s105840-007-9374-9

California Department of Water Resources (2003) Water desalination: findings and recommendations

Cayan DR, Bromirski PD, Hayhoe K, Tyree M, Dettinger MD, Flick RE (2008a) Climate change projections of sea level extremes along the California Coast: Special issue on California climate scenarios. Clim Chang 87(Suppl 1):S57–S73. doi:10.1007/s10584-007-9376-7

Cayan DR, Maurer EP, Dettinger M, Tyree M, Hayhoe K (2008b) Climate change scenarios for the California region. Clim Chang 87(suppl 1):S21–S42

Chao BF, Wu YH, Li YS (2008) Impact of artificial reservoir water impoundment on global sea level. Science 320:212–214

Chelton DB, Bernal PA, McGowan JA (1982) Large-scale interannual physical and biological interactions in the California current. J Mar Res 40:1095–1125

Christensen N, Lettenmaier DP (2007) A multi-model ensemble approach to assessment of climate change impacts on the hydrology and water resources of the Colorado River basin. Hydrol Earth SystSci 3:1–44

Christensen NS, Wood AW, Voisin N, Lettenmaier DP, Palmer RN (2004) Effects of climate change on the hydrology and water resources of the Colorado River basin. Clim Chang 62:337–363

Cohen R, Nelson B, Wolff G (2004) Energy down the drain: the hidden costs of California's water supply. National Resources Defense Council and The Pacific Institute

Colwell RR, Patz JA (1998) Climate, infectious disease and health: an interdisciplinary perspective. American Academy of Microbiology, Washington, D.C

County of San Diego Air Pollution Control District (2009) Annual Report. Air Quality in San Diego. Available online at http://sandiegohealth.org/air/sdapcd_annual/annual2009.pdf

Delfino RJ, Brummel S, Wu J, Stern H, Ostro B, Lipsett M, Winer A, Street DH, Zhang L, Tjoa T, Gillen DL (2009) The relationship of respiratory and cardiovascular hospital admissions to the southern California wildfires of 2003. Occup Environ Med 66:189–197

Department of Water Resources (2006) Progress on Incorporating Climate Change into Planning and Management of California's Water Resources—Technical Memorandum Report, California Department of Water Resources, Sacramento, Calif

Dockery DW, Pope CA 3rd, Xu X, Spengler JD, Ware JH, Fay ME et al (1993) An association between air pollution and mortality in six U.S. cities. New Engl J Med 329:1753–1759

Dudley JP, Gubler DJ, Enria DA, Morales MA, Pupo M, Bunning ML, Artsob H (2009) West Nile Virus in the New World: Trends in the spread and proliferation of West Nile Virus in the Western Hemisphere. Zoonoses and Public Health 56(6–7):357–369

Environment California Research and Policy Center (2007) Hot and smoggy: the ozone – hot weather connection in eight California cities. Los Angeles, CA

Feldhusen F (2000) The role of seafood in bacterial foodborne diseases. Microb Infect 2:1651–1660

Ferris B Jr, Higgins I, Higgins M, Peters J (1973) Chronic nonspecific respiratory disease in Berlin, New Hampshire, 1961 to 1967: a follow-up study. Am Rev Respir Dis 107:110–122

Galbraith H et al (2002) Global climate change and sea level rise: potential losses of intertidal habitat for shorebirds. Waterbirds 25:173–183

Gubler DJ, Reiter P, Ebi KL, Yap W, Nasci R, Patz JA (2001) Climate variability and change in the United States: potential impacts on vector- and rodent-borne diseases. Environ Health Perspect 109(Suppl 2):223–233. Available at: www.ehponline.org/docs/2001/suppl-2/223-233gubler/gubler.pdf

Hansen JE (2005) A slippery slope: how much global warming constitutes 'dangerous anthropogenic interferenced'? An editorial essay. Clim Change 68:269–279. doi:10.1007/s10584-005-4135-0

Harley CDG et al (2006) The impacts of climate change in coastal marine systems. Ecol Lett 9:228–241

Hodgkin JE, Abbey DE, Euler GL, Magie AR (1984) COPD prevalence in nonsmokers in high and low photochemical air pollution areas. Chest 86:830–838

Hoerling MP, Eischeid JK (2007) Past peak water in the southwest. In: Special Issue. Inconvenient hydrology? Southwest Hydrology 6(1)

Hughes M, Hall A, Kim J (2011) Human-induced changes in Wind, Temperature and Relative Humidity during Santa Ana events. Clim Change 109 (Suppl 1)

Iñiguez CF, Ballester J, Ferrandiz S, Pérez-Hoyos M, Sáez AL (2010) Relation between temperature and mortality in thirteen Spanish cities. Int J Environ Res Publ Health 7(8):3196–3210

IPCC (2007) In: Solomon S, Qin D, Manning M, Chen Z, Marquis M, Averyt KB, Tignor M, Miller HL (eds) Climate change 2007: the physical science basis. Contribution of Working Group I to the Fourth Assessment Report of the Intergovernmental Panel on Climate Change. Cambridge University Press, Cambridge

Kaiser R, Romieu I, Medina S, Schwartz J, Krzyzanowski M, Kunzli N (2004) Air pollution attributable postnatal infant mortality in U.S. metropolitan areas: a risk assessment study. Environ Health 3:4

Keith DW, DeCarolis JF, Denkenberger DC, Lenschow DH, Malyshev SL, Pacala S, Rasch PJ (2004) The influence of large-scale wind power on global climate. Proc Natl Acad Sci U S A 101(46):16115–16120

Knowlton K, Rotkin-Ellman M, King G, Margolis HG, Smith D, Solomon G, Trent R, English P (2009) The 2006 California heat wave: impacts on hospitalizations and emergency department visits. Environ Heal Perspect 117:1

Krewski D, Burnett RT, Goldberg MS, Hoover K, Siemiatycki J, Jerrett M, Abrahamonwicz M, White WH (2000) Reanalysis of the Harvard six cities study and the American Cancer Society study of particulate air pollution and mortality. Special Report. Health Effects Institute, Cambridge

Künzli N et al (2006) Health effects of the 2003 southern California wildfires on children. Am J RespirCrit Care Med 174:1221–1228

Lang JD (2004) Rodent-flea-plague relationships at the higher elevations of San Diego County, California. Journal Vector Ecology 29:236–247. Available at: www.sove.org/Journal%20PDF/December%202004/5Lang%2003-44.pdf

Lenton TM, Held H, Kriegler E, Hall JW, Lucht W, Rahmstorf S, Schnellnhuber HJ (2008) Tipping elements in the Earth's climate system. Proc Natl Acad Sci 105(6):1786–1793

MacCall A et al (2005) Report of the study group on the fisheries and ecosystem responses to recent regime shifts. In: King JR (ed) PICES Scientific Report No. 28, 162 pp. Available at: www.pices.int/publications/scientific_reports

MacDonald GM, Konstantine V, Hidalgo HG (2008) Southern California and the perfect drought: simultaneous prolonged drought in southern California and the Sacramento and Colorado River systems. Quarternary Int 188:11–23

Meehl GA, Washington WM, Collins WD, Arblaster JM, Hu A, Buja LE, Strand WG, Teng H (2005) How much more global warming and sea level rise? Science 307(5716):1769–1772

Miller NL, Schlegel NJ (2006) Climate change-projected Santa Ana fire weather occurrence. California Climate Change Center: Sacramento, California. CEC-500-2005-204-SF

Milly PCD, Dunne KA et al (2005) Global pattern of trends in streamflow and water availability in a changing climate. Nature 438(7066):347–350

Myers N, Mittermeier RA, Mittermeier CG, da Fonseca GAB, Kent J (2000) Biodiversity hotspots for conservation priorities. Nature 403:853–858

National Research Council (2001) Under the weather: climate, ecosystems, and infectious disease. National Academy Press, Washington, D.C

Pope CA III et al (1995) Particulate air pollution as a predictor of mortality in a prospective study of U.S. adults. Am J Respir Crit Care Med 151:669–674

Preston KL, Rotenberry JT (2007) California State Department of Transportation and Center for Conservation Biology: WRC MSHCP Niche Model Task Order. June 1, 2007. Center for Conservation Biology

Preston KL, Rotenberry JT, Redak R, Allen MF (2008) Habitat shifts of endangered species under altered climate conditions: importance of biotic interactions. Glob Chang Biol 14:2501–2515. doi:10.111/j.1365-2486.2008.01671.x

Rahmstorf S (2007) A semi-empirical approach to projecting future sea-level rise. Science 315(5810):368–370

Rodriguez JP, Roberts M, Dobson A (1997) Where are endangered species found in the United States? Endangered Species Update. University of Michigan School of Natural Resources and Environment

Rogers-Bennet L, Dondanville RF, Moore JD, Vilchis LI (2010) Response of red abalone reproduction to warm water starvation and disease stressors: implications of ocean warming. J Shellfish Res 29(3):599–611

Romieu I, Ramirez-Aguilar M, Moreno-Macias H, Barraza-Villarreal A, Miller P, Hernandez-Cadena I, Carbajal-Arroyo LA, Hernandez-Avila M (2004) Infant mortality and air pollution: modifying effect by social class. J Occup Environ Med 46:1210–1216

Rykaczewski RR, Checkley DM Jr (2008) Influence of ocean winds on the pelagic ecosystem in upwelling regions. Proc Natl Acad Sci 105:1965–1970

San Diego Regional Renewable Energy Group (2005) Potential for renewable energy in the San Diego Region. San Diego, California. Available at: www.renewablesg.org

SANDAG (2010a) Population and housing estimates: San Diego Region. San Diego, California. Available at: http://profilewarehouse.sandag.org/profiles/est/reg999est.pdf

SANDAG (2010b) Board Report – 2050 Regional Growth Forecast. San Diego, California. Available at: www.sandag.org/uploads/projectid/projectid_355_10794.pdf

Solomon S, Plattner G-K, Knutti R, Friedlingstein P (2009) Irreversible climate change due to carbon dioxide emissions. Proc Natl Acad Sci 106(6):1704–1709

Spracklen DV, Mickley LJ, Logan JA, Hudman RC, Yevich R, Flannigan MD, Westerling AL (2009) Impacts of climate change from 2000 to 2050 on wildfire activity and carbonaceous aerosol concentrations in the western United States. J Geophys Res 114:D20301. doi:10.1029/2008JD010966

State of California, Department of Finance (2007) Population Projections for California and Its Counties 2000–2050, Sacramento, California. July

Tamplin ML (2001) Coastal Vibrios: identifying relationships between environmental condition and human disease. Hum Ecol Risk Assess 7:1437–1445

The San Diego Foundation (2003) Community Needs Assessment Report. San Diego Regional Disaster Board After-the-Fire Fund, p. A2. (390,000 acres burned in 2003 fires) U.S. Climate Change Science Program. 2007. Effects of Climate Change on Energy Production and Use in the United States, Synthesis and Assessment Product 4.5. Available at: http://www.sc.doe.gov/ober/sap4-5-final-all.pdf

The San Diego Foundation (2007) Community needs assessment update. Helping the San Diego region recover and rebuild. Diego, p. 5. (369,000 acres burned in 2007 fires)

U.S. Climate Change Science Program (2007) Effects of climate change on energy production and use in the United States. Department of Energy, Office of Biological & Environmental Research, Washington, DC

Vicuna S (2006) Predictions of Climate Change Impacts on California Water Resources Using CALSIM II: A Technical Note, California Climate Change Center, Berkeley, CA

Westerling AL, Hidalgo HG, Cayan DR, Swetnam TW (2006) Warming and earlier spring increase in western U.S. forest wildfire activity. Science 313(5789):940–943

Woodruff TJ, Grillo J, Schoendorf KC (1997) The relationship between selected causes of postneonatal infant mortality and particulate air pollution in the United States. Environ Heal Perspect 105:608–612

Yates TL et al (2002) The ecology and evolutionary history of an emergent disease: hantavirus pulmonary syndrome. BioScience 52:989–998

Young IR, Zieger S, Babanin AV (2011) Global trends in wind speed and wave height. Science, published online March 24, 2011

Zhu T, Jenkins MW, Lund JR (2005) Estimated impacts of climate warming on California water availability under twelve future climate scenarios. J Am Water Resour Assoc 4(5):1027–1038

Index

A

Airborne LIDAR Assessment of Coastal Erosion (ALACE), 234
Akaike Information Criterion (AIC), 452
Albedo, snow, 97
    air temperature, 104
    anthropogenic emissions, 101
    fractional snow cover, 100
    Noah land-surface scheme, 99
    physical processes, 100
    satellite observations, 100
    sensible heat flux, 104, 107
    sensitivity study, 99
    snowfall and SWE, 101
    snowmelt, 105–108
    solar energy, 104
    surface insolation, 102–103
    SWE, 104, 105
    terrain elevations, 101
Atmospheric-oceanic general circulation models, 467–469

B

Base flood elevations (BFEs)
    coastal flood elevations and hazards, 254, 258
    FEMA, 259
    flood insurance rates, 259
    NGVD, 259
    TWL, 264–265
Beach choice and attendance model
    data estimation, 282
    demographic and economic projections, 284–285
    economic analysis, 278–279
    formulation, 282–284
    limitations, 295
    width projections, 284
Bias corrected spatial downscaling (BCSD), 3, 46–47
Big Creek system, 152

C

California agricultural landscape. *See also* Economic impacts; Extreme climate impact
    agricultural vulnerability, 409, 410
    agrobiodiversity, 415–416
    climate change responses
        chill hours computing, 412

        crop pets, 412–413
        DAYCENT model, 411
        decision tools and community strategies, 421–422
        elevated CO_2 effects, 412
        environmental factors, 412
        heat wave and drought effect, 411
        management practices, 410–411
        temperature effect, 409–410
        water needs, crop, 413–414
    climate response strategy
        biomass utilization, fuel production, 415
        carbon sequestration, 415
        cover cropping, 414
        crop breeding and diversity, 414
        farmscaping, 414–415
        irrigation and fertilizer use, 414
        manure management, 414
        organic production, 415
        tillage, 414
    GHG emissions and adaptation strategy, 408–409
    landscapes and land use options
        agrobiodiversity, 420
        climate change effects, 418–419
        crop diversification and practices, 419
        economic benefits, 420–421
        flood risk, 418
        GIS approach, 418
        horticulture development, 417–418
        mitigation benefits, 420
        soil erosion, 419–420
        wild species impacts, 421
    research goals, 409, 410
    soil and land management, climate mitigation, 416–417
    Yolo County
        location, 408, 409
        AB 32-Plus, precautionary change, 422–423
        A2 regional enterprise, 422
        B1 global sustainability, 422
        climate response, 422
        soil survey data, 419
California almond production
    February Tmin and state average, 328
    least square models, 328–329
    production, 318, 319
    statistical yield models, 323–324

temperature effect, 328
variety switching, 330–332
California Applications Program/California
Climate Change Center (CAP/CCCC), 431
California Borderlands, 213
California driving, future cost
cost, 36–37, 39
electricity prices
GHG estimation and regulation, 38, 39
oil price, 38
optimistic and pessimistic interpretation, 40
prices and demand, 37
statewide average residential, 38, 39
household income, 36
California economic growth
A2 and B1 scenarios, 28–29
historical and SRES projections, 26, 28
per capita terms, 31–32
California economic trend. *See* Socioeconomic and demographic trends
California ecosystem services
attributes, services and values
carbon sequestration, 474–476
ecosystem extent and distribution, 472–474
forage production, 475, 477
livestock, 477–478
social cost of carbon, 475
biophysical models
carbon sequestration, 468–469
climate data and projections, 467–468
ecosystem extent and distribution, 468
forage production, 469–470
climate stabilization, 466
cultural service, 466
definition, 466
economic
carbon sequestration valuation, 470
effects, 480–481
forage valuation, 470–471
impacts, 479–480
market value, 479
natural, non-irrigated forage, 477, 478
provisional services, 466
regulating services, 466
state, national and global economy, 467
supporting services, 466
terrestrial ecosystem, 476, 478
timing, magnitude and geographic distribution, 467
California Energy Commission (CEC), 253, 524
California Ocean Protection Council, 14
California perennial crops
adaptation options assessment, 319
agriculture, 317–318
almond
February Tmin and state average, 328
least square models, 328–329

production, 318, 319
statistical yield models, 323–324
temperature effect, 328
variety switching, 330–332
climate impact, 318, 329–330
climate projection, 319
crop and weather data analysis, 319
EPIC or CERES model, 318
fixed-effects, 326
Lasso models
temperature coefficients, 324–325
training and test datasets, 324
warming, 325
model beneficial effect, 329
pollination period, 318–319
predicted change, Lasso-regression tree model, 327
regression-Lasso tree models comparison, 326–327
statistical yield models
bootstrap analysis, 322
climate change impacts, 323
Lasso model, 321–322
regression tree modeling, 321
statewide gross cash income, 320, 321
steps, model yield response, 322–323
temperature and pressure measurement, 320
weather effect, 318
California population projections
climate-induced changes, 29
demographic component, 31
migration and fertility, 30
population forecasting, 29
statewide population, 29–30
and United States, 30, 31
California residential electricity consumption
adaptation simulation, 208–209
annual panel data, 193–194
anthropogenic climate change, 192
climate impacts, 193
demand estimation, 192
econometric estimation
bin definition and billing period, 199
climate change impacts, 198
consumption and temperature relationship, 199
maximum likelihood approach, 200
monthly household bill, 198
non-CARE households statistics, 197, 199
response functions, climate, 200, 201
temperature variables, 199
global warming, 193
large-scale bottom-up simulation models, 192–193
population and temperature simulation, 207–208
price and temperature simulation, 205–206
residential billing data

CEC climate zones, 195, 196
daily consumption, 196
econometric estimation, 195
households services, 194
temperature impact, 194–195
zip codes and weather station, 194, 195
temperature simulation
climate change effect, 201–202
GCM, 202
NCAR model simulation, 202
per household consumption, 204–205
response functions, climate, 201
weather-consumption and climate realtionship, 202–203
zip code, electricity consumption, 202–204
time series variation, 193
weather data, 196–198
California timberlands climate impact
A2 emissions scenarios
carbon market economic impact, 437–438
climate adaptation, 439
land use and policy implications, 439–441
PCM1 and GFDL scenario, 436–437
price effects, climate change, 437
production and price effects, 436
statewide economic impact, 438–439
domain and climatology, 431–432
ecological models, 432–433
economic and ecology consequences, 430
biological model integration, 434
carbon payment, 434–436
global timber prices, 434
management scenarios, 433
model description, 433
GCM data, 431
1989-2009 graph, timber production, 430
warming effect, 430
California urbanization projections
footprint layers, 33–34
population allocation, 34–35
selective metropolitan areas, 35–36
CALifornia Value Integrated Network (CALVIN) model, 7–8, 13
agricultural and urban water demands
hydrologic basins, 135–136
warm-dry and warm-only climate scenarios, 137–138
water scarcity, 136–137
groundwater storage, 139
hydro-economic optimization model, 135
permutation, 139
perturbed hydrology, 138
reservoir evaporation, 139
California water system
CALVIN model
hydrologic basins, 135–136

warm-dry and warm-only climate scenarios, 137–138
water scarcity, 136–137
groundwater storage, 139
hydro-economic optimization model, 135
permutation, 139
perturbed hydrology, 138
reservoir evaporation, 139
climate change impact, 133–134
hydrologic implication and records, 134–135, 147
hydropower, 145–146
limitations, 146–147
runoff reduction, 147
storage operations
climate affects, 141
filling surface water, 143
groundwater storage and use, 143–145
hydrology shift, 141–142
monthly aggregated statewide, 141–142
warm-only and warm-dry hydrology, 142–143
water supply
pay and percent water delivery, 139, 140
statewide water scarcity, 140
storage capacity, 140–141
water scarcity and costs estimation, 139–140
water supply, temperature and precipitation, 135
California wildfire
burned area, 457
CNRM CM3 and GFDL CM2.1 model, 459, 461
data and methods
domain and resolution, 447
fire history, 447–448
emissions scenarios, 457, 458
fire modeling
expected burned area, 454
Forest Service, 453
growth rates and allocation, 454
logistic regressions, 452
occurrence, 452, 453
probability, 452–453
relative humidity, 452
vegetation fraction and population, 452
fire severity, 446
flammability and fire frequency, 446
GPD, 455–456
growth and sprawl, 459
human-induced climatic change, 446
land surface characteristics
climate and hydrologic data, 450–451

ICLUS growth and development
scenarios, 449–450
protection responsibility, 449
topography, 448–449
vegetation, 448
lightning-caused ignitions, 456
logistic regression model, 454, 455
occurrence, 447
probabilistic models, 456
Santa Ana (SA) winds, 456
Sierra Nevada range, 458, 461
SRES A2 scenarios, 457, 458
wildland-urban interface, 446
Carbon payment, S434–S436
Carbon sequestration
carbon pool, 468–469
CCSM3, 474–475
climate stabilization, 466
economic value, 481
GFDL model, 474–475
market value, 479
MC1 model, 468–469
MTC, 470
natural ecosystems, 475
net economic value, 467
neutral climate future scenario, 474
PCM1 projects, 474
potential economic effects, 480
SCC, 470
terrestrial ecosystems, 478
vegetation distribution, 474
Center for Conservation Biology (CCB), 517
Central Valley agricultural water management
adaptation strategy, 304–305
California water resources, 300–301
CalSim-II and CALVIN, 303
climate sequences and analysis, 304, 306–307
cropping patterns, 305–306
flood conveyance systems, 302
hydrologic analysis
cropping pattern, 311, 313
crop water requirements, 314
demand analysis, 309–310
drought occurrence, 307–310
groundwater pumping, 310, 312, 315
Logit scenarios, 313
reservoir inflows, 307, 309
Sacramento valleys, 309, 310
San Joaquin Valley, 309, 310, 313
surface water delivery, 312, 314–315
hydrologic process, 303
irrigation efficiency, 305
potential implications, 315
system-wide water management, 302
water resources and contract impacts, 300–301, 303
watershed, 306
WEAP modeling system, 301
Central Valley crop production

A2 and B1 emission scenarios, 337
atmospheric CO_2 concentration, 336
climate change impacts, 336–337
cropland shift and extension, 336
data acquisition
climate data, 338
crop types and parameters, 339
management data, 339
soil data, 338–339
data analysis, 341–342
DAYCENT model, 337–338
high MSD, alfalfa and sunflower, 342
modeled *vs.* observed yields, 342
modeling strategies
crop rotations, 340–341
fertilization effect, 341
field crops, 339
rising CO_2 effect evaluation, 341
soil organic matter pools, 339–340
relative surface area, 340
temperatures effects, 342, 344
yield changes
county-level yield patterns, differences and changes, 348–350
elevated CO_2 effects, 350
between emission scenario, 346–348
within each emission scenario, 344–346
Central Valley Production Model (CVPM), 305, 391–392
Channel Island, 213–214
Climate Adaptation Strategy (CAS), 14
Climate change and mitigation policies, 486
cap-and-trade policies, health concern
AB 32 measures co-benefits, 495–496
fuel switching, 496–497
low-SES community, 494–495
mitigation measures, 496
SCAQMD, 495
climate gap
cumulative impacts, decision-making, 498–499
GIS mapping, 498
impacts and co-pollutants, 497
rates and impacts, 497–498
disproportionate economic impacts
agricultural employment, 492–493
basic necessities price, 491–492
infrastructure, SES, and insurance access, 493–494
economic costs and benefits, 494
extreme weather events-heat, 486–487 (*see also* Extreme weather events-heat)
households without access, 490
Climatic indicators
BCSD, 49, 53–54
CAD, 49
CDD, 49–54, 59
definition, 47

downscaled simulations and method, 51, 52
extremes and impact events, precipitation,
47, 48
frost days, 49–53, 55
heat wave, 49–53, 56–57
HPF, 50, 54
precipitation intensity, 49–53, 60
SOM, 49
SRES, 48
temperature extremes and impact events,
47, 48
warmest three nights, 49–53, 58
Coastal Data Information Program (CDIP), 510
Coastal flooding and erosion
analytical result
airports, 234–243
buildings and contents property,
243–245
emergency and healthcare facility,
237–238
hazardous materials sites, 238–239
limitations, 245–246
population at risk, 237
power plants, 240, 241
roads and railways, 239–240
seaports, 240–234
wastewater treatment plants, 240, 242
effects, sea-level rise, 230
GIS technique, 231
infrastructure impacts determination,
235–237
Pacific coast mapping, 233–234
population impacts estimation, 235
property losses and death, 230
protection, risk and cost, 231
San Francisco Bay mapping, 235
sea-level rise projections, 232–233
100-year flood, 231
Coastal Geomorphic Erosion Model (CGEM)
erosion/accretion portion, 225
longshore sediment transport formulation,
214
Santa Barbara, 221
Torrey Pines, 222–223
Coastal hazards. *See* Sea-level rise
Colorado River, 515
Community Climate System Model (CCSM),
121
Constant Elasticity of Supply (CES), 391
Constructed Analog Downscaling (CAD), 47, 49
Constructed Analogues (CA), 3
Cooling Degree Day (CDD), 523, 524, 526

D
Digital elevation models (DEMs), S234
Discrete continuous choice (DCC) model, 172
Dunes
deepwater wave sites, 261, 262
DHZ, 260
erosion response model, 260, 261

LIDAR data, 260
storm event, 260
toe extended inland, 261

E
Economic impacts
aggregate agricultural profits, 377,
379–380
agriculture
economics and agronomy, 366
production, 387–388
profits and crop yields, 366
base model, 392
beta and tau parameters, 390
CALVIN model, 389
caveats, 384–385
Central Valley, 388
CES production, 390–391
climate change threatens, 365
climate-related yield changes, 392,
394–396
CO_2 concentrations, 366
corn production, 366
county-level fixed effect, 368
crop
demands shifts, 393–394, 396
growth models, 366
prices and agricultural revenues
changes, 398–399
production per acre, 381–384
yields, 388
C3 species, 389
CVPM regions, 392
dairy herd silage, 391
data sources
B1 scenario, 376
climate change predictions, 374–375
crop production and yields, 373
degree-days and precipitation
variables, 371, 375
farm revenues, expenditures, and
profits, 370–373, 376
historical weather data, 373–374
precipitation variables, 370, 375
projected changes, 375
soil quality data, 373
economic costs, 388
farm profits, 377–381
geography and socio-economic conditions,
388
GFDL A2 scenario, 388
hedonic approach, 367
hydro-economic optimization model, 389
irrigated crops, pasture and grains, 392,
399
irrigation, 367
land availability changes, 392, 393, 396
land use surveys, 391
limitations and extensions, 401–402
livestock and dairy products, 381

mathematical programming model, 402
methodology
    agricultural profits and annual weather realizations, 368–369
    degree-days and precipitations, 370, 371
    estimation, 370
    IMPACT$_c$, 370
    temperature and precipitation, 368
    weather expectations, 369
net effect, 388
orchards, 399
per acre, farm profits, 366, 377, 378
PMP, 389–390
price, production, revenue percentage change, 398–399
production factor, 390
Ricardian approach, 367
runoff, agricultural land use, applied water and revenue, 400, 401
sensitivity analysis, 400–402
SWAP, 389, 391, 396
technological change, 393
total land use, 396–398
total water use, 397, 398
water supply and availability, 389, 395, 396
weather data, 367
Electricity and natural gas price impact
    block pricing, 173–174
    climate change effect, 172
    DCC model, 172
    discrete continuous choice models
        consumption analyzing, 186–187
        linked consumption, 184–186
        price elasticity demand, 183–184
    elasticity of demand
        DCC models, 174–175
        electricity, 175
        natural gas, 175
    household demand, 174
    linked demand estimation, 182
    price elasticity demand and estimation
        addressing linked consumption, 177–178
        GIS techniques, 179
        global warming, 179–180
        household-level data, 178
        joint demand, 180–181
        linear and log-log models, 179, 180, 181
        log-log demand function, 176–177
        negative and significant effect, 179
        price structure, 177
        statistics data, 179
    residential energy demand estimation
        household electricity and natural gas data, 187–188
        processing, 188
    residential use, 172–173
    water demand, 182

Elevated CO_2 effects, 350
El Niño/Southern Oscillation (ENSO), 509
ESRI's ArcGIS coastal mapping
    Pacific coastal mapping, 233–234
    San Francisco Bay, 235
Extreme climate impact
    adaptation options assessment, 362
    cold extremes, 362
    crop
        insurance system, 356–357
        shifting, 356
        and total indemnity payments, 357, 359
    disaster payment estimation, 357
    farming associated risk, 356
    global warming, 355–356
    indemnity payments and loss estimation, 358, 359
    NOAA dataset, 358–359
    NOAA Storm Event database, top 10 events, 360–361
    rainfall extremes, 362
    storm event, 357–358
    temperature or rainfall, 355
    warming, 362
Extreme impact events
    anthropogenic climate change, 44
    climate-related impact and events, 43–45
    events and projections, 44–45
    EVT analysis, 68
    formal vulnerability assessment, 69
    frequency and intensity, 68
    projections
        climatic indicators (see Climatic indicators)
        downscaled climate, 46–47
        PIER, 46
    return level analysis
        BCSD and CAD simulations, 64
        current climate simulations, 61–63
        downscaled model simulations, 59
        emissions scenarios, 63
        EVT, 56
        GEV distributions, 56–57, 60
        inter-model spread, 65, 66
        maximum likelihood calculations, 67
        precipitation trends, 59
        projected return periods, 63, 64
        statistical analysis, 57
        temperature values, 65–66
    temperature changes, 44, 68
    weather events, 44
Extreme weather events
    air pollution, 490–491
    drought, 43, 44
    extrinsic risk factors, 487–488
    floods, 43, 44, 336, 355, 486
    GHG emission scenarios, 489
    heat-wave mortality, 489, 490
    intrinsic risk factors, 487
    land cover characteristics, 487–489

material and socioeconomic deprivation, 489
storms, 355, 493
weather and climate fluctuations, 488
wildfire, 43, 44, 491, 493

F
Forest Service, 453

G
General circulation model (GCM), 301, 304, 431
Generalized Pareto distributions (GPDs), 454–456
Geographic information system (GIS), 179, 418
Global climate models and downscaling, 75–76
Gossypium hirsutum L., 339
Groundwater storage and use, 143–145

H
Heavy Precipitation Fraction (HPF), 50, 54
High-elevation hydropower systems
    Big Creek system, 152
    climate change effects, 152
    global greenhouse gas emissions effects, 151–152
    hydrologic conditions forecasting, 160–162
    hydropower system operations impact
        average monthly release, 163–164
        climate change impact, 165
        energy generation and revenues, 163
        flooding downstream, 166
        heat waves comparison, 165–166
        linear correlations, runoff timing affects, 163–164
        outputs comparison, 162
        runoff timing affects, 163, 164
        spills and storage, average monthly, 163–164
    UARP and Big Creek system operations, 162, 163
        annual generation, monthly percent, 157, 159–160
        annual runoff, 157, 159
        components, major characteristics, 155–157
        daily time series, 152
        electricity generation demand, 153
        energy-and-storage-driven model, 153
        future climate change scenarios, 154
        interannual variability, 157, 159
        ISO spot market price, 157, 159
        location, 154, 155
        monthly energy generation, 157
        multistep linear programming, 152–153
        reservoir inflows forecasting and construction, 153–155

I
Integrated Climate and Land Use Scenarios (ICLUS), 449–450

Intergovernmental Panel on Climate Change (IPCC), 22, 44, 232, 506

L
Land Data Assimilation System (LDAS), 448
Land use and policy implications, 439–441
Light Detection and Ranging (LIDAR), 234
Longshore sediment transport patterns
    angle of incidence, 217–218
    bathymetry, 212
    breaker zone, 214–215
    CERC formula, 216–217
    coastal landforms, 212
    drift divergence, 214, 219
    Earth's current climate trend, 226
    geomorphic and oceanographic setting
        California Borderlands, 213
        PDO, 214
        SCB, 212, 213
        tectonic processes, 212
        wave climate and energy, 213, 214
    oceanographic and geomorphic changes, 212
    ocean wave climate, 212
    SntBrb nest
        CDIP, 221
        CGEM results, 220, 221
        chronic sand loss, 220
        deep water wave direction, 219
        experiment, 221, 222
        Goleta Beach, 220
        SWAN transformed wave field, 217, 221
    TorPns nest
        beach, 222
        CGEM results, 220–222
        experiment, 224
        submarine canyons, 222
        SWAN transformed wave field, 218, 222
    wave energy flux, 218–219
    wave height and peak directions, 216
    wave transformation model, 214–218

M
Medicago sativa L., 339
Medieval Climate Anomaly (MCA), 509
Medieval Warm Period (MWP), 509
Metric ton of carbon (MTC), 470
Mission Beach, 510, 512–513

N
National Center for Atmospheric Research (NCAR), 121, 202
National Centers for Environmental Prediction (NCEP), 4
Natural and managed systems assessment
    CALVIN, 16
    CAS, 14

climate projections, 15
climate scenarios
    A2 emissions scenario, 3
    BCSD, 3
    B1 emissions scenario, 3
    CA, 3
    ENSO, 3
    GCMs, 2, 3
    greenhouse gases, 2
    precipitation changes, 4–6
    sea-level rise, 6
    warming, 3–4
executive order, 14
heterogeneous impacts, 15
impacts
    agriculture, 8–9
    climate change and policy, 13
    coastal resources, 9–10
    ecosystem services, 10–11
    energy demand and hydropower generation, 12–13
    public health, 12
    regional impacts foci, 13
    timber industry and wildfires, 11
    water supply, 6–8
policy decisions and development, 13, 14
sea-level rise and climate scenarios, 2
Special Issue, 2
SWAP model, 16
Natural forage production, livestock
    AOGCMs, 469
    AUM, 471
    B1 and A2 scenarios, 477
    dry matter, 471
    grasslands distribution, 466
    land cover model, 471
    natural, non-irrigated, 478
    non-irrigated, 470
    NRCS, 469
Natural Resources Conservation Service (NRCS), 469

O
Ocean bathymetry, 215
Oryza sativa L., 339

P
Pacific coast mapping, 233–234
Parallel Climate Model version 1 (PCM1), 432
Positive Mathematical Programming (PMP), 389–390
Potential economic impacts
    beach choice and attendance model
        data estimation, 282
        demographic and economic projections, 284–285
        formulation, 282–284
        limitations, 295
        width projections, 284
    beach economic analysis, 278–279

    climate change impact, 280–281
    coastal erosion and accretion, 278
    economic value, 279–280
    extremely stormy years, 290–291, 296
    inundation annual estimation
        expenditures and annual consumer, 285
        high growth *vs.* low growth, 286–290
        permanent beach, 285
        uneven impact, permanent beach loss, 286
    mean surface air temperature, 279
    mitigating beach loss, cost estimation
        annual impacts, 295
        beach nourishment, 293–294
        extremely stormy year, 294
        hopper dredge nourishment, 292
        nourishment parameters, 292
        sea-level rise, 291
    permanent inundation, 279
    sea-level rise effect, 278
    structure, 283
    tourism, 279
Public Interest Energy Research (PIER), 46

R
Residential Energy Consumption Survey (RECS), 178
Ricardian approach, 367
Risk Management Agency (RMA), 356

S
Sacramento Delta, 528
Sacramento Municipal Utility District (SMUD), 152
San Diego Association of Governments (SANDAG), 506
San Diego Gas and Electric (SDG&E), 523
San Diego region, 2050
    devastating wildfires, 506
    ecosystems
        biodiversity hotspot, 516
        climatic envelopes, 516
        coastal ocean, 518
        habitat, 516
        shrubland models, 517–518
    electricity
        annual consumption trends, 524, 526
        desalination demand, 527–528
        extreme temperature events and impact, 526–527
        peak demand trends, 524–526
        peak temperature and CDD, 524
        SDG&E, 523
        wind energy, 527
    energy use and greenhouse-gas emissions, 507
    ENSO, 509
    heat waves, 508
    IPCC, 506

precipitation, 508–509
public health
    aerosol fine particulate matter
        emissions, 519
    air pollution, 518
    air quality modeling, 519–521
    extreme heat events, 520, 522
    infectious disease, 523
    ozone air pollution, 520
    wildfire impacts, 522–523
research, 507
sea-level scenarios and coastal impacts
    modeling sea-level rise, 509–510
    wave activity, 510–513
temperature data, 506
warming, 507–508
water demand impacts, 513–515
water supply impact, 516
wildfires, 516
San Francisco Bay mapping, 235
San Joaquin Valley, 313
Santa Ana wind events
    actual and bivariate regression model, 124,
        125
    annual cycle
        fire incidence, 129
        number of days per month, 127
        relative humidity, 127–129
        wildfire activity, 127
    associated atmospheric dynamics, 123
    β and u contribution, 124, 126
    bivariate regression model, 124, 126
    climatic change response, 122–123
    coastal-ocean ecosystems impacts, 120
    desert and ocean, AMSL, 126
    ecological impacts, 120
    frequency and intensity response, 120
    katabatic offshore flow, 124
    land and ocean warming, 126–127
    predictor variables, 129–130
    regression model, 124
    societal implication, 130
    SRES-A1B emission scenario, 129
    wildfire region, 130
    WRF simulation, 121–122
Santa Barbara (SntBrb nest)
    CDIP, 221
    CGEM results, 220, 221
    chronic sand loss, 220
    deep water wave direction, 219
    experiment, 221, 222
    Goleta Beach, 220
    SWAN transformed wave field, 217, 221
Sea-level rise (SLR). *See also* Coastal flooding
    and erosion
    backshore characterization, 264
        erosion-resistant cliffs, 269
        geology and geomorphology, 269
        LIDAR data, 270
        topographic data, 270

BFEs (*see* Base flood elevations)
bluff and cliff morphology, 252
Bruun approach, 252–253
California seacliff, 253
coastal backshore response models
    cliffs, 262–264
    dunes (*see* Dunes)
coastal erosion response models, 271–272
coastal flood and erosion hazards, 253
    BFE, 258–259, 265, 270
    cliffs, 267–269
    dunes, 266, 268
    FEMA, 259
    flood elevations, 259
    GCM simulations, 270
    NGVD, 259
coastal response, 254, 255
downscaled regional global climate model,
    273
dune and coastal barrier erosion, 252
equilibrium profile approach, 252
flood component and elevations, 254, 273
geology and backshore type
    ArcGIS©, 256
    erosion models, 256
    filters, 257
    geologic unit, 255, 256
    geomorphic features, 255
    LIDAR topography, 255–256
    MATLAB©, 256
    morphologic diversity, 254
geology and tectonic movement, 253
geomorphology, 253
management implications, 262, 272–273
mass wasting and landslide processes, 254
regional downscaled global climate model,
    254
shoreline recession, 252
terrestrial and marine process, 253
TWL (*see* Total water levels)
wave impact approach, 253
Short for Simulating WAves Nearshore
    (SWAN), 214–216
Sierra Nevada floods
    climate change projections, 79–80
    datasets and models
        drainage areas, 77
        global climate models and
            downscaling, 75–76
        hydrological model, 76–77
        meteorological observation, 74–75
    downscaling and hydrological model
        simulations, 90
    drainage areas, 77, 78
    effects, climate change, 73
    evaporation and transpiration, 73, 74
    extreme precipitation changes, 80–81
    flood events, 3-day, 88–89
    frequencies and magnitudes, 73
    frequency analysis and factors, 81–82, 90

GCM simulation, 88
historical period simulation, 78
hydrological impacts, 73
log-transformed estimation, 91
NSN and SSN, 89–90
occurrences
    coupled ocean–atmosphere general circulation models, 85, 86
    "pineapple express" event, 82
    VIC streamflows simulation, 83, 85
rainfall, snowfall and streamfall, 72–73
risk mechanisms
    factors, 86
    flood-generating storms, 86–87
    precipitation and rainfall, 86–88
    rain-on-snow events, 87–88
river basin, rainfall, 89
runoff and baseflow simulation, 77
storm tracks, 74
streamflows simulation, 77
three-day annual maximum streamflows, 83
three-day maximum floods, 83, 84, 89
VIC streamflow simulation, 77–78
warming, 90
water-supply tightening, 72
winter-spring atmospheric circulations effects, 72
100-year floods estimation, 73
Sierra Nevada snowpack
aerosol depositions, 114
black carbon, 97–98
Colorado Rockies, 98
fine-resolution simulation, 99
hydrology, 114
hydropower generation, 96
insolation, air temperature, and orography, 97
melt, 98
orographic effects, 97
RCM, 98
RESM, 98
single-and multi-layer snow models
    Kain-Fritsch cumulus parameterization scheme, 100
    physical processes, 108–109
    runoff, 107, 109
    SAST model, 109
    solar and thermal energy, 110
    SWE, 110, 112
    WRF-SSiB model, 109–111
snow albedo, 97
    air temperature, 104
    anthropogenic emissions, 101
    fractional snow cover, 100
    Noah land-surface scheme, 99
    physical processes, 100
    satellite observations, 100
    sensible heat flux, 104, 107
    sensitivity study, 99
    snowfall and SWE, 101
    snowmelt, 105–108
    solar energy, 104
    surface, 102–103
    surface insolation, 103
    SWE, 104, 105
    terrain elevations, 101
surface air temperature, 96
surface energy balance, 113
WRF model version 2.2.1, 99
Social cost of carbon (SCC), 470, 475
Socioeconomic and demographic trends
anthropogenic GHG emissions, 22
California
    A2 and B1 scenarios interpretation, 26–28
    driving, future costs (see California driving, future cost)
    economic growth projections, 26, 28–29 (see also California economic growth)
    historical U. S. economic growth, 28
    labor productivity implications, 32–33
    methodology, 25–26
    population projections (see California population projections)
    urbanization projections (see California urbanization projections)
GHG mitigation efforts, 22
SRES A2 and B1 assessment
    anthropogenic GHG emissions, 23
    CO_2 emissions projections, 23–24
    differentiated world 24
    global sustainability, 24
    multi-model approach, 24
    United States projections, 25, 26
Solanum lycopersicum.L., 339
South Coast Air Quality Monitoring District (SCAQMD), 495
Southern California beach, economic impact. *See* Potential economic impacts
South Imperial Beach, 510, 511
Special Report on Emissions Scenarios (SRES), 21
Statewide Agricultural Production Model (SWAP), 8–9

T
Torrey pines (TorPns nest)
beach, 222
CGEM results, 222–223
experiment, 224
submarine canyons, 222
SWAN transformed wave field, 218, 222
Total water levels (TWL)
backshore characterization, 254
BFE, 264–265
coastal flood elevations, 270
cliff erosion rates, 262

coastal flood elevations, 254
erosion response models, 258
GCM, 257, 270
toe elevations, 260
wave effects and ocean water levels, 257
wave transformations, 271
Triticum aestivum L, 339

U
UARP and Big Creek system operations
    annual generation, monthly percent , 157,
        159–160
    annual runoff, 157, 159
    components, major characteristics,
        155–157
    daily time series, 152
    electricity generation demand, 153
    energy-and-storage-driven model, 153
    energy price, ISO spot market, 157, 159
    future climate change scenarios, 154
    interannual variability, 157, 159
    location, 154, 155
    monthly energy generation, 157

    multistep linear programming, 152–153
    reservoir inflows forecast estimation, 153
    reservoirs construction, 154–155
Upper American River Project (UARP), 152

V
Variable Infiltration Capacity (VIC), 5, 9, 71,
    514

W
Water Evaluation and Planning (WEAP)
    system, 299
Water Year Hydrological Classification Index,
    308
Weather Research and Forecast (WRF), 121

Y
100-year flood, 231

Z
Zea mays L., 339